General Fractional Derivatives

Theory, Methods and Applications

General Fractional Derivatives

Theory, Methods and Applications

Xiao-Jun Yang

CRC Press
Taylor & Francis Group
Boca Raton London New York

CRC Press is an imprint of the
Taylor & Francis Group, an **informa** business
A CHAPMAN & HALL BOOK

CRC Press
Taylor & Francis Group
6000 Broken Sound Parkway NW, Suite 300
Boca Raton, FL 33487-2742

Printed on acid-free paper

International Standard Book Number-13: 978-1-138-33616-2 (Hardback)

Library of Congress Cataloging-in-Publication Data

Names: Yang, Xiao-Jun (Mathematician), author.
Title: General fractional derivatives : theory, methods, and applications / Xiao Jun Yang
Description: Boca Raton : CRC Press, Taylor & Francis Group, 2019. | Includes bibliographical references and index.
Identifiers: LCCN 2019011932| ISBN 9781138336162 (hardback : alk. paper) | ISBN 9780429284083 (ebook)
Subjects: LCSH: Fractional calculus. | Calculus.
Classification: LCC QA314 .Y36 2019 | DDC 515/.83--dc23
LC record available at https://lccn.loc.gov/2019011932

Visit the Taylor & Francis Web site at
http://www.taylorandfrancis.com

and the CRC Press Web site at
http://www.crcpress.com

To the 110th Anniversary of the 1909 Founding of China University of Mining and Technology.

Contents

Preface

Modern developments in theoretical and applied science have widely depended on knowledge of the derivatives and integrals of the positive integer order, the properties of the mathematical functions, from the gamma and beta functions to the special functions, and the integro-differential operators where the integrals are of the convolution type and exist the singular, weakly singular and non-singular kernels of power-law type, which exhibit the derivatives and integrals of the variable order due to the appearance of the complexity in the natural phenomena. The aim of this monograph is mainly to investigate the fractional calculus and general fractional calculus of constant order, fractional calculus and general factional calculus of variable order, and their extended versions with respect to another function. Another aim is to present the anomalous relaxation and rheological models in light of the complexity in nature. The topics are important and interesting for scientists and engineers in the fields of mathematics, physics, chemistry and engineering.

In view of the above-mentioned avenues of their potential application to describe the numerous widespread real-world phenomena, we systematically investigate the special functions, the fractional calculus, fractional calculus with respect to another function, variable-order fractional calculus, variable-order fractional calculus with respect to another function, general fractional calculus, general fractional calculus with respect to another function, variable-order general fractional calculus, variable-order general fractional calculus with respect to another function. More specifically, we have clearly illustrated the basic theory of the general fractional calculus, the Fourier and Laplace transforms of general fractional integrals and general fractional derivatives and some illustrative examples.

The book is divided into five chapters with four appendices.

Chapter 1 investigates the history of fractional calculus of constant and variable orders, general fractional calculus of constant and variable orders and general calculus, for example, the Riemann-Liouville fractional derivative, Liouville-Grünwald fractional derivative, Marchaud fractional derivative, Hadamard fractional derivative, Hille-Tamarkin fractional derivative, Riesz fractional derivative, Hilfer fractional derivative, Erdélyi-Kober fractional derivative, Liouville-Sonine-Caputo fractional derivative and the corresponding inverse operators. Meanwhile, the special functions containing the Gamma function, Gauss hypergeometric function, Bessel function, Kummer function, Kohlrausch-Williams-Watts function, Bessel-Clifford function,

Macdonald function, Mittag-Leffler function, Wiman function, Wright function, Meijer G function, MacRobert E function, Rabotnov function, Agarwal function, Erdélyi-Magnus-Oberhettinger-Tricomi function, Fox H-function, Prabhakar function, Miller-Ross function, Gorenflo-Mainardi function and two-parametric Lorenzo-Hartley function. The special functions with respect to another function are proposed for the first time.

In Chapter 2 we report the theory of function space containing a family of the Lebesgue measurable functions and absolutely continuous functions. In view of the introductions of the fractional calculus and generalized fractional calculus in first chapter, we directly introduce the concepts of the fractional derivatives and fractional integrals of constant order within and without singular and non-singular; for example, Riemann-Liouville fractional calculus, Liouville-Sonine and Liouville-Sonine-Caputo fractional derivatives Liouville-Grünwald–Letnikov fractional derivatives, Tarasov type fractional derivatives, Riesz fractional calculus, Feller fractional calculus, Richard fractional calculus, Erdélyi-Kober type fractional calculus, Katugampola fractional calculus, Hadamard fractional calculus, Marchaud fractional derivatives, tempered fractional calculus within power-exponential, power-sine, power-cosine, power-Kohlrausch-Williams-Watts and power-Mittag-Leffler functions, tempered fractional calculus with respect to another function, tempered fractional derivatives of a purely imaginary order, Sabzikar-Meerschaert-Chen tempered fractional calculus, Torres tempered fractional derivatives, Liouville-Weyl fractional calculus, Kilbas-Srivastava-Trujillo fractional calculus, Sousa-de Oliveira fractional derivative with respect to another function, Liouville-Weyl fractional calculus with respect to another function in the sense of Riesz, Feller and Richard types, Hilfer derivatives, Riesz, Feller and Richard fractional derivatives with respect to another function, fractional calculus involving the exponential function, fractional derivatives involving the kernel of Prabhakar function, fractional derivatives involving the kernel of Wiman function and so on. The integral transforms of the fractional derivatives and fractional integrals of the constant order are discussed in detail. Some applications in engineering are also considered.

Chapter 3 addresses the general fractional derivatives of constant order involving the kernel of the special functions, such as the Mittag-Leffler function, Kohlrausch-Williams-Watts function, Wiman function, Prabhakar function, Gorenflo-Mainardi function, Miller-Ross function, Lorenzo-Hartley function, subcosine function and subsine function. The general fractional derivatives and the general fractional integrals in the complex and real orders are discussed in detail, and the Laplace transforms of the corresponding differential and integral operators are also given. The general fractional derivatives and the general fractional integrals with respect to another function are also presented. The general fractional derivatives are also considered to describe the relaxation and rheological models in complex phenomena.

Chapter 4 is devoted to the variable-order fractional calculus and the variable-order general fractional calculus involving the kernel of the special functions, such as the Mittag-Leffler function, weak power-law function, Lorenzo-Hartley function, Wiman function, Miller-Ross function and Prabhakar function. The definitions of the fractional derivatives and the general fractional integrals are presented. The Hilfer type fractional derivatives of variable order, the Riesz type fractional derivatives of variable order, the Feller type fractional derivatives of variable order and the Richard type fractional derivatives of variable order are also considered. The general fractional derivatives and general fractional integrals of the variable order involving the special functions are also discussed. The mathematical models for the relaxation and rheological models are addressed in detail.

In Chapter 5, we present the variable-order fractional calculus with respect to another function and the variable-order general fractional calculus involving the Mittag-Leffler function with respect to another function. The variable-order general fractional derivatives containing the kernel of the special functions, such as the Mittag-Leffler function and weak power-law function are presented. The definitions of the variable-order fractional derivatives and the variable-order general fractional integrals with respect to another function are introduced. The Hilfer type fractional derivatives of variable order with respect to another function and the Riesz type fractional derivatives of variable order with respect to another function are also proposed. The mathematical models for the relaxation and rheological models are discussed in detail.

It is my pleasure to express my grateful thanks to my friends and colleagues around the world who offered their suggestions of the preparation of the book. My special thanks go to my supervisor, Professor Feng Gao, and my colleagues, Professor Fu-Bao Zhou, Professor Yang Ju, Professor Hong-Wen Jing, Professor Zhan-Guo Ma and Professor Ting Li, in China University of Mining and Technology, who helped at various stages of the preparation of the book, and the financial support of the Yue-Qi Scholar of the China University of Mining and Technology (Grant No.04180004), the 333 Project of Jiangsu Province (Grant No.BRA2018320) and the State Key Research Development Program of the People's Republic of China (Grant No. 2016YFC0600705). Finally, I also wish to express my special thanks to Mrs. Aastha Sharma, senior acquisitions editor, Mrs. Shikha Garg, editorial assistant, and several staff members of CRC Press-Taylor & Francis Group for their cooperation in the production process of this book.

Xiao-Jun Yang

Professor

State Key Laboratory for Geomechanics and Deep Underground Engineering

China University of Mining and Technology

Xuzhou City, Jiangsu Province, China

Author

Dr. Xiao-Jun Yang is a full Professor of Applied Mathematics and Mechanic, at State Key Laboratory for Geomechanics and Deep Underground Engineering, China University of Mining and Technology, China. His scientific interests include: Viscoelasticity, Nonlinear Dynamics, Mathematical Physics, Analytical, Approximate, Numerical and Exact Solutions for ODEs and PDEs, Integral Transforms and Their Applications, Continuous Mechanics, Rock mechanics, Fluid Mechanics, Heat Transfer, and Traffic Flow, Wavelets, Signal Processing, Biomathematics, General Calculus and Applications, Fractional Calculus and Applications, Local Fractional Calculus and Applications, General Fractional Calculus and Applications, Variable Order Fractional Calculus and Applications and Variable Order Fractional Calculus and Applications with Respect to Another Function.

He has published over 160 journal articles and 5 books, as well as monographs, and edited volumes and 10 chapters. In 2017 and 2018, he was awarded the Elsevier Most Cited Chinese Researchers in Mathematics and the 2017 Chinese 100 Best Impact International Academic Paper. In 2018, he was selected as one of the "333" Program Talents of Jiangsu Province, China. He is currently an editor of several scientific journals, such as *Fractals-Complex Geometry, Patterns, and Scaling in Nature and Society, Applied Numerical Mathematics, Mathematical Modelling and Analysis, and Thermal Science.* He is also the referee of articles and books from Springer Nature, Elsevier, World Scientific and CRC Press publishers.

Symbol Description

$\mathbb{R}$	The set of real numbers	$Re\,(\alpha)$	The real part of the complex number–α
$\mathbb{C}$	The set of complex numbers	$Im\,(\alpha)$	The imaginary part of the complex number–α
$\mathbb{N}$	The set of natural numbers	i	Imaginary unit
$\mathbb{N}_0$	The set of positive integer numbers		

Chapter 1

Introduction

In this chapter, we introduce the histories of the fractional calculus of constant and variable orders, fractional calculus of constant and variable orders with respect to another function, general fractional calculus of constant and variable orders and general fractional calculus of constant and variable orders with respect to another function. Meanwhile, the special functions contain the Gamma function, Gauss hypergeometric function, Bessel function, Kummer function, Kohlrausch-Williams-Watts function, Bessel-Clifford function, Macdonald function, Mittag-Leffler function, Wiman function, Wright function, Meijer G function, MacRobert E function, Rabotnov function, Agarwal function, Erdélyi-Magnus-Oberhettinger-Tricomi function, Fox H function, Prabhakar function, Miller-Ross function, Gorenflo-Mainardi function, two-parametric Lorenzo-Hartley function, and so on. The special functions with respect to another function are proposed for the first time.

1.1 History of fractional calculus

The consideration for generalizing the notion of differentiation $d^n y/dx^n$ of order n from an integer number to a non-integer number was started at birth. The non-integer order, denoted by n, can be considered as follows: fractional, irrational, or complex, fractal dimension, real function or complex functions [1, 2, 3]. There are the fractional calculus (fractional-order calculus) of constant and variable orders, general fractional calculus, local fractional calculus (also called the fractal calculus), and general calculus, investigated by the researchers based on the differences of the kernel functions. To start with the ideas, we introduce their histories.

1.1.1 The contribution for fractional calculus and applications

On September 30, 1695, Leibniz replied to L'Hopital in the letter [4]: "*You can see by that, sir, that one can express by an infinite series a quantity such as* $d^{1/2}\overline{xy}$ *or* $d^{1:2}\overline{xy}$. *Although infinite series and geometry are distant relations, infinite series admits only the use of exponents that are positive and negative integers, and does not, as yet, know the use of fractional exponents.*" Later, in the same letter, Leibniz continues prophetically: "*Thus it follows that* $d^{1/2}x$ *will be equal to* $x\sqrt{dx : x}$. *This is an apparent paradox from which, one day, useful consequences will be drawn.*"

On December 28, 1695, in the correspondence with Johann Bernoulli, Leibniz mentioned derivatives of "general order." [5].

On May 28, 1697, in the Leibniz's correspondence with John Wallis, in which Wallis's infinite product for $\frac{1}{2}\pi$ is discussed, Leibniz states that differential calculus might have been used to achieve this result. He uses the notation $d^{\frac{1}{2}}y$ to denote the derivative of order [6].

In 1738, Euler took the first step that where he wrote: "*When n is a positive integer, the ratio* $d^n p$, *p is a function of x, to* dx^n *can always be expressed algebraically. Now it is asked: what kind of ratio can be made if n be a fraction? If n is a positive integer,* d^n *can be found by continued differentiation. Such a way, however, is not evident if n is a fraction. But the matter may be expedited with the help of the interpolation of series as explained earlier in this dissertation.*" (See [7]). He gave the definition of the fractional derivative as follows:

$$D_{a+}^{\alpha} f(x) = \frac{\Gamma(1-\beta)}{\Gamma(1-\alpha-\beta)} \frac{1}{(x-a)^{\beta+\alpha}}. \tag{1.1}$$

In 1772, Lagrange indirectly contributed to the law of exponents (indices) for differential operators of integer order by the expression [8]

$$\frac{d^m}{dx^m} \cdot \frac{d^n}{dx^n} y = \frac{d^{m+n}}{dx^{m+n}} y, \tag{1.2}$$

which is also true for the fractional calculus and proved by Letnikov in 1868 [9].

In 1812, Laplace proposed the idea of differentiation of non-integer order for the functions, which is represented by an integral [10]:

$$\int T(t) t^{-x} dt. \tag{1.3}$$

In 1819, Lacroix used the Euler idea, given by [6] $d^n x^a / dx^n$, and obtained by [11]

$$\frac{d^{\frac{1}{2}} \nu^a}{d\nu^{\frac{1}{2}}} = \frac{\Gamma(1+\alpha)}{\Gamma\left(1+\frac{1}{2}\right)} \nu^{a-\frac{1}{2}}, \tag{1.4}$$

which reduced for $\alpha = 1$ to

$$\frac{d^{\frac{1}{2}} \nu}{d\nu^{\frac{1}{2}}} = \frac{2\sqrt{\nu}}{\sqrt{\pi}}. \tag{1.5}$$

In 1822, Fourier contributed to the interesting results by the expression [12]:

$$\frac{d^\alpha f(x)}{dx^\alpha} = \frac{1}{2\pi} \int_{-\infty}^{+\infty} \lambda^\alpha d\lambda \int_{-\infty}^{+\infty} f(t) \cos(\lambda x - tx + p\pi/2)\, dt \tag{1.6}$$

in order to suggest the derivative for non-integer order.

In 1823, Abel considered the integral in the form [13, 14]

$$\int_{a}^{x} \frac{f(t)}{(x-t)^{\alpha}} dt = m(x), x > a, 0 < \alpha < 1 \quad (1.7)$$

to solve the tautochrone problem.

In 1832, Liouville derived the formula [15, 16]

$$D^{\alpha} f(x) = \sum_{k=0}^{\infty} C_k a_k^{\alpha} e^{a_k x}, \quad (1.8)$$

where $f(x) = \sum_{k=0}^{\infty} C_k a_k^{\alpha} e^{a_k x}$ and the fractional integration was defined as [15]

$$D^{-\alpha} f(x) = \frac{1}{(-1)^{\alpha} \Gamma(\alpha)} \int_{0}^{\infty} f(x+t) t^{\alpha-1} dt, x \in \mathbb{R}, Re(\alpha) > 0, \quad (1.9)$$

the fractional derivative as a limit of a difference quotient [16]

$$D^{\alpha} f = \lim_{h \to 0} \frac{\Delta_h^{\alpha} f}{h^{\alpha}}, \quad (1.10)$$

where $\Delta_h^{\alpha} f$ is a difference of fractional order, which is related to the sum of exponential functions, and the fractional derivative formula [15]

$$D^{\alpha} f(x) = \frac{1}{(-1)^{\alpha} \Gamma(\alpha)} \int_{0}^{\infty} \frac{d^n f(x+t)}{dx^n} t^{\alpha-1} dt, x \in \mathbb{R}, Re(\alpha) > 0, \quad (1.11)$$

and the fractional differential formula [15]

$$\frac{1}{\sqrt{2g}} \int_{0}^{h} \frac{1}{\sqrt{(x-t)}} \frac{df(t)}{dt} dt - m(t), x \in \mathbb{R}, \quad (1.12)$$

though not quite rigorously from the modern point of view.

In 1835, Liouville considered the fractional derivative of the function $f(x)$, denoted by $D^{\alpha} f(x)$, of order α, which is positive, negative, real and imaginary [17].

In 1847, Riemann derived the formula [18]

$$D^{-\alpha} f(x) = \frac{1}{\Gamma(\alpha)} \int_{c}^{x} (x-t)^{\alpha-1} f(t) dt, \quad (1.13)$$

which can be linked with the fractional integral formula of Liouville.

In 1869, Sonine called the Riemann-Liouville definition starting with the Cauchy's integral formula for the first time [19].

In 1867, Grünwald developed the Liouville's idea of the sum of the exponential functions [17], and proposed the following formula [20]:

$$D^{\alpha} f(x) = \lim_{h \to 0} \frac{\Delta_h^{\alpha} f(x)}{h^{\alpha}}, \tag{1.14}$$

where $\Delta_h^{\alpha} f$ is a difference of fractional order, given by

$$\Delta_h^{\alpha} f = \sum_{i=0}^{\infty} (-1)^i \binom{\alpha}{i} f(x - hi). \tag{1.15}$$

This operator is called the Liouville-Grünwald fractional operator [21, 22].

In 1868, Letnikov considered the above formula [23]

$$D^{\alpha} f(x) = \lim_{h \to 0} \frac{\Delta_h^{\alpha} f(x)}{h^{\alpha}}, \tag{1.16}$$

where $\Delta_h^{\alpha} f$ is a difference of fractional order, whose appropriate interpretation of the fractional difference is related with Riemann's and Liouville's expressions. This operator is also called the Grünwald-Letnikov fractional operator [3].

In 1872, Sonine introduced the following fractional derivative, given as [24]

$$D^{\alpha} f(x) = \frac{1}{\Gamma(p - \alpha + 1)} \int_a^x \frac{df(t)}{dt} (x - t)^{p-\alpha} dt. \tag{1.17}$$

In 1880, Sonine introduced the kernel of the form [25]

$$\Theta(t) = e^{\lambda t}, \tag{1.18}$$

which leads to the following expressions

$$\mathrm{M}(t) = \frac{1}{\Gamma(-\alpha)} \int_0^{\infty} e^{-\lambda t} f(t) dt \tag{1.19}$$

and

$$\mathrm{M}(t) = \int_a^b \frac{1}{t^{n+1}} e^{-\frac{\lambda}{2}\left(t - \frac{1}{t}\right)} f(t) dt. \tag{1.20}$$

In 1884, Sonine proposed a solution of the Abel type equation [26]

$$\int_a^x g(x - t) \Phi(x) dt = \int_a^x \Phi(x) dt, \tag{1.21}$$

where

$$g(h) = \int_0^h \psi(h-t)\varphi(t)\,dt = 1 \tag{1.22}$$

with

$$\psi(t) = t^{-p}\sum_{i=0}^{\infty}\frac{(ty)^n}{\Gamma(n+1)\Gamma(n-p+1)} \tag{1.23}$$

and

$$\varphi(t) = t^{-q}\sum_{i=0}^{\infty}\frac{(-ty)^n}{\Gamma(n+1)\Gamma(n-q+1)} \tag{1.24}$$

for $p+q=1$, which are called the Sonine conditions (for more details on the Sonine condition, see [3, 27, 28]), and introduced the kernel of the form

$$\Theta(t) = \frac{t^{\alpha}e^{\lambda t}}{\Gamma(1-\alpha)}, \tag{1.25}$$

which leads to the following expression

$$\mathrm{M}(t) = \frac{1}{\Gamma(-\alpha)}\int_0^{\infty}\frac{e^{-xt}}{t^{1+\alpha}}f(t)\,dt. \tag{1.26}$$

In 1890, Sonine introduced the fractional multiple integral of the form [29]

$$\begin{aligned}\mathrm{M}(t) = e^{-c_1 t}\int_a^x (x-t_1)^{\alpha_1}e^{b_1 t_1}dt_1 \ldots \int_a^{t_{i-1}} (x_{i-1}-t_i)^{\alpha_i}e^{b_i t_i}dt_i \\ \ldots \int_a^{t_{n-1}} f(t_n)(x_{n-1}-t_n)^{\alpha_n}e^{b_n t_n}dt_i\end{aligned} \tag{1.27}$$

In 1892, Hadamarod proposed the following fractional integral [30]

$$I^{\alpha}f(x) = \frac{z^{\alpha}}{\Gamma(\alpha)}\int_0^1\frac{f(zu)}{(1-u)^{1-\alpha}}du. \tag{1.28}$$

In 1896, Volterra introduced the general Abel's integral equation [31]

$$\int_a^x\frac{\kappa(x,t)f(t)}{(x-t)^{\alpha}}dt = g(x), \quad 0<\alpha<1, \tag{1.29}$$

where $\kappa(x,t)$ is a continuous function.

In 1909, Bocher considered that Abel's equation [32]

$$\frac{1}{\Gamma(\alpha)}\int_a^x\frac{f(t)}{(x-t)^{1-\alpha}}dt = g(x), \quad 0<\alpha<1, \tag{1.30}$$

is solvable in the set of all Lebesgue measurable functions if and only if there are

$$h(x) = \frac{1}{\Gamma(1-\alpha)} \int_a^x \frac{g(t)}{(x-t)^{1-\alpha}} dt \tag{1.31}$$

and

$$h(a) = 0, \tag{1.32}$$

where

$$h(t) = C + \int_a^x g(x)\, dx \tag{1.33}$$

and

$$\int_a^b |g(x)|\, dx < \infty. \tag{1.34}$$

In 1917, Weyl derived the fractional integrals by [33]:

$$I_+^\alpha f(x) = \frac{1}{\Gamma(\alpha)} \int_{-\infty}^x \frac{f(t)}{(x-t)^{1-\alpha}} dt, \tag{1.35}$$

$$I_-^\alpha f(x) = \frac{1}{\Gamma(\alpha)} \int_x^\infty \frac{f(t)}{(x-t)^{1-\alpha}} dt, \tag{1.36}$$

which are called the Weyl fractional integrals.

In 1927, Marchaud introduced the fractional differentiation of the form [34]:

$$D^\alpha f(x) = C\frac{1}{\Gamma(\alpha)} \int_0^\infty \frac{\Delta_t^l f(x)}{t^{\alpha+1}} f(t)\, dt, \tag{1.37}$$

where $\Delta_t^l f(x)$ is the finite difference of order l for $l > \alpha$ and $l \in \mathbb{N}$. This operator is now named Marchaud fractional derivatives. When $\Delta_t^l f(x)$, it is called the Weyl type finite difference for $l = 1$ and $1 > \alpha > 0$.

In 1927, Hadamard introduced the fractional integral expressed as [34]

$$I_a^{(\alpha)} f(x) = \frac{1}{\Gamma(\alpha)} \int_a^x \frac{f(t)}{\left(\ln \frac{x}{t}\right)^{1-\alpha}} \frac{dt}{t}, 0 < \alpha < 1, \tag{1.38}$$

which is called the Hadamard fractional integral, and the fractional derivative was given by

$$D_a^{(\alpha)} f(x) = \frac{1}{\Gamma(1-\alpha)} x \frac{d}{dx} \int_0^x \frac{f(t)}{\left(\ln \frac{x}{t}\right)^{\alpha}} \frac{dt}{t}, 0 < \alpha < 1, \tag{1.39}$$

which is called the Hadamard fractional derivative.

In 1928, Tonelli introduced the following result [35]:
If

$$f(t) = C + \int_a^x g(x)\,dx \tag{1.40}$$

and

$$\int_a^b |g(x)|\,dx < \infty, \tag{1.41}$$

where Abel's equation with $0 < \alpha < 1$,

$$\frac{1}{\Gamma(\alpha)} \int_a^x \frac{f(t)}{(x-t)^{1-\alpha}} dt = g(x), \tag{1.42}$$

has the solution [35]

$$f(x) = \frac{1}{\Gamma(1-\alpha)} \left[\frac{f(a)}{x^{\alpha}} + \int_a^x \frac{f^{(1)}(t)}{(x-t)^{\alpha}} dt \right], \tag{1.43}$$

which is called the Tonelli theorem.

In 1930, Tamarkin gave the expression as follows [36]:
The Abel's equation [36]

$$\frac{1}{\Gamma(\alpha)} \int_a^x \frac{f(t)}{(x-t)^{1-\alpha}} dt = g(x), \quad 0 < \alpha < 1, \tag{1.44}$$

is solvable in the set of all Lebesgue measurable functions if and only if there are [36]

$$h(x) = \frac{1}{\Gamma(1-\alpha)} \int_a^x \frac{g(t)}{(x-t)^{1-\alpha}} dt \tag{1.45}$$

and

$$h(a) = 0, \tag{1.46}$$

where

$$h(t) = C + \int_a^x g(x)\,dx \tag{1.47}$$

and

$$\int_a^b |g(x)|\,dx < \infty. \tag{1.48}$$

This was called the Tamarkin theorem.

In 1930, Hille and Tamarkin proposed the Abel type integral equation of the second kind [37]

$$g(x) - \frac{\lambda}{\Gamma(\alpha)} \int_0^x \frac{g(t)}{(x-t)^{1-\alpha}} dt = f(x), \quad 0 < \alpha < 1, \tag{1.49}$$

with the solution (also called the Hille-Tamarkin fractional derivative, see [37])

$$g(x) = \frac{d}{dx} \int_0^x E_\alpha \left[\lambda (x-t)^\alpha \right] f(t)\, dt, \tag{1.50}$$

the integral equation of the form

$$f(x) = g(x) - \lambda \int_0^x \kappa(x-t, \lambda)\, g(t)\, dt, \tag{1.51}$$

where

$$\kappa(x-t, \lambda) = e^{\lambda(t-x)}, \tag{1.52}$$

and the integral equation in the form [37]

$$f(x) = g(x) - \lambda \int_0^x \kappa(x-t, \lambda)\, g(t)\, dt, \tag{1.53}$$

where

$$\kappa(x, \lambda) = \frac{d}{dx} E_{1-\alpha} \left[\lambda \Gamma(1-\alpha) x^\alpha \right]. \tag{1.54}$$

In 1931, Watanabe considered the Leibniz' rule for analytic functions [38]:

$$D_a^\alpha \left[f(x) g(x) \right] = \sum_{k=-\infty}^{\infty} \binom{\alpha}{\beta+k} D_a^{\alpha-\beta-k} \left[f(x) \right] D_a^{\beta+k} \left[g(x) \right]. \tag{1.55}$$

In 1935, Sato considered an Abel type equation with a monotone increasing function, denoted by $\varphi(x)$, which is given as [39]:

$$\int_x^\infty \frac{f(u)\, du}{\left(\varphi(u) - \varphi(x) \right)^{\frac{1}{2}}} = \Omega(x), x > 0. \tag{1.56}$$

In 1938, Love introduced the convergent fractional integral by [40]:

$$I_+^\alpha f(x) = \frac{1}{\Gamma(\alpha)} \lim_{N \to \infty} \int_0^N \frac{f(x-t)}{t^{1-\alpha}} dt, \tag{1.57}$$

and Love and Young investigated the formula for fractional integration by parts in the form [41]

$$\int_a^b \left(D_{a+}^\alpha f(t) \right) g(t)\, dt = \int_a^b f(t) \left(D_{b-}^\alpha g(t) \right) dt \tag{1.58}$$

and the formula for the fractional integration by parts

$$\int_a^b \left(I_{a+}^{\alpha} f(t)\right) g(t)\, dt = \int_a^b f(t) \left(I_{b-}^{\alpha} g(t)\right) dt, \tag{1.59}$$

which is called the Love-Young formula for the fractional derivative.

In 1939, Hille introduced the following fractional differential operator [42]

$$R(\alpha, \lambda)\, g(x) = \lambda \frac{d}{dx} \int_0^x E_\alpha \left[\lambda (t-x)^\alpha\right] f(t)\, dt. \tag{1.60}$$

In 1940, Erdélyi and Kober derived the fractional integrals [43]

$$\frac{2x^{-2(\alpha+\beta)}}{\Gamma(\alpha)} \int_0^x \left(x^2 - t^2\right)^{\alpha-1} t^{2\beta+1} f(x)\, dt \tag{1.61}$$

and

$$\frac{2x^{-2\beta}}{\Gamma(\alpha)} \int_x^\infty \left(x^2 - t^2\right)^{\alpha-1} t^{1-2\beta-2\beta} f(x)\, dt, \tag{1.62}$$

which are called the Erdélyi-Kober fractional integrals.

In 1941, Cossar reported the following fractional integral [44]

$$-\frac{1}{\Gamma(1-\alpha)} \lim_{N\to\infty} \frac{d}{dx} \int_x^N \frac{f(t)}{(t-x)^\alpha} dt. \tag{1.63}$$

In 1949, Riesz considered the fractional form as [45]

$$(-\Delta)^{\frac{\alpha}{2}} f(x) = F^{-1} |x|^{-\alpha} F f(x) = \begin{cases} I_{R^n}^{\alpha} f(x), Re\alpha > 0, \\ D_{R^n}^{-\alpha} f(x), Re < 0, \end{cases} \tag{1.64}$$

where F is the Fourier transform operator and

$$I_{R^n}^{\alpha} f(x) = \frac{1}{\vartheta_n(\alpha)} \int_{R^n} \frac{f(t)}{|x-t|^{n-\alpha}} dt, \tag{1.65}$$

and the fractional derivative as

$$\frac{d^\alpha f(x)}{d|x|^\alpha} = \frac{1}{\vartheta_n(\alpha)} \frac{d^2}{dx^2} \int_{-\infty}^{\infty} |x-t|^{1-\alpha} f(t)\, dt \tag{1.66}$$

with

$$\frac{1}{\vartheta_n(\alpha)} = \frac{1}{2\cos\left(\frac{\pi\alpha}{2}\right) \Gamma(2-\alpha)}. \tag{1.67}$$

In 1951, Erdélyi introduced the fractional integrals by [46]

$$R_m^{\eta,\alpha} f(x) = \frac{m}{\Gamma(\alpha)} \int_0^x \frac{x^{-\eta-m\alpha+m-1}}{(x^m - t^m)^{1-\alpha}} f(t)\, dt \tag{1.68}$$

and

$$S_m^{\eta,\alpha} f(x) = \frac{m x^{-\eta-m\alpha+m-1}}{\Gamma(\alpha)} \int_0^x \frac{t^\eta}{(x^m - t^m)^{1-\alpha}} f(t)\, dt, \tag{1.69}$$

where $\alpha > 0$, $\eta > -\frac{1}{2}$ and $m > 0$.

In 1952, as the general version of the Riesz fractional calculus, Feller proposed [47]

$${}_F I_\vartheta^\alpha f(t) = \frac{\sin((\alpha-\vartheta)\pi/2)}{\sin(\pi\vartheta)} \frac{1}{\Gamma(\alpha)} \int_{-\infty}^t \frac{f(\tau)}{(\tau-t)^{1-\alpha}} d\tau + \frac{\sin((\alpha+\vartheta)\pi/2)}{\sin(\pi\vartheta)} \frac{1}{\Gamma(\alpha)} \int_t^\infty \frac{f(\tau)}{(t-\tau)^{1-\alpha}} d\tau, \tag{1.70}$$

and

$$\begin{aligned} {}_F D_\vartheta^\alpha f(t) = &-\frac{\sin((\alpha+\vartheta)\pi/2)}{\sin(\pi\vartheta)} \frac{1}{\Gamma(1-\alpha)} \frac{d}{dt} \int_{-\infty}^t \frac{f(\tau)}{(\tau-t)^\alpha} d\tau \\ &- \frac{\sin((\alpha-\vartheta)\pi/2)}{\sin(\pi\vartheta)} \frac{1}{\Gamma(1-\alpha)} \frac{d}{dt} \int_t^\infty \frac{f(\tau)}{(t-\tau)^\alpha} d\tau. \end{aligned} \tag{1.71}$$

In 1957, Hille and Phillips introduced the integral of the form [48]

$$I_\lambda^{(\alpha)} f(t) = \frac{\lambda^{\alpha+1}}{\Gamma(\alpha+1)} \int_0^\infty t^\alpha e^{\lambda(x-t)} f(t)\, dt. \tag{1.72}$$

In 1961, Chen proposed the fractional derivatives by [49]

$$D^\alpha f(x) = \frac{1}{\Gamma(n-\alpha)} \frac{d}{dx} \int_a^x \frac{(f(t) - f(a))}{|x-t|^{n-\alpha}} dt, \tag{1.73}$$

and

$$D^\alpha f(x) = \frac{1}{\Gamma(n-\alpha)} \int_a^x \frac{1}{|x-t|^\alpha} \frac{df(t)}{dt} dt, \tag{1.74}$$

and the fractional integrals by

$$I^\alpha f(x) = \frac{1}{\Gamma(\alpha)} \int_{-\infty}^x (t-x)^{\alpha-1} (f(t) - f(x))\, dt \tag{1.75}$$

and

$$I^\alpha f(x) = \frac{1}{\Gamma(\alpha)} \int_a^x |x-t|^{\alpha-1} f(t)\, dt. \tag{1.76}$$

In 1961, Peters considered the fractional integrals by [50]

$$Q_k^{\eta,\alpha} f(x) = \int_0^x t^{1-\eta} J_\alpha \left\{ k\sqrt{x^2 - t^2} \right\} \left(x^2 - t^2\right)^{\frac{\alpha}{2}} f(t)\, dt \tag{1.77}$$

and

$$P_k^{\eta,\alpha} f(x) = \int_0^x t^{\eta+1} I_\alpha \left\{ k\sqrt{x^2 - t^2} \right\} \left(x^2 - t^2\right)^{\frac{\alpha}{2}} f(t)\, dt, \tag{1.78}$$

which are called the Peters fractional operators, where $J_\alpha \left\{ k\sqrt{x^2 - t^2} \right\}$ and $I_\alpha \left\{ k\sqrt{x^2 - t^2} \right\}$ are the Bessel function and the modified Bessel function, respectively.

In 1962, Burlak introduced the fractional integral in the kernel of the Bessel function of the first kind as [51]

$$g(x) = \int_0^x J_\alpha \left\{ k\sqrt{x^2 - t^2} \right\} \left(x^2 - t^2\right)^{\frac{\alpha}{2}} f(t)\, dt \tag{1.79}$$

and the fractional derivative in the kernel of the modified Bessel function of the first kind as

$$f(x) = k \int_0^x t \left(x^2 - t^2\right)^{\frac{\alpha+1}{2}} I_{-(\alpha+1)} \left\{ k\sqrt{x^2 - t^2} \right\} g(t)\, dt, \tag{1.80}$$

which are called the Burlak fractional integral and derivative operators, respectively.

In 1962, Sneddon proposed the integral equation with trigonometric kernels [52]

$$f(x) = \int_0^\infty t^{-1} \cos(xt)\, g(t)\, dt. \tag{1.81}$$

In 1963, Srivastava introduced the integral equations with the kernel of the cosine function by [53]

$$f(x) = \int_0^\infty t^{2k-1} \cos(xt)\, g(t)\, dt, \tag{1.82}$$

$$f(x) = \int_0^\infty t^{2k-1} \sin(xt)\, g(t)\, dt, \tag{1.83}$$

$$m(x) = \int_0^\infty \cos(xt)\, g(t)\, dt \tag{1.84}$$

and

$$m(x) = \int_0^\infty \sin(xt)\, g(t)\, dt, \tag{1.85}$$

where $m(x) = 0$.

In 1964, Srivastava proposed the fractional integral in the kernel of the confluent hypergeometric function by [54]

$$I_a^{(\alpha)} f(x) = \int_0^x \frac{(x-t)^{\alpha-1}}{\Gamma(\alpha)} {}_1F_1(\beta;\alpha;x-t)\, f(t)\, dt, \tag{1.86}$$

where ${}_1F_1(\beta;\alpha;x-t)$ is the confluent hypergeometric function.

In 1965, Cooke introduced the fractional integral by [55]

$$I_{a;\sigma}^{\eta,\alpha} f(x) = \begin{cases} \frac{2x^{-2(\eta+\alpha)}}{\Gamma(\alpha)} \int_a^b \frac{t^{2(\eta+1)-1}}{(x^2-t^2)^{1-\alpha}} f(t)\, t^{2\eta-1} dt, \alpha > 0, \\ f(x), \alpha = 0, \\ \frac{2x^{-2(\eta+\alpha)-1}}{\Gamma(1+\alpha)} \frac{d}{dx} \int_a^b (x^2-t^2)^\alpha t^{2\eta+1} f(t)\, dt, -1 < \alpha < 0, \end{cases} \tag{1.87}$$

which is called the Cooke fractional operator.

In 1965, the solution of Abel's equation of the more general equation of the form [56]

$$\frac{1}{\Gamma(\alpha)} \int_a^x \frac{m(x-t)\, f(t)}{(x-t)^{1-\alpha}} dt = g(x), \quad 0 < \alpha < 1, \tag{1.88}$$

was considered by Sakalyuk, where $m(x-t)$ is a polynomial.

In 1966, Lowndes proposed the integral equations with the kernel of the trigonometric function [57]

$$f(x) = \int_0^\infty t^{-1} \cos(xt)\, g(t)\, dt \tag{1.89}$$

and

$$f(x) = \int_0^\infty t^{-1} \sin(xt)\, g(t)\, dt. \tag{1.90}$$

In 1967, Saxena introduced the fractional integral within the kernel of the Gauss hypergeometric function by [58]

$$I_a^{(\alpha)} f(x) = \frac{x^{-\gamma-1}}{\Gamma(\alpha)} \int_0^x {}_2F_1(1-\alpha, \beta+m; \beta; t/x)\, f(t)\, t^\gamma dt, \tag{1.91}$$

where ${}_2F_1(1-\alpha, \beta+m; \beta; t/x)$ is the Gauss hypergeometric function.

In 1967, Kalisch [59] proposed the fractional derivative of purely imaginary order α, where $\alpha = i\theta$, namely,

$$I_{a+}^{(i\theta)} f(x) = \frac{1}{\Gamma(1+i\theta)} \int_a^x (x-t)^{i\theta} f(t)\, dt \tag{1.92}$$

and

$$I_{b-}^{(i\theta)} f(x) = \frac{1}{\Gamma(1+i\theta)} \int_x^b (t-x)^{i\theta} f(t)\, dt. \tag{1.93}$$

In 1967, Parashar reported the fractional integral in the kernel of the Meijer G function (see [60]).

In 1967, Caputo introduced the fractional derivative of the form [61]

$$D^{\alpha} f(x) = \frac{1}{\Gamma(n-\alpha)} \int_0^x \frac{1}{(x-t)^{\alpha}} \frac{\partial^n f(t)}{\partial t^n} dt, \tag{1.94}$$

and did not report the results from Liouville and Sonine (see the report in Chapter 2).

In 1967, Bosanquet proposed that $\psi(t) = t^{\alpha-1}/\Gamma(\alpha)$ and $\varphi(t) = t^{-\alpha}/\Gamma(1-\alpha)$ in the theory of fractional calculus with the Sonine condition [62]:

$$\int_0^x \psi(x-t)\, \varphi(t)\, dt = 1. \tag{1.95}$$

In 1967, Dzherbashyan used the generalization of Hadamarod's idea and gave the following fractional integral [63]:

$$I^{\alpha} f(x) = \frac{1}{\Gamma(\alpha)} \int_0^1 \frac{f(zu)}{(-\ln u)^{1-\alpha}} du. \tag{1.96}$$

In 1968, Srivastava proposed the fractional operator of the form [64]

$$K_{\zeta,\alpha,n}^{(-)} f(x) = \frac{n}{\Gamma(\alpha)} x^{\zeta} \int_x^{\infty} (u^n - x^n)^{\alpha-1} u^{-\zeta-n\alpha+n-1} f(u)\, du, x > 0, \tag{1.97}$$

which is related to the generalized Whittaker transform.

In 1968, Flett suggested the Hadamard fractional integral with the principal value [65].

In 1968, Dzhrbashyan and Nersesyan proposed the fractional derivative [66]

$$D^{\alpha} f(x) = \frac{1}{\Gamma(n-\alpha)} \int_{0}^{\infty} \frac{1}{(x-t)^{n-\alpha}} \frac{\partial^n f(t)}{\partial t^n} dt. \tag{1.98}$$

In 1969, Kalla introduced the integral operators involving Fox's H-function (see [67]). In 1970, Smit and De Vries introduced the fractional derivative which is the same as the Caputo definition and did not report the results from Liouville, Sonine and Caputo.

In 1970, Lowndes proposed the fractional integral in the kernel of the Bessel function and modified Bessel function of the first kind in the following form [69]:

$$g(x) = 2^{-\alpha} x^{-2\eta} k^{\alpha+1} \int_{0}^{x} t^{2(\eta+\alpha)+1} \left(x^2 - t^2\right)^{\frac{\alpha+1}{2}} I_{-(\alpha+1)} \left\{k\sqrt{x^2-t^2}\right\} f(t)\, dt, \tag{1.99}$$

and

$$f(x) = 2^{\alpha} t^{-2(\eta+\alpha)-1} k^{-\alpha} \frac{d}{dx} \int_{0}^{x} t^{1+2\eta} \left(x^2 - t^2\right)^{\frac{\alpha}{2}} I_{-\alpha} \left\{k\sqrt{x^2-t^2}\right\} g(t)\, dt. \tag{1.100}$$

In 1970, Osler derived the fractional derivative with another function $h(z)$ in the complex domain by [70, 71]

$$D_{h(z)}^{(\alpha)} f(z) = \frac{1}{\Gamma(\alpha)} \int_{L} \frac{f(z)\, h^{(1)}(z)}{(h(z) - h(z))^{\alpha+1}} dz. \tag{1.101}$$

In 1970, Osler introduced the fractional derivative with respect to the continuously differentiable, increasing function in the form [72, 73, 74, 75]

$$D_{a;\kappa}^{(\alpha)} f(x) = \frac{1}{\Gamma(1-\alpha)} \frac{1}{\kappa^{(1)}(x)} \frac{d}{dx} \int_{a}^{x} \frac{f(t)}{(\kappa(x) - \kappa(t))^{\alpha}} \kappa^{(1)}(t)\, dt, 0 < \alpha < 1, \tag{1.102}$$

where $a = \kappa^{-1}(0)$, and, in 1971, the fractional integral of the form

$$I_{a;\kappa}^{(\alpha)} f(x) = \frac{1}{\Gamma(\alpha)} \int_{a}^{x} \frac{f(t)}{(\kappa(x) - \kappa(t))^{1-\alpha}} \kappa^{(1)}(t)\, dt, 0 < \alpha < 1, \tag{1.103}$$

and

$$D_{a;\kappa}^{(\alpha)} f(x) = \frac{1}{\Gamma(1-\alpha)} \frac{f(t)}{(\kappa(x) - \kappa(t))^{\alpha}} \frac{\alpha}{\Gamma(1-\alpha)} \int_{a}^{x} \frac{f(x) - f(t)}{(\kappa(x) - \kappa(t))^{1+\alpha}} \kappa^{(1)}(t)\, dt, \tag{1.104}$$

where $0 < \alpha < 1$.

In 1971, Love considered the fractional derivative of purely imaginary order as [76, 77]

$$D_a^{(i\alpha)} f(x) = \frac{1}{\Gamma(1-i\alpha)} \frac{d}{dx} \int_a^x \frac{f(t)}{(x-t)^{i\alpha}} dt, \tag{1.105}$$

and the fractional integral of pure imaginary order as [75, 76]

$$I^{(i\alpha)} f(t) = \frac{1}{\Gamma(i\alpha)} \int_0^\infty \frac{f(t)}{(x-t)^{1-i\alpha}} dt. \tag{1.106}$$

In 1971, Rafal'son introduced the Bessel fractional integration in the form (see [78])

$$I_-^{(\alpha)} f(t) = \frac{1}{\Gamma(\alpha)} \int_x^\infty (x-t)^{\alpha-1} e^{x-t} f(t)\, dt, \tag{1.107}$$

and the Bessel fractional derivative in the form

$$D_+^{(\alpha)} f(t) = \frac{1}{\Gamma(\alpha)} \int_x^\infty (x-t)^{\alpha-1} e^{x-t} f^{(\alpha)}(t)\, dt, \tag{1.108}$$

which is called the Rafal'son type Bessel fractional derivative.

In 1972, Prabhakar introduced the fractional integral in the kernel of the Humbert function by [79]

$$I_a^{(\alpha)} f(x) = \int_0^x \frac{(x-t)^{\alpha-1}}{\Gamma(\alpha)} \Theta_1 \left(\alpha, b; c; 1 - \frac{t}{x}; \lambda(x-t) \right) f(t)\, dt, \tag{1.109}$$

where $\Theta_1 \left(\alpha, b; c; 1 - \frac{t}{x}; \lambda(x-t)\right)$ is the Humbert function.

In 1973, based on the Sonine condition and Bosanquet's results, Choudhary proposed the following relationship [80]

$$f(t) = \int_a^t \psi(t-x) \left\{ -\frac{d}{dt} \int_u^t \varphi(x-u) f(u)\, du \right\} dx, \tag{1.110}$$

where $a > 0$ and $a > t$. It is called the Choudhary formula.

In 1975, Sneddon introduced the fractional integral by [81]

$$I_{a;\sigma}^{\eta,\alpha} f(x) = \frac{\sigma x^{-\sigma(\eta+\alpha)}}{\Gamma(\alpha)} \int_a^x \frac{t^{\sigma(\eta+1)-1}}{(x^\sigma - t^\sigma)^{1-\alpha}} f(t)\, dt, \tag{1.111}$$

which is called the Sneddon fractional integral.

In 1976, Malovichko reported the fractional integral in the kernel of the Bessel function [82]. In 1977, Krasnov presented the fractional derivative of a function with respect to another function in terms of fractional differences (see [83]).

In 1978, Saigo introduced the fractional integral operator involving the Gauss hypergeometric function in the form [84]

$$I_x^{\alpha,\beta,\eta} f(x) = \frac{x^{-\alpha-\beta}}{\Gamma(\alpha)} \int_0^x (x-t)^{\alpha-1} {}_2F_1(\alpha+\beta, -\eta; \alpha; 1-t/x) f(t)\, dt, \tag{1.112}$$

where ${}_2F_1(\alpha+\beta, -\eta; \alpha; 1-t/x)$ is the Gauss hypergeometric function.

In 1978, Srivastav introduced the fractional integral in the kernel of the Bessel function of the first kind in the following form [85]

$$Q_k^{\eta,\alpha} f(x) = 2^\alpha x^{-2(\alpha+\eta)} k^{\frac{1-\alpha}{2}} \int_0^x t^{2\eta+1} \left(x^2-t^2\right)^{\frac{1-\alpha}{2}} I_{\alpha-1}\left\{k\sqrt{x^2-t^2}\right\} f(t)\, dt, \tag{1.113}$$

which is called the Srivastav fractional integral.

In 1979, Gearhart reported the fractional integral [86]

$$I_-^{(\alpha)} f(t) = \frac{1}{\Gamma(\alpha)} \int_0^\infty (x-t)^{\alpha-1} e^{\lambda(x-t)} f(t)\, dt, \tag{1.114}$$

which is called the Rafal'son-Gearhart type Bessel fractional integral.

In 1979, McBride introduced the expression with respect to the continuously differentiable, increasing function by [87]

$$\frac{1}{\Gamma(\alpha)} \int_0^x \frac{f(t)}{(\kappa(x)-\kappa(t))^{1-\alpha}} \kappa^{(1)}(t)\, dt, \tag{1.115}$$

which yields the following

$$I_m^{(\alpha)} f(x) = \frac{m}{\Gamma(\alpha)} \int_0^x \frac{f(t)}{(x^m-t^m)^{1-\alpha}} x^{m-1} dt, \tag{1.116}$$

and the fractional integral in the form

$$I_a^{(\alpha)} f(x) = \int_0^x \frac{(x^m-t^m)^{\alpha-1}}{\Gamma(\alpha)} {}_2F_1\left(\alpha, b; c; 1-\frac{t^m}{x^m}\right) m t^{m-1} f(t)\, dt, \tag{1.117}$$

where ${}_2F_1\left(\alpha, b; c; 1-\frac{t}{x}\right)$ is the Gauss hypergeometric function.

In 1980, Skornik introduced the tempered fractional integral by [88, 89]

$$I_a^{(\alpha)} f(x) = e^{-\frac{x^2}{4}} \int_a^x \frac{(x-t)^{\alpha-1}}{\Gamma(\alpha)} f(t)\, e^{\frac{t^2}{4}} dt, 0 < \alpha < 1, \quad (1.118)$$

which is called the Skornik fractional integral, the tempered fractional derivative in the form

$$D_a^{(\alpha)} f(x) = \frac{e^{-\frac{x^2}{4}}}{\Gamma(1-\alpha)} \frac{d^n}{dx^n} \int_a^x \frac{e^{\frac{t^2}{4}}}{(x-t)^{\alpha}} f(t)\, dt, 0 < \alpha < 1, \quad (1.119)$$

and

$$D_a^{(\alpha)} I_a^{(\alpha)} f(x) = f(x). \quad (1.120)$$

In 1985, Estrada and Kanwal derived the integral equations including singular function with Cauchy kernel, given as (see [90])

$$\int_x^{\infty} \frac{f(u)\, du}{(h(x) - h(x))^{\alpha-1}} = \Omega(x), x > 0. \quad (1.121)$$

In 1986, Vu Kim Tuan reported the fractional integral in the kernel of the Meijer G function (see [91]).

In 1987, Samko, Kilbas and Marichev introduced (see [3])

$$I_a^{(\alpha)} f(x) = \int_0^x \frac{(h(x) - h(t))^{\alpha-1}}{\Gamma(\alpha)} \Theta_1 \left(\alpha, b; c; 1 - \frac{h(t)}{h(x)}; \lambda (h(x) - h(t))\right)$$
$$f(t)\, h^{(1)}(t)\, dt, \quad (1.122)$$

$$I^{(\alpha)} f(x) = \frac{1}{\Gamma(\alpha)} \int_a^x \frac{f(t) - f(a)}{(x-t)^{1-\alpha}} dt, \quad 0 < \alpha < 1 \quad (1.123)$$

$$D^{(m+\alpha)} f(x) = \frac{1}{\Gamma(\alpha)} \int_a^x \frac{f^{(m)}(t) - f^{(m)}(a)}{(x-t)^{1-\alpha}} dt, \ 0 < \alpha < 1, \ m \in \mathbb{N} \quad (1.124)$$

and

$$I_a^{(\alpha)} f(x) = \int_a^x \frac{\left(2sh\frac{x-t}{2}\right)^{\alpha-1}}{\Gamma(\alpha)} f(t)\, dt, \quad (1.125)$$

where $h^{(1)}(t) > 0$ and $\Theta_1 \left(\alpha, b; c; 1 - \frac{t}{x}; \lambda (x-t)\right)$ is the Humbert function.

In 1989, Peschanskii introduced the fractional integral operator involving the curvilinear convolution type in the form [92]

$$I_x^{\alpha,\beta,\eta} f(x) = \frac{1}{2\pi i} \int_{\Gamma} {}_2F_1(1, 1; 1+\alpha; t/x)\, f(t) \frac{dt}{t}, \quad (1.126)$$

where ${}_2F_1(1, 1; 1+\alpha; t/x)$ is the Gauss hypergeometric function.

In 1993, Samko and Ross proposed the variable-order fractional integral given by (see [93])

$$I_a^{(\alpha(x))} f(x) = \frac{1}{\Gamma(\alpha(x))} \int_a^x \frac{f(t)}{(x-t)^{\alpha(x)}} dt, \tag{1.127}$$

and the variable-order fractional derivative given as

$$D_a^{(\alpha(x))} f(x) = \frac{1}{\Gamma(1-\alpha(x))} \frac{d}{dx} \int_a^x \frac{f(t)}{(x-t)^{1+\alpha(x)}} dt. \tag{1.128}$$

In 2000, Hilfer introduced the fractional derivative by [94]

$$D^{\alpha,\beta} f(x) = I_0^{\beta(1-\alpha)} \cdot D^{(1)} \cdot I_0^{(1-\beta)(1-\alpha)} f(x), \tag{1.129}$$

where $D^{(1)} f(x) = \frac{df(x)}{dx}$, $I_0^{\beta(1-\alpha)}$ and $I_0^{(1-\beta)(1-\alpha)}$ are the Riemann-Liouville fractional integrals.

In 2003, Coimbra proposed the variable-order fractional integral in the form [95]

$$D_0^{(\alpha(x))} f(x) = \frac{1}{\Gamma(1-\alpha(x))} \int_0^x \frac{1}{(x-t)^{1+\alpha(x)}} \frac{df(t)}{dx} dt + \frac{f(0^+) - f(0^-)}{\Gamma(1-\alpha(x))(x-t)^{1+\alpha(x)}}. \tag{1.130}$$

In 2003, Samko and Cardoso introduced that Θ is a general operator, given by [96]

$$\Theta(\varpi) = \int_{-\infty}^x \psi(x-t) \varpi(t) dt = f(x). \tag{1.131}$$

Then, there exists the inverse operator expressed by [25]

$$\varpi(x) = \Theta^{-1} f = \frac{d}{dx} \int_{-\infty}^x \varphi(x-t) f(t) dt, \tag{1.132}$$

where

$$\int_0^x \psi(x-t) \varphi(t) dt = 1. \tag{1.133}$$

In 2004, Kilbas, Saigo and Saxena introduced the following general fractional derivative by [97]

$$D_+^{(\alpha)} f(x) = \frac{d^n}{dx^n} \int_a^x (\tau - t)^{\mu+n-\beta-1} E_{\alpha,\mu+n-\beta}^{\varphi} (\varpi (x-t)^{\alpha}) \Omega(t) dt. \tag{1.134}$$

In 2007, Agrawal proposed the fractional derivative and integral operators in terms of the Riesz fractional derivative, given by [98]

$$D^{\alpha} f(x) = \frac{1}{\Gamma(n-\alpha)} \frac{d^n}{dx^n} \int_a^b \frac{f(t)}{|x-t|^{n-\alpha}} dt, \tag{1.135}$$

$$D^{\alpha} f(x) = \frac{1}{\Gamma(n-\alpha)} \int_a^b \frac{1}{|x-t|^{\alpha}} \frac{d^n f(t)}{dt^n} dt \tag{1.136}$$

and

$$I^{\alpha} f(x) = \frac{1}{2\Gamma(\alpha)} \int_a^b |x-t|^{\alpha-1} f(t)\, dt. \tag{1.137}$$

In 2010, Gajda and Magdziarz introduced the fractional derivative in the form (see [99])

$$D_{+}^{(\alpha)} f(x) = \frac{d}{dx} \int_0^x M(x-t) f(t)\, dt, \tag{1.138}$$

where the Laplace transform of the function $M(x)$, denoted by $M(s)$, is

$$M(s) = \frac{1}{(s+\lambda)^{\alpha} + \lambda^{\alpha}}. \tag{1.139}$$

In 2014, Meerschaert, Sabzikar, Phanikumar and Zeleke suggested the fractional derivative with fractional difference in the form [100]

$$D_{+}^{(\alpha)} f(t) = \lim_{h \to 0} \frac{\Delta_h^{\alpha,\lambda} f(x)}{h^{\alpha}}, \tag{1.140}$$

where

$$\Delta_h^{\alpha,\lambda} f(x) = \sum_{i=0}^{\infty} \varpi_i e^{-\lambda i h} f(x-ih) \tag{1.141}$$

with

$$\varpi_i = \frac{(-1)^i \Gamma(1+\alpha)}{\Gamma(1+i)\Gamma(1+\alpha-i)}. \tag{1.142}$$

In 2014, Garra et al. introduced the fractional derivative in the form [101]

$$D_{+}^{(\alpha)} f(x) = \int_a^x (\tau-t)^{\mu+n-\beta-1} E_{\alpha,\mu+n-\beta}^{\varphi} \left(\varpi (x-t)^{\alpha}\right) \frac{d^n}{dt^n} \Omega(t)\, dt, \tag{1.143}$$

where $E_{\alpha,\mu+n-\beta}^{\varphi}\left(\varpi (x-t)^{\alpha}\right)$ is the generalized Mittag-Leffler function.

In 2015, Caputo and Fabrizio introduced the fractional derivative with the exponential function in the form [102]

$$D_x^{(\alpha)} f(x) = \frac{(2-\alpha)\Im(\alpha)}{2(1-\alpha)} \int_a^x \exp\left(-\frac{\alpha}{1-\alpha}(x-t)\right) \frac{df(t)}{dt} dt, \tag{1.144}$$

where $\Im(\alpha)$ is a parameter.

In 2015, Zayernouri, Ainsworth and Karniadakis proposed the fractional derivatives in the form [103]

$$D_+^{(\alpha)} f(x) = \frac{e^{\lambda x}}{\Gamma(1-\alpha)} \frac{d}{dx} \int_a^x \frac{f(t)}{(x-t)^{\alpha+1}} e^{-\lambda t} dt \tag{1.145}$$

and

$$D_-^{(\alpha)} f(x) = \frac{e^{\lambda x}}{\Gamma(1-\alpha)} \frac{d}{dx} \int_x^b \frac{f(t)}{(x-t)^{\alpha+1}} e^{-\lambda t} dt. \tag{1.146}$$

In 2015, Yang, Srivastava and Machado introduced the fractional derivative with the exponential function by [104]

$$D_x^{(\alpha)} f(x) = \frac{(2-\alpha)\Im(\alpha)}{2(1-\alpha)} \frac{d}{dx} \int_a^x \exp\left(-\frac{\alpha}{1-\alpha}(x-t)\right) f(t)\, dt. \tag{1.147}$$

In 2016, Atangana and Baleanu proposed the fractional derivative with the Mittag-Leffler function [105]

$$D_x^{(\alpha)} f(x) = \frac{\Im(\alpha)}{1-\alpha} \int_a^x \mathrm{E}_\alpha\left(-\frac{\alpha}{1-\alpha}(x-t)^\alpha\right) \frac{df(t)}{dt} dt. \tag{1.148}$$

In 2015, Sabzikar, Meerschaert and Chen introduced the fractional integration by [106, 107, 108, 109]

$$I_-^{(\alpha)} f(t) = \frac{1}{\Gamma(\alpha)} \int_x^\infty (x-t)^{\alpha-1} e^{-\lambda(x-t)} f(t)\, dt, \tag{1.149}$$

and the fractional derivative by

$$D_+^{(\alpha)} f(t) = \frac{1}{\Gamma(\alpha)} \int_x^\infty (x-t)^{\alpha-1} e^{-\lambda(x-t)} f^{(\alpha)}(t)\, dt. \tag{1.150}$$

In 2016, Yang proposed the fractional derivatives as [110]

$$D_x^{(\alpha)} f(x) = \frac{1+\alpha^2}{\sqrt{\pi^\alpha (1-\alpha)}} \int_a^x \exp\left(-\frac{\alpha}{1-\alpha}(x-t)^{2\alpha}\right) \frac{df(t)}{dt} dt, \tag{1.151}$$

$$D_x^{(\alpha)} f(x) = \frac{\Im(\alpha(x))}{1-\alpha(x)} \int_a^x \exp\left(-(x-t)^{\alpha(x)}\right) \frac{df(t)}{dt} dt \tag{1.152}$$

and

$$D_x^{(\alpha(x))} f(x) = \frac{1}{\Gamma(1-\alpha(x))} \int_a^x \mathrm{E}_{\alpha(x)}\left(-(x-t)^{\alpha(x)}\right) \frac{df(t)}{dt} dt. \tag{1.153}$$

In 2016 Li and Deng proposed the fractional derivatives [111]

$$D_+^{(\alpha)} f(x) = \frac{e^{-\lambda x}}{\Gamma(n-\alpha)} \frac{d^n}{dx^n} \int_a^x \frac{f(t)}{(x-t)^{\alpha+1}} e^{\lambda t} dt \tag{1.154}$$

and

$$D_-^{(\alpha)} f(x) = \frac{(-1)^n e^{-\lambda x}}{\Gamma(n-\alpha)} \frac{d^n}{dx^n} \int_x^b \frac{f(t)}{(x-t)^{\alpha+1}} e^{\lambda t} dt. \tag{1.155}$$

In 2017, Torres introduced the fractional derivatives [112]

$$D_+^{(\alpha)} f(x) = \lambda^\alpha f(x) + \frac{\alpha}{\Gamma(1-\alpha)} \int_{-\infty}^x \frac{f(x)-f(t)}{(x-t)^{\alpha+1}} e^{-\lambda(x-t)} dt \tag{1.156}$$

and

$$D_-^{(\alpha)} f(x) = \lambda^\alpha f(x) + \frac{\alpha}{\Gamma(1-\alpha)} \int_x^\infty \frac{f(x)-f(t)}{(x-t)^{\alpha+1}} e^{-\lambda(x-t)} dt. \tag{1.157}$$

In 2017, Sun, Hao, Zhang and Baleanu reported the fractional derivative [113]

$$D_x^{(\alpha)} f(x) = \frac{\Gamma(1+\alpha)}{(1-\alpha)^{1/\alpha}} \int_0^x \exp\left(-\frac{\alpha}{1-\alpha}(x-t)^\alpha\right) \frac{df(t)}{dt} dt. \tag{1.158}$$

In 2017, Yang and Machado introduced the variable-order fractional derivative with the another function by [114]

$$D_a^{(\xi(x),\varphi)} f(x) = \frac{1}{\Gamma(1-\xi(t))} \int_a^x \frac{f_\varphi^{(1)}(t)}{(\varphi(x)-\varphi(t))^{\xi(x)}} dt. \tag{1.159}$$

In 2017, Dehghan, Abbaszadeh and Deng presented the fractional derivative in the form [115]

$$D_{+}^{(\alpha)} f(x) = \frac{1}{\Gamma(\alpha-1)} \int_{0}^{x} (x-t)^{\alpha+1} e^{-\lambda(x-t)} \frac{d^{\alpha} f(t)}{d|t|^{\alpha}} dt, \tag{1.160}$$

where $\frac{d^{\alpha} f(t)}{d|t|^{\alpha}}$ is the Riesz fractional derivative.

In 2017, Yang, Machado and Baleanu introduced [116]

$$D_{+}^{(\alpha)} f(x) = \int_{a}^{\tau} E_{\alpha,\upsilon}^{\varphi,\phi} \left(-(\tau-t)^{\alpha}\right) \frac{df(t)}{dt} dt, \tag{1.161}$$

$$D_{+}^{(\alpha)} f(x) = \frac{d}{d\tau} \int_{a}^{\tau} E_{\alpha,\upsilon}^{\varphi,\phi} \left(-(\tau-t)^{\alpha}\right) f(t) dt, \tag{1.162}$$

$$D_{+}^{(\alpha)} f(x) = \int_{a}^{\tau} E_{\alpha,\upsilon}^{\varphi} \left(-(\tau-t)^{\alpha}\right) \frac{df(t)}{dt} dt, \tag{1.163}$$

$$D_{+}^{(\alpha)} f(x) = \frac{d}{d\tau} \int_{a}^{\tau} E_{\alpha,\upsilon}^{\varphi} \left(-(\tau-t)^{\alpha}\right) f(t) dt, \tag{1.164}$$

$$D_{+}^{(\alpha)} f(x) = \int_{a}^{\tau} E_{\alpha,\upsilon} \left(-(\tau-t)^{\alpha}\right) \frac{df(t)}{dt} dt, \tag{1.165}$$

$$D_{+}^{(\alpha)} f(x) = \frac{d}{d\tau} \int_{a}^{\tau} E_{\alpha,\upsilon} \left(-(\tau-t)^{\alpha}\right) f(t) dt, \tag{1.166}$$

$$D_{+}^{(\alpha)} f(x) = \int_{a}^{\tau} E_{\alpha} \left(-(\tau-t)^{\alpha}\right) \frac{df(t)}{dt} dt, \tag{1.167}$$

$$D_{+}^{(\alpha)} f(x) = \frac{d}{d\tau} \int_{a}^{\tau} E_{\alpha} \left(-(\tau-t)^{\alpha}\right) f(t) dt, \tag{1.168}$$

$$D_{+}^{(\alpha)} f(x) = \int_{a}^{\tau} Exp_{\alpha} \left(-(\tau-t)^{\alpha}\right) \frac{df(t)}{dt} dt, \tag{1.169}$$

$$D_{+}^{(\alpha)} f(x) = \frac{d}{d\tau} \int_{a}^{\tau} Exp_{\alpha} \left(-(\tau-t)^{\alpha}\right) f(t) dt, \tag{1.170}$$

$$D_{+}^{(\alpha)} f(x) = \int_{a}^{\tau} Exp \left(-(\tau-t)\right) \frac{df(t)}{dt} dt, \tag{1.171}$$

and

$$D_{+}^{(\alpha)} f(x) = \frac{d}{d\tau} \int_{a}^{\tau} Exp \left(-(\tau-t)\right) f(t) dt. \tag{1.172}$$

In 2017, Yang, Gao, Machado and Baleanu proposed [117]

$$D_x^{(\varpi)}\Pi(x) = \frac{\alpha\wp(\alpha)}{1-\alpha}\int_a^x \sin c\left(-\frac{\alpha(x-t)}{1-\alpha}\right)\frac{df(t)}{dt}dt, \tag{1.173}$$

where

$$\sin c(x) = \frac{\sin(\pi x)}{\pi x}. \tag{1.174}$$

In 2018, Sousa and de Oliveira introduced the fractional derivative by [118]

$$D_+^{(\alpha)} f(x) = \left(\frac{\varpi(x)}{\frac{d\varphi(x)}{dx}}\frac{d}{dx}\right)\int_a^x E_\alpha\left(\omega(t)(\varphi(x)-\varphi(t))^\alpha\right)\Omega(t)\frac{d\varphi(t)}{dt}dt, \tag{1.175}$$

where $\varpi(x) = \frac{M(\alpha(x))}{1-\alpha(x)}$, $\omega(t) = -\frac{\alpha(t)}{1-\alpha(t)}$, $\varphi^{(1)}(t) = \frac{d\varphi(t)}{dt}$ and $E_\alpha(\varpi(x-t)^\alpha)$ is the generalized Mittag-Leffler function.

1.1.2 The contribution for generalized fractional calculus and applications

The fractional calculus has found the important applications in fields of mathematics, science and applied engineering, including fluid flow, heat transfer, rheology, electrical circuit, networks, electromagnetic theory, control theory and probability.

We would like to refer to the important works for fractional derivatives and integrals and related works from Holmgren [119], Davis [120], Hardy and Littlewood [121], Burkill [122], Copson [123], Zygmund [124], Bassam [125], Kuttner[126], Kesarwani [127], Kalisch [128], O'Neil [129], Kober [130], Gaer and Rubel [131], Love [132], Prabhakar [133], Rubin[134], Adams [135], Koeller [136], Ahern and Jevti [137], Glöckle and Nonnenmacher [138], Bagley and Torvik [139], Koeller [140], Srivastava, Saigo and Owa [141], Debnath [142], Ross [143], Agarwal, Lakshmikantham and Nieto [144], Magin [145], Engheia [146], Kilicman and Al Zhour [147], Machado, Kiryakova and Mainardi [148] and so on.

Meanwhile, I would like to refer to the books for the fractional calculus from Oldham and Spanier [149], Podlubny [150], Magin [151], Kilbas, Srivastava and Trujillo [152], Sabatier, Agrawal and Machado [153], Diethelm [154], Monje et al. [155], Tarasov [156], Ortigueira [157], Baleanu, Machado and Luo [158], Anastassiou [159], Klafter, Lim and Metzler [160], Meerschaert and Sikorskii [161], Baleanu, Diethelm, Scalas and Trujillo [162], Malinowska and Torres [163], Abbas, Benchohra and N'Guérékata [164], Yang [165], Uchaikin [167], Richard [168], Carpinteri and Mainardi [169], Atanackovic, Pilipovic, Stankovic and Zorica [170], Li and Zeng [171], Yong, Jinrong and Lu [172], Guo, Xueke and Fenghui [173], Lorenzo and Hartley [174], West [175], Ahmad et al. [176] and so on. More important contributions were noted in [176, 177, 178, 179, 180].

The generalized fractional calculus have the increasing interests for scientists and engineers to model the complex phenomena in science and engineering practice. In 1884, Sonine proposed a solution of the Abel type equation (see [26]) and derived the Sonine conditions, it is an important contribution to the generalized fractional derivatives and integrals. In 1930, Hille and Tamarkin proposed the Abel type integral equation of the second kind (see [37]) and derived the generalized fractional derivative involving the Mittag-Leffler function. Erdélyi and Kober introduced the generalized fractional derivative and the generalized fractional integral (see [43]) (which are called the Erdélyi-Kober fractional derivative and integral, respectively). Since Samko, Kilbas and Marichev and Kiryakova published their works on general fractional calculus (see [3, 181]), the generalized fractional calculus has played an important role in modelling the complex phenomena with the power-law behaviors in mathematical physics and engineering. For the details of the history of the generalized fractional calculus, readers refer to the results [182, 183, 184, 185, 186, 187, 188].

Where will fractional calculus go? Seemly, it is an open problem (see [189]).

1.2 History of special functions

The special functions are used in modern engineering and physical science, such as communication systems, electro-optics, nonlinear wave, propagation, electromagnetic theory, electric circuit theory, and quantum mechanics. To start with, let us introduce the history investigation of the special functions.

- **The Euler gamma function**

In 1729, Euler showed that (see [190]; also see [190, 194])

$$\Gamma(x) = \int_0^\infty e^{-t} t^{x-1} dt, Re(x) \geq -1, \tag{1.176}$$

which is called the Euler gamma function and it was introduced in 1809 by Legendre as (see [191, 239]):

$$\Gamma(2x) = \frac{2^{2x-1}}{\sqrt{\pi}} \Gamma(2x) \Gamma\left(x + \frac{1}{2}\right), 2x \neq -1, -2, ...$$

In 1877, Prym considered (see [240])

$$\Gamma(x) = \int_0^1 e^{-t} t^{x-1} dt + \int_1^\infty e^{-t} t^{x-1} dt,$$

which is called the Prym decomposition, and there are

$$\chi(x,a) = \int_0^a t^{x-1} e^{-t} dt \ \ (a \geq 0;\ Re(x) > 0)$$

and

$$\Gamma(x,a) = \int_a^\infty t^{x-1} e^{-t} dt \ \ (a \geq 0;\ Re(x) > 0 \text{ when } a = 0).$$

For more details of the history investigation of the Euler gamma functions, see [190, 191, 192, 193, 194, 195, 241, 242, 243, 244].

The properties of the incomplete gamma functions are presented as follows (see [233]):

1. For $Re(x) > -1$ there is the difference equation (see [190]; also see [245])

$$\Gamma(x+1) = \Gamma(x)\, x.$$

2. For $x \in \mathbb{C}$ there is the Euler reflection formula (see [192]; also see [246])

$$\Gamma(1-x)\,\Gamma(x) = \frac{\pi}{\sin(\pi x)}.$$

3. For $x \in \mathbb{C}$ there is (see [245, 248], also see [249])

$$\frac{1}{\Gamma(x+1)} = x e^{-\gamma x} \prod_{n=1}^{\infty} \left(1 + \frac{x}{n}\right) e^{-\frac{x}{n}},$$

   where γ is Euler's constant.

4. For $x, p \in \mathbb{C}$, $Re(x) \geq 0$ and $Re(p) \geq 0$, there is [250]

$$\int_0^\infty e^{-pt} t^{x-1} dt = \frac{\Gamma(p)}{p^x}.$$

5. For $x, p \in \mathbb{C}$, $Re(x) \geq 0$ and $Re(p) \geq 0$, there are (for the real parameters, see [232, 238, 250])

$$\int_0^\infty e^{-t^p} dt = \Gamma\left(\frac{1}{p} + 1\right),$$

$$\int_0^\infty e^{-t^p} t^{x-1} dt = \frac{1}{p}\Gamma\left(\frac{x}{p}\right)$$

   and

$$\int_0^\infty e^{-pt^2} t^{x-1} dt = \frac{p^{-\frac{x}{2}}}{2}\Gamma\left(\frac{x}{2}\right).$$

6. For $p, q \in \mathbb{C}$, $Re(q) \geq 0$ and $Re(p) \geq 0$, there is (for real parameters, see [232, 251]; also see [181])

$$\int_0^\infty e^{-pt^q} dt = p^{-\frac{1}{q}} \Gamma\left(1 + \frac{1}{q}\right).$$

7. For $x, p \in \mathbb{C}$, $Re(q) \geq 0$, $Re(x) \geq 0$ and $Re(p) \geq 0$, there is [212]

$$\int_0^\infty e^{-pt^q} t^{x-1} dt = \frac{p^{-\frac{x}{q}}}{q} \Gamma\left(\frac{x}{q}\right).$$

8. For $x \in \mathbb{C}$ there is (see [252]; also see [253])

$$\frac{1}{\Gamma(x)} = \frac{1}{2\pi i} \int_{\mathrm{E}} s^{-x} e^s ds,$$

where the contour of integration E is the Hankel contour.

9. For $x \in \mathbb{C}$ and $Re(x) > 1$, there is (see [253, 254, 255])

$$\Gamma(x) = \frac{1}{\cos(2\pi x)} \int_0^\infty t^{x-1} \cos t dt, -\frac{1}{2} + n < x < \frac{1}{2} + n, n \in \mathbb{N}_0.$$

10. For $x \in \mathbb{C}$ and $Re(x) > 1$, there is (see [253, 254, 255])

$$\Gamma(x) = \frac{1}{\sin(2\pi x)} \int_0^\infty t^{x-1} \sin t dt, -1 + n < x < 1 + n, n \in \mathbb{N}_0.$$

- **Euler beta function**

In 1772, Euler gave the beta function defined as [247]

$$B(x, y) = \int_0^1 e^{-t} t^{x-1} (1-t)^{y-1} dt, Re(x) > 0, Re(y) > 0, \tag{1.177}$$

which is called the Euler beta function.

In 1763, Bayes considered the incomplete beta function defined by [256]

$$B(x, y; m) = \int_0^m t^{x-1} (1-t)^{y-1} dt, x > 0, y > 0. \tag{1.178}$$

For more details of the history investigation of the incomplete gamma functions, see [257].

The property of the Euler beta function is presented as follows [253]:

$$B(x, y) = \frac{\Gamma(x)\Gamma(y)}{\Gamma(x+y)}.$$

- **The Gauss hypergeometric function**

The major development of the theory of the hypergeometric function was carried out by Gauss and published in 1812 (see [196]).

In 1812, Gauss introduced the series by [196]

$$1 + \frac{ab}{c}t + \frac{a(a+1)b(b+1)}{c(c+1)}\frac{t^2}{2!} + \cdots = \sum_{n=0}^{\infty} \frac{(a)_n (b)_n}{(c)_n}\frac{t^n}{n!}, \tag{1.179}$$

which is called the Gauss hypergeometric function (GHF), where the Pochhammer symbol, which was introduced by Pochhammer in 1870, is given as [197]

$$(a)_n = \frac{\Gamma(a+n)}{\Gamma(a)}, n = 0, 1, 2, \cdots. \tag{1.180}$$

- **The Bessel function**

In 1824, Bessel proposed the function (see [198]; also see [199, 200, 201])

$$J_p(t) = \sum_{n=0}^{\infty} \frac{(-1)^n}{\Gamma(k+p+1)}\frac{(t/2)^{2n-p}}{n!}, \tag{1.181}$$

which is called the Bessel function of the first kind of order.

- **The Kummer function**

In 1836, Kummer proposed the confluent hypergeometric function defined as follows [202]

$$M(a; c; t) = \sum_{n=0}^{\infty} \frac{(a)_n}{(c)_n}\frac{t^n}{n!}, \tag{1.182}$$

which is called the Kummer function or Kummer confluent hypergeometric function.

- **The Kohlrausch-Williams-Watts function**

In 1854, Kohlrausch introduced the stretched exponential function defined as [203]

$$e_\alpha(t) = e^{-t^\alpha}, \tag{1.183}$$

which is called the Kohlrausch function, Kohlrausch-Williams-Watts function or the Williams-Watts function due to the repeated work in 1970 [204].

- **The Bessel-Clifford function**

In 1882, Clifford introduced a Bessel type function in the form [205]

$$J_\alpha(t) = \sum_{k=0}^{\infty} \frac{\left(-t^2/4\right)^k}{(\alpha+1)_k \Gamma(k+1)}, \tag{1.184}$$

which is called the Bessel-Clifford function.

- **The Macdonald function**

In 1898, Macdonald introduced the function by [206]

$$K_\nu(t) = \frac{\pi}{2\sin(\nu\pi)} \left(I_{-\nu}(t) - I_\nu(t)\right), \tag{1.185}$$

which is called the Macdonald function, where

$$I_\nu(t) = \sum_{n=0}^{\infty} \frac{(t/2)^{\nu+2m}}{n!\Gamma(n+1+\nu)}. \tag{1.186}$$

- **The Mittag-Leffler function**

In 1903, the Mittag-Leffler function, introduced by Swedish mathematician Gosta Mittag-Leffler, is defined as [207]

$$E_\nu(t) = \sum_{n=0}^{\infty} \frac{t^n}{\Gamma(n\nu+1)}. \tag{1.187}$$

- **The Wiman function**

In 1905, Wiman introduced the function by [208]

$$E_{\nu,\upsilon}(t) = \sum_{n=0}^{\infty} \frac{t^n}{\Gamma(n\nu+\upsilon)}, \tag{1.188}$$

which is called the Wiman function.

In 1907, Barnes reported the hypergeometric function [209]

$${}_2F_1(a_1, a_2; c_1; t) = \sum_{n=0}^{\infty} \frac{(a_1)_n (a_2)_n}{(c_1)_n} \frac{t^n}{n!} = \frac{1}{2\pi i} \frac{\Gamma(c)}{\Gamma(a)\Gamma(b)} \int_L \frac{\Gamma(a+s)\Gamma(b+s)\Gamma(-s)}{\Gamma(c+s)} (-t)^s \, ds, \tag{1.189}$$

where the numbers 2 and 1 refer to the number of the numerator and denominator parameters, respectively.

- **The Wright function**

In 1933, Maitland Wright introduced the function defined by [210]

$$W_{\nu,\mu}(t)=\sum_{n=0}^{\infty}\frac{1}{\Gamma(n\nu+\mu)}\frac{t^n}{n!},\nu>-1,\mu\in\mathbb{C}, \tag{1.190}$$

which is called the Wright function.

- **The generalized Bailey hypergeometric function**

In 1935, Bailey introduced the generalized hypergeometric function by [211]

$$\begin{aligned}&{}_pF_q(a_1,a_2,\cdots,a_p;c_1,c_2,\cdots,c_q;t)\\&={}_pF_q\left(\begin{array}{c}a_1,a_1,\cdots,a_p;t\\c_1,c_1,\cdots,c_q\end{array}\right)\\&={}_pF_q\left(\begin{array}{c}\mathbf{a_p}\\\mathbf{b_q}\end{array}\middle|t\right)\\&=\sum_{n=0}^{\infty}\frac{(a_1)_n\cdots(a_p)_n}{(c_1)_n\cdots(c_q)_n}\frac{t^n}{n!},\end{aligned} \tag{1.191}$$

which is called the generalized Bailey hypergeometric function.

- **The Meijer G-function**

In 1936, Meijer addressed the function [212]

$$G_{p,q}^{m,n}\left(t\middle|\begin{array}{c}a_1,\cdots,a_p\\b_1,\cdots,b_q\end{array}\right)=\frac{1}{2\pi i}\int_L\frac{\prod_{j=1}^{m}\Gamma(b_j-s)\prod_{j=1}^{n}\Gamma(1-a_j+s)}{\prod_{j=n+1}^{p}\Gamma(a_j-s)\prod_{j=m+1}^{q}\Gamma(1-b_j+s)}t^s ds, \tag{1.192}$$

which is called the Meijer G-function function.

- **The MacRobert E-function**

In 1939, MacRobert reported the function [213]

$$E\left(\begin{array}{c}\mathbf{a_p}\\\mathbf{b_q}\end{array}\middle|t\right)=\begin{cases}\dfrac{\prod_{j=1}^{p}\Gamma(a_j)}{\prod_{j=1}^{q}\Gamma(b_j)}\,{}_pF_q\left(\begin{array}{c}\mathbf{a_p}\\\mathbf{b_q}\end{array}\middle|-t^{-1}\right)\ (p=q+1/and/\,|\mathrm{t}|>1),\\\dfrac{\prod_{j=1}^{p}\Gamma(a_j-a_h)}{\prod_{j=1}^{q}\Gamma(b_j-a_h)}\Gamma(a_h)t^{a_h}\\\times{}_{p+1}F_{q-1}\left(\begin{array}{c}a_h,1+a_h-b_1,\cdots,1+a_h-b_q;(-1)^{p-q}t\\1+a_h-a_1,\cdots,1+a_h-a_p\end{array}\right)\\(p=q+1\ and\ |\mathrm{t}|<1),\end{cases} \tag{1.193}$$

which is called the MacRobert E-function.

Remark that the generalized hypergeometric function and the MacRobert E-function had the same aim, but Meijer's G-function was able to include those as particular cases as well. Meijer's G-function is actually a special case of Meijer's G-function (see [214, 215, 216]).

- **The Rabotnov function**

In 1948, Rabotnov introduced the fractional exponential function by [217]

$$\Re_{\nu}(t) = \sum_{n=0}^{\infty} \frac{t^{(n+1)(\nu+1)-1}}{\Gamma((n+1)(\nu+1))}, \tag{1.194}$$

which is called the Rabotnov function.

- **The Agarwal function**

In 1953, Agarwal proposed the function [218]

$$A_{\nu,\upsilon}(t) = \sum_{n=0}^{\infty} \frac{t^{n\upsilon+\nu-1}}{\Gamma(n\upsilon+\nu)}, \tag{1.195}$$

which is called the Agarwal function.

- **The Erdélyi-Magnus-Oberhettinger-Tricomi function**

In 1954, Erdélyi, Magnus, Oberhettinger and Tricomi introduced the function by [219]

$$\mathrm{A}_{\nu,\upsilon}(t) = t^{\nu-1} \sum_{n=0}^{\infty} \frac{t^{n\upsilon}}{\Gamma(n\upsilon+\nu)}, \tag{1.196}$$

which is called the Erdélyi-Magnus-Oberhettinger-Tricomi function.

- **The Fox H-function**

In 1961, Charles Fox reported the function [220]

$$\begin{aligned} &H_{p,q}^{m,n}\left(t\left|\begin{matrix}(a_1,A_1),\cdots,(a_p,A_p)\\(b_1,B_1),\cdots,(b_q,B_p)\end{matrix}\right.\right)\\ &= \frac{1}{2\pi i}\int_L \frac{\prod_{j=1}^{m}\Gamma(b_j+\mathrm{B}_j s)\prod_{j=1}^{n}\Gamma(1-a_j-A_j s)}{\prod_{j=n+1}^{p}\Gamma(a_j+A_j s)\prod_{j=m+1}^{q}\Gamma(1-b_j-B_j s)} t^{-s} ds, \end{aligned} \tag{1.197}$$

which is called the Fox H-function.

- **The Prabhakar function**

In 1971, Prabhakar introduced the function by (see [221]; also see [222])

$$E_{\nu,\upsilon}^{\varphi}(t) = \sum_{n=0}^{\infty} \frac{(\varphi)_n}{\Gamma(n\nu+\upsilon)} \frac{t^n}{\Gamma(n+1)}, \tag{1.198}$$

which is called the Prabhakar function.

- **The Miller-Ross function**

In 1993, Miller and Ross introduced the function by [223]

$$\mathrm{M}_{\nu}\left(a,t\right)=\mathrm{M}_{\nu}\left(at^{\nu}\right)=t^{\nu}\sum_{n=0}^{\infty}\frac{a^{n}t^{n}}{\Gamma\left(n+1+\nu\right)},\tag{1.199}$$

which is called the Miller-Ross function.

- **The Gorenflo-Mainardi function**

In 1996, Gorenflo and Mainardi proposed the function [224]

$$A_{\nu,\upsilon}\left(a,t\right)=G_{\nu,\upsilon}\left(at^{\nu}\right)=\sum_{n=0}^{\infty}\frac{a^{n}t^{n\upsilon+\nu-1}}{\Gamma\left(n\upsilon+\nu\right)},\tag{1.200}$$

which is called the Gorenflo-Mainardi function.

- **The Lorenzo-Hartley R-function**

In 1999, Hartley and Lorenzo introduced the function by [225]

$$R_{\nu,\upsilon}\left(a,t\right)=R_{\nu,\upsilon}\left(at^{\nu}\right)=\sum_{n=0}^{\infty}\frac{a^{n}t^{(n+1)\upsilon-\nu-1}}{\Gamma\left((n+1)\,\upsilon-\nu\right)},\tag{1.201}$$

which is called the Lorenzo-Hartley R-function.

For more details of the theory of the special functions, see [226, 227, 228, 229, 230, 231, 232, 233, 234, 236, 237, 238].

1.3 Special functions with respect to another function

We now introduce the special functions with respect to another function, for example, the Gamma function with respect to another function and the Pochhammer symbol with respect to another function. The Gamma function with respect to another function was used to present the variable-order fractional derivative and integral by Samko and Ross in 1993 (see [93]). The definition has not been presented since 25 years ago. Here, we first give the definitions of the special functions.

The Gamma function with respect to another function, denoted by $\Gamma\left(\gamma\left(x\right)\right)$, is defined as

$$\Gamma\left(\gamma\left(x\right)\right)=\int_{0}^{\infty}e^{-t}t^{\gamma(x)-1}dt,\ \gamma\left(x\right)\geq-1.\tag{1.202}$$

More generally, the Gamma function with respect to another function denoted by $\Gamma(\gamma(x))$, is defined as

$$\Gamma(\gamma(x)) = \int_0^{\infty} e^{-t} t^{\gamma(x)-1} dt, \ \mathrm{Re}(\gamma(x)) \geq -1. \tag{1.203}$$

With the use of the motivation of the Prym decomposition, we define the lower Gamma function with respect to another function as

$$\chi(\gamma(x), a) = \int_0^{a} t^{\gamma(x)-1} e^{-t} dt \ \ (a \geq 0; \ Re(\gamma(x)) > 0) \tag{1.204}$$

and the upper Gamma function with respect to another function as

$$\Gamma(\gamma(x), a) = \int_a^{\infty} t^{\gamma(x)-1} e^{-t} dt \ \ (a \geq 0; \ Re(\gamma(x)) > 0 \text{ when } a = 0), \tag{1.205}$$

respectively.

As the direct results, we have

$$\Gamma(\gamma(x) + 1) = \Gamma(\gamma(x)) \gamma(x) \tag{1.206}$$

and

$$\chi(\gamma(x), a) + \Gamma(\gamma(x), a) = \Gamma(\gamma(x)). \tag{1.207}$$

The Pochhammer symbol with respect to another function is defined as

$$(\gamma(x))_\kappa = \frac{\Gamma(\gamma(x) + \kappa)}{\Gamma(\gamma(x))}. \tag{1.208}$$

The corresponding lower Srivastava-Chaudhry-Agarwal symbol with respect to another function is defined as

$$(\gamma(x); a)_\kappa = \frac{\chi(\gamma(x) + \kappa, a)}{\Gamma(\gamma(x))}, \tag{1.209}$$

where $x \in \mathbb{C}$, $\kappa \in \mathbb{C}$ and $a \geq 0$, and the upper Srivastava-Chaudhry-Agarwal symbol with respect to another function as

$$[\gamma(x); a]_\kappa = \frac{\Gamma(\gamma(x) + \kappa, a)}{\Gamma(\gamma(x))} \ \ (x \in \mathbb{C}, \ \gamma \in \mathbb{C}, \ a \geq 0). \tag{1.210}$$

The Gauss hypergeometric function with respect to another function is defined as

$$_2F_1(a(x), b(x); c(x); t) = \sum_{\kappa=0}^{\infty} \frac{(a(x))_\kappa (b(x))_\kappa}{(c(x))_\kappa} \frac{t^\kappa}{\kappa!}, \tag{1.211}$$

which can be represented in the form:

$$
\begin{aligned}
&{}_2F_1\left(a\left(x\right),b\left(x\right);c\left(x\right);t\right)\\
&=\tfrac{1}{2\pi i}\tfrac{\Gamma(c(x))}{\Gamma(a(x))\Gamma(b(x))}\int\limits_{L}\tfrac{\Gamma(a(x)+s)\Gamma(b(x)+s)\Gamma(-s)}{\Gamma(c(x)+s)}\left(-t\right)^{s}ds,
\end{aligned}
\tag{1.212}
$$

where the Pochhammer symbol with respect to another is denoted as

$$
\left(a\left(x\right)\right)_{\kappa}=\frac{\Gamma\left(a\left(x\right)+\kappa\right)}{\Gamma\left(a\left(x\right)\right)},\kappa=0,1,2,\cdots. \tag{1.213}
$$

The Bessel function with respect to another function of the first kind of order is defined as

$$
J_{p(x)}\left(t\right)=\sum_{\kappa=0}^{\infty}\frac{\left(-1\right)^{\kappa}}{\Gamma\left(k+p\left(x\right)+1\right)}\frac{\left(t/2\right)^{2\kappa-p(x)}}{\kappa!},p\left(x\right)\geq0,-\infty<x<\infty. \tag{1.214}
$$

The Kummer confluent hypergeometric function with respect to another function is defined as

$$
M\left(a\left(x\right);c\left(x\right);t\right)=\sum_{\kappa=0}^{\infty}\frac{\left(a\left(x\right)\right)_{\kappa}}{\left(c\left(x\right)\right)_{\kappa}}\frac{t^{\kappa}}{\kappa!},-\infty<x<\infty. \tag{1.215}
$$

The Bessel-Clifford function with respect to another function is defined as

$$
J_{\alpha(x)}\left(t\right)=\sum_{k=0}^{\infty}\frac{\left(-t^{2}/4\right)^{k}}{\left(\alpha\left(x\right)+1\right)_{k}\Gamma\left(k+1\right)}. \tag{1.216}
$$

The Macdonald function with respect to another function is defined as

$$
K_{\nu(x)}\left(t\right)=\frac{\pi}{2\sin\left(\nu\left(x\right)\pi\right)}\left(I_{-\nu(x)}\left(t\right)-I_{\nu(x)}\left(t\right)\right), \tag{1.217}
$$

where

$$
I_{\nu(x)}\left(t\right)=\sum_{\kappa=0}^{\infty}\frac{\left(t/2\right)^{\nu(x)+2\kappa}}{\kappa!\Gamma\left(\kappa+1+\nu\left(x\right)\right)}. \tag{1.218}
$$

The Mittag-Leffler function with respect to another function is defined as

$$
E_{\nu(x)}\left(t\right)=\sum_{\kappa=0}^{\infty}\frac{t^{\kappa}}{\Gamma\left(\kappa\nu\left(x\right)+1\right)}, \tag{1.219}
$$

where $t,\nu\left(x\right)\in\mathbb{C}$, $\Re\left(\nu\left(x\right)\right),\Re\left(\upsilon\left(x\right)\right)\in\mathbb{R}_{0}^{+}$, $\kappa\in\mathbb{N}$ and $\Gamma\left(\cdot\right)$ is the gamma function with respect to another function.

The Wiman function with respect to another function is defined as

$$
E_{\nu(x),\upsilon(x)}\left(t\right)=\sum_{\kappa=0}^{\infty}\frac{t^{\kappa}}{\Gamma\left(\kappa\nu\left(x\right)+\upsilon\left(x\right)\right)}, \tag{1.220}
$$

where $t, \nu(x), \upsilon(x) \in \mathbb{C}$, $\Re(\nu(x)), \Re(\upsilon(x)) \in \mathbb{R}_0^+$ and $\kappa \in \mathbb{N}$.

In particular, when $\upsilon(x) = \upsilon$, the Wiman function with respect to another function is defined as

$$E_{\nu(x),\upsilon}(t) = \sum_{\kappa=0}^{\infty} \frac{t^{\kappa}}{\Gamma(\kappa\nu(x) + \upsilon)}, \tag{1.221}$$

where $t, \nu(x), \upsilon \in \mathbb{C}$, $\Re(\nu(x)), \Re(\upsilon) \in \mathbb{R}_0^+$ and $\kappa \in \mathbb{N}$.

The Wright function with respect to another function is defined as

$$\mathrm{W}_{\nu(x),\mu(x)}(t) = \sum_{\kappa=0}^{\infty} \frac{1}{\Gamma(\kappa\nu(x) + \mu(x))} \frac{t^{\kappa}}{\kappa!}, \tag{1.222}$$

where $\nu(x) > -1$ and $\mu(x) \in \mathbb{C}$.

In particular, the Wright function with respect to another function is defined as

$$\mathrm{W}_{\nu(x),\mu}(t) = \sum_{\kappa=0}^{\infty} \frac{1}{\Gamma(\kappa\nu(x) + \mu)} \frac{t^{\kappa}}{\kappa!}, \tag{1.223}$$

where $\nu(x) > -1$ and $\mu \in \mathbb{C}$.

The Rabotnov function with respect to another function is defined as

$$\Re_{\nu(x)}(t) = \sum_{\kappa=0}^{\infty} \frac{t^{(\kappa+1)(\nu(x)+1)-1}}{\Gamma((\kappa+1)(\nu(x)+1))}. \tag{1.224}$$

The Agarwal function with respect to another function is defined as

$$A_{\nu(x),\upsilon(x)}(t) = \sum_{\kappa=0}^{\infty} \frac{t^{\kappa\upsilon(x)+\nu(x)-1}}{\Gamma(\kappa\upsilon(x) + \nu(x))}. \tag{1.225}$$

In particular, the Agarwal function with respect to another function is expressed as

$$A_{\nu(x),\upsilon}(t) = \sum_{\kappa=0}^{\infty} \frac{t^{\kappa\upsilon(x)+\nu-1}}{\Gamma(\kappa\upsilon(x) + \nu)}. \tag{1.226}$$

The Erdélyi-Magnus-Oberhettinger-Tricomi function with respect to another function is defined as

$$\mathrm{A}_{\nu(x),\upsilon(x)}(t) = \sum_{\kappa=0}^{\infty} \frac{t^{\kappa}}{\Gamma(\kappa\upsilon(x) + \nu(x))}. \tag{1.227}$$

The Fox H-function with respect to another function, denoted as

$$H_{p,q}^{m,n}\left(t\left|\begin{matrix}(\mathbf{a}(x), \mathbf{A}(x))\\(\mathbf{b}(x), \mathbf{B}(x))_{p,q}\end{matrix}\right.\right) = H_{p,q}^{m,n}\left(t\left|\begin{matrix}(a_1(x), A_1(x)), \cdots, (a_p(x), A_p(x))\\(b_1(x), B_1(x)), \cdots, (b_q(x), B_p(x))\end{matrix}\right.\right),$$

is defined as

$$
\begin{aligned}
&H_{p,q}^{m,n}\left(t\left|\begin{array}{l}(\mathbf{a}(x),\mathbf{A}(x))\\(\mathbf{b}(x),\mathbf{B}(x))_{p,q}\end{array}\right.\right)\\
&=H_{p,q}^{m,n}\left(t\left|\begin{array}{l}(a_1(x),A_1(x)),\cdots,(a_p(x),A_p(x))\\(b_1(x),B_1(x)),\cdots,(b_q(x),B_p(x))\end{array}\right.\right)\\
&=\frac{1}{2\pi i}\int_L\frac{\prod_{j=1}^{m}\Gamma\left(b_j(x)+\mathrm{B}_j(x)s\right)\prod_{j=1}^{n}\Gamma(1-a_j(x)-A_js(x))}{\prod_{j=n+1}^{p}\Gamma(a_j(x)+A_j(x)s)\prod_{j=m+1}^{q}\Gamma(1-b_j(x)-B_j(x)s)}t^{-s}ds.
\end{aligned}
\tag{1.228}
$$

The Prabhakar function with respect to another function is defined as

$$E_{\nu(x),\upsilon(x)}^{\varphi(x)}(t)=\sum_{\kappa=0}^{\infty}\frac{(\varphi(x))_\kappa}{\Gamma(n\nu(x)+\upsilon(x))}\frac{t^\kappa}{\Gamma(\kappa+1)}, \tag{1.229}$$

where $t,\nu(x),\upsilon(x),\varphi(x)\in\mathbb{C}$, $\Re(\nu(x)),\Re(\upsilon(x)),\Re(\varphi(x))\in\mathbb{R}_0^+$, and $\kappa\in\mathbb{N}$.

In particular, we have the following cases: (1) when $\varphi(x)=\varphi$,

$$E_{\nu(x),\upsilon(x)}^{\varphi}(t)=\sum_{\kappa=0}^{\infty}\frac{(\varphi)_\kappa}{\Gamma(n\nu(x)+\upsilon(x))}\frac{t^\kappa}{\Gamma(\kappa+1)}; \tag{1.230}$$

(2) when $\nu(x)=\nu$,

$$E_{\nu,\upsilon(x)}^{\varphi(x)}(t)=\sum_{\kappa=0}^{\infty}\frac{(\varphi(x))_\kappa}{\Gamma(n\nu+\upsilon(x))}\frac{t^\kappa}{\Gamma(\kappa+1)}; \tag{1.231}$$

(3) when $\upsilon(x)=\upsilon$,

$$E_{\nu(x),\upsilon}^{\varphi(x)}(t)=\sum_{\kappa=0}^{\infty}\frac{(\varphi(x))_\kappa}{\Gamma(n\nu(x)+\upsilon)}\frac{t^\kappa}{\Gamma(\kappa+1)}; \tag{1.232}$$

(4) when $\varphi(x)=\varphi$ and $\nu(x)=\nu$,

$$E_{\nu,\upsilon(x)}^{\varphi}(t)=\sum_{\kappa=0}^{\infty}\frac{(\varphi)_\kappa}{\Gamma(n\nu+\upsilon(x))}\frac{t^\kappa}{\Gamma(\kappa+1)}; \tag{1.233}$$

(5) when $\varphi(x)=\varphi$ and $\upsilon(x)=\upsilon$,

$$E_{\nu(x),\upsilon}^{\varphi}(t)=\sum_{\kappa=0}^{\infty}\frac{(\varphi)_\kappa}{\Gamma(n\nu(x)+\upsilon)}\frac{t^\kappa}{\Gamma(\kappa+1)}; \tag{1.234}$$

(6) when $\nu(x)=\nu$ and $\upsilon(x)=\upsilon$,

$$E_{\nu,\upsilon}^{\varphi(x)}(t)=\sum_{\kappa=0}^{\infty}\frac{(\varphi(x))_\kappa}{\Gamma(n\nu+\upsilon)}\frac{t^\kappa}{\Gamma(\kappa+1)}; \tag{1.235}$$

(7) when $\varphi(x) = 1$,

$$E_{\nu(x),\upsilon(x)}(t) = E^{1}_{\nu(x),\upsilon(x)}(t) = \sum_{\kappa=0}^{\infty} \frac{t^{\kappa}}{\Gamma(n\nu(x) + \upsilon(x))}; \tag{1.236}$$

(8) when $\varphi(x) = 1$ and $\upsilon(x) = 1$,

$$E_{\nu(x)}(t) = E^{1}_{\nu(x),1}(t) = \sum_{\kappa=0}^{\infty} \frac{t^{\kappa}}{\Gamma(n\nu(x) + 1)}. \tag{1.237}$$

The generalized Bailey hypergeometric function with respect to another function is defined as

$$\begin{aligned} &{}_pF_q(a_1(x), a_2(x), \cdots, a_p(x); c_1(x), c_2(x), \cdots, c_q(x); t) \\ &= {}_pF_q\left(\begin{matrix} a_1(x), a_2(x), \cdots, a_p(x); \\ c_1(x), c_2(x), \cdots, c_q(x); \end{matrix} \quad t \right) \\ &= {}_pF_q\left(\begin{matrix} a_1(x), a_2(x), \cdots, a_p(x) \\ c_1(x), c_2(x), \cdots, c_q(x) \end{matrix} \middle| \; t \right) \\ &= \sum_{n=0}^{\infty} \frac{(a_1(x))_{\kappa} \cdots (a_p(x))_{\kappa}}{(c_1(x))_{\kappa} \cdots (c_q(x))_{\kappa}} \frac{t^{\kappa}}{\kappa!}. \end{aligned} \tag{1.238}$$

The Miller-Ross function with respect to another function is defined as

$$\mathrm{M}_{\nu(x)}(a, t) = t^{\nu(x)} \sum_{\kappa=0}^{\infty} \frac{a^{\kappa} t^{\kappa}}{\Gamma(\kappa + 1 + \nu(x))}.$$

The Gorenflo-Mainardi function with respect to another function is defined as

$$G_{\nu(x),\upsilon(x)}(a, t) = \sum_{\kappa=0}^{\infty} \frac{a^{n} t^{n\upsilon(x)+\nu(x)-1}}{\Gamma(n\upsilon(x) + \nu(x))}. \tag{1.239}$$

In particular, when $\upsilon(x) = \upsilon$, the Gorenflo-Mainardi function with respect to another function can be written as

$$G_{\nu(x),\upsilon}(a, t) = \sum_{\kappa=0}^{\infty} \frac{a^{n} t^{n\upsilon+\nu(x)-1}}{\Gamma(n\upsilon + \nu(x))}, \tag{1.240}$$

and when $\nu(x) = \nu$, the Gorenflo-Mainardi function with respect to another function is given as

$$G_{\nu,\upsilon(x)}(a, t) = \sum_{\kappa=0}^{\infty} \frac{a^{n} t^{n\upsilon(x)+\nu-1}}{\Gamma(n\upsilon(x) + \nu)}. \tag{1.241}$$

The two-parametric Lorenzo-Hartley function with respect to another function is defined as

$$R_{\nu(x),\upsilon(x)}(a, t) = \sum_{\kappa=0}^{\infty} \frac{a^{n} t^{(n+1)\upsilon(x)-\nu(x)-1}}{\Gamma((n+1)\upsilon(x) - \nu(x))}. \tag{1.242}$$

In particular, when $\upsilon(x) = \upsilon$, two-parametric Lorenzo-Hartley function with respect to another function can be represented in the form:

$$R_{\nu(x),\upsilon}(a,t) = \sum_{\kappa=0}^{\infty} \frac{a^n t^{(n+1)\upsilon-\nu(x)-1}}{\Gamma((n+1)\upsilon-\nu(x))}, \tag{1.243}$$

and when $\nu(x) = \nu$, two-parametric Lorenzo-Hartley function with respect to another function can be represented as

$$R_{\nu,\upsilon(x)}(a,t) = \sum_{\kappa=0}^{\infty} \frac{a^n t^{(n+1)\upsilon(x)-\nu-1}}{\Gamma((n+1)\upsilon(x)-\nu)}. \tag{1.244}$$

Chapter 2

Fractional Derivatives of Constant Order and Applications

In this chapter, we start with the theory of function space containing a family of the Lebesgue measurable functions and absolutely continuous functions. We directly introduce the concepts of the fractional derivatives and fractional integrals within and without singular and non-singular kernels, and the fractional derivative and fractional integrals with respect to another function. The integral transforms of the fractional derivatives and fractional integrals will be discussed in detail. Finally, some applications in engineering are also considered.

2.1 Fractional derivatives within power-law kernel

To begin with the concepts of the fractional derivatives and fractional integrals, we introduce the following function spaces used in the book.

Definition 2.1 *Let* $\mathrm{M} = [a,b]$ $(-\infty \leq a < b \leq \infty)$ *be a finite or infinite interval of the real axis* $\mathbb{R} = (-\infty, \infty)$. *The space* $L_\kappa(a,b)\,(1 \leq \kappa \leq \infty)$ *is denoted as the set of those Lebesgue measurable functions* $f(t)$ *on* M *for which* $\|f\|_\kappa$, *where*

$$\|f\|_\kappa = \left(\int_a^b |f(t)|^\kappa\, dt\right)^{1/\kappa} \quad (1 \leq \kappa \leq \infty) \tag{2.1}$$

and

$$\|f\|_\infty = esssup_{a<t<b}\,|f(t)|\,. \tag{2.2}$$

Here $esssup_{t\in\mathrm{M}}\,|f(t)|$ *is the essential maximum of the function on* M *(see [258]; also see [3, 75, 152]).*

Definition 2.2 *(see [152]) We also need the weighted* L_κ*-space with the weight involving the function* $h(t)$ *defined in the interval* $\mathrm{M} = [a,b]$. *Such a space, which we denote by*

$$L^\kappa_{h(\cdot)}(a,b)\,(h(t) \in \mathbb{R}; 1 \leq \kappa \leq \infty)\,, \tag{2.3}$$

consists of those Lebesgue measurable functions $f(t)$ *on* (a,b) *for which* $\|f\|_{L^{\kappa}_{h(\cdot)}} < \infty$ *with*

$$\|f\|_{L^{\kappa}_{h(\cdot)}} = \left(\int_a^b |h(t) f(t)|^{\kappa} \frac{dt}{t} \right)^{1/\kappa} \quad (1 \leq \kappa \leq \infty) \tag{2.4}$$

and

$$\|f\|_{L^{\infty}_{h(\cdot)}} = esssup_{a<t<b} \left[|h(t) f(t)| \right]. \tag{2.5}$$

Definition 2.3 *(see [152]) We also need the weighted* L_{κ}*-space with the power weight. Such a space, which we denote as*

$$L^{\kappa}_{h}(a,b)\left(h \in \mathbb{R}; 1 \leq \kappa \leq \infty\right), \tag{2.6}$$

consists of those Lebesgue measurable functions $f(t)$ *on* (a,b) *for which* $\|f\|_{L^{\kappa}_{h}} < \infty$ *with*

$$\|f\|_{L^{\kappa}_{h}} = \left(\int_a^b \left|t^h f(t)\right|^{\kappa} \frac{dt}{t} \right)^{1/\kappa} \quad (1 \leq \kappa \leq \infty) \tag{2.7}$$

and

$$\|f\|_{L^{\infty}_{h}} = esssup_{a<t<b} \left[\left|t^h f(t)\right| \right]. \tag{2.8}$$

Definition 2.4 *(see [152]) We need the weighted* L_{κ}*-space with the weight involving the function* $g(t)$ *defined in the interval* $\mathrm{M} = [a,b]$*. Such a space, which we denote by*

$$L^{\kappa}_{g(\cdot)}(a,b)\left(g(t) \in \mathbb{R}; 1 \leq \kappa \leq \infty\right), \tag{2.9}$$

consists of those Lebesgue measurable functions $f(t)$ *on* (a,b) *for which* $\|f\|_{L^{\kappa}_{g(\cdot)}} < \infty$ *with*

$$\|f\|_{L^{\kappa}_{g(\cdot)}} = \left(\int_a^b |g(t) f(t)|^{\kappa} dt \right)^{1/\kappa} \quad (1 \leq \kappa \leq \infty) \tag{2.10}$$

and

$$\|f\|_{L^{\infty}_{g(\cdot)}} = esssup_{a<t<b} \left[|g(t) f(t)| \right]. \tag{2.11}$$

Definition 2.5 *(see [152]) We also need the weighted* L_{κ}*-space with the power weight. Such a space, which we denote as*

$$L^{\kappa}_{g}(a,b)\left(g \in \mathbb{R}; 1 \leq \kappa \leq \infty\right), \tag{2.12}$$

consists of those Lebesgue measurable functions $f(t)$ *on* (a,b) *for which* $\|f\|_{L_h^\kappa} < \infty$ *with*

$$\|f\|_{L_g^\kappa} = \left(\int_a^b |t^g f(t)|^\kappa \, dt \right)^{1/\kappa} \quad (1 \leq \kappa \leq \infty) \tag{2.13}$$

and

$$\|f\|_{L_g^\infty} = esssup_{a<t<b} \left[|t^g f(t)| \right]. \tag{2.14}$$

In particular, we have the following cases:
(1) When $h(t) = t^h$, there is

$$L_{h(\cdot)}^\kappa (a,b) = L_h^\kappa (a,b); \tag{2.15}$$

(2) When $g = 1/\kappa$, there is

$$L_{g(\cdot)}^\kappa (a,b) = L_\kappa (a,b); \tag{2.16}$$

(3) When $g(t) = 1/\kappa$, there is

$$L_{g(\cdot)}^\kappa (a,b) = L_\kappa (a,b); \tag{2.17}$$

(4) When $g(t) = t^g$, there is

$$L_{g(\cdot)}^\kappa (a,b) = L_g^\kappa (a,b). \tag{2.18}$$

For more details on the above spaces, see [3, 75, 152].

Definition 2.6 *(see [152]) Let* $\mathrm{M} = [a,b]$. *A function* $f(t)$ *is called absolutely continuous in an interval* M, *if for any* $\varepsilon > 0$ *there exists* $\delta > 0$ *such that for any finite set of the pairwise nonintersecting intervals* $[a_k, b_k] \in \mathrm{M}$, $\kappa = 1, 2, \cdots, n$, *the inequality*

$$\sum_{\iota=1}^{n} |f(g_i) - f(h_i)| < \varepsilon \tag{2.19}$$

holds, where

$$\sum_{\kappa=1}^{n} (b_k - a_k) < \delta. \tag{2.20}$$

The space of these functions is denoted as $AC(\mathrm{M})$ *(see [3, 75, 152]).*

Let now $\mathrm{M} = [a,b]$ $(-\infty \leq a < b \leq \infty)$ be a finite interval and let $AC[a,b]$ be the space of functions $f(t)$ which are absolutely continuous on $[a,b]$ (see [3, 75, 152]).

In 1968, Kolmogorov and Fomin observed that the space $AC(\mathrm{M})$ is in agreement with the space of the primitives of the Lebesgue summable functions (see [259]; also see [3, 75]):

$$f(t) \in AC(\mathrm{M}) \Leftrightarrow f(t) = h + \int_a^t \phi(t)\,dt, \tag{2.21}$$

where $\phi(t) \in L_\kappa$, which is called the Kolmogorov-Fomin condition.

It is to say that, if an absolutely continuous function $f(t)$ has a derivative $f^{(1)}(t) = \phi(t)$ almost everywhere on $[a,b]$, then there are $h = f(a)$ and $f(t) \in AC(\mathrm{M})$.

Definition 2.7 *(see [3, 75]; also see [152]) Let $AC^\kappa(\mathrm{M})$, where $\kappa = 1, 2, \cdots$, be the space of functions $f(t)$, which has the continuous derivatives up to order $\kappa - 1$ in the interval* M *with the condition given as follows:*

$$f^{(\kappa-1)}(t) \in AC^\kappa(\mathrm{M}). \tag{2.22}$$

It is to say that, if $f(t)$ has the continuous derivatives up to order $\kappa - 1$ in the interval such that $f^{(\kappa-1)}(t) \in AC^\kappa[a,b]$, then the space is represented in the form:

$$AC^\kappa[a,b] = \left\{f : [a,b] \to \mathbb{C} \text{ and } \left(D^{(\kappa-1)}f\right)(t) \in AC[a,b] \quad \left(D = \frac{d}{dt}\right)\right\}, \tag{2.23}$$

which is called the Samko-Kilbas-Marichev condition.

Lemma 2.1 *(see [152]) The space $AC^\kappa[a,b]$ consists of those and only those functions $f(t)$ which can be represented as*

$$f(t) = \left(I_{a+}^\kappa \phi\right)(t) + \sum_{j=0}^{\kappa} h_j (t-a)^j, \tag{2.24}$$

where $\phi(t) \in L(a,b)$ and $h_j \ (j \in \mathbb{N}_0)$ are arbitrary constants, and

$$\left(I_{a+}^\kappa \psi\right)(t) = \frac{1}{\Gamma(\kappa)} \int_a^t (\tau - t)^{\kappa-1} \phi(\tau)\,d\tau. \tag{2.25}$$

When $\phi(t) = f^{(\kappa)}(t)$, we have (see [152])

$$h_j = \frac{f^{(\kappa)}(a)}{\Gamma(1+\kappa)}. \tag{2.26}$$

We use a weighted modification of the space $AC^\kappa[a,b]$ $(-\infty \le a < b \le \infty)$, where the usual derivative $D = d/dt$ is replaced by the so-called $\delta-$derivative, defined as

$$\delta = tD \quad \left(D = \frac{d}{dt}\right). \tag{2.27}$$

In the case, we denote as $AC^{\kappa}_{\delta,g(\cdot)}[a,b]$ $(-\infty \le a < b \le \infty)$, involves the Lebesgue measurable functions φ on (a,b) such that $g(t)f(t)$ has $\delta-$derivatives up to order $\kappa - 1$ on $[a,b]$ and $\delta^{\kappa-1}[g(t)f(t)]$ is absolutely continuous on $[a,b]$:

$$\begin{aligned} &AC^{\kappa}_{\delta,g(\cdot)}[a,b] \\ &= \left\{f : [a,b] \to \mathbb{C} \text{ and } \delta^{(\kappa-1)}[g(t)f(t)] \in AC[a,b], g(t) \in \mathbb{R},\ \delta = t\tfrac{d}{dt}\right\}, \end{aligned} \tag{2.28}$$

we denote as $AC^{\kappa}_{\delta,h(\cdot)}[a,b]$ $(-\infty \le a < b \le \infty)$, involves the Lebesgue measurable functions φ on (a,b) such that $h(t)f(t)$ has $\delta-$derivatives up to order $\kappa - 1$ on $[a,b]$ and $\delta^{\kappa-1}[h(t)f(t)]$ is absolutely continuous on $[a,b]$:

$$\begin{aligned} &AC^{\kappa}_{\delta,h(\cdot)}[a,b] \\ &= \left\{f : [a,b] \to \mathbb{C} \text{ and } \delta^{(\kappa-1)}[h(t)f(t)] \in AC[a,b], h(t) \in \mathbb{R},\ \delta = t\tfrac{d}{dt}\right\}, \end{aligned} \tag{2.29}$$

we denote as $AC^{\kappa}_{\delta,g}[a,b]$ $(-\infty \le a < b \le \infty)$, involves the Lebesgue measurable functions φ on (a,b) such that $t^g f(t)$ has $\delta-$derivatives up to order $\kappa-1$ on $[a,b]$ and $\delta^{\kappa-1}[t^g f(t)]$ is absolutely continuous on $[a,b]$:

$$AC^{\kappa}_{\delta,g}[a,b] = \left\{f : [a,b] \to \mathbb{C} \text{ and } \delta^{(\kappa-1)}[t^g f(t)] \in AC[a,b], g \in \mathbb{R},\ \delta = t\frac{d}{dt}\right\}, \tag{2.30}$$

and we denote as $AC^{\kappa}_{\delta,h}[a,b]$ $(-\infty \le a < b \le \infty)$, involves the Lebesgue measurable functions φ on (a,b) such that $t^h f(t)$ has $\delta-$derivatives up to order $\kappa - 1$ on $[a,b]$ and $\delta^{\kappa-1}\left[t^h f(t)\right]$ is absolutely continuous on $[a,b]$:

$$AC^{\kappa}_{\delta,h}[a,b] = \left\{f : [a,b] \to \mathbb{C} \text{ and } \delta^{(\kappa-1)}\left[t^h f(t)\right] \in AC[a,b], h \in \mathbb{R},\ \delta = t\frac{d}{dt}\right\}. \tag{2.31}$$

We also use a weighted modification of the space $AC^{\kappa}[a,b]$, where the usual derivative $D = d/dt$ is replaced by the so-called $\mu-$derivative, defined as

$$\mu = e^{-\eta t} D \left(D = \frac{d}{dt}\right), \tag{2.32}$$

where η is the constant.

Similarly, we denote as $AC^{\kappa,\eta}_{\mu,g(\cdot)}[a,b]$ $(-\infty \le a < b \le \infty)$, involves the Lebesgue measurable functions φ on (a,b) such that $g(t)f(t)$ has $\mu-$derivatives up to order $\kappa - 1$ on $[a,b]$ and $\mu^{\kappa-1}[g(t)f(t)]$ is absolutely continuous on $[a,b]$:

$$\begin{aligned} &AC^{\kappa,\eta}_{\mu,g(\cdot)}[a,b] \\ &= \left\{f : [a,b] \to \mathbb{C} \text{ and } \mu^{(\kappa-1)}[g(t)f(t)] \in AC[a,b], g(t) \in \mathbb{R},\ \delta = e^{-\eta t}\tfrac{d}{dt}\right\}, \end{aligned} \tag{2.33}$$

we denote as $AC^{\kappa,\eta}_{\mu,h(\cdot)}[a,b]$ $(-\infty \le a < b \le \infty)$, involves the Lebesgue measurable functions φ on (a,b) such that $h(t)f(t)$ has $\mu-$derivatives up to order

$\kappa - 1$ on $[a, b]$ and $\mu^{\kappa-1}[h(t)f(t)]$ is absolutely continuous on $[a, b]$:

$$AC^{\kappa,\eta}_{\mu,h(\cdot)}[a,b] \\ = \left\{f:[a,b]\to\mathbb{C} \text{ and } \mu^{(\kappa-1)}[h(t)f(t)]\in AC[a,b], h(t)\in\mathbb{R},\ \delta=e^{-\eta t}\tfrac{d}{dt}\right\}, \tag{2.34}$$

we denote as $AC^{\kappa,\eta}_{\mu,g}[a,b]$ $(-\infty\le a<b\le\infty)$, involves the Lebesgue measurable functions φ on (a, b) such that $t^g f(t)$ has $\mu-$derivatives up to order $\kappa-1$ on $[a, b]$ and $\mu^{\kappa-1}[t^g f(t)]$ is absolutely continuous on $[a, b]$:

$$AC^{\kappa,\eta}_{\mu,g}[a,b] = \left\{f:[a,b]\to\mathbb{C} \text{ and } \mu^{(\kappa-1)}[t^g f(t)]\in AC[a,b], g\in\mathbb{R},\ \mu=e^{-\eta t}\frac{d}{dt}\right\}, \tag{2.35}$$

and we denote as $AC^{\kappa,\eta}_{\mu,h}[a,b]$ $(-\infty\le a<b\le\infty)$, involves the Lebesgue measurable functions φ on (a, b) such that $t^h f(t)$ has $\mu-$derivatives up to order $\kappa-1$ on $[a, b]$ and $\mu^{\kappa-1}\left[t^h f(t)\right]$ is absolutely continuous on $[a, b]$:

$$AC^{\kappa,\eta}_{\mu,h}[a,b] = \left\{f:[a,b]\to\mathbb{C} \text{ and } \mu^{(\kappa-1)}\left[t^h f(t)\right]\in AC[a,b], h\in\mathbb{R}, \mu=e^{-\eta t}\frac{d}{dt}\right\}. \tag{2.36}$$

In particular, we have the following cases:
(1) When $h(t) = t^h$, there is

$$AC^{\kappa}_{\delta,h(\cdot)}[a,b] = AC^{\kappa}_{\delta,h}[a,b]; \tag{2.37}$$

(2) When $g(t) = t^g$, there is

$$AC^{\kappa}_{\delta,g(\cdot)}[a,b] = AC^{\kappa}_{\delta,g}[a,b]; \tag{2.38}$$

(3) When $h(t) = t^h$, there is

$$AC^{\kappa,\eta}_{\mu,h(\cdot)}[a,b] = AC^{\kappa,\eta}_{\mu,h}[a,b]; \tag{2.39}$$

(4) When $g(t) = t^g$, there is

$$AC^{\kappa,\eta}_{\mu,g(\cdot)}[a,b] = AC^{\kappa,\eta}_{\mu,g}[a,b]; \tag{2.40}$$

(5) When $h(t) = 1$, there is (see [152])

$$AC^{\kappa}_{\delta,h(\cdot)}[a,b]\Big|_{h(t)=1} = AC^{\kappa}_{\delta,h}[a,b]\Big|_{h=0} = AC^{\kappa}_{\delta,1}[a,b] = AC^{\kappa}_{\delta}[a,b]; \tag{2.41}$$

(6) When $g(t) = 1$, there is

$$AC^{\kappa}_{\delta,g(\cdot)}[a,b]\Big|_{g(t)=1} = AC^{\kappa}_{\delta,g}[a,b]\Big|_{g=0} = AC^{\kappa}_{\delta,1}[a,b] = AC^{\kappa}_{\delta}[a,b]; \tag{2.42}$$

(7) When $h(t) = 1$, there is

$$AC^{\kappa,\eta}_{\mu,h(\cdot)}[a,b]\Big|_{h(t)=1} = AC^{\kappa,\eta}_{\mu,h}[a,b]\Big|_{h=0} = AC^{\kappa,\eta}_{\mu,1}[a,b] = AC^{\kappa,\eta}_{\mu}[a,b]; \tag{2.43}$$

(8) When $g(t) = 1$, there is

$$AC_{\mu,g(\cdot)}^{\kappa,\eta}[a,b]\Big|_{g(t)=1} = AC_{\mu,g(\cdot)}^{\kappa,\eta}[a,b]\Big|_{g=0} = AC_{\mu,1}^{\kappa,\eta}[a,b] = AC_{\mu}^{\kappa,\eta}[a,b]; \tag{2.44}$$

(9) When $h(t) = 1$ and $\kappa = 1$, there is (see [152])

$$AC_{\delta,h(\cdot)}^{1}[a,b]\Big|_{h(t)=1} = AC_{\delta,h}^{1}[a,b]\big|_{h=0} = AC_{\delta,1}^{1}[a,b] = AC_{\delta}^{1}[a,b] = AC_{\delta}[a,b]; \tag{2.45}$$

(10) When $g(t) = 1$ and $\kappa = 1$, there is

$$AC_{\delta,g(\cdot)}^{1}[a,b]\Big|_{g(t)=1} = AC_{\delta,g}^{1}[a,b]\big|_{g=0} = AC_{\delta,1}^{1}[a,b] = AC_{\delta}^{1}[a,b] = AC_{\delta}[a,b]; \tag{2.46}$$

(11) When $h(t) = 1$ and $\kappa = 1$, there is

$$AC_{\mu,h(\cdot)}^{1,\eta}[a,b]\Big|_{h(t)=1} = AC_{\mu,h}^{1,\eta}[a,b]\Big|_{h=0} = AC_{\mu,1}^{1,\eta}[a,b] = AC_{\mu}^{1,\eta}[a,b] = AC_{\mu}^{\eta}[a,b]; \tag{2.47}$$

(12) When $g(t) = 1$ and $\kappa = 1$, there is

$$AC_{\mu,g(\cdot)}^{1,\eta}[a,b]\Big|_{g(t)=1} = AC_{\mu,g(\cdot)}^{1,\eta}[a,b]\Big|_{g=0} = AC_{\mu,1}^{1,\eta}[a,b] = AC_{\mu}^{1,\eta}[a,b] = AC_{\mu}^{\eta}[a,b]. \tag{2.48}$$

Lemma 2.2 *(see [152]) Let $-\infty \le a < b \le \infty$, $\kappa \in \mathbb{N}$ and $h \in \mathbb{R}$. The space $AC_{\delta,h}^{\kappa}[a,b]$ consists of those and only those functions $f(t)$ which are represented as*

$$f(t) = t^{h}\left[\left(I_{a+}^{\kappa}\phi\right)(t) + \sum_{j=0}^{\kappa} h_j \left(\log\frac{\tau}{t}\right)^{j}\right], \tag{2.49}$$

where $\phi(t) \in L(a,b)$ and h_j $(j \in \mathbb{N}_0)$ are arbitrary constants, and

$$\left(I_{a+}^{\kappa}\phi\right)(t) - \frac{1}{\Gamma(\kappa)}\int_{a}^{t}\left(\log\frac{\tau}{t}\right)^{\kappa-1}\phi(\tau)\frac{d\tau}{\tau}. \tag{2.50}$$

In particular, $f(t) \in AC_{\delta}^{\kappa}[a,b]$ if and only if

$$f(t) = \left(I_{a+}^{\kappa}\phi\right)(t) + \sum_{j=0}^{\kappa} h_j \left(\log\frac{\tau}{t}\right)^{j}, \tag{2.51}$$

where $\phi(t) \in L(a,b)$ and h_j $(j \in \mathbb{N}_0)$ are arbitrary constants, and

$$\left(I_{a+}^{\kappa}\phi\right)(t) = \frac{1}{\Gamma(\kappa)}\int_{a}^{t}\left(\log\frac{\tau}{t}\right)^{\kappa-1}\phi(\tau)\,d\tau. \tag{2.52}$$

Moreover, in Eq. (2.47) there are

$$\phi(t) = \frac{d}{dt}\left[\delta^{\kappa-1}\left(t^h f(t)\right)\right] \tag{2.53}$$

and

$$h_j = \frac{\delta^{\kappa}\left[a^h f(a)\right]}{\Gamma(1+\kappa)}, \tag{2.54}$$

while, in Eq. (2.47), there are

$$\phi(t) = \frac{d}{dt}\left[\delta^{\kappa-1} f(t)\right] \tag{2.55}$$

and

$$h_j = \frac{\delta^{\kappa} f(a)}{\Gamma(1+\kappa)}. \tag{2.56}$$

Let $\mathrm{M} = [a,b]$ $(-\infty \le a < b \le \infty)$ and $\kappa \in \mathbb{N}_0$. We denote $H^{\kappa}(\mathrm{M})$ as a space of functions $f(t)$ which are κ times continuously differentiate on M with the norm (see [152])

$$\|f\|_{H^{\kappa}} = \sum_{j=1}^{\kappa}\left\|f^{(j)}\right\|_H = \sum_{j=1}^{\kappa}\max_{t\in \mathrm{M}}\left|f^{(j)}(t)\right|, \ \kappa \in \mathbb{N}_0. \tag{2.57}$$

In particular, for $\kappa = 1$, $H^0(\mathrm{M}) = H(\mathrm{M})$ is the space of continuous functions f on M with the norm (see [152])

$$\|f\|_H = \max_{t\in \mathrm{M}}|f(t)|. \tag{2.58}$$

When $\mathrm{M} = [a,b]$ is a finite interval and $h \in \mathbb{C}$ $(0 \le Re(h) < 1)$, we introduce the weighted space, denoted as $H_h[a,b]$, of function $f(t)$ on $(a,b]$, such that the function $(t-a)^h f(t) \in H[a,b]$, and (see [152])

$$\|f\|_{H_h} = \left\|(t-a)^h f(t)\right\|_H, \ H_0[a,b] = H[a,b]. \tag{2.59}$$

For $\kappa \in \mathbb{N}$ we denote by $H_h^{\kappa}[a,b]$ the Banach space of functions $f(t)$ which are continuously differentiable on $[a,b]$ up to order $\kappa - 1$ and have the derivative $f^{(\kappa)}(t)$ of order κ on $(a,b]$ such that $f^{(\kappa)}(t) \in H_h[a,b]$ (see [152]):

$$H_h^{\kappa}[a,b] = \left\{ f : \|f\|_{H_h^{\kappa}} = \sum_{j=0}^{\kappa-1}\left\|f^{(j)}\right\|_H + \|f\|_{H_h}\right\}. \tag{2.60}$$

When $\kappa = 0$, there is (see [152])

$$H_h^0[a,b] = H_h[a,b] = \left\{ f : \|f\|_{H_h} = \left\|(t-a)^h f(t)\right\|_H \right\}. \tag{2.61}$$

Lemma 2.3 *(see [152]) Let $\kappa \in \mathbb{N}_0$ and $h \in \mathbb{C}$ $(0 \le Re(h) < 1)$. The space $H_h^\kappa[a,b]$ consists of those and only those functions $f(t)$ which are represented as*

$$f(t) = \left(I_{a+}^{\kappa}\phi\right)(t) + \sum_{j=0}^{\kappa-1} h_j (t-a)^j, \tag{2.62}$$

where $\phi(t) \in H_h[a,b]$ and h_j $(j \in \mathbb{N}_0)$ are arbitrary constants, and

$$\left(I_{a+}^{\kappa}\phi\right)(t) = \frac{1}{\Gamma(\kappa)} \int_a^t (\tau - t)^{\kappa-1} \phi(\tau)\, d\tau. \tag{2.63}$$

Moreover, there are (see [152])

$$\phi(t) = f^{(\kappa)}(t) \tag{2.64}$$

and

$$h_j = \frac{f^{(\kappa)}(a)}{\Gamma(\kappa+1)}. \tag{2.65}$$

Let $h \in \mathbb{C}$ $(0 \le Re(h) < 1)$ and $\mathrm{M} = [a,b]$ $(-\infty \le a < b \le \infty)$ be a finite interval. We introduce the weighted space, denoted as $H_{h,\log}[a,b]$, of functions $f(t)$ on $(a,b]$ and such that $(\log(t/a))^h f(t) \in H[a,b]$ (see [152]):

$$\|f\|_{H_{h,\log}} = \left\| \left(\log\left(\frac{t}{a}\right)\right)^h f(t) \right\|_H$$

and

$$H_{0,\log}[a,b] = H[a,b].$$

For $\kappa \in \mathbb{N}$ we denote by $H_{\delta,h}^{\kappa}[a,b]$ the Banach space of functions $f(t)$ which have continuous δ−derivatives on $[a,b]$ up to order $\kappa - 1$ and derivative $\delta^\kappa f(t)$ on $(a,b]$ of order κ such that $\delta^\kappa f(t) \in H_{h,\log}[a,b]$: (see [152])

$$H_{h,\log}^{\kappa}[a,b] = \left\{ f : \|f\|_{H_{h,\log}^{\kappa}} = \sum_{j=0}^{\kappa-1} \left\|\delta^{(j)} f\right\|_H + \left\|\delta^{(\kappa)} f\right\|_{H_{h,\log}} \right\}. \tag{2.66}$$

Lemma 2.4 *(see [152]) Let $-\infty \le a < b \le \infty$, $\kappa \in \mathbb{N}_0$ and $h \in \mathbb{C}$ $(0 \le Re(h) < 1)$. The space, denoted by $H_{h,\log}^{\kappa}[a,b]$, consists of those and only those functions $f(t)$ which are represented as*

$$f(t) = \left(I_{a+}^{\kappa}\phi\right)(t) + \sum_{j=0}^{\kappa} h_j \left(\log\frac{\tau}{t}\right)^j, \tag{2.67}$$

where $\phi(t) \in H_{h,\log}[a,b]$ and h_j $(j \in \mathbb{N}_0)$ are arbitrary constants, and

$$\left(I_{a+}^{\kappa}\phi\right)(t) = \frac{1}{\Gamma(\kappa)} \int_a^t \left(\log\frac{\tau}{t}\right)^{\kappa-1} \phi(\tau)\, \frac{d\tau}{\tau}. \tag{2.68}$$

Moreover,

$$\phi(t) = \delta^{\kappa} f(t) \tag{2.69}$$

and

$$h_j = \frac{\delta^{\kappa} f(a)}{\Gamma(1+\kappa)}. \tag{2.70}$$

Generally, we have the following results.

Let $h \in \mathbb{C}$ $(0 \leq Re(h) < 1)$ and $\mathrm{M} = [a,b]$ $(-\infty \leq a < b \leq \infty)$ be a finite interval. We introduce the weighted space, denoted as $H_{h,g(\cdot)}[a,b]$, of functions $f(t)$ on $(a,b]$ and such that $g^h(t) f(t) \in H[a,b]$ (see [152]):

$$\|f\|_{H_{h,g(\cdot)}} = \left\|g^h(t) f(t)\right\|_H$$

and

$$H_{0,g(\cdot)}[a,b] = H[a,b].$$

For $\kappa \in \mathbb{N}$ we denote by $H^{\kappa}_{\delta,h}[a,b]$ the Banach space of functions $f(t)$ which have continuous δ–derivatives on $[a,b]$ up to order $\kappa - 1$ and derivative $\delta^{\kappa} f(t)$ on $(a,b]$ of order κ such that $\delta^{\kappa} f(t) \in H_{h,g(\cdot)}[a,b]$:

$$H^{\kappa}_{h,g(\cdot)}[a,b] = \left\{ f : \|f\|_{H^{\kappa}_{h,g(\cdot)}} = \sum_{j=0}^{\kappa-1} \left\|\delta^{(j)} f\right\|_H + \left\|\delta^{(\kappa)} f\right\|_{H_{h,g(\cdot)}} \right\}. \tag{2.71}$$

Lemma 2.5 *Let $-\infty \leq a < b \leq \infty$, $\kappa \in \mathbb{N}_0$ and $h \in \mathbb{C}$ $(0 \leq Re(h) < 1)$. The space, denoted by $H^{\kappa}_{h,g(\cdot)}[a,b]$, consists of those and only those functions $f(t)$ which are represented as*

$$f(t) = \left(I^{\kappa}_{a+}\phi\right)(t) + \sum_{j=0}^{\kappa} h_j g^j(t), \tag{2.72}$$

where $\phi(t) \in H_{h,g(\cdot)}[a,b]$ and h_j $(j \in \mathbb{N}_0)$ are arbitrary constants, and

$$\left(I^{\kappa}_{a+}\phi\right)(t) = \frac{1}{\Gamma(\kappa)} \int_a^t g^{\kappa-1}(\tau)\,\phi(\tau)\,d\tau. \tag{2.73}$$

In particular, we consider the following case:

Let $h \in \mathbb{C}$ $(0 \leq Re(h) < 1)$ and $\mathrm{M} = [a,b]$ $(-\infty \leq a < b \leq \infty)$ be a finite interval. We introduce the weighted space, denoted as $H_{h,g(\cdot)}[a,b]$, of functions $f(t)$ on $(a,b]$ and such that $(g(t) - g(a))^h f(t) \in H[a,b]$ (see [176]):

$$\|f\|_{H_{h,g(\cdot)}} = \left\|(g(t) - g(a))^h f(t)\right\|_H$$

and

$$H_{0,g(\cdot)}[a,b] = H[a,b].$$

For $\kappa \in \mathbb{N}$ we denote by $H^{\kappa}_{\delta,h}[a,b]$ the Banach space of functions $f(t)$ which have continuous $\delta-$derivatives on $[a,b]$ up to order $\kappa-1$ and derivative $\delta^{\kappa} f(t)$ on $(a,b]$ of order κ such that $\delta^{\kappa} f(t) \in H_{h,g(\cdot)}[a,b]$:

$$H^{\kappa}_{h,g(\cdot)}[a,b] = \left\{ f : \|f\|_{H^{\kappa}_{h,g(\cdot)}} = \sum_{j=0}^{\kappa-1} \left\|\delta^{(j)} f\right\|_{H} + \left\|\delta^{(\kappa)} f\right\|_{H_{h,g(\cdot)}} \right\}. \tag{2.74}$$

Lemma 2.6 *Let $-\infty \le a < b \le \infty$, $\kappa \in \mathbb{N}_0$ and $h \in \mathbb{C}$ $(0 \le Re(h) < 1)$. The space, denoted by $H^{\kappa}_{h,g(\cdot)}[a,b]$, consists of those and only those functions $f(t)$ which are represented as*

$$f(t) = \left(I^{\kappa}_{a+}\phi\right)(t) + \sum_{j=0}^{\kappa} h_j \left(g(t) - g(a)\right)^j, \tag{2.75}$$

where $\phi(t) \in H_{h,g(\cdot)}[a,b]$ and h_j $(j \in \mathbb{N}_0)$ are arbitrary constants, and

$$\left(I^{\kappa}_{a+}\phi\right)(t) = \frac{1}{\Gamma(\kappa)} \int_{a}^{t} \left(g(\tau) - g(a)\right)^{\kappa-1} \phi(\tau)\, g^{(1)}(\tau)\, d\tau. \tag{2.76}$$

Moreover,

$$\phi(t) = h g^{(h)}(t) \frac{d}{dt}\left[\delta^{\kappa-1}\left(\left(g(t) - g(a)\right)^h f(t)\right)\right] \tag{2.77}$$

and

$$h_j = \frac{\delta^{\kappa}\left[g^h(a)\, f(a)\right]}{\Gamma(1+\kappa)}. \tag{2.78}$$

When $g(t) = t^r$, where $r \in \mathbb{R}_+$, we give the following results:

Lemma 2.7 *Let $-\infty \le a < b \le \infty$, $\kappa \in \mathbb{N}_0$ and $h \in \mathbb{C}$ $(0 \le Re(h) < 1)$. The space, denoted by $H^{\kappa}_{h,(\cdot)^r}[a,b]$, consists of those and only those functions $f(t)$ which are represented as*

$$f(t) = \left(I^{\kappa}_{a+}\phi\right)(t) + \sum_{j=0}^{\kappa} h_j \left(t^r - a^r\right)^j, \tag{2.79}$$

where $\phi(t) \in H_{h,(\cdot)^r}[a,b]$ and h_j $(j \in \mathbb{N}_0)$ are arbitrary constants, and

$$\left(I^{\kappa}_{a+}\phi\right)(t) = \frac{r}{\Gamma(\kappa)} \int_{a}^{t} \left(t^r - a^r\right)^{\kappa-1} \phi(\tau)\, \tau^{r-1} d\tau. \tag{2.80}$$

Moreover,

$$\phi(t) = r t^{r-1} \frac{d}{dt}\left[\delta^{\kappa-1}\left(\left(t^r - a^r\right) f(t)\right)\right] \tag{2.81}$$

and

$$h_j = \frac{\delta^{\kappa}\left[a^r f(a)\right]}{\Gamma(1+\kappa)}. \tag{2.82}$$

When $g(t) = e^{rt}$, where $r \in \mathbb{R}_+$, we have the following results:

Lemma 2.8 *Let $-\infty \leq a < b \leq \infty$, $\kappa \in \mathbb{N}_0$ and $h \in \mathbb{C}$ $(0 \leq Re(h) < 1)$. The space, denoted by $H^{\kappa}_{h,e^{r(\cdot)}}[a,b]$, consists of those and only those functions $f(t)$ which are represented as*

$$f(t) = \left(I^{\kappa}_{a+}\phi\right)(t) + \sum_{j=0}^{\kappa} h_j \left(e^{rt} - e^{ra}\right)^j, \tag{2.83}$$

where $\phi(t) \in H_{h,e^{r(\cdot)}}[a,b]$ and h_j $(j \in \mathbb{N}_0)$ are arbitrary constants, and

$$\left(I^{\kappa}_{a+}\phi\right)(t) = \frac{r}{\Gamma(\kappa)} \int_a^t \left(e^{rt} - e^{ra}\right)^{\kappa-1} \phi(\tau) e^{r\tau} d\tau. \tag{2.84}$$

Moreover, there are

$$\phi(t) = re^{rt}\frac{d}{dt}\left[\delta^{\kappa-1}\left(\left(e^{rt} - e^{ra}\right) f(t)\right)\right] \tag{2.85}$$

and

$$h_j = \frac{\delta^{\kappa}\left[e^{ra} f(a)\right]}{\Gamma(1+\kappa)}. \tag{2.86}$$

More generally, we have the following results.

Let $h \in \mathbb{C}$ $(0 \leq Re(h) < 1)$ and $\mathrm{M} = [a,b]$ $(-\infty \leq a < b \leq \infty)$ be a finite interval. We introduce the weighted space, denoted as $H_{g(\cdot)}[a,b]$, of functions $f(t)$ on $(a,b]$ and such that $g(t) f(t) \in H[a,b]$:

$$\|f\|_{H_{g(\cdot)}} = \|g(t) f(t)\|_H$$

and

$$H_{g(\cdot)}[a,b] = H[a,b].$$

For $\kappa \in \mathbb{N}$ we denote by $H^{\kappa}_{\delta}[a,b]$ the Banach space of functions $f(t)$ which have continuous δ–derivatives on $[a,b]$ up to order $\kappa - 1$ and derivative $\delta^{\kappa} f(t)$ on $(a,b]$ of order κ such that $\delta^{\kappa} f(t) \in H_{g(\cdot)}[a,b]$:

$$H^{\kappa}_{g(\cdot)}[a,b] = \left\{ f : \|f\|_{H^{\kappa}_{g(\cdot)}} = \sum_{j=0}^{\kappa-1} \left\|\delta^{(j)} f\right\|_H + \left\|\delta^{(\kappa)} f\right\|_{H_{g(\cdot)}} \right\}. \tag{2.87}$$

Lemma 2.9 *Let $-\infty \leq a < b \leq \infty$, $\kappa \in \mathbb{N}_0$ and $h \in \mathbb{C}$ $(0 \leq Re(h) < 1)$.*

The space, denoted by $H^{\kappa}_{g(\cdot)}[a,b]$, *consists of those and only those functions* $f(t)$ *which are represented as*

$$f(t) = \left(I^{\kappa}_{a+}\phi\right)(t) + \sum_{j=0}^{\kappa} h_j g^j(t), \tag{2.88}$$

where $\phi(t) \in H_{g(\cdot)}[a,b]$ *and* h_j $(j \in \mathbb{N}_0)$ *are arbitrary constants, and*

$$\left(I^{\kappa}_{a+}\phi\right)(t) = \frac{1}{\Gamma(\kappa)} \int_a^t g^{\kappa-1}(\tau)\,\phi(\tau)\,g^{(1)}(\tau)\,d\tau. \tag{2.89}$$

Moreover, there are

$$\phi(t) = g^{(1)}(t)\frac{d}{dt}\left[\delta^{\kappa-1}\left(g(t)\,f(t)\right)\right] \tag{2.90}$$

and

$$h_j = \frac{\delta^{\kappa}\left[g(a)\,f(a)\right]}{\Gamma(1+\kappa)}. \tag{2.91}$$

For the Banach spaces X and Y, we denote by the continuous embedding $X \to Y$(see [152]):

(1) if $f \in X$, then $f \in Y$;

(2) $\|f\|_Y \le \theta \|f\|_X$;

where the constant $\theta > 0$ does not depend on f.

Property 2.1 *Let* $\kappa \in \mathbb{N}_0$, *and let* σ_1 *and* σ_2 *be complex numbers such that*

$$0 \le Re(\sigma_1) \le Re(\sigma_2) < 1. \tag{2.92}$$

Then, the following embedding formulas hold [152]:

(1)

$$H^{\kappa}[a,b] \to H^{\kappa}_{\sigma_1}[a,b] \to H^{\kappa}_{\sigma_2}[a,b] \tag{2.93}$$

with $\|f\|_{H^{\kappa}_{\sigma_2}} \le \theta \|f\|_{H^{\kappa}_{\sigma_1}}$, *where*

$$\theta = \min\left\{1, (b-a)^{Re(\sigma_2-\sigma_1)}\right\}.$$

(2)

$$H^{\kappa}_{\delta}[a,b] \to H^{\kappa}_{\delta,\sigma_1}[a,b] \to H^{\kappa}_{\delta,\sigma_2}[a,b] \tag{2.94}$$

with $\|f\|_{H^{\kappa}_{\delta,\sigma_2}} \le \theta \|f\|_{H^{\kappa}_{\delta,\sigma_1}}$, *where*

$$\theta = \min\left\{1, \left(\log\frac{b}{a}\right)^{Re(\sigma_2-\sigma_1)}\right\}.$$

2.2 Riemann-Liouville fractional calculus

Today, as the first view, the derivatives of arbitrary order are usually defined by way of the so-called Riemann-Liouville definition (Riemann-Liouville definition of fractional integral) suggested by Holmgren in 1866 (see [119]; also see [260]). There is another view that Sonine first presents the definition of the Riemann–Liouville operator (see [261]; also see [261, 262]). We now present the two equations by Krug in 1890 represented in the forms (see [263]; also see [168]).

2.2.1 Riemann-Liouville fractional integrals

Let $\mathrm{M} = [a, b]$ be a finite interval $h \in \mathbb{C}$ $(0 \leq Re(h) < 1)$ and let $[Re(\alpha)]$ be the integer part of $Re(\alpha)$.

Definition 2.8 *Let* $\mathrm{M} = [a, b]$ $(-\infty \leq a < b \leq \infty)$ *be a finite interval on the real axis* $\mathbb{R}$. *The Riemann-Liouville fractional integrals* $\left(I_{a+}^{\alpha} f\right)(t)$ *and* $\left(I_{b-}^{\alpha} f\right)(t)$ *of order* $\alpha \in \mathbb{C}$ $(Re(\alpha) > 0)$ *are defined as (see [2, 19, 143, 261, 262, 263])*

$$\left(I_{a+}^{\alpha} f\right)(t) = \frac{1}{\Gamma(\alpha)} \int_{a}^{t} \frac{f(\tau)}{(\tau - t)^{1-\alpha}} d\tau \tag{2.95}$$

and

$$\left(I_{b-}^{\alpha} f\right)(t) = \frac{1}{\Gamma(\alpha)} \int_{t}^{b} \frac{f(\tau)}{(t - \tau)^{1-\alpha}} d\tau, \tag{2.96}$$

respectively, which are called the left-sided and the right-sided Riemann-Liouville fractional integrals.

2.2.2 Riemann-Liouville fractional derivatives

Definition 2.9 *The Riemann-Liouville fractional derivatives* $\left(D_{a+}^{\alpha} f\right)(t)$ *and* $\left(D_{b-}^{\alpha} f\right)(t)$ *of order* $\alpha \in \mathbb{C}$ $(Re(\alpha) \geq 0)$ *are defined as (see [119]; also see [2, 19, 143, 261, 262, 263])*

$$\left(D_{a+}^{\alpha} f\right)(t) = \left(\tfrac{d}{dt}\right)^{\kappa} \left(I_{a+}^{\kappa-\alpha} f\right)(t) = \tfrac{1}{\Gamma(n-\alpha)} \left(\tfrac{d}{dt}\right)^{\kappa} \int_{a}^{t} \tfrac{f(\tau)}{(\tau-t)^{\alpha-\kappa+1}} d\tau \tag{2.97}$$

and

$$\left(D_{b-}^{\alpha} f\right)(t) = \left(-\tfrac{d}{dt}\right)^{\kappa} \left(I_{b-}^{\alpha} f\right)(t) = \tfrac{1}{\Gamma(n-\alpha)} \left(-\tfrac{d}{dt}\right)^{\kappa} \int_{t}^{b} \tfrac{f(\tau)}{(t-\tau)^{\alpha-\kappa+1}} d\tau, \tag{2.98}$$

respectively, which are called the left-sided and the right-sided Riemann-Liouville fractional integrals.

If $0 < Re(\alpha) < 1$, then we have the left-sided and the right-sided Riemann-Liouville fractional derivatives as follows (see [263]; also see [3, 75, 143, 152])

$$\left(D_{a+}^{\alpha} f\right)(t) = \tfrac{d}{dt}\left(I_{a+}^{1-\alpha} f\right)(t) = \tfrac{1}{\Gamma(1-\alpha)} \tfrac{d}{dt} \int\limits_a^t \tfrac{f(\tau)}{(\tau-t)^{\alpha}} d\tau \tag{2.99}$$

and

$$\left(D_{b-}^{\alpha} f\right)(t) = -\tfrac{d}{dt}\left(I_{b-}^{\alpha} f\right)(t) = -\tfrac{1}{\Gamma(1-\alpha)} \tfrac{d}{dt} \int\limits_t^b \tfrac{f(\tau)}{(t-\tau)^{\alpha}} d\tau. \tag{2.100}$$

When $\alpha \in \mathbb{R}$ $(\alpha \geq 0)$, we have the left-sided and the right-sided Riemann-Liouville fractional integrals in the interval $[a, b]$ as follows (see [263]; also see [3, 75, 152])

$$\left(I_{a+}^{\alpha} f\right)(t) = \frac{1}{\Gamma(\alpha)} \int\limits_a^t \frac{f(\tau)}{(\tau - t)^{1-\alpha}} d\tau \;\; (t > a; \alpha > 0), \tag{2.101}$$

$$\left(I_{b-}^{\alpha} f\right)(t) = \frac{1}{\Gamma(\alpha)} \int\limits_t^b \frac{f(\tau)}{(t - \tau)^{1-\alpha}} d\tau \;\; (t < b; \alpha > 0), \tag{2.102}$$

$$\left(D_{a+}^{\alpha} f\right)(t) = \left(\tfrac{d}{dt}\right)^{\kappa}\left(I_{a+}^{\kappa-\alpha} f\right)(t) = \tfrac{1}{\Gamma(\kappa-\alpha)}\left(\tfrac{d}{dt}\right)^{\kappa} \int\limits_a^t \tfrac{f(\tau)}{(\tau-t)^{\alpha-\kappa+1}} d\tau \tag{2.103}$$

and

$$\left(D_{b-}^{\alpha} f\right)(t) = \left(-\tfrac{d}{dt}\right)^{\kappa}\left(I_{b-}^{\alpha} f\right)(t) = \tfrac{1}{\Gamma(\kappa-\alpha)}\left(-\tfrac{d}{dt}\right)^{\kappa} \int\limits_t^b \tfrac{f(\tau)}{(t-\tau)^{\alpha-\kappa+1}} d\tau. \tag{2.104}$$

If $0 < \alpha < 1$, then we have the left-sided and the right-sided Riemann-Liouville fractional integrals and derivatives in the interval $[a, b]$ as follows (see [263]; also see [3, 75, 152])

$$\left(I_{a+}^{\alpha} f\right)(t) = \frac{1}{\Gamma(\alpha)} \int\limits_a^t \frac{f(\tau)}{(\tau - t)^{1-\alpha}} d\tau, \tag{2.105}$$

$$\left(I_{b-}^{\alpha} f\right)(t) = \frac{1}{\Gamma(\alpha)} \int\limits_t^b \frac{f(\tau)}{(t - \tau)^{1-\alpha}} d\tau, \tag{2.106}$$

$$\left(D_{a+}^{\alpha} f\right)(t) = \tfrac{d}{dt}\left(I_{a+}^{1-\alpha} f\right)(t) = \tfrac{1}{\Gamma(1-\alpha)} \tfrac{d}{dt} \int\limits_a^t \tfrac{f(\tau)}{(\tau-t)^{\alpha}} d\tau \tag{2.107}$$

and

$$\left(D_{b-}^{\alpha} f\right)(t) = -\tfrac{d}{dt}\left(I_{b-}^{\alpha} f\right)(t) = -\tfrac{1}{\Gamma(1-\alpha)} \tfrac{d}{dt} \int\limits_t^b \tfrac{f(\tau)}{(t-\tau)^{\alpha}} d\tau. \tag{2.108}$$

If $0 < \alpha < 1$, then we have the left-sided and the right-sided Liouville fractional integrals and derivatives in the domain $\mathbb{R} = [-\infty, \infty]$ as follows (see [16, 17]; also see [152, 168]):

$$\left({}_{L}I_{+}^{\alpha} f\right)(t) = \frac{1}{\Gamma(\alpha)} \int_{-\infty}^{t} \frac{f(\tau)}{(\tau - t)^{1-\alpha}} d\tau, \tag{2.109}$$

$$\left({}_{L}I_{-}^{\alpha} f\right)(t) = \frac{1}{\Gamma(\alpha)} \int_{t}^{\infty} \frac{f(\tau)}{(t - \tau)^{1-\alpha}} d\tau, \tag{2.110}$$

$$\left({}_{L}D_{+}^{\alpha} f\right)(t) = \frac{d}{dt}\left(I_{+}^{1-\alpha} f\right)(t) = \frac{1}{\Gamma(1-\alpha)} \frac{d}{dt} \int_{-\infty}^{t} \frac{f(\tau)}{(\tau-t)^{\alpha}} d\tau \tag{2.111}$$

and

$$\left({}_{L}D_{-}^{\alpha} f\right)(t) = -\frac{d}{dt}\left(I_{-}^{\alpha} f\right)(t) = -\frac{1}{\Gamma(1-\alpha)} \frac{d}{dt} \int_{t}^{\infty} \frac{f(\tau)}{(t-\tau)^{\alpha}} d\tau. \tag{2.112}$$

We note that the left-sided and the right-sided Riemann fractional integrals and derivatives are presented by Liouville in 1832 with the complement functions (see [16, 17]; also see [151, 168]).

If $0 < \alpha < 1$, then we have the Riemann fractional derivatives as follows (see [151, 168]):

$$\left({}_{R}I_{a+}^{\alpha} f\right)(t) = \frac{1}{\Gamma(\alpha)} \int_{0}^{t} \frac{f(\tau)}{(\tau - t)^{1-\alpha}} d\tau, \tag{2.113}$$

$$\left({}_{R}I_{0-}^{\alpha} f\right)(t) = \frac{1}{\Gamma(\alpha)} \int_{t}^{0} \frac{f(\tau)}{(t - \tau)^{1-\alpha}} d\tau, \tag{2.114}$$

$$\left({}_{R}D_{a+}^{\alpha} f\right)(t) = \frac{d}{dt}\left(I_{a+}^{1-\alpha} f\right)(t) = \frac{1}{\Gamma(1-\alpha)} \frac{d}{dt} \int_{a}^{t} \frac{f(\tau)}{(\tau-t)^{\alpha}} d\tau \tag{2.115}$$

and

$$\left({}_{R}D_{b-}^{\alpha} f\right)(t) = -\frac{d}{dt}\left(I_{b-}^{\alpha} f\right)(t) = -\frac{1}{\Gamma(1-\alpha)} \frac{d}{dt} \int_{t}^{b} \frac{f(\tau)}{(t-\tau)^{\alpha}} d\tau. \tag{2.116}$$

It is seen that the Riemann fractional derivatives were presented with the complement functions [18, 151, 168].

2.2.3 Riemann-Liouville fractional derivatives of a purely imaginary order

If $Re(\alpha) = 0$ $(\alpha \neq 0)$, the left-sided and the right-sided Riemann-Liouville fractional integrals and derivatives in the interval $[a, b]$ are defined as (see [59];

also see [3, 75, 152])

$$\left(I_{a+}^{i\theta}f\right)(t)=\frac{1}{\Gamma(i\theta)}\int_a^t\frac{f(\tau)}{(\tau-t)^{1-i\theta}}d\tau, \tag{2.117}$$

$$\left(I_{b-}^{i\theta}f\right)(t)=\frac{1}{\Gamma(i\theta)}\int_t^b\frac{f(\tau)}{(t-\tau)^{1-i\theta}}d\tau, \tag{2.118}$$

$$\left(D_{a+}^{i\theta}f\right)(t)=\tfrac{d}{dt}\left(I_{a+}^{1-i\theta}f\right)(t)=\tfrac{1}{\Gamma(i\theta)}\tfrac{d}{dt}\int_a^t\tfrac{f(\tau)}{(\tau-t)^{i\theta}}d\tau \tag{2.119}$$

and

$$\left(D_{b-}^{i\theta}f\right)(t)=-\tfrac{d}{dt}\left(I_{b-}^{i\theta}f\right)(t)=-\tfrac{1}{\Gamma(1-i\theta)}\tfrac{d}{dt}\int_t^b\tfrac{f(\tau)}{(t-\tau)^{i\theta}}d\tau. \tag{2.120}$$

Property 2.2 *If $Re(\beta)\geq 0$, then we have*

$$\left(I_{a+}^{\alpha}\frac{(t-a)^{\beta-1}}{\Gamma(\beta)}\right)(t)=\frac{(t-a)^{\beta+\alpha-1}}{\Gamma(\beta+\alpha)}, \tag{2.121}$$

$$\left(D_{a+}^{\alpha}\frac{(t-a)^{\beta-1}}{\Gamma(\beta)}\right)(t)=\frac{(t-a)^{\beta-\alpha-1}}{\Gamma(\beta-\alpha)}, \tag{2.122}$$

$$\left(I_{b-}^{\alpha}\frac{(b-t)^{\beta-1}}{\Gamma(\beta)}\right)(t)=\frac{(b-t)^{\beta+\alpha-1}}{\Gamma(\beta+\alpha)} \tag{2.123}$$

and

$$\left(D_{b-}^{\alpha}\frac{(b-t)^{\beta-1}}{\Gamma(\beta)}\right)(t)=\frac{(b-t)^{\beta-\alpha-1}}{\Gamma(\beta-\alpha)}. \tag{2.124}$$

In particular, if $0<Re(\alpha)<1$, then we have

$$\left(D_{a+}^{\alpha}1\right)(t)=\frac{(t-a)^{-\alpha}}{\Gamma(1-\alpha)} \tag{2.125}$$

and

$$\left(D_{b-}^{\alpha}1\right)(t)=\frac{(b-t)^{-\alpha}}{\Gamma(1-\alpha)}. \tag{2.126}$$

For $\kappa=1,2,\cdots,[Re(\alpha)]+1$, there are

$$\left(D_{a+}^{\alpha}\frac{(t-a)^{\beta}}{\Gamma(1+\beta)}\right)(t)=0 \tag{2.127}$$

and

$$\left(D_{b-}^{\alpha}\frac{(b-t)^{\beta}}{\Gamma(1+\beta)}\right)(t)=0. \tag{2.128}$$

In particular, we give

$$\left(D_{a+}^{\alpha}\frac{(t-a)^{\alpha-1}}{\Gamma(\alpha)}\right)(t)=0, \tag{2.129}$$

$$\left(D_{b-}^{\alpha}\frac{(b-t)^{\alpha-1}}{\Gamma(\alpha)}\right)(t)=0, \tag{2.130}$$

$$\left(D_{a+}^{\alpha}\frac{(t-a)^{\alpha-[Re(\alpha)]-1}}{\Gamma(\alpha)}\right)(t)=0, \tag{2.131}$$

and

$$\left(D_{b-}^{\alpha}\frac{(b-t)^{\alpha-[Re(\alpha)]-1}}{\Gamma(\alpha)}\right)(t)=0. \tag{2.132}$$

Thus, we have the following results (see [152]).

Property 2.3 *(see [152]) Let $Re(\alpha)>0$. Then we have:*
(1) For $0<Re(\alpha)<1$, there exists

$$\left(D_{a+}^{\alpha}f\right)(t)=0 \tag{2.133}$$

if and only if

$$f(t)=h(t-a)^{\alpha-1}, \tag{2.134}$$

where h is any constant and $h\in\mathbb{R}$.
(2) For $0<Re(\alpha)<1$, there exists

$$\left(D_{b-}^{\alpha}f\right)(t)=0 \tag{2.135}$$

if and only if

$$f(t)=g(b-t)^{\alpha-1}, \tag{2.136}$$

where g is any constant and $g\in\mathbb{R}$.
(3) For $0<Re(\alpha)$ and $\kappa=[Re(\alpha)]+1$, there exists

$$\left(D_{a+}^{\alpha}f\right)(t)=0 \tag{2.137}$$

if and only if

$$f(t)=\sum_{j=1}^{\kappa}h_j(t-a)^{\alpha-j}, \tag{2.138}$$

where h_j $(j=1,2,\cdots,\kappa)$ are arbitrary constants and $h_j\in\mathbb{R}$.

(4) For $0 < Re(\alpha)$ *and* $\kappa = [Re(\alpha)] + 1$*, there exists*

$$\left(D_{b-}^{\alpha} f\right)(t) = 0 \tag{2.139}$$

if and only if

$$f(t) = \sum_{j=1}^{\kappa} g_j (b-t)^{\alpha-j}, \tag{2.140}$$

where g_j $(j = 1, 2, \cdots, \kappa)$ *are arbitrary constants and* $g_j \in \mathbb{R}$.

Here, we present the boundedness of the fractional integration operators $\left(I_{a+}^{\alpha} f\right)(t)$ and $\left(I_{b-}^{\alpha} f\right)(t)$.

Theorem 2.1 *(Samko-Kilbas-Marichev Theorem) (see [3, 75]; also see [152])*
Let the fractional integration operators $\left(I_{a+}^{\alpha} f\right)(t)$ *and* $\left(I_{b-}^{\alpha} f\right)(t)$ *with* $Re(\alpha) > 0$ *be bounded in* $L_{\kappa}(a, b)$ $(1 \le \kappa \le \infty)$*. Then we have*

$$\left\|I_{a+}^{\alpha} f\right\|_{\kappa} \le \theta \left\|f\right\|_{\kappa} \tag{2.141}$$

and

$$\left\|I_{b-}^{\alpha} f\right\|_{\kappa} \le \theta \left\|f\right\|_{\kappa} \tag{2.142}$$

where

$$\theta = \frac{(b-a)^{Re(\alpha)}}{Re(\alpha)\left|\Gamma(\alpha)\right|}. \tag{2.143}$$

When $\alpha \in \mathbb{R}_{+}$, the Samko-Kilbas-Marichev theorem can be represented in the form:

Let the fractional integration operators $\left(I_{a+}^{\alpha} f\right)(t)$ and $\left(I_{b-}^{\alpha} f\right)(t)$ with $\alpha \in \mathbb{R}_{+}$ be bounded in $L_{\kappa}(a, b)$ $(1 \le \kappa \le \infty)$. Then we have

$$\left\|I_{a+}^{\alpha} f\right\|_{\kappa} \le \theta \left\|f\right\|_{\kappa} \tag{2.144}$$

and

$$\left\|I_{b-}^{\alpha} f\right\|_{\kappa} \le \theta \left\|f\right\|_{\kappa} \tag{2.145}$$

where

$$\theta = \frac{(b-a)^{\alpha}}{\Gamma(1+\alpha)}. \tag{2.146}$$

Theorem 2.2 *(Hardy-Littlewood Theorem) (see [121]; also see [3, 75, 152])*
If $0 < \alpha < 1$ *and* $1 < \kappa < 1/\alpha$*, then the operators* $\left(I_{a+}^{\alpha} f\right)(t)$ *and* $\left(I_{b-}^{\alpha} f\right)(t)$ *are bounded from* $L_{\kappa}(a, b)$ *into* $L_{\ell}(a, b)$*, where* $\ell = \kappa / (1 - \alpha\kappa)$.

When $0 < Re(\alpha) < 1$ *and* $1 < \kappa < 1/Re(\alpha)$*, the Hardy-Littlewood theorem can be represented as follows:*

If $0 < Re(\alpha) < 1$ *and* $1 < \kappa < 1/Re(\alpha)$*, then the operators* $\left(I_{a+}^{\alpha} f\right)(t)$ *and* $\left(I_{b-}^{\alpha} f\right)(t)$ *are bounded from* $L_{\kappa}(a, b)$ *into* $L_{\ell}(a, b)$*, where* $\ell = \kappa / (1 - Re(\alpha)\kappa)$.

We now consider the conditions for the existence of the fractional derivatives $\left(D_{a+}^{\alpha} f\right)(t)$ *and* $\left(D_{b-}^{\alpha} f\right)(t)$ *in the space* $AC^{\kappa}[a, b]$.

Theorem 2.3 *(see [3, 75]; also see [152])*

Let $Re(\alpha) \geq 0$, $\kappa = [Re(\alpha)] + 1$ and $f(t) \in AC^{\kappa}(a,b)$. Then fractional derivatives $\left(D_{a+}^{\alpha} f\right)(t)$ and $\left(D_{b-}^{\alpha} f\right)(t)$ exist almost everywhere on (a,b) and can be represented as follows:

$$\left(D_{a+}^{\alpha} f\right)(t) = \sum_{j=0}^{\kappa-1} \frac{f^{(j)}(a)}{\Gamma(1+j-\alpha)}(t-a)^{j-\alpha} + \frac{1}{\Gamma(\kappa-\alpha)} \int_{a}^{t} \frac{f^{(\kappa)}(\tau)}{(t-\tau)^{\alpha-\kappa+1}} d\tau \tag{2.147}$$

and

$$\left(D_{b-}^{\alpha} f\right)(t) = \sum_{j=0}^{\kappa-1} \frac{(-1)^{\kappa} f^{(j)}(b)}{\Gamma(1+j-\alpha)}(b-t)^{j-\alpha} + \frac{(-1)^{\kappa}}{\Gamma(\kappa-\alpha)} \int_{t}^{b} \frac{f^{(\kappa)}(\tau)}{(\tau-t)^{\alpha-\kappa+1}} d\tau, \tag{2.148}$$

respectively.

Theorem 2.4 *(see [3]; also see [75, 152])*

Let $0 \leq Re(\alpha) < 1$ $(\alpha \neq 0)$ and $f(t) \in AC(a,b)$. Then we have

$$\left(D_{a+}^{\alpha} f\right)(t) = \frac{1}{\Gamma(1-\alpha)} \left(\frac{f(a)}{(t-a)^{\alpha}} + \int_{a}^{t} \frac{f^{(1)}(\tau)}{(t-\tau)^{\alpha}} d\tau \right) \tag{2.149}$$

and

$$\left(D_{b-}^{\alpha} f\right)(t) = \frac{1}{\Gamma(1-\alpha)} \left(\frac{f(b)}{(b-t)^{\alpha}} - \int_{t}^{b} \frac{f^{(1)}(\tau)}{(\tau-t)^{\alpha}} d\tau \right). \tag{2.150}$$

Theorem 2.5 *(see [3, 75]; also see [152])*

Let $\alpha > 0$, $\kappa = \alpha + 1$ and $f(t) \in AC^{\kappa}(a,b)$. Then fractional derivatives $\left(D_{a+}^{\alpha} f\right)(t)$ and $\left(D_{b-}^{\alpha} f\right)(t)$ exist almost everywhere on (a,b) and can be represented as follows:

$$\left(D_{a+}^{\alpha} f\right)(t) = \sum_{j=0}^{\kappa-1} \frac{f^{(j)}(a)}{\Gamma(1+j-\alpha)}(t-a)^{j-\alpha} + \frac{1}{\Gamma(\kappa-\alpha)} \int_{a}^{t} \frac{f^{(\kappa)}(\tau)}{(t-\tau)^{\alpha-\kappa+1}} d\tau \tag{2.151}$$

and

$$\left(D_{b-}^{\alpha} f\right)(t) = \sum_{j=0}^{\kappa-1} \frac{(-1)^{\kappa} f^{(j)}(b)}{\Gamma(1+j-\alpha)}(b-t)^{j-\alpha} + \frac{(-1)^{\kappa}}{\Gamma(\kappa-\alpha)} \int_{t}^{b} \frac{f^{(\kappa)}(\tau)}{(\tau-t)^{\alpha-\kappa+1}} d\tau, \tag{2.152}$$

respectively.

Theorem 2.6 *(Tamarkin Theorem) (see [166]; also see [3, 75, 152])*

Let $0 < \alpha < 1$ *and* $f(t) \in AC(a,b)$*. Then we have*

$$\left(D_{a+}^{\alpha} f\right)(t) = \frac{1}{\Gamma(1-\alpha)} \left(\frac{f(a)}{(t-a)^{\alpha}} + \int_{a}^{t} \frac{f^{(1)}(\tau)}{(t-\tau)^{\alpha}} d\tau \right) \tag{2.153}$$

and

$$\left(D_{b-}^{\alpha} f\right)(t) = \frac{1}{\Gamma(1-\alpha)} \left(\frac{f(b)}{(b-t)^{\alpha}} - \int_{t}^{b} \frac{f^{(1)}(\tau)}{(\tau-t)^{\alpha}} d\tau \right). \tag{2.154}$$

The semigroup property of the fractional derivative and integration operators $\left(D_{a+}^{\alpha} f\right)(t)$, $\left(D_{b-}^{\alpha} f\right)(t)$, $\left(I_{a+}^{\alpha} f\right)(t)$ and $\left(I_{b-}^{\alpha} f\right)(t)$ are given as follows:

Property 2.4 *(1) Let* $Re(\alpha) > 0$, $Re(\beta) > 0$, $t \in [a,b]$ *and* $f(t) \in L_{\kappa}(a,b)$ $(1 \leq \kappa \leq \infty)$*. Then we have (see [3, 75]; also see [152])*

$$\left(I_{a+}^{\alpha} I_{a+}^{\beta} f\right)(t) = \left(I_{a+}^{\alpha+\beta} f\right)(t) \tag{2.155}$$

and

$$\left(I_{b-}^{\alpha} I_{b-}^{\beta} f\right)(t) = \left(I_{b-}^{\alpha+\beta} f\right)(t). \tag{2.156}$$

(2) Let $Re(\beta) > Re(\alpha) > 0$ *,* $t \in [a,b]$ *and* $f(t) \in L_{\kappa}(a,b)$ $(1 \leq \kappa \leq \infty)$*. Then we have (see [166]; also see [3, 75, 152])*

$$\left(D_{a+}^{\alpha} I_{a+}^{\beta} f\right)(t) = \left(I_{a+}^{\beta-\alpha} f\right)(t) \tag{2.157}$$

and

$$\left(D_{b-}^{\alpha} I_{b-}^{\beta} f\right)(t) = \left(I_{b-}^{\beta-\alpha} f\right)(t). \tag{2.158}$$

(3) Let $Re(\alpha) > 0$, $t \in [a,b]$ *and* $f(t) \in L_{\kappa}(a,b)$ $(1 \leq \kappa \leq \infty)$*. Then we have (see [3, 75]; also see [152])*

$$\left(D_{a+}^{\alpha} I_{a+}^{\alpha} f\right)(t) = f(t) \tag{2.159}$$

and

$$\left(D_{b-}^{\alpha} I_{b-}^{\alpha} f\right)(t) = f(t). \tag{2.160}$$

(4) Let $Re(\alpha) > 0$ *,* $\kappa \in \mathbb{N}$ *and* $t \in [a,b]$*. Then we have (see [3, 75]; also see [152])*

$$\left(D^{\alpha} D_{a+}^{\kappa} f\right)(t) = \left(D_{a+}^{\alpha+\kappa} f\right)(t) \tag{2.161}$$

and

$$\left(D^{\kappa} D_{b-}^{\kappa} f\right)(t) = (-1)^{\kappa} \left(D_{b-}^{\alpha+\kappa} f\right)(t). \tag{2.162}$$

(5) Let $Re(\alpha) > 0$, $g = [Re(\alpha)] + 1$, $t \in [a,b]$ *and let* $f_{g-\alpha}(t) = \left(I_{a+}^{g-\alpha} f\right)(t)$*. If* $f(t) \in I_{a+}^{\alpha}(L_{\kappa})$*, where* $I_{a+}^{\alpha}(L_{\kappa}) = \left\{ f : f = I_{a+}^{\alpha} l,\ l \in L_{\kappa}(a,b) \right\}$*, then we have (see [3, 75]; also see [152])*

$$\left(I_{a+}^{\beta} D_{a+}^{\alpha} f\right)(t) = f(t). \tag{2.163}$$

(6) Let $Re(\alpha) > 0$, $g = [Re(\alpha)] + 1$, $t \in [a, b]$ *and let* $f_{g-\alpha}(t) = \left(I_{a+}^{g-\alpha} f\right)(t)$. *If* $f(t) \in I_{b-}^{\alpha}(L_\kappa)$, *where* $I_{b-}^{\alpha}(L_\kappa) = \left\{f : f = I_{b-}^{\alpha} l,\ l \in L_\kappa(a, b)\right\}$, *then we have (see [3, 75]; also see [152])*

$$\left(I_{b-}^{\alpha} D_{b-}^{\alpha} f\right)(t) = f(t). \tag{2.164}$$

(7) Let $Re(\alpha) > 0$, $\kappa = [Re(\alpha)] + 1$, $t \in [a, b]$ *and let* $f_{\kappa-\alpha}(t) = \left(I_{a+}^{\kappa-\alpha} f\right)(t)$. *If* $f_{\kappa-\alpha}(t) \in AC^\kappa[a, b]$, *then we have (see [3, 75]; also see [152])*

$$\left(I_{a+}^{\alpha} D_{a+}^{\alpha} f\right)(t) = f(t) - \sum_{j=0}^{\kappa-1} \frac{f_{\kappa-\alpha}^{(\kappa-j)}(a)}{\Gamma(\alpha - 1 - j)} (t - a)^{\alpha-j}. \tag{2.165}$$

Theorem 2.7 *(Tamarkin Theorem) (see [166]; also see [3, 75, 152]) Let* $\alpha > 0$, $\kappa = [\alpha]+1$, $t \in [a, b]$ *and let* $f_{\kappa-\alpha}(t) = \left(I_{a+}^{\kappa-\alpha} f\right)(t)$. *If* $f_{\kappa-\alpha}(t) \in AC^\kappa[a, b]$, *then we have (see [166]; also see [3, 75, 152])*

$$\left(I_{a+}^{\alpha} D_{a+}^{\alpha} f\right)(t) = f(t) - \sum_{j=0}^{\kappa-1} \frac{f_{\kappa-\alpha}^{(\kappa-j)}(a)}{\Gamma(\alpha - 1 - j)} (t - a)^{\alpha-j}. \tag{2.166}$$

In particular, if $1 > Re(\alpha) > 0$, then we have the following result:

Let $1 > Re(\alpha) > 0$, $t \in [a, b]$ and let $f_{1-\alpha}(t) = \left(I_{a+}^{1-\alpha} f\right)(t)$. Then, we have (see [166]; also see [3, 75, 152])

$$\left(I_{a+}^{\alpha} D_{a+}^{\alpha} f\right)(t) = f(t) - \frac{(t - a)^{\alpha-1}}{\Gamma(\alpha)} f_{1-\alpha}(a), \tag{2.167}$$

where $f_{1-\alpha}(a) = \left(I_{a+}^{1-\alpha} f\right)(t)$.

In particular, we have the following result.

Let $Re(\alpha) > \kappa$, $t \in [a, b]$ and $f(t) \in L_\kappa(a, b)$ $(1 \le \kappa \le \infty)$. Then we have (see [3, 75]; also see [152])

$$\left(D^{\alpha} I_{a+}^{\kappa} f\right)(t) = \left(I_{a+}^{\alpha-\kappa} f\right)(t) \tag{2.168}$$

and

$$\left(D^{\kappa} I_{b-}^{\kappa} f\right)(t) = \left(I_{b-}^{\alpha-\kappa} f\right)(t). \tag{2.169}$$

Property 2.5 *(see [3, 75]; also see [152])*

Let $\alpha > 0$, $\beta > 0$, $g > \alpha > g-1$ $(g \in \mathbb{N})$, $\kappa > \beta > \kappa-1$ $(\kappa \in \mathbb{N})$, $\alpha+\beta > g$ *and* $f_{\kappa-\alpha}(t) \in AC^\kappa[a, b]$. *Then we have*

$$\left(D_{a+}^{\alpha} D_{a+}^{\beta} f\right)(t) = \left(D_{a+}^{\alpha+\beta} f\right)(t) - \sum_{j=1}^{\kappa} \left(D_{a+}^{(\beta-j)} f\right)(a+) \frac{(t - a)^{-\alpha-j}}{\Gamma(1 - a - j)}. \tag{2.170}$$

For the more details of the semigroup property of the fractional derivative and integration operators, see [166, 3, 75]; also see [152, 154].

We now consider the rules for fractional integration by parts (see [3, 75]; also see [152]).

Property 2.6 *(see [3, 75]; also see [152])*

Let $\alpha > 0$, $g_1 \geq 1$, $g_2 \geq 1$, and $(1/g_1)+(1/g_2) \leq 1+\alpha$. If $f_1(t) \in L_{g_1}(a,b)$ and $f_2(t) \in L_{g_2}(a,b)$, then we have

$$\int_a^b f_1(\tau)\left(I_{a+}^{\alpha} f_2\right)(\tau)\, d\tau = \int_a^b f_2(\tau)\left(I_{b-}^{\alpha} f_1\right)(\tau)\, d\tau. \tag{2.171}$$

Property 2.7 *(see [3, 75]; also see [152])*

Let $\alpha > 0$, $g_1 \geq 1$, $g_2 \geq 1$, and $(1/g_1)+(1/g_2) \leq 1+\alpha$. If $f_1(t) \in I_{a+}^{\alpha}(L_{\kappa})$, where $I_{a+}^{\alpha}(L_{\kappa}) = \{f : f = I_{a+}^{\alpha} l,\ l \in L_{\kappa}(a,b)\}$, and $f_2(t) \in I_{b-}^{\alpha}(L_{\kappa})$, where $I_{b-}^{\alpha}(L_{\kappa}) = \{f : f = I_{b-}^{\alpha} l,\ l \in L_{\kappa}(a,b)\}$, then we have

$$\int_a^b f_1(\tau)\left(D_{a+}^{\alpha} f_2\right)(\tau)\, d\tau = \int_a^b f_2(\tau)\left(D_{b-}^{\alpha} f_2\right)(\tau)\, d\tau. \tag{2.172}$$

Let $0 < Re(\alpha) < 1$. Then we have

$$\left(I_{+}^{\alpha} f\right)(t) = \frac{1}{\Gamma(\alpha)} \int_{-\infty}^{t} \frac{f(\tau)}{(t-\tau)^{1-\alpha}} d\tau, \tag{2.173}$$

$$\left(I_{-}^{\alpha} f\right)(t) = \frac{1}{\Gamma(\alpha)} \int_{t}^{\infty} \frac{f(\tau)}{(t-\tau)^{1-\alpha}} d\tau, \tag{2.174}$$

$$\left(D_{+}^{\alpha} f\right)(t) = \frac{1}{\Gamma(1-\alpha)} \frac{d}{dt} \int_{-\infty}^{t} \frac{f^{(1)}(\tau)}{(t-\tau)^{\alpha}} d\tau \tag{2.175}$$

and

$$\left(D_{-}^{\alpha} f\right)(t) = \frac{1}{\Gamma(1-\alpha)} \frac{d}{dt} \int_{t}^{\infty} \frac{f^{(1)}(\tau)}{(t-\tau)^{\alpha}} d\tau. \tag{2.176}$$

Property 2.8 *(see [3, 75])*

If $0 < Re(\alpha) < 1$ and $f(t) \in H_1(\mathbb{R})$, then the following relations (see [3, 75])

$$F\left(I_{+}^{\beta} f(t)\right)(\omega) = \frac{f(\omega)}{(-i\omega)^{\alpha}}$$

and

$$F\left(I_{-}^{\beta} f(t)\right)(\omega) = \frac{f(\omega)}{(i\omega)^{\alpha}}$$

hold, where $(\mp i\omega)^{\alpha} = |\omega|^{\alpha} e^{\mp \alpha\pi i\ sgn(\alpha)/2}$ and $F\{f(t)\}(\omega) = f(\omega)$.

Property 2.9 *(see [3, 75]) If $0 < Re(\alpha) < 1$, then the following relations (see [3, 75])*

$$F\left(D_{+}^{\alpha} f(t)\right)(\omega) = (-i\omega)^{\alpha} f(\omega)$$

and

$$F\left(D_{-}^{\alpha} f(t)\right)(\omega) = (i\omega)^{\alpha} f(\omega)$$

hold, where $(\mp i\omega)^{\alpha} = |\omega|^{\alpha} e^{\mp \alpha\pi i\ sgn(\alpha)/2}$ and $F\{f(t)\}(\omega) = f(\omega)$.

Let $0 < Re(\alpha) < 1$. Then we have (see [152])

$$\left(I_{0+}^{\alpha} f\right)(t) = \frac{1}{\Gamma(\alpha)} \int_{0}^{t} \frac{f(\tau)}{(t-\tau)^{1-\alpha}} d\tau, \tag{2.177}$$

and

$$\left(D_{0+}^{\alpha} f\right)(t) = \frac{1}{\Gamma(1-\alpha)} \frac{d}{dt} \int_{0}^{t} \frac{f^{(1)}(\tau)}{(t-\tau)^{\alpha}} d\tau. \tag{2.178}$$

Property 2.10 *(see [150]) If $0 < Re(\alpha) < 1$, then the following relations (see [150])*

$$L\left(I_{+}^{\beta} f(t)\right)(s) = \frac{(f)(s)}{s^{\alpha}} \tag{2.179}$$

and

$$L\left(D_{+}^{\alpha} f(t)\right)(\omega) = s^{\alpha}(f)(s), \tag{2.180}$$

hold, where $L\{f(t)\}(s) = f(s)$.

Property 2.11 *(see [150]) If $\kappa < Re(\alpha) < 1 + \kappa$, then we have (see [150])*

$$L\left(D_{+}^{\kappa-\alpha} f(t)\right)(\omega) = s^{\kappa-\alpha}(f)(s), \tag{2.181}$$

where $L\{f(t)\}(s) = f(s)$.

Find that, when $D^{\kappa}\left(\left(I_{0+}^{\alpha} f\right)(t)\right)|_{t=0} = 0$, there is (see [150])

$$L\left\{\left(D_{0+}^{\alpha} f\right)(t)\right\} = s^{u} f(s), \tag{2.182}$$

and when $D^{\kappa}\left(\left(I_{0+}^{\alpha,\lambda} f\right)(t)\right)|_{t=0} \neq 0$, we have (see [150])

$$L\left\{\left(D_{0+}^{\alpha} f\right)(t)\right\} = s^{\alpha} f(s) - \sum_{\mu=0}^{\kappa-1} s^{\kappa-\mu-1}\left(D^{\mu}\left[I_{0+}^{\kappa-\alpha} f(+0)\right]\right). \tag{2.183}$$

In particular, when $\kappa = 1$, there is (see [150])

$$L\left\{D_{0+}^{\alpha} f(t)\right\} = s^{\alpha} f(s) - I_{0+}^{1-\alpha} f(+0), \tag{2.184}$$

and when $\kappa = 1$ and $D^{\kappa}\left(\left(I_{0+}^{\alpha,\lambda} f\right)(t)\right)|_{t=0} = 0$, we have (see [150])

$$L\left\{D_{0+}^{\alpha} f(t)\right\} = s^{\alpha} f(s). \tag{2.185}$$

Property 2.12 *(see [150]) For* $0 < Re(\alpha)$ *and* $\kappa = [Re(\alpha)] + 1$*, there exists (see [150])*

$$M\{D_{0+}^{\alpha} f(t)\} = \frac{\Gamma(1+\alpha-\varpi)}{\Gamma(1-\varpi)} f(\varpi-\alpha) + \sum_{j=0}^{\kappa-1} \frac{\Gamma(1+j-\varpi)}{\Gamma(1-\varpi)} \left[t^{\varpi-\kappa-1}\left(I_{0+}^{\kappa-\alpha} f\right)(t)\right]_0^{\infty}. \tag{2.186}$$

In particular, when $0 < Re(\alpha) < 1$, there is (see [150])

$$M\left\{D_{0+}^{\alpha} f(t)\right\} = \frac{\Gamma(1+\alpha-\varpi)}{\Gamma(1-\varpi)} f(\varpi-\alpha) + \left[t^{\varpi-1}\left(I_{0+}^{1-\alpha} f\right)(t)\right]_0^{\infty}. \tag{2.187}$$

For more results on the Riemann-Liouville fractional integrals and fractional derivatives on a finite interval of the real line, see (see [149, 150, 152, 153]).

2.3 Liouville-Sonine-Caputo fractional derivatives

2.3.1 Motivations

The left-sided fractional differential operator is defined as (see [152])

$$\left({}_{LC}D_{a+}^{\alpha} f\right)(t) = \frac{1}{\Gamma(\kappa-\alpha)} \left(\frac{d}{dt}\right)^{\kappa} \int_a^t \frac{1}{(\tau-t)^{\alpha-\kappa+1}} \left[f(t) - \sum_{j=0}^{\kappa-1} \frac{f^{(j)}(a)}{\Gamma(1+j)} (t-a)^j\right] d\tau, \tag{2.188}$$

where $Re(\alpha) \geq 0$, $t > a$ and $\kappa = [Re(\alpha)] + 1$, and the right-sided fractional differential operator as (see [152])

$$\left({}_{LC}D_{b-}^{\alpha} f\right)(t) = \frac{1}{\Gamma(\kappa-\alpha)} \left(-\frac{d}{dt}\right)^{\kappa} \int_t^b \frac{1}{(t-\tau)^{\alpha-\kappa+1}} \left[f(t) - \sum_{j=0}^{\kappa-1} \frac{(-1)^j f^{(j)}(b)}{\Gamma(1+j)} (b-t)^j\right] d\tau, \tag{2.189}$$

where $Re(\alpha) \geq 0$, $t < b$ and $\kappa = [Re(\alpha)] + 1$, which are due to the Kilbas-Srivastava-Trujillo ideas (see [152]). The operators are called as the left-sided and right-sided Liouville-Sonine-Caputo fractional derivatives, respectively.

Liouville introduced the fractional derivative operator in the form [15]

$${}_{L}D_{0+}^{\alpha} f(t) = \frac{1}{(-1)^{\alpha}\Gamma(\alpha)} \int_0^{\infty} \frac{d^{\kappa} f(t+\tau)}{dt^{\kappa}} \tau^{\alpha-1} d\tau, \tag{2.190}$$

where $Re(\alpha) > 0$ and $t \in \mathbb{R}$, which, by taking

$$\frac{d^{\kappa} f(t+\tau)}{dt^{\kappa}} = \frac{d^{\kappa} f(t+\tau)}{d(t+\tau)^{\kappa}}, \tag{2.191}$$

can be rewritten as

$$ {}_{L}D_{0+}^{\alpha} f(t) = \frac{1}{(-1)^{\alpha}\Gamma(\alpha)} \int_{0}^{\infty} \frac{d^{\kappa} f(t+\tau)}{d(t+\tau)^{\kappa}} \tau^{\alpha-1} d\tau. \tag{2.192} $$

Setting $t+\tau = h$, which leads to

$$ d(t+\tau) = d\tau = dh, \tag{2.193} $$

we have from Eqs. (2.192) and (2.193) that

$$ {}_{L}D_{t+}^{\alpha} f(t) = \frac{1}{(-1)^{\alpha}\Gamma(\alpha)} \int_{t}^{\infty} \frac{d^{\kappa} f(h)}{dh^{\kappa}} (h-t)^{\alpha-1} dh, \tag{2.194} $$

which yields

$$ {}_{L}D_{t+}^{\alpha} f(t) = \frac{1}{(-1)^{\alpha}\Gamma(\alpha)} \int_{t}^{\infty} \frac{\frac{d^{\kappa} f(h)}{dh^{\kappa}}}{(h-t)^{\alpha}} dh = \frac{1}{(-1)^{\alpha}\Gamma(\alpha)} \int_{t}^{\infty} \frac{f^{(\kappa)}(h)}{(h-t)^{\alpha}} dh, \tag{2.195} $$

From Eq. (2.194), we have [3, 75, 261]

$$ \frac{\Gamma(z+\kappa)}{\Gamma(z)} = (z)_{\kappa} $$

which leads to [75]

$$ {}_{L}D_{t+}^{\alpha} f(t) = \frac{1}{\Gamma(\kappa-\alpha)} \int_{t}^{\infty} \frac{f^{(\kappa)}(h)}{(h-t)^{\alpha-\kappa+1}} dh, \tag{2.196} $$

which can be used to consider the problems in mathematical physics (see [264, 265, 266, 267, 268]).

Liouville introduced the special case of the fractional derivative represented in the form [15]

$$ \lambda\!\!\!{}^{-} \int_{0}^{h} \frac{1}{\sqrt{(h-t)}} \frac{df(t)}{dt} dt = \lambda\!\!\!{}^{-} \int_{0}^{h} \frac{f^{(1)}(t)}{(h-t)^{\frac{1}{2}}} dt = \lambda\!\!\!{}^{-}\Gamma\left(\frac{1}{2}\right) {}_{L}D_{0+}^{\frac{1}{2}} f(t), \tag{2.197} $$

where $\lambda\!\!\!{}^{-} = 1/\sqrt{2g}$.

When $g = \frac{1}{2\pi}$, we easily obtain that

$$ \frac{1}{\sqrt{\pi}} \int_{0}^{h} \frac{1}{\sqrt{(h-t)}} \frac{df(t)}{dt} dt = \frac{1}{\Gamma\left(\frac{1}{2}\right)} \int_{0}^{h} \frac{1}{\sqrt{(h-t)}} \frac{df(t)}{dt} dt = {}_{L}D_{0+}^{\frac{1}{2}} f(t). \tag{2.198} $$

Sonine introduced the fractional derivative presented as follows (see [24]):

$$\left({}_{S}D_{a+}^{\alpha}f\right)(t)=\frac{1}{\Gamma(p+1-\alpha)}\int_{a}^{t}\frac{df(t)}{dt}(h-t)^{p-\alpha}\,dt, \tag{2.199}$$

where $Re(p)<\alpha<Re(p+1)$, which can be rewritten as follows:

$$\left({}_{S}D_{a+}^{\alpha}f\right)(t)=\frac{1}{\Gamma(p+1-\alpha)}\int_{a}^{t}\frac{df(t)}{dt}\frac{1}{(h-t)^{\alpha-p}}dt, \tag{2.200}$$

which, by taking $p=0$, becomes

$$\left({}_{S}D_{a+}^{\alpha}f\right)(t)=\frac{1}{\Gamma(1-\alpha)}\int_{a}^{t}\frac{df(t)}{dt}\frac{1}{(h-t)^{\alpha}}dt. \tag{2.201}$$

Caputo introduced the fractional derivative in the form (see [61])

$${}_{C}D_{0+}^{\kappa+\alpha}f(x)=\frac{1}{\Gamma(1-\alpha)}\int_{0}^{t}\frac{1}{(h-t)^{\alpha}}\frac{d^{\kappa+1}f(t)}{dt^{\kappa+1}}dt, \tag{2.202}$$

which is, by taking $p=\kappa+1$ and $h=\kappa+\alpha$, given as

$${}_{C}D_{0+}^{h}f(x)=\frac{1}{\Gamma(p-h)}\int_{0}^{t}\frac{1}{(h-t)^{h-p+1}}\frac{d^{p}f(t)}{dt^{p}}dt. \tag{2.203}$$

Thus, we get [61]

$${}_{C}D_{0+}^{\alpha}f(x)=\frac{1}{\Gamma(\kappa-\alpha)}\int_{0}^{t}\frac{1}{(h-t)^{\alpha-\kappa+1}}\frac{d^{\kappa}f(t)}{dt^{\kappa}}dt=\frac{1}{\Gamma(\kappa-\alpha)}\int_{0}^{t}\frac{f^{(\kappa)}(t)}{(h-t)^{\alpha-\kappa+1}}dt. \tag{2.204}$$

Thus, we have

$${}_{S}D_{a+}^{\alpha}f(x)={}_{C}D_{a+}^{\alpha}f(x)=\frac{1}{\Gamma(1-\alpha)}\int_{0}^{t}\frac{f^{(1)}(t)}{(h-t)^{\alpha}}dt, \tag{2.205}$$

where $\kappa=1$ and $a=0$,

$${}_{S}D_{a+}^{\frac{1}{2}}f(x)={}_{C}D_{a+}^{\frac{1}{2}}f(x)=\frac{1}{\Gamma\left(\frac{1}{2}\right)}\int_{0}^{t}\frac{f^{(1)}(t)}{(h-t)^{\frac{1}{2}}}dt={}_{L}D_{0+}^{\frac{1}{2}}f(t), \tag{2.206}$$

where $\kappa=1$ and $\alpha=\frac{1}{2}$,

$${}_{S}D_{+\infty}^{\alpha}f(x)=\frac{1}{\Gamma(1-\alpha)}\int_{t}^{\infty}\frac{f^{(1)}(t)}{(h-t)^{\alpha}}dt=-\frac{1}{\Gamma(1-\alpha)}\int_{\infty}^{t}\frac{f^{(1)}(t)}{(h-t)^{\alpha}}dt=-{}_{C}D_{+\infty}^{\alpha}f(x), \tag{2.207}$$

where $a = -\infty$,

$$ {}_{L}D^{\alpha}_{-\infty}f(x) = \lim_{a\to-\infty}\left({}_{LC}D^{\alpha}_{a+}f\right)(t) \tag{2.208}$$

and

$$\lim_{a\to 0}\left({}_{LC}D^{\alpha}_{a+}f\right)(t) = -{}_{C}D^{\alpha}_{0+}f(x). \tag{2.209}$$

Thus, we have

$$ {}_{LS}D^{\alpha}_{a+}f(x) = \frac{1}{\Gamma(1-\alpha)}\int_{a}^{t}\frac{f^{(1)}(\tau)}{(t-\tau)^{\alpha}}d\tau, \tag{2.210}$$

which is due to Liouville and Sonine,

$$ {}_{LSC}D^{\alpha}_{+\infty}f(x) = \frac{1}{\Gamma(\kappa-\alpha)}\int_{\infty}^{t}\frac{f^{(\kappa)}(t)}{(h-t)^{\alpha-\kappa+1}}dt, \tag{2.211}$$

which is due to Liouville (for the case $a = \infty$), Sonine (for the case $\kappa = 1$) and Gerasimov [269, 265],

$$ {}_{LSC}D^{\alpha}_{+\infty}f(x) = \frac{1}{\Gamma(\kappa-\alpha)}\int_{0}^{t}\frac{f^{(\kappa)}(t)}{(h-t)^{\alpha-\kappa+1}}dt, \tag{2.212}$$

which is due to Liouville [15], Sonine (for the case $\kappa = 1$), Caputo (for the case $a = 0$, see [61, 270, 271]) and Smit and De Vries [68]

$$ {}_{LSC}D^{\alpha}_{a+}f(x) = \frac{1}{\Gamma(\kappa-\alpha)}\int_{a}^{t}\frac{f^{(\kappa)}(t)}{(h-t)^{\alpha-\kappa+1}}dt, \tag{2.213}$$

which is due to the detailed information of the books by Podlubny [150], Kilbas, Srivastava, Trujillo [152] and Diethelm [154].

2.3.2 Liouville-Sonine-Caputo fractional derivatives

Definition 2.10 *Let $Re(\alpha) \geq 0$ and $\kappa = [Re(\alpha)] + 1$. The left-sided and right-sided fractional derivatives are defined as*

$$ {}_{LSC}D^{\alpha}_{a+}f(x) = I^{\alpha}_{a+}\left[f^{(\kappa)}(x)\right] = \frac{1}{\Gamma(\kappa-\alpha)}\int_{a}^{t}\frac{f^{(\kappa)}(t)}{(h-t)^{\alpha-\kappa+1}}dt, \tag{2.214}$$

and

$$ {}_{LSC}D^{\alpha}_{b-}f(x) = I^{\alpha}_{b-}\left[f^{(\kappa)}(x)\right] = \frac{(-1)^{\kappa}}{\Gamma(\kappa-\alpha)}\int_{t}^{b}\frac{f^{(\kappa)}(t)}{(t-h)^{\alpha-\kappa+1}}dt, \tag{2.215}$$

which are called the left-sided and right-sided Liouville-Sonine-Caputo fractional derivatives (due to the honors of Liouville, Sonine and Caputo), respectively.

See that many authors called them as the Liouville-Caputo fractional derivatives [265, 272, 273] or Caputo fractional derivatives [150, 152, 154, 274, 275].

In particular, when $0 < Re(\alpha) < 1$, Eq.(2.214) and Eq.(2.215) can be represented in the forms [150, 152, 154]:

$$ {}_{LS}D^{\alpha}_{a+} f(x) = I^{\alpha}_{a+} \left[f^{(1)}(x) \right] = \frac{1}{\Gamma(1-\alpha)} \int_a^t \frac{f^{(1)}(t)}{(h-t)^{\alpha}} dt, \tag{2.216} $$

and

$$ {}_{LS}D^{\alpha}_{b-} f(x) = I^{\alpha}_{b-} \left[f^{(1)}(x) \right] = -\frac{1}{\Gamma(1-\alpha)} \int_t^b \frac{f^{(1)}(t)}{(t-h)^{\alpha}} dt, \tag{2.217} $$

which are called as the left-sided and right-sided Liouville-Sonine fractional derivatives in honor of Liouville and Sonine.

The relationships among the Riemann-Liouville fractional derivatives and Liouville-Sonine-Caputo fractional derivatives are presented as follows [152].

Property 2.13 *Let $Re(\alpha) \geq 0$, $\kappa = [Re(\alpha)] + 1$ and $f(t) \in AC^{\kappa}(a,b)$. Then fractional derivatives $\left({}_{LSC}D^{\alpha}_{a+} f\right)(t)$ and $\left({}_{LSC}D^{\alpha}_{b-} f\right)(t)$ exist almost everywhere on (a,b) and can be represented as follows (see [150, 151, 154]):*

$$ {}_{LSC}D^{\alpha}_{a+} f(x) = \sum_{j=0}^{\kappa-1} \frac{f^{(j)}(a)}{\Gamma(1+j-\alpha)} (t-a)^{j-\alpha} + \left(D^{\alpha}_{a+} f\right)(t) \tag{2.218} $$

and

$$ {}_{LSC}D^{\alpha}_{b-} f(x) = \sum_{j=0}^{\kappa-1} \frac{(-1)^{\kappa} f^{(j)}(b)}{\Gamma(1+j-\alpha)} (b-t)^{j-\alpha} + \left(D^{\alpha}_{b-} f\right)(t), \tag{2.219} $$

respectively.

Property 2.14 *Let $\alpha \geq 0$, $\kappa = \alpha + 1$ and $f(t) \in AC^{\kappa}(a,b)$. Then fractional derivatives $\left({}_{LSC}D^{\alpha}_{a+} f\right)(t)$ and $\left({}_{LSC}D^{\alpha}_{b-} f\right)(t)$ exist almost everywhere on (a,b) and can be represented as follows (see [150, 151, 154]):*

$$ {}_{LSC}D^{\alpha}_{a+} f(x) = \sum_{j=0}^{\kappa-1} \frac{f^{(j)}(a)}{\Gamma(1+j-\alpha)} (t-a)^{j-\alpha} + \left(D^{\alpha}_{a+} f\right)(t) \tag{2.220} $$

and

$$ {}_{LSC}D^{\alpha}_{b-} f(x) = \sum_{j=0}^{\kappa-1} \frac{(-1)^{\kappa} f^{(j)}(b)}{\Gamma(1+j-\alpha)} (b-t)^{j-\alpha} + \left(D^{\alpha}_{b-} f\right)(t), \tag{2.221} $$

respectively.

Property 2.15 *Let* $\alpha \geq 0$ *and* $\kappa = \alpha + 1$. *Then fractional derivatives* $\left({}_{LSC}D_{0+}^{\alpha}f\right)(t)$ *and* $\left({}_{LSC}D_{+}^{\alpha}f\right)(t)$ *can be represented as follows (see [152]):*

$$
{}_{LSC}D_{0+}^{\alpha}f\left(x\right)=\sum_{j=0}^{\kappa-1}\frac{f^{(j)}\left(0\right)}{\Gamma\left(1+j-\alpha\right)}t^{j-\alpha}+\left(D_{0+}^{\alpha}f\right)\left(t\right), \tag{2.222}
$$

and

$$
{}_{LSC}D_{+}^{\alpha}f\left(x\right)=\frac{1}{\Gamma\left(\kappa-\alpha\right)}\int_{-\infty}^{t}\frac{f^{(\kappa)}\left(t\right)}{\left(h-t\right)^{\alpha-\kappa+1}}dt. \tag{2.223}
$$

The relations among the Riemann-Liouville fractional derivatives and Liouville-Sonine-Caputo fractional derivatives are given as follows (see [152]):

Property 2.16 *Let* $0 < Re\left(\alpha\right) < 1$ *and* $f\left(t\right) \in AC\left(a,b\right)$. *Then the Riemann-Liouville and Liouville-Sonine fractional derivatives* $\left({}_{LS}D_{a+}^{\alpha}f\right)(t)$ *and* $\left({}_{LS}D_{b-}^{\alpha}f\right)(t)$ *exist almost everywhere on* (a,b) *and can be represented as follows (see [150, 151, 154]):*

$$
{}_{LS}D_{a+}^{\alpha}f\left(x\right)=\left(D_{a+}^{\alpha}f\right)\left(t\right)-\frac{\left(t-a\right)^{-\alpha}}{\Gamma\left(1-\alpha\right)} \tag{2.224}
$$

and

$$
{}_{LS}D_{b-}^{\alpha}f\left(x\right)=\left(D_{b-}^{\alpha}f\right)\left(t\right)-\frac{\left(b-t\right)^{-\alpha}}{\Gamma\left(1-\alpha\right)}, \tag{2.225}
$$

respectively.

Property 2.17 *Let* $0 < \alpha < 1$ *and* $f\left(t\right) \in AC\left(a,b\right)$. *Then fractional derivatives* $\left(D_{a+}^{\alpha}f\right)(t)$ *and* $\left(D_{b-}^{\alpha}f\right)(t)$ *exist almost everywhere on* (a,b) *and can be represented as follows (see [150, 151, 154]):*

$$
{}_{LS}D_{a+}^{\alpha}f\left(x\right)=\left(D_{a+}^{\alpha}f\right)\left(t\right)-\frac{\left(t-a\right)^{-\alpha}}{\Gamma\left(1-\alpha\right)} \tag{2.226}
$$

and

$$
{}_{LS}D_{b-}^{\alpha}f\left(x\right)=\left(D_{b-}^{\alpha}f\right)\left(t\right)-\frac{\left(b-t\right)^{-\alpha}}{\Gamma\left(1-\alpha\right)}, \tag{2.227}
$$

respectively.

Property 2.18 *Let* $0 < \alpha < 1$. *Then fractional derivatives* $\left(D_{0+}^{\alpha}f\right)(t)$ *and* $\left(D_{+}^{\alpha}f\right)(t)$*exists almost everywhere on* (a,b) *and can be represented as follows (see [152]):*

$$
{}_{LS}D_{0+}^{\alpha}f\left(x\right)=\left(D_{0+}^{\alpha}f\right)\left(t\right)-\frac{t^{-\alpha}}{\Gamma\left(1-\alpha\right)} \tag{2.228}
$$

and

$$
{}_{LS}D_{+}^{\alpha}f\left(x\right)=\frac{1}{\Gamma\left(1-\alpha\right)}\int_{-\infty}^{t}\frac{f^{(1)}\left(t\right)}{\left(h-t\right)^{\alpha}}dt. \tag{2.229}
$$

In the above, we have the following:

Property 2.19 *(1) If $f^{(\kappa-1)}(a)=0$, where $\kappa=[Re(\alpha)]+1$, we have (see [152])*

$$_{LSC}D_{a+}^{\alpha}f(x)=D_{a+}^{\alpha}f(x). \tag{2.230}$$

(2) If $f^{(\kappa-1)}(b)=0$, where $\kappa=[Re(\alpha)]+1$, we have (see [152])

$$_{LSC}D_{b-}^{\alpha}f(x)=D_{b-}^{\alpha}f(x). \tag{2.231}$$

Let $Re(\alpha)\geq 0$ and $\kappa=[Re(\alpha)]+1$. If $f(t)\in H^{\kappa}[a,b]$, then there are (see [152])

$$_{LSC}D_{a+}^{\alpha}f(x)\in H[a,b], \tag{2.232}$$

$$_{LSC}D_{b-}^{\alpha}f(x)\in H[a,b], \tag{2.233}$$

$$_{LS}D_{a+}^{\alpha}f(x)\in H[a,b], \tag{2.234}$$

$$_{LS}D_{b-}^{\alpha}f(x)\in H[a,b]. \tag{2.235}$$

From the definitions of the Riemann-Liouville fractional derivatives and Liouville-Sonine-Caputo fractional derivatives, for $0<Re(\beta)$ we have the results as follows (see [152]):

Property 2.20 *The following relations (see [152])*

$$\left({}_{LSC}D_{a+}^{\alpha}\frac{(t-a)^{\beta-1}}{\Gamma(\beta)}\right)(t)=\frac{(t-a)^{\beta-\alpha-1}}{\Gamma(\beta-\alpha)}, \tag{2.236}$$

$$\left({}_{LSC}D_{b-}^{\alpha}\frac{(b-t)^{\beta-1}}{\Gamma(\beta)}\right)(t)=\frac{(b-t)^{\beta-\alpha-1}}{\Gamma(\beta-\alpha)}, \tag{2.237}$$

$$\left({}_{LSC}D_{a+}^{\alpha}(t-a)^{\kappa-1}\right)(t)=0, \tag{2.238}$$

and

$$\left({}_{LSC}D_{b-}^{\alpha}(b-t)^{\kappa-1}\right)(t)=0, \tag{2.239}$$

are valid.

Property 2.21 *For $0<Re(\beta)$ we have (see [152])*

$$\left({}_{LS}D_{a+}^{\alpha}\frac{(t-a)^{\beta-1}}{\Gamma(\beta)}\right)(t)=\frac{(t-a)^{\beta-\alpha-1}}{\Gamma(\beta-\alpha)}, \tag{2.240}$$

and

$$\left({}_{LSC}D_{b-}^{\alpha}\frac{(b-t)^{\beta-1}}{\Gamma(\beta)}\right)(t)=\frac{(b-t)^{\beta-\alpha-1}}{\Gamma(\beta-\alpha)}. \tag{2.241}$$

Property 2.22 *The following relations (see [3, 75, 150, 152])*

$$F\left({}_{LSC}D_{+}^{\alpha}f(t)\right)(\omega)=(-i\omega)^{\alpha}(Ff)(\omega)$$

and

$$F\left({}_{LS}D_{+}^{\alpha}f(t)\right)(\omega)=(-i\omega)^{\alpha}(Ff)(\omega)$$

holds, where $F\{f(t)\}(\omega)=(Ff)(\omega)$.

Property 2.23 *There are the following relations (see [150])*

$$L\left({}_{LSC}D_{0+}^{\alpha}f(t)\right)(s)=s^{\alpha}\left((Lf)(s)-\sum_{j=0}^{\kappa-1}\frac{f^{(j)}(0)}{s^{j-\alpha+1}}\right), \tag{2.242}$$

and

$$L\left({}_{LS}D_{0+}^{\alpha}f(t)\right)(s)=s^{\alpha}\left((Lf)(s)-f(0)\right), \tag{2.243}$$

where $L\{f(t)\}(s)=(Lf)(s)$.

2.4 Liouville-Grünwald–Letnikov fractional derivatives

2.4.1 Motivations

Liouville first derived in 1832 the fractional difference derivative formula represented in the form (see [15, 16])

$${}_{d}D^{\alpha}f(t)=\lim_{h\to 0}\frac{\Delta_{h}^{\alpha}f(t)}{h^{\alpha}}, \tag{2.244}$$

where $\Delta_{h}^{\alpha}f(t)$ is the fractional difference (see [15]), which is expressible as the sum of the exponentials (see [276]).

In fact, Riemann (in the case $\alpha=2$) derived the fractional difference derivative in the form (see [18, 22, 276, 277])

$${}_{d}D^{[\alpha]}f(t)=\lim_{h\to 0}\frac{\Delta_{h}^{\alpha}f(t)}{h^{\alpha}}, \tag{2.245}$$

where

$$\Delta_{h}^{\alpha}f(t)=\sum_{i=0}^{\infty}(-1)^{i}\begin{pmatrix}\alpha\\ i\end{pmatrix}f(x-hi). \tag{2.246}$$

It is to say that the Riemann work is important for us to present the basic formula for the difference derivative operator in the integer-order sense.

Based on the restrictions of Liouville fractional difference operator (see [17]), Grünwald developed in the following formula (see [20]):

$$ {}_dD^{\alpha} f(t) = \lim_{h\to 0} \frac{\Delta_h^{\alpha} f(t)}{h^{\alpha}}, \tag{2.247} $$

where $\Delta_h^{\alpha} f(t)$ is a difference of fractional order, given by (see [20])

$$ \Delta_h^{\alpha} f = \sum_{i=0}^{\infty} (-1)^i \left(\begin{array}{c} \alpha \\ i \end{array} \right) f(x - hi). \tag{2.248} $$

In fact, Eq.(2.247) is called the Liouville-Grünwald fractional derivative in the book (see [276]).

Letnikov considered the above formula (see [23])

$$ {}_dD^{\alpha} f(t) = \lim_{h\to 0} \frac{\Delta_h^{\alpha} f(t)}{h^{\alpha}}, \tag{2.249} $$

where $\Delta_h^{\alpha} f(t)$ is a difference of fractional order, whose appropriate interpretation of the fractional difference are related with Riemann's and Liouville's expressions [15, 18]. Due to Letnikov, the fractional difference derivative is called the Grünwald–Letnikov fractional derivative [3, 75]. For more details on the fractional difference derivative operators, see [22, 278].

Definition 2.11 *For $\alpha > 0$ and $h > 0$, the left-sided and right-sided fractional differences are given as follows (see [3, 20, 152]):*

$$ \Delta_h^{\alpha} f(t) = \sum_{i=0}^{\infty} (-1)^i \left(\begin{array}{c} \alpha \\ i \end{array} \right) f(t - hi), \tag{2.250} $$

$$ \Delta_{-h}^{\alpha} f(t) = \sum_{i=0}^{\infty} (-1)^i \left(\begin{array}{c} \alpha \\ i \end{array} \right) f(x + hi) \ . \tag{2.251} $$

Property 2.24 *(see [3, 152]) If $\alpha > 0$ and $\beta > 0$, then the semigroup properties (see [3, 152])*

$$ \left(\Delta_h^{\alpha} \Delta_h^{\beta} f \right)(t) = \Delta_h^{\alpha+\beta} f(t) \tag{2.252} $$

and

$$ \left(\Delta_{-h}^{\alpha} \Delta_{-h}^{\beta} f \right)(t) = \Delta_{-h}^{\alpha+\beta} f(t), \tag{2.253} $$

hold for any bounded function $f(t)$.

Property 2.25 *(see [22, 152]) If $\alpha > 0$ and $f(t) \in L_1(\mathbb{R})$, then the Fourier transform of the fractional difference $\Delta_h^{\alpha} f(t)$ is represented in the form (see [22, 152]):*

$$ \left(F\Delta_h^{\alpha} f(t)\right)(\omega) = \left(1 - e^{ith}\right)^{\alpha} (Ff)(\omega). \tag{2.254} $$

2.4.2 Liouville-Grünwald-Letnikov fractional derivatives

Definition 2.12 *Following the above results, the left- and right-sided fractional difference derivatives, denoted by* ${}_dD_+^\alpha f(t)$ *and* ${}_dD_-^\alpha f(t)$*, are defined as follows:*

$$ {}_dD_+^\alpha f(t) = \lim_{h \to +0} \frac{\Delta_h^\alpha f(t)}{h^\alpha} \quad (\alpha > 0), \tag{2.255} $$

and

$$ {}_dD_-^\alpha f(t) = \lim_{h \to +0} \frac{\Delta_{-h}^\alpha f(t)}{h^\alpha} \quad (\alpha > 0), \tag{2.256} $$

respectively, which, in honor of Liouville, Grünwald and Letnikov, are called the Liouville-Grünwald-Letnikov fractional derivatives [3, 22, 75, 152].

Definition 2.13 *When*

$$ f_\iota(t) = \begin{cases} f(t), t \in [a,b], \\ 0, t \notin [a,b], \end{cases} \tag{2.257} $$

we have

$$ \Delta_{h,+a}^\alpha f(t) = \sum_{i=0}^{\left[\frac{t-a}{h}\right]} (-1)^i \binom{\alpha}{i} f(t-hi) \quad (t \in \mathbb{R}, h > 0, \alpha > 0) \tag{2.258} $$

and

$$ \Delta_{-h,-b}^\alpha f(t) = \sum_{i=0}^{\left[\frac{b-t}{h}\right]} (-1)^i \binom{\alpha}{i} f(x+hi) \quad (t \in \mathbb{R}, h > 0, \alpha > 0), \tag{2.259} $$

such that the left-sided and right-sided Liouville-Grünwald-Letnikov fractional derivatives on the interval $[a,b]$ *can be expressed as follows:*

$$ {}_dD_{a+}^\alpha f(t) = \lim_{h \to +0} \frac{\Delta_{h,+a}^\alpha f(t)}{h^\alpha} \tag{2.260} $$

and

$$ {}_dD_{b-}^\alpha f(t) = \lim_{h \to +0} \frac{\Delta_{-h,-h}^\alpha f(t)}{h^\alpha}, \tag{2.261} $$

respectively.

Definition 2.14 *(see [279]) Let* $f(t) \in L_\kappa(\mathbb{R})$*. Then we have*

$$ {}_dD_{a+}^\alpha f(t) = \frac{\alpha}{\Gamma(1-\alpha)} \int_a^t \frac{f(t) - f(\tau)}{(\tau - t)^{1+\alpha}} d\tau \tag{2.262} $$

and

$$ {}_dD_{b-}^\alpha f(t) = \frac{\alpha}{\Gamma(1-\alpha)} \int_t^b \frac{f(t) - f(\tau)}{(t - \tau)^{1+\alpha}} d\tau. \tag{2.263} $$

where $0 < \alpha < 1$ *and* $1 < \kappa < 1/\alpha$*.*

2.4.3 Kilbas-Srivastava-Trujillo fractional derivatives

Definition 2.15 *(see [152]) For $\alpha > 0$ there are*

$$M_{a+}^{\alpha} f(t) = \frac{f(t)(t-a)^{-\alpha}}{\Gamma(1-\alpha)} +_d D_{a+}^{\alpha} f(t) \tag{2.264}$$

and

$$M_{-b}^{\alpha} f(t) = \frac{f(t)(b-t)^{-\alpha}}{\Gamma(1-\alpha)} +_d D_{b-}^{\alpha} f(t), \tag{2.265}$$

where $M_{a+}^{\alpha} f(t)$ and $M_{-b}^{\alpha} f(t)$ are the left-sided and right-sided Kilbas-Srivastava-Trujillo fractional derivatives, respectively.

Definition 2.16 *(see [3, 75]) For $\alpha > 0$ there are*

$$M_{0+}^{\alpha} f(t) = \frac{1}{\Gamma(1-\alpha)} \left(f(t) t^{-\alpha} + \alpha \int_0^t \frac{f(t) - f(\tau)}{(\tau - t)^{1+\alpha}} d\tau \right) \tag{2.266}$$

and

$$M_{-0}^{\alpha} f(t) = \frac{1}{\Gamma(1-\alpha)} \left(f(t)(-t)^{-\alpha} + \alpha \int_t^0 \frac{f(t) - f(\tau)}{(t - \tau)^{1+\alpha}} d\tau \right) \tag{2.267}$$

where $M_{0+}^{\alpha} f(t)$ and $M_{-0}^{\alpha} f(t)$ are the left-sided and right-sided Kilbas-Srivastava-Trujillo fractional derivatives, respectively.

Property 2.26 *(see [279]) For $1 > \alpha > 0$ there are*

$${}_dD_{a+}^{\alpha} l = 0, \ {}_dD_{b-}^{\alpha} l = 0, M_{a+}^{\alpha} l = \frac{l(t-a)^{-\alpha}}{\Gamma(1-\alpha)}, \tag{2.268}$$

$$M_{-b}^{\alpha} l = \frac{l(b-t)^{-\alpha}}{\Gamma(1-\alpha)}, \tag{2.269}$$

where l is any constant.

Property 2.27 *(see [279]) For $1 > \alpha > 0$ there are*

$${}_dD_{+}^{\alpha} f(t) = \lim_{h \to +0} \frac{\Delta_h^{\alpha} f(t)}{|h|^{\alpha}} \tag{2.270}$$

and

$${}_dD_{-}^{\alpha} f(t) = \lim_{h \to +0} \frac{\Delta_{-h}^{\alpha} f(t)}{|h|^{\alpha}}, \tag{2.271}$$

which leads to

$$\begin{aligned} {}_{Ta}D_t^{\alpha} f(t) &= \frac{1}{2cos(\alpha\pi/2)} \lim_{h \to +0} \frac{\Delta_h^{\alpha} f(t) + \Delta_{-h}^{\alpha} f(t)}{|h|^{\alpha}} \\ &= \frac{\alpha}{2\Gamma(1-\alpha)cos(\alpha\pi/2)} \left[\int_{-\infty}^{t} \frac{f(t)-f(\tau)}{(\tau-t)^{1+\alpha}} d\tau + \int_t^{\infty} \frac{f(t)-f(\tau)}{(t-\tau)^{1+\alpha}} d\tau \right]. \end{aligned} \tag{2.272}$$

For the information of the Liouville-Grünwald-Letnikov fractional derivatives, see [20, 279].

2.5 Tarasov type fractional derivatives

We introduce the fractional derivative of Liouville–Grünwald–Letnikov-Riesz type, introduced by Tarasov (see [279]), which is called the Tarasov fractional derivative in honor of Tarasov.

2.5.1 Tarasov type fractional derivatives

Definition 2.17 *Let $1 > \alpha > 0$. The fractional derivative of Liouville–Grünwald–Letnikov-Riesz type given as (see [279])*

$$ {}_{Ta}D_t^{\alpha} f(t) = \frac{\alpha}{2\Gamma(1-\alpha)\cos(\alpha\pi/2)} \left[\int_{-\infty}^{t} \frac{f(t)-f(\tau)}{(\tau - t)^{1+\alpha}} d\tau + \int_{t}^{\infty} \frac{f(t)-f(\tau)}{(t-\tau)^{1+\alpha}} d\tau \right], \tag{2.273}$$

is called the Tarasov fractional derivative, which is a family of the fractional derivative of the Grünwald–Letnikov–Riesz type.

Since there are (see [279])

$$ {}_{d}D_{+}^{\alpha} f(t) = \lim_{h\to+0} \frac{\Delta_h^{\alpha} f(t)}{|h|^{\alpha}} = \frac{\alpha}{\Gamma(1-\alpha)} \int_{-\infty}^{t} \frac{f(t)-f(\tau)}{(\tau-t)^{1+\alpha}} d\tau = \frac{\alpha}{\Gamma(1-\alpha)} \int_{0}^{\infty} \frac{f(\tau)-f(t-\tau)}{\tau^{1+\alpha}} d\tau \tag{2.274}$$

and

$$ {}_{d}D_{-}^{\alpha} f(t) = \lim_{h\to+0} \frac{\Delta_{-h}^{\alpha} f(t)}{|h|^{\alpha}} = \frac{\alpha}{\Gamma(1-\alpha)} \int_{t}^{\infty} \frac{f(t)-f(\tau)}{(t-\tau)^{1+\alpha}} d\tau = \frac{\alpha}{\Gamma(1-\alpha)} \int_{0}^{\infty} \frac{f(\tau)-f(t+\tau)}{\tau^{1+\alpha}} d\tau, \tag{2.275}$$

we have (see [279])

$$ {}_{Ta}D_t^{\alpha} f(t) = -\frac{\alpha}{2\Gamma(1-\alpha)\cos(\alpha\pi/2)} \int_{0}^{\infty} \frac{f(\tau)-f(t+\tau)}{\tau^{1+\alpha}} d\tau. \tag{2.276}$$

Making use of

$$ \frac{\alpha}{\Gamma(1-\alpha)} = \Gamma(1+\alpha) \frac{\sin(\alpha\pi)}{\pi} \tag{2.277}$$

and

$$ \sin(\alpha\pi) = 2\sin(\alpha\pi/2)\cos(\alpha\pi/2), \tag{2.278}$$

the Tarasov fractional derivative can be rewritten as follows:

$$ \begin{aligned} {}_{Ta}D_t^{\alpha} f(t) &= -\frac{\alpha}{2\Gamma(1-\alpha)\cos(\alpha\pi/2)} \int_{0}^{\infty} \frac{f(\tau)-f(t+\tau)}{\tau^{1+\alpha}} d\tau \\ &= \frac{\Gamma(1+\alpha)\sin(\alpha\pi/2)}{\pi} \int_{0}^{\infty} \frac{f(t+\tau)-f(\tau)}{\tau^{1+\alpha}} d\tau. \end{aligned} \tag{2.279}$$

2.5.2 Extended Tarasov type fractional derivatives

Definition 2.18 *Let* $1 > \alpha > 0$. *The extended Tarasov type fractional derivative of Liouville–Grünwald–Letnikov-Feller type is defined as*

$$ {}_{FV}D^{\alpha}_{t;\vartheta} f(t) = H_{+}(\vartheta,\alpha) \times \lim_{h \to +0} \frac{\Delta^{\alpha}_{h} f(t)}{|h|^{\alpha}} + H_{-}(\vartheta,\alpha) \times \lim_{h \to +0} \frac{\Delta^{\alpha}_{-h} f(t)}{|h|^{\alpha}}, \tag{2.280} $$

where two components can be expressed as follows:

$$ H_{-}(\vartheta,\alpha) = \frac{\sin((\alpha-\vartheta)\pi/2)}{\sin(\pi\vartheta)} \tag{2.281} $$

and

$$ H_{+}(\vartheta,\alpha) = \frac{\sin((\alpha+\vartheta)\pi/2)}{\sin(\pi\vartheta)}. \tag{2.282} $$

With the use of

$$ \frac{\alpha}{\Gamma(1-\alpha)} = \Gamma(1+\alpha)\frac{\sin(\alpha\pi)}{\pi}, \tag{2.283} $$

we obtain

$$ \begin{aligned} {}_{FV}D^{\alpha}_{t;\vartheta} f(t) &= H_{+}(\vartheta,\alpha) \times \lim_{h \to +0} \tfrac{\Delta^{\alpha}_{h} f(t)}{|h|^{\alpha}} + H_{-}(\vartheta,\alpha) \times \lim_{h \to +0} \tfrac{\Delta^{\alpha}_{-h} f(t)}{|h|^{\alpha}} \\ &= \tfrac{\sin((\alpha+\vartheta)\pi/2)}{\sin(\pi\vartheta)} \times \tfrac{\Gamma(1+\alpha)\sin(\pi\alpha)}{\pi} \times \int_0^{\infty} \tfrac{f(\tau)-f(t-\tau)}{\tau^{1+\alpha}} d\tau \\ &- \tfrac{\sin((\alpha-\vartheta)\pi/2)}{\sin(\pi\vartheta)} \times \tfrac{\Gamma(1+\alpha)\sin(\pi\alpha)}{\pi} \int_0^{\infty} \tfrac{f(\tau)-f(t+\tau)}{\tau^{1+\alpha}} d\tau. \end{aligned} \tag{2.284} $$

Thus, the extended Tarasov type fractional derivative of Liouville–Grünwald–Letnikov-Feller type can be expressed as follows:

$$ \begin{aligned} {}_{FV}D^{\alpha}_{t;\vartheta} f(t) &= H_{+}(\vartheta,\alpha) \times \lim_{h \to +0} \tfrac{\Delta^{\alpha}_{h} f(t)}{|h|^{\alpha}} + H_{-}(\vartheta,\alpha) \times \lim_{h \to +0} \tfrac{\Delta^{\alpha}_{-h} f(t)}{|h|^{\alpha}} \\ &= \tfrac{\sin(\pi\alpha)}{\sin(\pi\vartheta)} \times \tfrac{\Gamma(1+\alpha)\sin((\alpha+\vartheta)\pi/2)}{\pi} \times \int_{\infty}^{t} \tfrac{f(t)-f(\tau)}{(\tau-t)^{1+\alpha}} d\tau \\ &+ \tfrac{\sin(\pi\alpha)}{\sin(\pi\vartheta)} \times \tfrac{\Gamma(1+\alpha)\sin((\alpha-\vartheta)\pi/2)}{\pi} \times \int_{\infty}^{b} \tfrac{f(t)-f(\tau)}{(t-\tau)^{1+\alpha}} d\tau. \end{aligned} \tag{2.285} $$

Definition 2.19 *Let* $1 > \alpha > 0$. *The extended Tarasov type fractional derivative of Liouville–Grünwald–Letnikov-Richard type is defined as*

$$ {}_{L}D^{\alpha}_{\tau} f(t) = \frac{\alpha}{2\Gamma(1-\alpha)\, sin(\alpha\pi/2)} \left[\int_{-\infty}^{t} \frac{f(t)-f(\tau)}{(\tau-t)^{1+\alpha}} d\tau + \int_{t}^{\infty} \frac{f(t)-f(\tau)}{(t-\tau)^{1+\alpha}} d\tau \right]. \tag{2.286} $$

The relations among the Tarasov type fractional derivatives of the Liouville–Grünwald–Letnikov-Riesz, Liouville–Grünwald–Letnikov-Feller and Liouville–Grünwald–Letnikov-Richard types in the sense of Tarasov type are presented as follows:

(1) When $\vartheta = 0$, we have (see [168])

$$H_+(0,\alpha) = H_-(0,\alpha) = \frac{1}{2cos(\alpha\pi/2)}, \tag{2.287}$$

which yields from Eq. (5.63) and Eq. (5.64) that the Tarasov fractional derivative of two forms:

$$\begin{aligned}
&{}_{FV}D^{\alpha}_{t;0}f(t) = H_+(0,\alpha) \times \lim_{h\to+0} \frac{\Delta^{\alpha}_{h}f(t)}{|h|^{\alpha}} + H_-(0,\alpha) \times \lim_{h\to+0} \frac{\Delta^{\alpha}_{-h}f(t)}{|h|^{\alpha}} \\
&= \frac{1}{2cos(\alpha\pi/2)} \times \left(\lim_{h\to+0} \frac{\Delta^{\alpha}_{h}f(t)}{|h|^{\alpha}} + \lim_{h\to+0} \frac{\Delta^{\alpha}_{-h}f(t)}{|h|^{\alpha}} \right) \\
&= {}_{Ta}D^{\alpha}_{\tau}f(t).
\end{aligned} \tag{2.288}$$

(2) When $\vartheta = 1$, we have (see [279])

$$H_+(1,\alpha) = H_-(1,\alpha) = \frac{1}{2sin(\alpha\pi/2)}, \tag{2.289}$$

such that

$$\begin{aligned}
&{}_{FV}D^{\alpha}_{t;1}f(t) = H_+(1,\alpha) \times \lim_{h\to+0} \frac{\Delta^{\alpha}_{h}f(t)}{|h|^{\alpha}} + H_-(1,\alpha) \times \lim_{h\to+0} \frac{\Delta^{\alpha}_{-h}f(t)}{|h|^{\alpha}} \\
&= \frac{1}{2sin(\alpha\pi/2)} \times \left(\lim_{h\to+0} \frac{\Delta^{\alpha}_{h}f(t)}{|h|^{\alpha}} + \lim_{h\to+0} \frac{\Delta^{\alpha}_{-h}f(t)}{|h|^{\alpha}} \right) \\
&= {}_{L}D^{\alpha}_{\tau}f(t).
\end{aligned} \tag{2.290}$$

Analogous to the above considerations, we have the new family of the Tarasov type fractional derivatives of the Liouville–Grünwald–Letnikov-Riesz, Liouville–Grünwald–Letnikov-Feller and Liouville–Grünwald–Letnikov-Richard types as follows.

Definition 2.20 *Let $1 > \alpha > 0$. The Tarasov type fractional derivatives of the Liouville–Grünwald–Letnikov-Riesz on $[a,b]$ is defined by*

$$ {}_{Ta}D^{\alpha}_{t;[a,b]}f(t) = \frac{1}{2cos(\alpha\pi/2)} \lim_{h\to+0} \frac{\Delta^{\alpha}_{h,+a}f(t) + \Delta^{\alpha}_{-h,-b}f(t)}{|h|^{\alpha}}. \tag{2.291}$$

Definition 2.21 *Let $1 > \alpha > 0$. The Tarasov type fractional derivatives of the Liouville–Grünwald–Letnikov-Feller on $[a,b]$ is defined by*

$$ {}_{FV}D^{\alpha}_{t;[a,b];\vartheta}f(t) = H_+(\vartheta,\alpha) \times \lim_{h\to+0} \frac{\Delta^{\alpha}_{h,+a}f(t)}{|h|^{\alpha}} + H_-(\vartheta,\alpha) \times \lim_{h\to+0} \frac{\Delta^{\alpha}_{-h,-b}f(t)}{|h|^{\alpha}}. \tag{2.292}$$

Definition 2.22 *Let $1 > \alpha > 0$. The Tarasov type fractional derivatives of the Liouville–Grünwald–Letnikov-Richard on $[a,b]$ is defined by*

$$ {}_{L}D^{\alpha}_{t;[a,b]}f(t) = \frac{1}{2sin(\alpha\pi/2)} \lim_{h\to+0} \frac{\Delta^{\alpha}_{h,+a}f(t) + \Delta^{\alpha}_{-h,-b}f(t)}{|h|^{\alpha}}. \tag{2.293}$$

We now give the expressions as follows

$$\begin{aligned} {}_{Ta}D^{\alpha}_{t;[a,b]}f(t) &= \frac{1}{2cos(\alpha\pi/2)} \lim_{h\to+0} \frac{\Delta^{\alpha}_{h,+a}f(t)+\Delta^{\alpha}_{-h,-b}f(t)}{|h|^{\alpha}} \\ &= \frac{\Gamma(1+\alpha)\sin(\alpha\pi/2)}{\pi} \times \left[\int_a^t \frac{f(t)-f(\tau)}{(\tau-t)^{1+\alpha}}d\tau + \int_t^b \frac{f(t)-f(\tau)}{(t-\tau)^{1+\alpha}}d\tau\right], \end{aligned} \tag{2.294}$$

$$\begin{aligned} &{}_{FV}D^{\alpha}_{t;[a,b];\vartheta}f(t) = H_{+}(\vartheta,\alpha) \times \lim_{h\to+0} \frac{\Delta^{\alpha}_{h,+a}f(t)}{|h|^{\alpha}} + H_{-}(\vartheta,\alpha) \times \lim_{h\to+0} \frac{\Delta^{\alpha}_{-h,-b}f(t)}{|h|^{\alpha}} \\ &= \frac{\alpha}{\Gamma(1-\alpha)} \times \left[\frac{\sin((\alpha+\vartheta)\pi/2)}{\sin(\pi\vartheta)} \int_a^t \frac{f(t)-f(\tau)}{(\tau-t)^{1+\alpha}}d\tau + \frac{\sin((\alpha-\vartheta)\pi/2)}{\sin(\pi\vartheta)} \int_t^b \frac{f(t)-f(\tau)}{(t-\tau)^{1+\alpha}}d\tau\right]. \end{aligned} \tag{2.295}$$

and

$$\begin{aligned} {}_{Ta}D^{\alpha}_{t;[a,b]}f(t) &= \frac{1}{2sin(\alpha\pi/2)} \lim_{h\to+0} \frac{\Delta^{\alpha}_{h,+a}f(t)+\Delta^{\alpha}_{-h,-b}f(t)}{|h|^{\alpha}} \\ &= \frac{1}{2sin(\alpha\pi/2)} \times \frac{\alpha}{\Gamma(1-\alpha)} \times \left[\int_a^t \frac{f(t)-f(\tau)}{(\tau-t)^{1+\alpha}}d\tau + \int_t^b \frac{f(t)-f(\tau)}{(t-\tau)^{1+\alpha}}d\tau\right] \\ &= \frac{\Gamma(1+\alpha)\cos(\alpha\pi/2)}{\pi} \times \left[\int_a^t \frac{f(t)-f(\tau)}{(\tau-t)^{1+\alpha}}d\tau + \int_t^b \frac{f(t)-f(\tau)}{(t-\tau)^{1+\alpha}}d\tau\right]. \end{aligned} \tag{2.296}$$

In particular, we take the following forms:

$$\begin{aligned} &{}_{FV}D^{\alpha}_{t;[a,b];0}f(t) = H_{+}(0,\alpha) \times \lim_{h\to+0} \frac{\Delta^{\alpha}_{h,+a}f(t)}{|h|^{\alpha}} + H_{-}(0,\alpha) \times \lim_{h\to+0} \frac{\Delta^{\alpha}_{-h,-b}f(t)}{|h|^{\alpha}} \\ &= \frac{1}{2cos(\alpha\pi/2)} \times \frac{\alpha}{\Gamma(1-\alpha)} \times \left[\int_a^t \frac{f(t)-f(\tau)}{(\tau-t)^{1+\alpha}}d\tau + \int_t^b \frac{f(t)-f(\tau)}{(t-\tau)^{1+\alpha}}d\tau\right] \\ &= \frac{\Gamma(1+\alpha)\sin(\alpha\pi/2)}{\pi} \times \left[\int_a^t \frac{f(t)-f(\tau)}{(\tau-t)^{1+\alpha}}d\tau + \int_t^b \frac{f(t)-f(\tau)}{(t-\tau)^{1+\alpha}}d\tau\right] \\ &= {}_{Ta}\,D^{\alpha}_{t;[a,b]}f(t) \end{aligned} \tag{2.297}$$

and

$$\begin{aligned} &{}_{FV}D^{\alpha}_{t;[a,b];1}f(t) = H_{+}(\vartheta,\alpha) \times \lim_{h\to+0} \frac{\Delta^{\alpha}_{h,+a}f(t)}{|h|^{\alpha}} + H_{-}(\vartheta,\alpha) \times \lim_{h\to+0} \frac{\Delta^{\alpha}_{-h,-b}f(t)}{|h|^{\alpha}} \\ &= \frac{1}{2sin(\alpha\pi/2)} \times \frac{\alpha}{\Gamma(1-\alpha)} \times \left[\int_a^t \frac{f(t)-f(\tau)}{(\tau-t)^{1+\alpha}}d\tau + \int_t^b \frac{f(t)-f(\tau)}{(t-\tau)^{1+\alpha}}d\tau\right] \\ &= \frac{\Gamma(1+\alpha)\cos(\alpha\pi/2)}{\pi} \times \left[\int_a^t \frac{f(t)-f(\tau)}{(\tau-t)^{1+\alpha}}d\tau + \int_t^b \frac{f(t)-f(\tau)}{(t-\tau)^{1+\alpha}}d\tau\right] \\ &= {}_{L}\,D^{\alpha}_{t;[a,b]}f(t) \end{aligned} \tag{2.298}$$

Note that ${}_{Ta}D_t^{\alpha}f(t)$, ${}_{Ta}D_{t;[a,b]}^{\alpha}f(t)$, ${}_{L}D_t^{\alpha}f(t)$, ${}_{L}D_{t;[a,b]}^{\alpha}f(t)$, ${}_{FV}D_{t;\vartheta}^{\alpha}f(t)$ and ${}_{FV}D_{t;[a,b];\vartheta}^{\alpha}f(t)$ are the family of the Tarasov type fractional derivatives.

2.6 Riesz fractional calculus

The definition of the Riesz fractional derivative in the one-dimensional and three-dimensional spaces are given as follows [44, 67].

Definition 2.23 *The Riesz fractional integration of functions of many variables as potential type operators is defined as [44, 168]*

$$(-\Delta)^{\frac{\alpha}{2}} f(t) = F^{-1}|t|^{-\alpha} Ff(t) = \begin{cases} {}_{Rz}I_{\mathbb{R}^n}^{\alpha} f(t), Re(\alpha) > 0, \\ {}_{Rz}D_{\mathbb{R}^n}^{\alpha} f(t), Re(\alpha) > 0. \end{cases} \tag{2.299}$$

where ${}_{Rz}I_{\mathbb{R}^n}^{\alpha} f(t)$ $(Re(\alpha) > 0)$ is the Riesz fractional integral in the three-dimensional space and ${}_{Rz}D_{\mathbb{R}^n}^{\alpha} f(t)$ is the Riesz fractional derivative in the three-dimensional space.

Definition 2.24 *The Riesz fractional integral in the one-dimensional space is defined as [44, 168]*

$${}_{Rz}I_{\mathbb{R}}^{\alpha} f(t) = \frac{\left({}_{L}I_{+}^{\alpha}f + {}_{L}I_{-}^{\alpha}f\right)(t)}{2\cos(\pi\alpha/2)} = \frac{1}{2\Gamma(\alpha)\cos(\pi\alpha/2)} \int_{-\infty}^{\infty} \frac{f(\tau)}{|\tau - t|^{1-\alpha}} d\tau. \tag{2.300}$$

Definition 2.25 *The Riesz fractional derivative in the one-dimensional space is defined as [44, 168]*

$${}_{Rz}D_{\mathbb{R}}^{\alpha} f(t) = \frac{\left({}_{L}D_{+}^{\alpha}f + {}_{L}D_{-}^{\alpha}f\right)(t)}{2\cos(\pi\alpha/2)} = \frac{\Gamma(1+\alpha)\sin(\pi\alpha/2)}{\pi} \times \int_{0}^{\infty} \frac{f(t+\tau) - 2f(t) + f(t-\tau)}{\tau^{1-\alpha}} d\tau, \tag{2.301}$$

where

$${}_{L}D_{+}^{\alpha} f(t) = \frac{\alpha}{\Gamma(1-\alpha)} \int_{-\infty}^{t} \frac{f(t) - f(\tau)}{(t-\tau)^{1+\alpha}} d\tau \tag{2.302}$$

and

$${}_{L}D_{-}^{\alpha} f(t) = \frac{\alpha}{\Gamma(1-\alpha)} \int_{t}^{\infty} \frac{f(t) - f(\tau)}{(\tau - t)^{1+\alpha}} d\tau. \tag{2.303}$$

With the help of

$$\cos(\pi\alpha/2) = \frac{e^{i\pi\alpha/2} + e^{-i\pi\alpha/2}}{2}, \tag{2.304}$$

we have

$$_{Rz}I_{\mathbb{R}}^{\alpha}f(t) = \frac{\left({}_{L}I_{+}^{\alpha}f + {}_{L}I_{-}^{\alpha}f\right)(t)}{2\cos(\pi\alpha/2)} = \frac{\left({}_{L}I_{+}^{\alpha}f + {}_{L}I_{-}^{\alpha}f\right)(t)}{e^{i\pi\alpha/2} + e^{-i\pi\alpha/2}} \tag{2.305}$$

and

$$_{Rz}D_{\mathbb{R}}^{\alpha}f(t) = \frac{\left({}_{L}D_{+}^{\alpha}f + {}_{L}D_{-}^{\alpha}f\right)(t)}{2\cos(\pi\alpha/2)} = \frac{\left({}_{L}D_{+}^{\alpha}f + {}_{L}D_{-}^{\alpha}f\right)(t)}{e^{i\pi\alpha/2} + e^{-i\pi\alpha/2}}. \tag{2.306}$$

2.7 Feller fractional calculus

The conceptions of the Feller fractional derivative and Feller fractional integral are given as follows [47].

Definition 2.26 *Let* $1 > Re(\alpha) > 0$. *We define the Feller fractional integral by (see [47]; for real number of the order, see [168])*

$$\begin{aligned} &_{F}I_{\mathbb{R},\vartheta}^{\alpha}f(t) = H_{-}(\vartheta,\alpha)_{L}\,I_{+}^{\alpha}f(t) + H_{+}(\vartheta,\alpha)_{L}\,I_{-}^{\alpha}f(t) \\ &= \frac{\sin((\alpha-\vartheta)\pi/2)}{\sin(\pi\vartheta)} \times \frac{1}{\Gamma(\alpha)}\int_{-\infty}^{t}\frac{f(\tau)}{(\tau-t)^{1-\alpha}}d\tau + \frac{\sin((\alpha+\vartheta)\pi/2)}{\sin(\pi\vartheta)} \times \frac{1}{\Gamma(\alpha)}\int_{t}^{\infty}\frac{f(\tau)}{(t-\tau)^{1-\alpha}}d\tau, \end{aligned} \tag{2.307}$$

where two components can be expressed as follows:

$$H_{-}(\vartheta,\alpha) = \frac{\sin((\alpha-\vartheta)\pi/2)}{\sin(\pi\vartheta)}, \tag{2.308}$$

and

$$H_{+}(\vartheta,\alpha) = \frac{\sin((\alpha+\vartheta)\pi/2)}{\sin(\pi\vartheta)}. \tag{2.309}$$

Definition 2.27 *Let* $1 > Re(\alpha) > 0$. *The fractional Feller derivative can be represented in the form (see [47]; for real number of the order, see [168]):*

$$\begin{aligned} &_{F}D_{\mathbb{R},\vartheta}^{\alpha}f(t) = -\left(H_{+}(\vartheta,\alpha)_{L}\,D_{+}^{\alpha}f(t) + H_{-}(\vartheta,\alpha)_{L}\,D_{-}^{\alpha}f(t)\right) \\ &= -\frac{\sin((\alpha+\vartheta)\pi/2)}{\sin(\pi\vartheta)} \times \frac{1}{\Gamma(1-\alpha)}\frac{d}{dt}\int_{-\infty}^{t}\frac{f(\tau)}{(\tau-t)^{\alpha}}d\tau - \frac{\sin((\alpha-\vartheta)\pi/2)}{\sin(\pi\vartheta)} \times \frac{1}{\Gamma(1-\alpha)}\frac{d}{dt}\int_{t}^{\infty}\frac{f(\tau)}{(t-\tau)^{\alpha}}d\tau. \end{aligned} \tag{2.310}$$

The formulations are called the Feller fractional calculus in honor of Feller. For $1 > Re(\alpha) > 0$, we have, by taking (see [168])

$$H_{+}(0,\alpha) = H_{-}(0,\alpha) = \frac{1}{2cos(\alpha\pi/2)}, \tag{2.311}$$

that

$$\begin{aligned} &_{F}I_{\mathbb{R},0}^{\alpha}f(t) = H_{-}(0,\alpha)_{L}\,I_{+}^{\alpha}f(t) + H_{+}(0,\alpha)_{L}\,I_{-}^{\alpha}f(t) \\ &= \frac{\left({}_{L}I_{+}^{\alpha}f + {}_{L}I_{-}^{\alpha}f\right)(t)}{2\cos(\pi\alpha/2)} = {}_{Rz}\,I_{\mathbb{R}}^{\alpha}f(t) \end{aligned} \tag{2.312}$$

and

$$\begin{aligned} {}_F D^{\alpha}_{\mathbb{R},0} f(t) &= -\left(H_+(0,\alpha)_L\, D^{\alpha}_+ f(t) + H_-(0,\alpha)_L\, D^{\alpha}_- f(t)\right) \\ &= -\frac{{}_L D^{\alpha}_+ f(t) + {}_L D^{\alpha}_- f(t)}{2cos(\alpha\pi/2)} = {}_{Rz} D^{\alpha}_{\mathbb{R}} f(t). \end{aligned} \quad (2.313)$$

2.8 Richard fractional calculus

In 2011, Richard proposed the fractional derivative and fractional integral (see [280]; also see [168]).

For $1 > Re(\alpha) > 0$, we have, by taking

$$H_+(1,\alpha) = H_-(1,\alpha) = \frac{1}{2sin(\alpha\pi/2)}, \quad (2.314)$$

that (see [280])

$${}_F I^{\alpha}_{\mathbb{R},1} f(t) = H_-(1,\alpha)_L\, I^{\alpha}_+ f(t) + H_+(1,\alpha)_L\, I^{\alpha}_- f(t) = \frac{\left({}_L I^{\alpha}_+ f + {}_L I^{\alpha}_- f\right)(t)}{2sin(\alpha\pi/2)} \quad (2.315)$$

and

$${}_F D^{\alpha}_{\mathbb{R},1} f(t) = -\left(H_+(1,\alpha)_L D^{\alpha}_+ f(t) + H_-(1,\alpha)_L\, D^{\alpha}_- f(t)\right) = \frac{\left({}_L D^{\alpha}_+ f + {}_L D^{\alpha}_- f\right)(t)}{2sin(\alpha\pi/2)}, \quad (2.316)$$

which are deduced in 2011 by Richard (see [168]).

Definition 2.28 *For $1 > Re(\alpha) > 0$, Richard defines (see [168])*

$${}_{Ri} I^{\alpha}_{\mathbb{R}} f(t) = \frac{\left({}_L I^{\alpha}_+ f + {}_L I^{\alpha}_- f\right)(t)}{2sin(\alpha\pi/2)} = \frac{1}{2\Gamma(\alpha)\sin(\pi\alpha/2)} \int_{-\infty}^{\infty} \frac{f(\tau)}{|\tau - t|^{1-\alpha}} d\tau \quad (2.317)$$

and

$${}_{Ri} D^{\alpha}_{\mathbb{R}} f(t) = \frac{\left({}_L D^{\alpha}_+ f + {}_L D^{\alpha}_- f\right)(t)}{2sin(\alpha\pi/2)} = \frac{\Gamma(1+\alpha)\cos(\pi\alpha/2)}{\pi} \times \int_0^{\infty} \frac{f(t+\tau) - f(t-\tau)}{\tau^{1-\alpha}} d\tau, \quad (2.318)$$

which are called the Richard fractional integral and the Richard fractional derivative, respectively.

In fact, according to the expression of the Riesz fractional derivative, we write Richard fractional derivative as

$${}_{Ri} D^{\alpha}_{\mathbb{R}} f(t) = \frac{\left({}_L D^{\alpha}_+ f + {}_L D^{\alpha}_- f\right)(t)}{2sin(\alpha\pi/2)} = \frac{\Gamma(1+\alpha)\cos(\pi\alpha/2)}{\pi} \times \int_0^{\infty} \frac{f(t+\tau) - 2f(t) + f(t-\tau)}{\tau^{1-\alpha}} d\tau. \quad (2.319)$$

They are called the Richard fractional calculus in honor of Richard.

2.9 Erdélyi-Kober type fractional calculus

2.9.1 Erdélyi-Kober type operators of fractional integration and fractional derivative

The conceptions of the Erdélyi-Kober type operators of fractional integration and fractional derivative, as the extensions of the Riemann-Liouville left-sided and right-sided fractional integrals, are given as follows [3, 105, 181].

Let $\mathrm{M} = (a, b)$ $(-\infty \leq a < b \leq \infty)$ be a finite interval on the real axis $\mathbb{R}$.

Definition 2.29 *Let $\alpha \in \mathbb{C}$, $Re(\alpha) > 0$, $\sigma > 0$, $t > 0$ and $\eta \in \mathbb{C}$. We define the left-sided and the right-sided Erdélyi-Kober type fractional integrals as (see [3, 152])*

$$\left(I^{\alpha}_{a+;\sigma,\eta} f\right)(t) = \frac{\sigma t^{-\sigma(\alpha+\eta)}}{\Gamma(\alpha)} \int_{a}^{t} \frac{\tau^{\sigma(\eta+1)-1} f(\tau)}{\left(t^{\sigma} - \tau^{\sigma}\right)^{1-\alpha}} d\tau \tag{2.320}$$

and

$$\left(I^{\alpha}_{b-;\sigma,\eta} f\right)(t) = \frac{\sigma t^{-\sigma\eta}}{\Gamma(\alpha)} \int_{t}^{b} \frac{\tau^{\sigma(1-\alpha-\eta)-1} f(\tau)}{\left(\tau^{\sigma} - t^{\sigma}\right)^{1-\alpha}} d\tau, \tag{2.321}$$

respectively.

Definition 2.30 *When $a = -\infty$ and $b = -\infty$, we should present the following notations (see [3, 152])*

$$\left({}_{E}I^{\alpha}_{+;\sigma,\eta} f\right)(t) = \frac{\sigma t^{-\sigma(\alpha+\eta)}}{\Gamma(\alpha)} \int_{-\infty}^{t} \frac{\tau^{\sigma(\eta+1)-1} f(\tau)}{\left(t^{\sigma} - \tau^{\sigma}\right)^{1-\alpha}} d\tau \tag{2.322}$$

and

$$\left({}_{E}I^{\alpha}_{-;\sigma,\eta} f\right)(t) = \frac{\sigma t^{-\sigma\eta}}{\Gamma(\alpha)} \int_{t}^{\infty} \frac{\tau^{\sigma(1-\alpha-\eta)-1} f(\tau)}{\left(\tau^{\sigma} - t^{\sigma}\right)^{1-\alpha}} d\tau, \tag{2.323}$$

which are called the left-sided and the right-sided Erdélyi-Kober type fractional integrals.

In particular, when $\sigma = 1$, we should obtain the following notations in the forms:

$$\left(I^{\alpha}_{a+;1,\eta} f\right)(t) = \frac{t^{-(\alpha+\eta)}}{\Gamma(\alpha)} \int_{a}^{t} \frac{\tau^{(\eta+1)-1} f(\tau)}{\left(t - \tau\right)^{1-\alpha}} d\tau \tag{2.324}$$

and

$$\left(I^{\alpha}_{b-;1,\eta} f\right)(t) = \frac{t^{-\eta}}{\Gamma(\alpha)} \int_{t}^{b} \frac{\tau^{-(\alpha+\eta)} f(\tau)}{\left(\tau - t\right)^{1-\alpha}} d\tau, \tag{2.325}$$

which are called the left-sided and the right-sided Erdélyi-Kober type fractional integrals, respectively.

In particular, when $\sigma = 1$, $a = -\infty$ and $b = \infty$ we represent (see [3, 152])

$$\left(I_{\eta,\alpha}^{+} f\right)(t) = \left(I_{0+;1,\eta}^{\alpha} f\right)(t) = \frac{t^{-(\alpha+\eta)}}{\Gamma(\alpha)} \int_{0}^{t} \frac{\tau^{\eta} f(\tau)}{(t-\tau)^{1-\alpha}} d\tau \tag{2.326}$$

and

$$\left(K_{\eta,\alpha}^{-} f\right)(t) = \left(I_{-;1,\eta}^{\alpha} f\right)(t) = \frac{t^{-\eta}}{\Gamma(\alpha)} \int_{t}^{\infty} \frac{\tau^{-(\alpha+\eta)} f(\tau)}{(\tau-t)^{1-\alpha}} d\tau, \tag{2.327}$$

which are known as the Kober or Erdélyi-Kober integral operators.

In particular, when $\eta = 0$, we should obtain the following notations in the forms:

$$\left(I_{a+;\sigma,0}^{\alpha} f\right)(t) = \frac{\sigma t^{-\sigma\alpha}}{\Gamma(\alpha)} \int_{a}^{t} \frac{\tau^{\sigma-1} f(\tau)}{\left(t^{\sigma} - \tau^{\sigma}\right)^{1-\alpha}} d\tau \tag{2.328}$$

and

$$\left(I_{b-;\sigma,0}^{\alpha} f\right)(t) = \frac{\sigma}{\Gamma(\alpha)} \int_{t}^{b} \frac{\tau^{\sigma(1-\alpha)-1} f(\tau)}{\left(\tau^{\sigma} - t^{\sigma}\right)^{1-\alpha}} d\tau, \tag{2.329}$$

which are called the left-sided and the right-sided Erdélyi-Kober type fractional integrals, respectively.

In particular, when $\eta = 0$, $a = -\infty$ and $b = -\infty$, we should present the following notations

$$\left({}_{Y}I_{+;\sigma,0}^{\alpha} f\right)(t) = \frac{\sigma t^{-\sigma\alpha}}{\Gamma(\alpha)} \int_{-\infty}^{t} \frac{\tau^{\sigma-1} f(\tau)}{\left(t^{\sigma} - \tau^{\sigma}\right)^{1-\alpha}} d\tau \tag{2.330}$$

and

$$\left({}_{Y}I_{-;\sigma,0}^{\alpha} f\right)(t) = \frac{\sigma}{\Gamma(\alpha)} \int_{t}^{\infty} \frac{\tau^{\sigma(1-\alpha)-1} f(\tau)}{\left(\tau^{\sigma} - t^{\sigma}\right)^{1-\alpha}} d\tau, \tag{2.331}$$

which are called the left-sided and the right-sided Erdélyi-Kober type fractional integrals, respectively.

In particular, when $\eta = 0$ and $\sigma = 1$, we should obtain the following notations in the forms (see [280]):

$$\left(I_{a+;1,0}^{\alpha} f\right)(t) = \frac{t^{-\alpha}}{\Gamma(\alpha)} \int_{a}^{t} \frac{f(\tau)}{(t-\tau)^{1-\alpha}} d\tau \tag{2.332}$$

and

$$\left(I_{b-;1,0}^{\alpha} f\right)(t) = \frac{1}{\Gamma(\alpha)} \int_{t}^{b} \frac{\tau^{-\alpha} f(\tau)}{(\tau-t)^{1-\alpha}} d\tau, \tag{2.333}$$

which are called the left-sided and the right-sided Erdélyi-Kober type fractional integrals, respectively.

In particular, when $\eta = 0$, $\sigma = 1$, $a = -\infty$ and $b = -\infty$, we should obtain the following notations in the forms (see [280]):

$$\left({}_LI^{\alpha}_{+;1,0}f\right)(t) = \frac{t^{-\alpha}}{\Gamma(\alpha)} \int\limits_{-\infty}^{t} \frac{f(\tau)}{(t-\tau)^{1-\alpha}} d\tau \tag{2.334}$$

and

$$\left({}_LI^{\alpha}_{-;1,0}f\right)(t) = \frac{1}{\Gamma(\alpha)} \int\limits_{t}^{\infty} \frac{\tau^{-\alpha} f(\tau)}{(\tau-t)^{1-\alpha}} d\tau, \tag{2.335}$$

which are called the left-sided and the right-sided Erdélyi-Kober type fractional integrals, respectively.

Let $\mathrm{M} = (a, b)$ $(-\infty \leq a < b \leq \infty)$ be a finite interval on the real axis $\mathbb{R}$.

Definition 2.31 *Let $\alpha \in \mathbb{C}$, $Re(\alpha) > 0$, $\kappa = [Re(\alpha)] + 1$, $\sigma > 0$, $t > 0$ and $\eta \in \mathbb{C}$. We define the left-sided and the right-sided Erdélyi-Kober type fractional derivatives as (see [3, 152])*

$$\left(D^{\alpha}_{a+;\sigma,\eta}f\right)(t) = t^{-\sigma\eta} \left(\tfrac{1}{\sigma t^{\sigma-1}} D\right)^{\kappa} t^{\sigma(\alpha+\eta)} \left(I^{\alpha}_{a+;\sigma,\eta+\alpha}f\right)(t) \tag{2.336}$$

and

$$\left(D^{\alpha}_{b-;\sigma,\eta}f\right)(t) = t^{-\sigma(\eta+\alpha)} \left(\tfrac{1}{\sigma t^{\sigma-1}} D\right)^{\kappa} t^{\sigma(\kappa+\eta-\alpha)} \left(I^{\alpha}_{b-;\sigma,\eta+\alpha-\kappa}f\right)(t), \tag{2.337}$$

respectively, where $D = d/dt$.

Definition 2.32 *Let $\alpha \in \mathbb{C}$, $1 > Re(\alpha) > 0$, $\sigma > 0$, $t > 0$ and $\eta \in \mathbb{C}$. We define the left-sided and the right-sided Erdélyi-Kober type fractional derivatives as (see [3, 152])*

$$\left(D^{\alpha}_{a+;\sigma,\eta}f\right)(t) = t^{-\sigma\eta} \left(\tfrac{1}{\sigma t^{\sigma-1}} D\right) t^{\sigma(\alpha+\eta)} \left(I^{\alpha}_{a+;\sigma,\eta+\alpha}f\right)(t) \tag{2.338}$$

and

$$\left(D^{\alpha}_{b-;\sigma,\eta}f\right)(t) = t^{-\sigma(\eta+\alpha)} \left(\tfrac{1}{\sigma t^{\sigma-1}} D\right) t^{\sigma(1+\eta-\alpha)} \left(I^{\alpha}_{b-;\sigma,\eta+\alpha-1}f\right)(t), \tag{2.339}$$

respectively, where $D = d/dt$.

Definition 2.33 *Let $\alpha \in \mathbb{C}$, $Re(\alpha) > 0$, $\kappa = [Re(\alpha)] + 1$, $\sigma = 1$, $t > 0$ and $\eta \in \mathbb{C}$. We define the left-sided and the right-sided Erdélyi-Kober type fractional derivatives as (see [3, 152])*

$$\left(D^{\alpha}_{a+;1,\eta}f\right)(t) = t^{-\eta} \left(\tfrac{1}{\sigma} D\right)^{\kappa} t^{\alpha+\eta} \left(I^{\alpha}_{a+;1,\eta+\alpha}f\right)(t) \tag{2.340}$$

and

$$\left(D^{\alpha}_{b-;1,\eta}f\right)(t) = t^{-(\eta+\alpha)} \left(\tfrac{1}{\sigma} D\right)^{\kappa} t^{\kappa+\eta-\alpha} \left(I^{\alpha}_{b-;1,\eta+\alpha-\kappa}f\right)(t), \tag{2.341}$$

respectively, where $D = d/dt$.

Definition 2.34 *Let $\alpha \in \mathbb{C}$, $Re(\alpha) > 0$, $\kappa = [Re(\alpha)] + 1$, $\sigma = 1$, $t > 0$ and $\eta \in \mathbb{C}$. We define the left-sided and the right-sided Erdélyi-Kober type fractional derivatives as (see [3, 152])*

$$\left(D^{\alpha}_{+;1,\eta} f\right)(t) = t^{-\eta}\left(\frac{1}{\sigma} D\right)^{\kappa} t^{\alpha+\eta}\left(I^{\alpha}_{+;1,\eta+\alpha} f\right)(t) \tag{2.342}$$

and

$$\left(D^{\alpha}_{-;1,\eta} f\right)(t) = t^{-(\eta+\alpha)}\left(\frac{1}{\sigma} D\right)^{\kappa} t^{\kappa+\eta-\alpha}\left(I^{\alpha}_{-;1,\eta+\alpha-\kappa} f\right)(t), \tag{2.343}$$

respectively, where $D = d/dt$.

Definition 2.35 *Let $\alpha \in \mathbb{C}$, $1 > Re(\alpha) > 0$, $\sigma = 1$, $t > 0$ and $\eta \in \mathbb{C}$. We define the left-sided and the right-sided Erdélyi-Kober type fractional derivatives as (see [3, 152])*

$$\left(D^{\alpha}_{a+;1,\eta} f\right)(t) = t^{-\eta}(D)\, t^{\alpha+\eta}\left(I^{\alpha}_{a+;1,\eta+\alpha} f\right)(t) \tag{2.344}$$

and

$$\left(D^{\alpha}_{b-;\sigma,\eta} f\right)(t) = t^{-\sigma(\eta+\alpha)}\left(\tfrac{1}{\sigma t^{\sigma-1}} D\right) t^{\sigma(1+\eta-\alpha)}\left(I^{\alpha}_{b-;\sigma,\eta+\alpha-1} f\right)(t), \tag{2.345}$$

respectively, where $D = d/dt$.

Definition 2.36 *Let $\alpha \in \mathbb{C}$, $1 > Re(\alpha) > 0$, $\sigma = 1$, $t > 0$ and $\eta \in \mathbb{C}$. We define the left-sided and the right-sided Erdélyi-Kober type fractional derivatives as (see [3, 152])*

$$\left(D^{\alpha}_{+;1,\eta} f\right)(t) = t^{-\eta}(D)\, t^{\alpha+\eta}\left(I^{\alpha}_{+;1,\eta+\alpha} f\right)(t) \tag{2.346}$$

and

$$\left(D^{\alpha}_{-;\sigma,\eta} f\right)(t) = t^{-\sigma(\eta+\alpha)}\left(\frac{1}{\sigma t^{\sigma-1}} D\right) t^{\sigma(1+\eta-\alpha)}\left(I^{\alpha}_{-;\sigma,\eta+\alpha-1} f\right)(t), \tag{2.347}$$

respectively, where $D = d/dt$.

In particular, we have the following cases:

(1) When $\kappa = [Re(\alpha)] + 1$, $a = -\infty$ and $b = -\infty$, we should present the following notations (see [3, 152])

$$\left({}_{E}D^{\alpha}_{+;\sigma,\eta} f\right)(t) = t^{-\sigma\eta}\left(\frac{1}{\sigma t^{\sigma-1}} D\right)^{\kappa} t^{\sigma(\alpha+\eta)}\left(I^{\alpha}_{+;\sigma,\eta+\alpha} f\right)(t) \tag{2.348}$$

and

$$\left({}_{E}D^{\alpha}_{-;\sigma,\eta} f\right)(t) = t^{-\sigma(\eta+\alpha)}\left(\frac{1}{\sigma t^{\sigma-1}} D\right)^{\kappa} t^{\sigma(\kappa+\eta-\alpha)}\left(I^{\alpha}_{-;\sigma,\eta+\alpha-\kappa} f\right)(t). \tag{2.349}$$

(2) When $\kappa = [Re(\alpha)] + 1$, $\eta = 0$, there are the following notations (see [3, 152])

$$\left(D^{\alpha}_{a+;\sigma,0}f\right)(t) = \left(\frac{1}{\sigma t^{\sigma-1}}D\right)^{\kappa} t^{\sigma\alpha}\left(I^{\alpha}_{a+;\sigma,\alpha}f\right)(t) \tag{2.350}$$

and

$$\left(D^{\alpha}_{b-;\sigma,0}f\right)(t) = t^{-\sigma\alpha}\left(\frac{1}{\sigma t^{\sigma-1}}D\right)^{\kappa} t^{\sigma(\kappa-\alpha)}\left(I^{\alpha}_{b-;\sigma,\alpha-\kappa}f\right)(t). \tag{2.351}$$

(3) When $\kappa = [Re(\alpha)] + 1$ and $\eta = 0$, there are the following notations (see [3, 152])

$$\left(D^{\alpha}_{+;\sigma,0}f\right)(t) = \left(\frac{1}{\sigma t^{\sigma-1}}D\right)^{\kappa} t^{\sigma\alpha}\left(I^{\alpha}_{+;\sigma,\alpha}f\right)(t) \tag{2.352}$$

and

$$\left(D^{\alpha}_{-;\sigma,0}f\right)(t) = t^{-\sigma\alpha}\left(\frac{1}{\sigma t^{\sigma-1}}D\right)^{\kappa} t^{\sigma(\kappa-\alpha)}\left(I^{\alpha}_{-;\sigma,\alpha-\kappa}f\right)(t). \tag{2.353}$$

(4) When $0 < Re(\alpha) < 1$ and $\eta = 0$, there are the following notations (see [3, 152])

$$\left(D^{\alpha}_{a+;\sigma,0}f\right)(t) = \left(\frac{1}{\sigma t^{\sigma-1}}D\right) t^{\sigma\alpha}\left(I^{\alpha}_{a+;\sigma,\alpha}f\right)(t) \tag{2.354}$$

and

$$\left(D^{\alpha}_{b-;\sigma,0}f\right)(t) = t^{-\sigma\alpha}\left(\frac{1}{\sigma t^{\sigma-1}}D\right) t^{\sigma(1-\alpha)}\left(I^{\alpha}_{b-;\sigma,\alpha-1}f\right)(t). \tag{2.355}$$

(5) When $0 < Re(\alpha) < 1$, $\eta = 0$, $a = -\infty$ and $b = -\infty$, there are the following notations (see [3, 152])

$$\left(D^{\alpha}_{+;\sigma,0}f\right)(t) = \left(\frac{1}{\sigma t^{\sigma-1}}D\right) t^{\sigma\alpha}\left(I^{\alpha}_{+;\sigma,\alpha}f\right)(t) \tag{2.356}$$

and

$$\left(D^{\alpha}_{-;\sigma,0}f\right)(t) = t^{-\sigma\alpha}\left(\frac{1}{\sigma t^{\sigma-1}}D\right) t^{\sigma(1-\alpha)}\left(I^{\alpha}_{-;\sigma,\alpha-1}f\right)(t). \tag{2.357}$$

(6) When $0 < Re(\alpha) < 1$, $\eta = 0$ and $\sigma = 1$, there are the following notations (see [3, 152])

$$\left(D^{\alpha}_{a+;1,0}f\right)(t) = (D)\, t^{\alpha}\left(I^{\alpha}_{a+;1,\alpha}f\right)(t) \tag{2.358}$$

and

$$\left(D^{\alpha}_{b-;1,0}f\right)(t) = t^{-\alpha}(D)\, t^{1-\alpha}\left(I^{\alpha}_{b-;1,\alpha-1}f\right)(t). \tag{2.359}$$

(7) When $0 < Re(\alpha) < 1$, $\eta = 0$, $\sigma = 1$, $a = -\infty$ and $b = -\infty$, there are the following notations (see [3, 152])

$$\left(D^{\alpha}_{+;1,0}f\right)(t) = (D)\, t^{\alpha}\left(I^{\alpha}_{+;1,\alpha}f\right)(t) \tag{2.360}$$

and

$$\left(D^{\alpha}_{-;1,0}f\right)(t) = t^{-\alpha}(D)\,t^{1-\alpha}\left(I^{\alpha}_{-;1,\alpha-1}f\right)(t). \tag{2.361}$$

The above operators are also called the left-sided and the right-sided Erdélyi-Kober type fractional derivatives.

Property 2.28 *(see [3, 152]) Let $Re(\alpha) > 0$ and $-\infty \leq a < b \leq \infty$. Then we have*

$$\left(D^{\alpha}_{a+;\sigma,\eta}I^{\alpha}_{a+;\sigma,\eta}f\right)(t) = f(t), \tag{2.362}$$

$$\left(D^{\alpha}_{b-;\sigma,\eta}I^{\alpha}_{b-;\sigma,\eta}f\right)(t) = f(t), \tag{2.363}$$

$$\left(D^{\alpha}_{+;\sigma,\eta}I^{\alpha}_{+;\sigma,\eta}f\right)(t) = f(t), \tag{2.364}$$

and

$$\left(D^{\alpha}_{-;\sigma,\eta}I^{\alpha}_{-;\sigma,\eta}f\right)(t) = f(t). \tag{2.365}$$

With the motivation of the Riesz idea, we define the following fractional integrals and fractional derivatives of the Erdélyi-Kober-Riesz, Erdélyi-Kober-Feller and Erdélyi-Kober-Richard types as follows.

2.9.2 Fractional integrals and fractional derivatives of the Erdélyi-Kober-Riesz, Erdélyi-Kober-Feller and Erdélyi-Kober-Richard types

Definition 2.37 *Let $\alpha \in \mathbb{C}$, $\sigma \in \mathbb{C}$, $a < t < b$ and $Re(\alpha) > 0$. The fractional integral in the sense of the Erdélyi-Kober-Riesz type on $\mathbb{R}$ as*

$$\begin{aligned}{}_{Rw}I^{\alpha}_{\mathbb{R}}f(t) &= \frac{\left(\left({}_{E}I^{\alpha}_{+;\sigma,\eta}f\right)+\left({}_{E}I^{\alpha}_{-;\sigma,\eta}f\right)\right)(t)}{2\cos(\pi\alpha/2)} \\ &= \frac{\sigma t^{-\sigma\eta}}{2\cos(\pi\alpha/2)\Gamma(\alpha)} \times \left(t^{-\sigma\alpha}\int\limits_{-\infty}^{t}\frac{\tau^{\sigma(\eta+1)-1}f(\tau)}{(t^{\sigma}-\tau^{\sigma})^{1-\alpha}}d\tau + \int\limits_{t}^{\infty}\frac{\tau^{\sigma(1-\alpha-\eta)-1}f(\tau)}{(\tau^{\sigma}-t^{\sigma})^{1-\alpha}}d\tau\right),\end{aligned} \tag{2.366}$$

and the fractional derivative in the sense of the Erdélyi-Kober-Riesz type on $\mathbb{R}$ as

$$\begin{aligned}&{}_{Rw}D^{\alpha}_{\mathbb{R}}f(t) \\ &= \frac{\left(\left({}_{E}D^{\alpha}_{+;\sigma,\eta}f\right)f+\left({}_{E}D^{\alpha}_{-;\sigma,\eta}f\right)\right)(t)}{2\cos(\pi\alpha/2)} \\ &= \frac{t^{-\sigma\eta}}{2\cos(\pi\alpha/2)} \times \left[\left(\frac{1}{\sigma t^{\sigma-1}}D\right)^{\kappa} t^{\sigma(\alpha+\eta)}\left(I^{\alpha}_{a+;\sigma,\eta+\alpha}f\right)(t) + t^{-(\eta+\alpha)}\left(\frac{1}{\sigma}D\right)^{\kappa}\right. \\ &\left. + t^{-\sigma\alpha}\left(\frac{1}{\sigma t^{\sigma-1}}D\right)^{\kappa} t^{\sigma(\kappa+\eta-\alpha)}\left(I^{\alpha}_{b-;\sigma,\eta+\alpha-\kappa}f\right)(t)\right],\end{aligned} \tag{2.367}$$

respectively.

Definition 2.38 *Let $\alpha \in \mathbb{C}$, $\sigma \in \mathbb{C}$, $a < t < b$ and $Re(\alpha) > 0$. The fractional integral in the sense of the Erdélyi-Kober-Feller type on $\mathbb{R}$ defined as*

$$\begin{aligned} {}_{Rv}I_{\mathbb{R}}^{\alpha} f(t) &= \frac{\left(\left({}_E I_{+;\sigma,\eta}^{\alpha} f\right) + \left({}_E I_{-;\sigma,\eta}^{\alpha} f\right)\right)(t)}{2\sin(\pi\alpha/2)} \\ &= \frac{\sigma t^{-\sigma\eta}}{2\sin(\pi\alpha/2)\Gamma(\alpha)} \times \left(t^{-\sigma\alpha} \int_{-\infty}^{t} \frac{\tau^{\sigma(\eta+1)-1} f(\tau)}{(t^{\sigma}-\tau^{\sigma})^{1-\alpha}} d\tau + \int_{t}^{\infty} \frac{\tau^{\sigma(1-\alpha-\eta)-1} f(\tau)}{(\tau^{\sigma}-t^{\sigma})^{1-\alpha}} d\tau \right) \end{aligned} \tag{2.368}$$

and the fractional derivative in the sense of the Erdélyi-Kober-Feller type on $\mathbb{R}$ as

$$\begin{aligned} {}_{Rv}D_{\mathbb{R}}^{\alpha} f(t) &= \frac{\left(\left({}_E D_{+;\sigma,\eta}^{\alpha} f\right) f + \left({}_E D_{-;\sigma,\eta}^{\alpha} f\right)\right)(t)}{2\sin(\pi\alpha/2)} \\ &= \frac{t^{-\sigma\eta}}{2\sin(\pi\alpha/2)} \times \left[\left(\frac{1}{\sigma t^{\sigma-1}} D\right)^{\kappa} t^{\sigma(\alpha+\eta)} \left(I_{a+;\sigma,\eta+\alpha}^{\alpha} f\right)(t) + t^{-(\eta+\alpha)} \left(\frac{1}{\sigma} D\right)^{\kappa} \right. \\ &\left. + t^{-\sigma\alpha} \left(\frac{1}{\sigma t^{\sigma-1}} D\right)^{\kappa} t^{\sigma(\kappa+\eta-\alpha)} \left(I_{b-;\sigma,\eta+\alpha-\kappa}^{\alpha} f\right)(t) \right], \end{aligned} \tag{2.369}$$

respectively.

Definition 2.39 *Let $\alpha \in \mathbb{C}$, $\sigma \in \mathbb{C}$, $a < t < b$ and $Re(\alpha) > 0$. The fractional integral in the sense of the Erdélyi-Kober-Richard type on $\mathbb{R}$ as*

$$\begin{aligned} {}_{Ro}I_{\mathbb{R}}^{\alpha} f(t) &= \frac{\left(\left({}_E I_{+;\sigma,\eta}^{\alpha} f\right) + \left({}_E I_{-;\sigma,\eta}^{\alpha} f\right)\right)(t)}{2\sin(\pi\alpha/2)} \\ &= \frac{\sigma t^{-\sigma\eta}}{2\sin(\pi\alpha/2)\Gamma(\alpha)} \times \left(t^{-\sigma\alpha} \int_{-\infty}^{t} \frac{\tau^{\sigma(\eta+1)-1} f(\tau)}{(t^{\sigma}-\tau^{\sigma})^{1-\alpha}} d\tau + \int_{t}^{\infty} \frac{\tau^{\sigma(1-\alpha-\eta)-1} f(\tau)}{(\tau^{\sigma}-t^{\sigma})^{1-\alpha}} d\tau \right), \end{aligned} \tag{2.370}$$

and the fractional derivative in the sense of the Erdélyi-Kober-Richard type on $\mathbb{R}$ as

$$\begin{aligned} {}_{Ro}D_{\mathbb{R}}^{\alpha} f(t) &= \frac{\left(\left({}_E D_{+;\sigma,\eta}^{\alpha} f\right) f + \left({}_E D_{-;\sigma,\eta}^{\alpha} f\right)\right)(t)}{2\sin(\pi\alpha/2)} \\ &= \frac{t^{-\sigma\eta}}{2\sin(\pi\alpha/2)} \times \left[\left(\frac{1}{\sigma t^{\sigma-1}} D\right)^{\kappa} t^{\sigma(\alpha+\eta)} \left(I_{a+;\sigma,\eta+\alpha}^{\alpha} f\right)(t) + t^{-(\eta+\alpha)} \left(\frac{1}{\sigma} D\right)^{\kappa} \right. \\ &\left. + t^{-\sigma\alpha} \left(\frac{1}{\sigma t^{\sigma-1}} D\right)^{\kappa} t^{\sigma(\kappa+\eta-\alpha)} \left(I_{b-;\sigma,\eta+\alpha-\kappa}^{\alpha} f\right)(t) \right], \end{aligned} \tag{2.371}$$

respectively.

2.10 Katugampola fractional calculus

2.10.1 Katugampola fractional integrals and Katugampola fractional derivatives

We present the definitions and properties of the fractional integrals and fractional derivatives involving the Riemann-Liouville and the Hadamard fractional calculus, introduced by Katugampola in 2011 and in 2014, respectively.

Definition 2.40 *The generalized versions of the well-known Riemann-Liouville and the Hadamard fractional integrals are defined by (see [281])*

$$\left({}_K I^{\alpha}_{a+;\sigma} f\right)(t) = \frac{\sigma^{1-\alpha}}{\Gamma(\alpha)} \int_a^t \frac{\tau^{\sigma-1} f(\tau)}{(t^\sigma - \tau^\sigma)^{1-\alpha}} d\tau \tag{2.372}$$

and (see [281])

$$\left({}_K I^{\alpha}_{b-;\sigma} f\right)(t) = \frac{\sigma^{1-\alpha}}{\Gamma(\alpha)} \int_t^b \frac{\tau^{\sigma-1} f(\tau)}{(\tau^\sigma - t^\sigma)^{1-\alpha}} d\tau, \tag{2.373}$$

which are called the left-sided and the right-sided Katugampola fractional integrals in honor of Katugampola, where $\alpha \in \mathbb{C}$, $\sigma \in \mathbb{C}$, $a < t < b$ and $Re(\alpha) > 0$.

In fact, we rewrite Eq. (2.373) as

$$\left({}_K I^{\alpha}_{a+;\sigma} f\right)(t) = \frac{\sigma^{-\alpha}}{\Gamma(\alpha)} \int_a^t \frac{f(\tau)}{(t^\sigma - \tau^\sigma)^{1-\alpha}} (d\tau^\sigma). \tag{2.374}$$

In a similar way, we also rewrite Eq. (2.374) as

$$\left(I^{\alpha}_{a+;\sigma,0} f\right)(t) = \frac{t^{-\sigma\alpha}}{\Gamma(\alpha)} \int_a^t \frac{f(\tau)}{(t^\sigma - \tau^\sigma)^{1-\alpha}} (d\tau^\sigma), \tag{2.375}$$

which, compared between Eq. (2.374) and Eq. (2.375), implies that

$$\left({}_K I^{\alpha}_{a+;\sigma} f\right)(t) = \sigma^{-\alpha} t^{\sigma\alpha} \left(I^{\alpha}_{a+;\sigma,0} f\right)(t). \tag{2.376}$$

Definition 2.41 *The fractional derivatives involving the generalized version of the well-known Riemann-Liouville and the Hadamard fractional integrals defined as (see [282])*

$$\left({}_K D^{\alpha}_{a+;\sigma} f\right)(t) = \left(t^{1-\sigma} \tfrac{d}{dt}\right)^{\kappa} \left({}_K I^{\alpha}_{a+;\sigma} f\right)(t) \tag{2.377}$$

and

$$\left({}_K D^{\alpha}_{b-;\sigma} f\right)(t) = \left(t^{1-\sigma} \tfrac{d}{dt}\right)^{\kappa} \left({}_K I^{\alpha}_{b-;\sigma} f\right)(t), \tag{2.378}$$

which are called the left-sided and the right-sided Katugampola fractional derivatives, respectively, where $\alpha \in \mathbb{C}$, $\sigma \in \mathbb{C}$, $a < t < b$ and $\kappa = [Re(\alpha)] + 1$.

The properties of the Katugampola fractional derivatives are given as follows.

Property 2.29 *(see [282]) Let $\alpha \in \mathbb{R}$, $\sigma \in \mathbb{C}$, $\lambda \in \mathbb{C}$ and $0 < \alpha < 1$. Then, for $f(t) \in L_\kappa(a,b)$ $(-\infty < a < b < \infty)$ we have (see [282])*

$$\left({}_K D^{\alpha}_{a+;\sigma,\lambda} {}_K I^{\alpha}_{a+;\sigma,\lambda} f\right)(t) = f(t), \tag{2.379}$$

$$\left({}_K D^{\alpha}_{b-;\sigma,\lambda} {}_K I^{\alpha}_{b-;\sigma,\lambda} f\right)(t) = f(t), \tag{2.380}$$

$$\left({}_K D^{\alpha}_{+;\sigma,\lambda} {}_K I^{\alpha}_{+;\sigma,\lambda} f\right)(t) = f(t), \tag{2.381}$$

$$\left({}_K D^{\alpha}_{-;\sigma,\lambda} {}_K I^{\alpha}_{-;\sigma,\lambda} f\right)(t) = f(t), \tag{2.382}$$

$$\left({}_K D^{\alpha}_{+;\sigma,\lambda} f \, {}_K I^{\alpha}_{+;\sigma} f\right)(t) = f(t) \tag{2.383}$$

and

$$\left({}_K D^{\alpha}_{-;\sigma,\lambda} {}_K I^{\alpha}_{-;\sigma} f\right)(t) = f(t). \tag{2.384}$$

By the motivation of the idea, we have the following results.

Definition 2.42 *Let $\alpha \in \mathbb{C}$, $\sigma \in \mathbb{C}$,$\lambda \in \mathbb{C}$, $a < t < b$ and $Re(\alpha) > 0$. The left-sided fractional integral involving the exponential function in the interval $[a,b]$, denoted by $\left({}_{Kv} I^{\alpha}_{a+;\sigma} f\right)(t)$, is defined as*

$$\left({}_{Kv} I^{\alpha}_{a+;\sigma,\lambda} f\right)(t) = \frac{e^{\lambda t}\sigma^{1-\alpha}}{\Gamma(\alpha)} \int_a^t \frac{\tau^{\sigma-1} e^{-\lambda\tau}}{\left(t^\sigma - \tau^\sigma\right)^{1-\alpha}} f(\tau)\, d\tau, \tag{2.385}$$

and the right-sided fractional integral involving the exponential function in the interval $[a,b]$, denoted by $\left({}_{Kv} I^{\alpha}_{b-;\sigma} f\right)(t)$, as

$$\left({}_{Kv} I^{\alpha}_{b-;\sigma,\kappa} f\right)(t) = \frac{e^{\lambda t}\sigma^{1-\alpha}}{\Gamma(\alpha)} \int_t^b \frac{\tau^{\sigma-1} e^{-\lambda\tau} f(\tau)}{\left(\tau^\sigma - t^\sigma\right)^{1-\alpha}} d\tau, \tag{2.386}$$

where $\kappa = [Re(\alpha)] + 1$.

2.10.2 Katugampola type fractional integrals and Katugampola type fractional derivatives involving the exponential function

Definition 2.43 *Let $\alpha \in \mathbb{C}$, $\sigma \in \mathbb{C}$, $\lambda \in \mathbb{C}$, $a < t < b$, $Re(\alpha) > 0$ and $\kappa = [Re(\alpha)] + 1$. The left-sided Katugampola type fractional derivative involving the exponential function in the interval $[a,b]$, denoted by $\left({}_{Kv} D^{\alpha}_{a+;\sigma,\lambda} f\right)(t)$, is defined as*

$$\left({}_{Kv} D^{\alpha}_{a+;\sigma,\lambda} f\right)(t) = \frac{\sigma^{\alpha-\kappa+1} e^{\lambda t}}{\Gamma(\kappa-\alpha)} \left(t^{1-\sigma}\frac{d}{dt}\right)^{\kappa} \int_a^t \frac{\tau^{\sigma-1} e^{-\lambda\tau} f(\tau)}{\left(t^\sigma - \tau^\sigma\right)^{\alpha-\kappa+1}} d\tau, \tag{2.387}$$

and the right-sided Katugampola type fractional derivative involving the exponential function in the interval $[a,b]$ *denoted by* $\left({}_{Kv}D^{\alpha}_{b-;\sigma,\lambda}f\right)(t)$, *as*

$$\left({}_{Kv}D^{\alpha}_{b-;\sigma,\lambda}f\right)(t)=\frac{\sigma^{\alpha-\kappa+1}e^{\lambda t}}{\Gamma(\kappa-\alpha)}\left(t^{1-\sigma}\frac{d}{dt}\right)^{\kappa}\int\limits_{t}^{b}\frac{\tau^{\sigma-1}e^{-\lambda\tau}f(\tau)}{(\tau^{\sigma}-t^{\sigma})^{\alpha-\kappa+1}}d\tau. \quad (2.388)$$

Property 2.30 *Let* $\alpha\in\mathbb{R}$, $\sigma\in\mathbb{C}$,$\lambda\in\mathbb{C}$ *and* $0<\alpha<1$. *Then, for* $f(t)\in L_{\kappa}(a,b)$ $(-\infty<a<b<\infty)$ *we have*

$$\left({}_{Kv}D^{\alpha}_{a+;\sigma,\lambda}{}_{Kv}I^{\alpha}_{a+;\sigma,\lambda}f\right)(t)=f(t), \quad (2.389)$$

$$\left({}_{Kv}D^{\alpha}_{b-;\sigma,\lambda}{}_{Kv}I^{\alpha}_{b-;\sigma,\lambda}f\right)(t)=f(t), \quad (2.390)$$

$$\left({}_{Kv}D^{\alpha}_{+;\sigma,\lambda}{}_{Kv}I^{\alpha}_{+;\sigma,\lambda}f\right)(t)=f(t), \quad (2.391)$$

$$\left({}_{Kv}D^{\alpha}_{-;\sigma,\lambda}{}_{Kv}I^{\alpha}_{-;\sigma,\lambda}f\right)(t)=f(t), \quad (2.392)$$

$$\left({}_{Kv}D^{\alpha}_{+;\sigma,\lambda}f\,{}_{Kv}I^{\alpha}_{+;\sigma}f\right)(t)=f(t) \quad (2.393)$$

and

$$\left({}_{Kv}D^{\alpha}_{-;\sigma,\lambda}{}_{Kv}I^{\alpha}_{-;\sigma}f\right)(t)=f(t). \quad (2.394)$$

2.11 Hadamard fractional calculus

2.11.1 Hadamard fractional integrals and fractional derivatives

The definitions and properties of the Hadamard fractional integrals and fractional derivatives are introduced as follows [3, 30, 152].

Let $\mathrm{M}=(a,b)$ $(0\le a<b\le\infty)$ be a finite interval on the real axis $\mathbb{R}_{+}$.

Definition 2.44 *Let* $\alpha\in\mathbb{C}$ *and* $Re(\alpha)>0$. *The left-sided and the right-sided Hadamard type fractional integrals are defined as (see [3, 152])*

$$\left({}_{H}I^{\alpha}_{a+}f\right)(t)=\frac{1}{\Gamma(\alpha)}\int\limits_{a}^{t}\left(\log\frac{t}{\tau}\right)^{\sigma}f(\tau)\frac{d\tau}{\tau}, \quad (2.395)$$

and

$$\left({}_{H}I^{\alpha}_{b-}f\right)(t)=\frac{1}{\Gamma(\alpha)}\int\limits_{t}^{b}\left(\log\frac{\tau}{t}\right)^{\sigma}f(\tau)\frac{d\tau}{\tau}, \quad (2.396)$$

respectively.

Definition 2.45 *Let $\alpha \in \mathbb{C}$, $Re(\alpha) > 0$ and $\delta = tD$, where $D = \frac{d}{dt}$. The left- and right-sided Hadamard fractional derivatives are defined as*

$$\begin{aligned}\left({}_H D^{\alpha}_{a+} f\right)(t) &= \tfrac{1}{\Gamma(\alpha)} (tD)^{\kappa} \int\limits_a^t \left(\log \tfrac{t}{\tau}\right)^{\sigma} f(\tau) \tfrac{d\tau}{\tau} \\ &= \tfrac{1}{\Gamma(\alpha)} \left(t\tfrac{d}{dt}\right)^{\kappa} \int\limits_a^t \left(\log \tfrac{t}{\tau}\right)^{\sigma} f(\tau) \tfrac{d\tau}{\tau},\end{aligned} \tag{2.397}$$

and

$$\begin{aligned}\left({}_H D^{\alpha}_{b-} f\right)(t) &= \tfrac{1}{\Gamma(\alpha)} (tD)^{\kappa} \int\limits_t^b \left(\log \tfrac{\tau}{t}\right)^{\sigma} f(\tau) \tfrac{d\tau}{\tau} \\ &= \tfrac{1}{\Gamma(\alpha)} \left(t\tfrac{d}{dt}\right)^{\kappa} \int\limits_t^b \left(\log \tfrac{\tau}{t}\right)^{\sigma} f(\tau) \tfrac{d\tau}{\tau},\end{aligned} \tag{2.398}$$

respectively.

Property 2.31 *(see [152]) Let $0 \le a < b \le \infty$, $Re(\alpha) > 0$ and $Re(\beta) > 0$. Then we have*

$$\left({}_H I^{\alpha}_{a+} \left(\log \frac{t}{a}\right)^{\beta}\right)(t) = \frac{\Gamma(1+\beta)}{\Gamma(\alpha+\beta)} \left(\log \frac{t}{a}\right)^{\beta+\alpha-1}, \tag{2.399}$$

$$\left({}_H I^{\alpha}_{b-} \left(\log \frac{b}{t}\right)^{\beta}\right)(t) = \frac{\Gamma(1+\beta)}{\Gamma(\alpha+\beta)} \left(\log \frac{b}{t}\right)^{\beta+\alpha-1}, \tag{2.400}$$

$$\left({}_H D^{\alpha}_{a+} \left(\frac{t}{a}\right)^{\beta}\right)(t) = \frac{\Gamma(1+\beta)}{\Gamma(\beta-\alpha)} \left(\log \frac{t}{a}\right)^{\beta-\alpha-1} \tag{2.401}$$

and

$$\left({}_H D^{\alpha}_{b-} \left(\log \frac{b}{t}\right)^{\beta}\right)(t) = \frac{\Gamma(1+\beta)}{\Gamma(\beta-\alpha)} \left(\log \frac{b}{t}\right)^{\beta-\alpha-1}. \tag{2.402}$$

Property 2.32 *(see [152]) Let $0 \le a < b \le \infty$, $Re(\alpha) > 0$ and $L_{\kappa}(a,b)$ $(1 \le \kappa \le \infty)$. Then we have*

$$\left({}_H D^{\alpha}_{a+} {}_H I^{\alpha}_{a+}\right)(t) = f(t) \tag{2.403}$$

and

$$\left({}_H D^{\alpha}_{b-} {}_H I^{\alpha}_{b-}\right)(t) = f(t). \tag{2.404}$$

2.11.2 Hadamard type fractional integrals and fractional derivatives

Let $\alpha \in \mathbb{C}$, $Re(\alpha) > 0$ and $\ell_{\lambda} = e^{-\lambda t} D$, where $\lambda \in \mathbb{C}$ and $D = \frac{d}{dt}$. Following the motivation of the idea, we give the definitions of the Hadamard type fractional derivatives as follows:

Definition 2.46 *Let $\alpha \in \mathbb{C}$, $Re(\alpha) > 0$ and $\kappa = [Re(\alpha)] + 1$. The left-sided and the right-sided Hadamard type fractional derivatives are defined as*

$$\left({}_{Hv}D^{\alpha}_{a+}f\right)(t) = \frac{1}{\Gamma(\alpha)}\left(e^{-\lambda t}\frac{d}{dt}\right)^{\kappa}\int_{a}^{t}\left(\log\frac{t}{\tau}\right)^{\sigma}f(\tau)\frac{d\tau}{\tau}, \tag{2.405}$$

and

$$\left({}_{Hv}D^{\alpha}_{b-}f\right)(t) = \frac{1}{\Gamma(\alpha)}\left(e^{-\lambda t}\frac{d}{dt}\right)^{\kappa}\int_{t}^{b}\left(\log\frac{\tau}{t}\right)^{\sigma}f(\tau)\frac{d\tau}{\tau}, \tag{2.406}$$

respectively.

Definition 2.47 *Let $\alpha \in \mathbb{C}$ and $1 > Re(\alpha) > 0$. The left-sided and the right-sided Hadamard type fractional derivatives are defined as*

$$\left({}_{Hv}D^{\alpha}_{a+}f\right)(t) = \frac{1}{\Gamma(\alpha)}\left(e^{-\lambda t}\frac{d}{dt}\right)\int_{a}^{t}\left(\log\frac{t}{\tau}\right)^{\sigma}f(\tau)\frac{d\tau}{\tau}, \tag{2.407}$$

and

$$\left({}_{Hv}D^{\alpha}_{b-}f\right)(t) = \frac{1}{\Gamma(\alpha)}\left(e^{-\lambda t}\frac{d}{dt}\right)\int_{t}^{b}\left(\log\frac{\tau}{t}\right)^{\sigma}f(\tau)\frac{d\tau}{\tau}, \tag{2.408}$$

respectively.

Definition 2.48 *Let $\alpha \in \mathbb{C}$, $Re(\alpha) > 0$ and $\kappa = [Re(\alpha)] + 1$. We define the left-sided and the right-sided Hadamard type fractional integrals as*

$$\left({}_{Hv}I^{\alpha}_{a+}f\right)(t) = \frac{1}{\Gamma(\alpha)}\left(t\frac{d}{dt}\right)^{\kappa}\int_{a}^{t}\left(\log\frac{t}{\tau}\right)^{\sigma}e^{\kappa\lambda\tau}f(\tau)\frac{d\tau}{\tau}, \tag{2.409}$$

and

$$\left({}_{Hv}I^{\alpha}_{b-}f\right)(t) = \frac{1}{\Gamma(\alpha)}\left(t\frac{d}{dt}\right)^{\kappa}\int_{t}^{b}\left(\log\frac{\tau}{t}\right)^{\sigma}e^{\kappa\lambda\tau}f(\tau)\frac{d\tau}{\tau}, \tag{2.410}$$

respectively.

Definition 2.49 *Let $\alpha \in \mathbb{C}$ and $1 > Re(\alpha) > 0$. We define the left-sided Hadamard type fractional integral as*

$$\left({}_{Hv}I^{\alpha}_{a+}f\right)(t) = \frac{1}{\Gamma(\alpha)}\left(t\frac{d}{dt}\right)\int_{a}^{t}\left(\log\frac{t}{\tau}\right)^{\sigma}e^{\lambda\tau}f(\tau)\frac{d\tau}{\tau}, \tag{2.411}$$

and the right-sided Hadamard type fractional integral as

$$\left({}_{Hv}I^{\alpha}_{b-}f\right)(t) = \frac{1}{\Gamma(\alpha)}\left(t\frac{d}{dt}\right)^{\kappa}\int_{t}^{b}\left(\log\frac{\tau}{t}\right)^{\sigma}e^{\lambda\tau}f(\tau)\frac{d\tau}{\tau}, \tag{2.412}$$

respectively.

Property 2.33 *Let $0 \leq a < b \leq \infty$, $1 > Re(\alpha) > 0$ and $Re(\beta) > 0$. Then we have*

$$\left({}_{Hv}I^{\alpha}_{a+}\left(\log\frac{t}{a}\right)^{\beta}\right)(t) = \frac{\Gamma(1+\beta)}{\Gamma(\alpha+\beta)}e^{-\lambda t}\left(\log\frac{t}{a}\right)^{\beta+\alpha-1}, \tag{2.413}$$

$$\left({}_{Hv}I^{\alpha}_{b-}\left(\log\frac{b}{t}\right)^{\beta}\right)(t) = \frac{\Gamma(1+\beta)}{\Gamma(\alpha+\beta)}e^{-\lambda t}\left(\log\frac{b}{t}\right)^{\beta+\alpha-1}, \tag{2.414}$$

$$\left({}_{Hv}D^{\alpha}_{a+}e^{-\lambda t}\left(\frac{t}{a}\right)^{\beta}\right)(t) = \frac{\Gamma(1+\beta)}{\Gamma(\beta-\alpha)}t\left(\log\frac{t}{a}\right)^{\beta-\alpha-1} \tag{2.415}$$

and

$$\left({}_{Hv}D^{\alpha}_{b-}e^{-\lambda t}\left(\log\frac{b}{t}\right)^{\beta}\right)(t) = \frac{\Gamma(1+\beta)}{\Gamma(\beta-\alpha)}t\left(\log\frac{b}{t}\right)^{\beta-\alpha-1}. \tag{2.416}$$

Property 2.34 *Let $0 \leq a < b \leq \infty$, $Re(\alpha) > 0$ and $L_{\kappa}(a,b)$ $(1 \leq \kappa \leq \infty)$. Then we have*

$$\left({}_{Hv}D^{\alpha}_{a+}{}_{Hv}I^{\alpha}_{a+}\right)(t) = f(t) \tag{2.417}$$

and

$$\left({}_{Hv}D^{\alpha}_{b-}{}_{Hv}I^{\alpha}_{b-}\right)(t) = f(t). \tag{2.418}$$

Definition 2.50 *Let $\alpha \in \mathbb{C}$, $Re(\alpha) > 0$ and $\kappa = [Re(\alpha)] + 1$. We define the left-sided fractional derivative involving the exponential function as*

$$\left({}_{Sv}D^{\alpha}_{a+}f\right)(t) = \tfrac{1}{\Gamma(\alpha)}\left(e^{-\lambda t}\tfrac{d}{dt}\right)^{\kappa}\int_a^t \tfrac{f(\tau)}{(\tau-t)^{\alpha-\kappa+1}}d\tau, \tag{2.419}$$

and the right-sided fractional derivative involving the exponential function as

$$\left({}_{Sv}D^{\alpha}_{b-}f\right)(t) = \tfrac{1}{\Gamma(\alpha)}\left(e^{-\lambda t}\tfrac{d}{dt}\right)^{\kappa}\int_t^b \tfrac{f(\tau)}{(t-\tau)^{\alpha-\kappa+1}}d\tau, \tag{2.420}$$

respectively.

In particular, when $\alpha \in \mathbb{C}$ and $1 > Re(\alpha) > 0$, we define the left-sided and the right-sided fractional derivatives involving the exponential function as

$$\left({}_{Sv}D^{\alpha}_{a+}f\right)(t) = \tfrac{1}{\Gamma(\alpha)}\left(e^{-\lambda t}\tfrac{d}{dt}\right)\int_a^t \tfrac{f(\tau)}{(\tau-t)^{\alpha-\kappa+1}}d\tau, \tag{2.421}$$

and

$$\left({}_{Sv}D^{\alpha}_{b-}f\right)(t) = \tfrac{1}{\Gamma(\alpha)}\left(e^{-\lambda t}\tfrac{d}{dt}\right)\int_t^b \tfrac{f(\tau)}{(t-\tau)^{\alpha-\kappa+1}}d\tau, \tag{2.422}$$

respectively.

Definition 2.51 *Let $\alpha \in \mathbb{C}$, $Re(\alpha) > 0$ and $\kappa = [Re(\alpha)] + 1$. We define the left-sided fractional derivative involving the kernel of the power-exponential function as*

$$\left({}_{Sv}I_{a+}^{\alpha} f\right)(t) = \frac{1}{\Gamma(\alpha)} \int\limits_a^t \frac{e^{\lambda\tau}}{(\tau - t)^{\alpha-\kappa+1}} f(\tau)\, d\tau, \tag{2.423}$$

and the right-sided fractional derivative involving the kernel of the power-exponential function as

$$\left({}_{Sv}I_{b-}^{\alpha} f\right)(t) = \frac{1}{\Gamma(\alpha)} \int\limits_t^b \frac{e^{\lambda\tau}}{(t-\tau)^{\alpha-\kappa+1}} f(\tau)\, d\tau. \tag{2.424}$$

In particular, when $\alpha \in \mathbb{C}$ and $1 > Re(\alpha) > 0$, we define the left-sided fractional integral involving the kernel of the power-exponential function as

$$\left({}_{Sv}I_{a+}^{\alpha} f\right)(t) = \frac{1}{\Gamma(\alpha)} \int\limits_a^t \frac{e^{\lambda\tau}}{(\tau - t)^{\alpha}} f(\tau)\, d\tau, \tag{2.425}$$

and the right-sided fractional integral involving the kernel of the power-exponential function as

$$\left({}_{Sv}I_{b-}^{\alpha} f\right)(t) = \frac{1}{\Gamma(\alpha)} \int\limits_t^b \frac{e^{\lambda\tau}}{(t-\tau)^{\alpha}} f(\tau)\, d\tau. \tag{2.426}$$

Property 2.35 *Let $0 \le a < b \le \infty$, $1 > Re(\alpha) > 0$ and $Re(\beta) > 0$. Then we have*

$$\left({}_{Sv}I_{a+}^{\alpha}\left(e^{-\lambda\tau}\frac{(t-a)^{\beta-1}}{\Gamma(\beta)}\right)\right)(t) = \frac{(t-a)^{\beta+\alpha-1}}{\Gamma(\beta+\alpha)}, \tag{2.427}$$

$$\left({}_{Sv}I_{b-}^{\alpha}\left(e^{-\lambda\tau}\frac{(t-a)^{\beta-1}}{\Gamma(\beta)}\right)\right)(t) = \frac{(t-a)^{\beta-\alpha-1}}{\Gamma(\beta-\alpha)}, \tag{2.428}$$

$$\left({}_{Sv}D_{a+}^{\alpha}\frac{(t-a)^{\beta-1}}{\Gamma(\beta)}\right)(t) = e^{-\lambda t}\frac{(b-t)^{\beta+\alpha-1}}{\Gamma(\beta+\alpha)} \tag{2.429}$$

and

$$\left({}_{Sv}D_{b-}^{\alpha}\frac{(t-a)^{\beta-1}}{\Gamma(\beta)}\right)(t) = e^{-\lambda t}\frac{(b-t)^{\beta-\alpha-1}}{\Gamma(\beta-\alpha)}. \tag{2.430}$$

Property 2.36 *Let $0 \le a < b \le \infty$, $Re(\alpha) > 0$ and $L_\kappa(a,b)$ $(1 \le \kappa \le \infty)$. Then we have*

$$\left({}_{Sv}D_{a+}^{\alpha}\, {}_{Sv}I_{a+}^{\alpha}\right)(t) = f(t), \tag{2.431}$$

and

$$\left({}_{Sv}D_{b-}^{\alpha}\, {}_{Sv}I_{b-}^{\alpha}\right)(t) = f(t). \tag{2.432}$$

2.12 Marchaud fractional derivatives

We introduce the concepts and properties of the Marchaud fractional derivatives in honor of Marchaud [3, 75].

Definition 2.52 *Let* $1 > Re(\alpha) > 0$. *The left-sided Marchaud fractional derivative is defined by (for real number of order, see [3, 75]):*

$$\left({}_{M}D_{+}^{\alpha} f\right)(t) = \frac{\alpha}{\Gamma(1-\alpha)} \int_{0}^{\infty} \frac{f(\tau)-f(t-\tau)}{\tau^{\alpha+1}} d\tau = \frac{\alpha}{\Gamma(1-\alpha)} \int_{-\infty}^{t} \frac{f(t)-f(\tau)}{(t-\tau)^{\alpha+1}} d\tau, \tag{2.433}$$

and the right-sided Marchaud fractional derivative by

$$\left({}_{M}D_{-}^{\alpha} f\right)(t) = \frac{\alpha}{\Gamma(1-\alpha)} \int_{0}^{\infty} \frac{f(\tau)-f(t+\tau)}{\tau^{\alpha+1}} d\tau = \frac{\alpha}{\Gamma(1-\alpha)} \int_{t}^{\infty} \frac{f(t)-f(\tau)}{(t-\tau)^{\alpha+1}} d\tau, \tag{2.434}$$

where $-\infty < t < \infty$.

In fact, for $1 > Re(\alpha) > 0$ the left-sided Marchaud fractional derivative is defined as

$$\frac{\alpha}{\Gamma(1-\alpha)} \int_{-\infty}^{t} \frac{f(t)-f(\tau)}{(t-\tau)^{\alpha+1}} d\tau = \frac{1}{\Gamma(1-\alpha)} \frac{d}{dt} \int_{0}^{\infty} \frac{f(t-\tau)}{\tau^{\alpha}} d\tau = \frac{a}{\Gamma(1-\alpha)} \int_{0}^{\infty} \frac{f(t)-f(t-\tau)}{\tau^{\alpha+1}} d\tau, \tag{2.435}$$

and the right-sided Marchaud fractional derivative as

$$\frac{\alpha}{\Gamma(1-\alpha)} \int_{t}^{\infty} \frac{f(t)-f(\tau)}{(t-\tau)^{\alpha+1}} d\tau = \frac{1}{\Gamma(1-\alpha)} \frac{d}{dt} \int_{0}^{\infty} \frac{f(t+\tau)}{\tau^{\alpha}} d\tau = \frac{a}{\Gamma(1-\alpha)} \int_{0}^{\infty} \frac{f(t)-f(t+\tau)}{\tau^{\alpha+1}} d\tau. \tag{2.436}$$

In particular, when $1 > \alpha > 0$, the left-sided Marchaud fractional derivative is defined by (see [3, 75]):

$$\left({}_{M}D_{+}^{\alpha} f\right)(t) = \frac{\alpha}{\Gamma(1-\alpha)} \int_{0}^{\infty} \frac{f(\tau)-f(t-\tau)}{\tau^{\alpha+1}} d\tau = \frac{\alpha}{\Gamma(1-\alpha)} \int_{-\infty}^{t} \frac{f(t)-f(\tau)}{(t-\tau)^{\alpha+1}} d\tau, \tag{2.437}$$

and the right-sided Marchaud fractional derivative as

$$\left({}_{M}D_{-}^{\alpha} f\right)(t) = \frac{\alpha}{\Gamma(1-\alpha)} \int_{0}^{\infty} \frac{f(\tau)-f(t+\tau)}{\tau^{\alpha+1}} d\tau = \frac{\alpha}{\Gamma(1-\alpha)} \int_{t}^{\infty} \frac{f(t)-f(\tau)}{(t-\tau)^{\alpha+1}} d\tau. \tag{2.438}$$

For $1 > \alpha > 0$ we have (see [3, 75])

$$\frac{\alpha}{\Gamma(1-\alpha)} \int_{-\infty}^{t} \frac{f(t)-f(\tau)}{(t-\tau)^{\alpha+1}} d\tau = \frac{1}{\Gamma(1-\alpha)} \frac{d}{dt} \int_{0}^{\infty} \frac{f(t-\tau)}{\tau^{\alpha}} d\tau = \frac{a}{\Gamma(1-\alpha)} \int_{0}^{\infty} \frac{f(t)-f(t-\tau)}{\tau^{\alpha+1}} d\tau, \tag{2.439}$$

and

$$\frac{\alpha}{\Gamma(1-\alpha)} \int_{t}^{\infty} \frac{f(t)-f(\tau)}{(t-\tau)^{\alpha+1}} d\tau = \frac{1}{\Gamma(1-\alpha)} \frac{d}{dt} \int_{0}^{\infty} \frac{f(t+\tau)}{\tau^{\alpha}} d\tau = \frac{a}{\Gamma(1-\alpha)} \int_{0}^{\infty} \frac{f(t)-f(t+\tau)}{\tau^{\alpha+1}} d\tau. \tag{2.440}$$

2.13 Tempered fractional calculus

2.13.1 Motivations

In 1697, Bernoulli was of great interest to finding the following number given as (see [283])

$$e=\lim_{\kappa\to\infty}\left(1+\frac{1}{\kappa}\right)^{\kappa}, \tag{2.441}$$

which is now known as e, called as the Euler number. In 1748, Euler represented the exponential function in the form (see [284, 285, 286])

$$e^{t}=\lim_{\kappa\to\infty}\left(1+\frac{t}{\kappa}\right)^{\kappa}. \tag{2.442}$$

Euler also gives the following expression (see [284])

$$e^{i\theta}=\cos\theta+i\sin\theta, \tag{2.443}$$

which leads to (see [284])

$$\cos\theta=\frac{e^{i\theta}+e^{-i\theta}}{2} \tag{2.444}$$

and

$$\sin\theta=\frac{e^{i\theta}-e^{-i\theta}}{2i}. \tag{2.445}$$

2.13.2 Tempered fractional derivatives

The conceptions of the tempered fractional calculus containing the Bessel fractional integrals and Bessel fractional derivatives are given as follows.

Definition 2.53 *Let $\alpha\in\mathbb{C}$, $1>Re(\alpha)>0$, $-\infty<a<b<\infty$ and $\lambda\in\mathbb{C}$. The left-sided tempered fractional derivative in the sense of Riemann-Liouville type is defined as*

$$\left({}_{Cp}^{RL}D_{a+}^{\alpha,\lambda}f\right)(t)=\frac{1}{\Gamma(1-\alpha)}\frac{d}{dt}\int_{a}^{t}\frac{f(\tau)}{(t-\tau)^{\alpha}}e^{-\lambda(t-\tau)}d\tau, \tag{2.446}$$

and the right-sided tempered fractional derivative in the sense of Riemann-Liouville type as

$$\left({}_{Cp}^{RL}D_{b-}^{\alpha,\lambda}f\right)(t)=\frac{-1}{\Gamma(1-\alpha)}\frac{d}{dt}\int_{t}^{b}\frac{f(\tau)}{(\tau-t)^{\alpha}}e^{-\lambda(\tau-t)}d\tau, \tag{2.447}$$

respectively.

Definition 2.54 *Let $\alpha \in \mathbb{C}$, $Re(\alpha) > 0$, $\kappa = [Re(\alpha)] + 1$, $-\infty < a < b < \infty$ and $\lambda \in \mathbb{C}$. The left-sided tempered fractional derivative in the sense of Riemann-Liouville type is defined as*

$$\left({}_{Cp}^{RL}D_{a+}^{\alpha,\lambda} f\right)(t) = \frac{1}{\Gamma(\kappa-\alpha)} \frac{d^\kappa}{dt^\kappa} \int_a^t \frac{f(\tau)}{(t-\tau)^{\alpha-\kappa+1}} e^{-\lambda(t-\tau)} d\tau, \tag{2.448}$$

and the right-sided tempered fractional derivative in the sense of Riemann-Liouville type as

$$\left({}_{Cp}^{RL}D_{b-}^{\alpha,\lambda} f\right)(t) \frac{(-1)^\kappa}{\Gamma(\kappa-\alpha)} \frac{d^\kappa}{dt^\kappa} \int_t^b \frac{f(\tau)}{(\tau-t)^{\alpha-\kappa+1}} e^{-\lambda(\tau-t)} d\tau, \tag{2.449}$$

respectively.

In particular, when $\alpha \in \mathbb{R}$, $1 > \alpha > 0$, $-\infty < a < b < \infty$ and $\lambda \in \mathbb{C}$, the left-sided tempered fractional derivative in the sense of Riemann-Liouville type is defined as

$$\left({}_{Cp}^{RL}D_{a+}^{\alpha,\lambda} f\right)(t) = \frac{1}{\Gamma(1-\alpha)} \frac{d}{dt} \int_a^t \frac{f(\tau)}{(t-\tau)^\alpha} e^{-\lambda(t-\tau)} d\tau, \tag{2.450}$$

and the right-sided tempered fractional derivative in the sense of Riemann-Liouville type as

$$\left({}_{Cp}^{RL}D_{b-}^{\alpha,\lambda} f\right)(t) = -\frac{1}{\Gamma(1-\alpha)} \frac{d}{dt} \int_t^b \frac{f(\tau)}{(\tau-t)^\alpha} e^{-\lambda(\tau-t)} d\tau. \tag{2.451}$$

Similarly, when $\alpha \in \mathbb{C}$, $Re(\alpha) > 0$, $\kappa = [Re(\alpha)] + 1$, $-\infty < a < b < \infty$ and $\lambda \in \mathbb{C}$, the left-sided tempered fractional derivative in the sense of Riemann-Liouville type is defined as

$$\left({}_{Cp}^{RL}D_{a+}^{\alpha,\lambda} f\right)(t) = \frac{1}{\Gamma(\kappa-\alpha)} \frac{d^\kappa}{dt^\kappa} \int_a^t \frac{f(\tau)}{(t-\tau)^{\alpha-\kappa+1}} e^{-\lambda(t-\tau)} d\tau, \tag{2.452}$$

and the right-sided tempered fractional derivative in the sense of Riemann-Liouville type as

$$\left({}_{Cp}^{RL}D_{b-}^{\alpha,\lambda} f\right)(t) = \frac{(-1)^\kappa}{\Gamma(\kappa-\alpha)} \frac{d^\kappa}{dt^\kappa} \int_t^b \frac{f(\tau)}{(\tau-t)^{\alpha-\kappa+1}} e^{-\lambda(\tau-t)} d\tau. \tag{2.453}$$

Definition 2.55 *Let $\alpha \in \mathbb{R}$, $\kappa < \alpha < \kappa + 1$, $a = -\infty$, $b = \infty$ and $\lambda \in \mathbb{C}$. The left-sided tempered fractional derivative in the sense of Riemann-Liouville type on the real line $\mathbb{R}$ is defined as*

$$\left({}_{Cp}^{RL}D_{+}^{\alpha,\lambda} f\right)(t) = \frac{1}{\Gamma(\kappa-\alpha)} \frac{d^\kappa}{dt^\kappa} \int_{-\infty}^t \frac{f(\tau)}{(t-\tau)^{\alpha-\kappa+1}} e^{-\lambda(t-\tau)} d\tau, \tag{2.454}$$

and the right-sided tempered fractional derivative in the sense of Riemann-Liouville type on the real line $\mathbb{R}$ *as*

$$\left({}_{Cp}^{RL}D_{-}^{\alpha,\lambda}f\right)(t)=\frac{(-1)^{\kappa}}{\Gamma(\kappa-\alpha)}\frac{d^{\kappa}}{dt^{\kappa}}\int_{t}^{\infty}\frac{f(\tau)}{(\tau-t)^{\alpha-\kappa+1}}e^{-\lambda(\tau-t)}d\tau. \tag{2.455}$$

In particular, when $\alpha\in\mathbb{R}$,$1>\alpha>0$, $a=-\infty$, $b=\infty$ and $\lambda\in\mathbb{C}$, the left -sided tempered fractional derivative in the sense of Riemann-Liouville type on the real line $\mathbb{R}$ is defined as

$$\left({}_{Cp}^{RL}D_{+}^{\alpha,\lambda}f\right)(t)=\frac{1}{\Gamma(1-\alpha)}\frac{d}{dt}\int_{-\infty}^{t}\frac{f(\tau)}{(t-\tau)^{\alpha}}e^{-\lambda(t-\tau)}d\tau, \tag{2.456}$$

and the right-sided tempered fractional derivative in the sense of Riemann-Liouville type on the real line $\mathbb{R}$ as

$$\left({}_{Cp}^{RL}D_{-}^{\alpha,\lambda}f\right)(t)=\frac{-1}{\Gamma(1-\alpha)}\frac{d}{dt}\int_{t}^{\infty}\frac{f(\tau)}{(\tau-t)^{\alpha}}e^{-\lambda(\tau-t)}d\tau. \tag{2.457}$$

2.13.3 Tempered fractional derivatives with respect to another function

Definition 2.56 *Let* $\alpha\in\mathbb{C}$, $1>Re(\alpha)>0$, $-\infty<a<b<\infty$ *and* $\lambda\in\mathbb{C}$. *The left-sided tempered fractional derivative in the sense of Riemann-Liouville type with respect to another function* $\phi(t)$ $\left(h^{(1)}(t)>0;t\in[a,b]\right)$ *is defined as*

$$\left({}_{Cp}^{RL}D_{a+;h}^{\alpha,\lambda}f\right)(t)=\frac{1}{\Gamma(1-\alpha)}\frac{1}{h^{(1)}(t)}\frac{d}{dt}\int_{a}^{t}\frac{e^{-\lambda(h(t)-h(\tau))}h^{(1)}(\tau)}{(h(t)-h(\tau))^{\alpha}}f(\tau)\,d\tau, \tag{2.458}$$

and the right-sided tempered fractional derivative in the sense of Riemann-Liouville type with respect to another function as

$$\left({}_{Cp}^{RL}D_{b-;h}^{\alpha,\lambda}f\right)(t)=\frac{-1}{\Gamma(1-\alpha)}\frac{1}{h^{(1)}(t)}\frac{d}{dt}\int_{t}^{b}\frac{e^{-\lambda(h(\tau)-h(t))}h^{(1)}(\tau)}{(h(\tau)-h(t))^{\alpha}}f(\tau)\,d\tau, \tag{2.459}$$

respectively.

Definition 2.57 *Let* $\alpha\in\mathbb{C}$, $Re(\alpha)>0$, $\kappa=[Re(\alpha)]+1$, $-\infty<a<b<\infty$ *and* $\lambda\in\mathbb{C}$. *The left sided tempered fractional derivative in the sense of Riemann-Liouville type with respect to another function is defined as*

$$\left({}_{Cp}^{RL}D_{a+}^{\alpha,\lambda}f\right)(t)=\frac{1}{\Gamma(\kappa-\alpha)}\left(\frac{1}{h^{(1)}(t)}\frac{d}{dt}\right)^{\kappa}\int_{a}^{t}\frac{e^{-\lambda(h(t)-h(\tau))}h^{(1)}(\tau)}{(h(t)-h(\tau))^{\alpha}}f(\tau)\,d\tau, \tag{2.460}$$

and the right-sided tempered fractional derivative in the sense of Riemann-Liouville type with respect to another function as

$$\left({}_{Cp}^{RL}D_{b-}^{\alpha,\lambda}f\right)(t)=\frac{(-1)^{\kappa}}{\Gamma(\kappa-\alpha)}\left(\frac{1}{h^{(1)}(t)}\frac{d}{dt}\right)^{\kappa}\int_{t}^{b}\frac{e^{-\lambda(h(t)-h(\tau))}h^{(1)}(\tau)}{(h(\tau)-h(t))^{\alpha-\kappa+1}}df(\tau)\,\tau, \tag{2.461}$$

respectively.

In particular, when $\alpha \in \mathbb{R}$, $1 > \alpha > 0$, $-\infty < a < b < \infty$ and $\lambda \in \mathbb{C}$, the left-sided tempered fractional derivative in the sense of Riemann-Liouville type with respect to another function is defined as

$$\left({}_{Cp}^{RL}D_{a+}^{\alpha,\lambda} f\right)(t) = \frac{1}{\Gamma(1-\alpha)} \frac{1}{h^{(1)}(t)} \frac{d}{dt} \int_a^t \frac{e^{-\lambda(h(t)-h(\tau))} h^{(1)}(\tau)}{(h(t)-h(\tau))^{\alpha-\kappa+1}} f(\tau)\, d\tau, \tag{2.462}$$

and the right-sided tempered fractional derivative in the sense of Riemann-Liouville type with respect to another function as

$$\left({}_{Cp}^{RL}D_{b-}^{\alpha,\lambda} f\right)(t) = -\frac{1}{\Gamma(1-\alpha)} \frac{1}{h^{(1)}(t)} \frac{d}{dt} \int_t^b \frac{e^{-\lambda(h(\tau)-h(t))} h^{(1)}(\tau)}{(h(\tau)-h(t))^{\alpha}} f(\tau)\, d\tau\ . \tag{2.463}$$

Similarly, when $\alpha \in \mathbb{C}$, $Re(\alpha) > 0$, $\kappa = [Re(\alpha)] + 1$, $-\infty < a < b < \infty$ and $\lambda \in \mathbb{C}$, the left-sided tempered fractional derivative in the sense of Riemann-Liouville type with respect to another function is defined as

$$\left({}_{Cp}^{RL}D_{a+}^{\alpha,\lambda} f\right)(t) = \frac{1}{\Gamma(\kappa-\alpha)} \left(\frac{1}{h^{(1)}(t)} \frac{d}{dt}\right)^{\kappa} \int_a^t \frac{e^{-\lambda(h(t)-h(\tau))} h^{(1)}(\tau)}{(h(t)-h(\tau))^{\alpha-\kappa+1}} f(\tau)\, d\tau, \tag{2.464}$$

and the right-sided tempered fractional derivative in the sense of Riemann-Liouville type with respect to another function as

$$\left({}_{Cp}^{RL}D_{b-}^{\alpha,\lambda} f\right)(t) = \frac{(-1)^{\kappa}}{\Gamma(\kappa-\alpha)} \left(\frac{1}{h^{(1)}(t)} \frac{d}{dt}\right)^{\kappa} \int_t^b \frac{e^{-\lambda(h(\tau)-h(t))} h^{(1)}(\tau)}{(h(\tau)-h(t))^{\alpha-\kappa+1}} f(\tau)\, d\tau. \tag{2.465}$$

Definition 2.58 *Let $\alpha \in \mathbb{R}$, $\kappa < \alpha < \kappa + 1$, $a = -\infty$, $b = \infty$ and $\lambda \in \mathbb{C}$. The left-sided tempered fractional derivative in the sense of Riemann-Liouville type with respect to another function on the real line $\mathbb{R}$ is defined as*

$$\left({}_{Cp}^{RL}D_{+}^{\alpha,\lambda} f\right)(t) = \frac{1}{\Gamma(\kappa-\alpha)} \left(\frac{1}{h^{(1)}(t)} \frac{d}{dt}\right)^{\kappa} \int_{-\infty}^t \frac{e^{-\lambda(h(t)-h(\tau))} h^{(1)}(\tau)}{(h(t)-h(\tau))^{\alpha-\kappa+1}} f(\tau)\, d\tau, \tag{2.466}$$

and the right-sided tempered fractional derivative in the sense of Riemann-Liouville type with respect to another function on the real line $\mathbb{R}$ as

$$\left({}_{Cp}^{RL}D_{-}^{\alpha,\lambda} f\right)(t) = \frac{(-1)^{\kappa}}{\Gamma(\kappa-\alpha)} \left(\frac{1}{h^{(1)}(t)} \frac{d}{dt}\right)^{\kappa} \int_t^{\infty} \frac{e^{-\lambda(h(\tau)-h(t))} h^{(1)}(\tau)}{(h(\tau)-h(t))^{\alpha-\kappa+1}} f(\tau)\, d\tau. \tag{2.467}$$

2.13.4 Tempered fractional derivatives of a purely imaginary order

In particular, when $\alpha \in \mathbb{R}$, $1 > \alpha > 0$, $a = -\infty$, $b = \infty$ and $\lambda \in \mathbb{C}$, the left-sided tempered fractional derivative in the sense of Riemann-Liouville type with respect to another function on the real line $\mathbb{R}$ is defined as

$$\left({}_{Cp}^{RL}D_{+}^{\alpha,\lambda} f\right)(t) = \frac{1}{\Gamma(1-\alpha)} \frac{1}{h^{(1)}(t)} \frac{d}{dt} \int_{-\infty}^t \frac{e^{-\lambda(h(t)-h(\tau))} h^{(1)}(\tau)}{(h(t)-h(\tau))^{\alpha-\kappa+1}} f(\tau)\, d\tau, \tag{2.468}$$

and the right-sided tempered fractional derivative in the sense of Riemann-Liouville type with respect to another function on the real line $\mathbb{R}$ as

$$\left({}_{Cp}^{RL}D_{-}^{\alpha,\lambda}f\right)(t)=\frac{-1}{\Gamma(1-\alpha)}\frac{1}{h^{(1)}(t)}\frac{d}{dt}\int_{t}^{\infty}\frac{e^{-\lambda(h(\tau)-h(t))}h^{(1)}(\tau)}{(h(\tau)-h(t))^{\alpha-\kappa+1}}f(\tau)\,d\tau. \tag{2.469}$$

In particular, when $\alpha=i\chi$, $\chi>0$, $-\infty<a<b<\infty$ and $\lambda\in\mathbb{C}$, the left-sided tempered fractional derivatives of a purely imaginary order in the sense of Riemann-Liouville type are defined as

$$\left({}_{Cp}^{RL}D_{a+}^{i\chi,\lambda}f\right)(t)=\frac{1}{\Gamma(1-i\chi)}\frac{d}{dt}\int_{a}^{t}\frac{f(\tau)}{(t-\tau)^{i\chi}}e^{-\lambda(t-\tau)}d\tau, \tag{2.470}$$

and

$$\left({}_{Cp}^{RL}D_{+}^{i\chi,\lambda}f\right)(t)=\frac{1}{\Gamma(1-i\chi)}\frac{d}{dt}\int_{-\infty}^{t}\frac{f(\tau)}{(t-\tau)^{i\chi}}e^{-\lambda(t-\tau)}d\tau, \tag{2.471}$$

and the right-sided tempered fractional derivatives of a purely imaginary order in the sense of Riemann-Liouville type as

$$\left({}_{Cp}^{RL}D_{b-}^{i\chi,\lambda}f\right)(t)=\frac{-1}{\Gamma(1-i\chi)}\frac{d}{dt}\int_{t}^{b}\frac{f(\tau)}{(\tau-t)^{i\chi}}e^{-\lambda(t-\tau)}d\tau, \tag{2.472}$$

and

$$\left({}_{Cp}^{RL}D_{-}^{i\chi,\lambda}f\right)(t)=\frac{-1}{\Gamma(1-i\chi)}\frac{d}{dt}\int_{t}^{\infty}\frac{f(\tau)}{(\tau-t)^{i\chi}}e^{-\lambda(t-\tau)}d\tau. \tag{2.473}$$

2.13.5 Tempered fractional integrals

Definition 2.59 *Let $\alpha\in\mathbb{C}$, $1>Re(\alpha)>0$, $-\infty<a<b<\infty$ and $\lambda\in\mathbb{C}$. The left-sided tempered fractional integral of Riemann-Liouville type is defined as*

$$\left({}_{Cp}I_{a+}^{\alpha,\lambda}f\right)(t)=\int_{a}^{t}\frac{1}{(t-\tau)^{\alpha+1}}E_{1,-\alpha}^{1-\alpha}\left(-\lambda\left(t-\tau\right)\right)f\left(\tau\right)d\tau, \tag{2.474}$$

and the right-sided tempered fractional integral of Riemann-Liouville type as

$$\left({}_{Cp}I_{b-}^{\alpha,\lambda}f\right)(t)=\int_{t}^{b}\frac{1}{(\tau-t)^{\alpha+1}}E_{1,-\alpha}^{1-\alpha}\left(-\lambda\left(\tau-t\right)\right)f\left(\tau\right)d\tau, \tag{2.475}$$

where $E_{1,-\alpha}^{1-\alpha}(t)$ is the Prabhakar function (for more details of the special functions, see Chapter 1 and Appendix D).

Generally, when $\alpha \in \mathbb{C}$, $\kappa + 1 > Re(\alpha) > \kappa$, $-\infty < a < b < \infty$ and $\lambda \in \mathbb{C}$, the left-sided tempered fractional integral of Riemann-Liouville type is defined as

$$\left({}_{Cp}^{RL}I_{a+}^{\alpha,\lambda} f\right)(t) = \int_a^t \frac{1}{(t-\tau)^{\alpha+1}} E_{1,-\alpha}^{\kappa-\alpha}\left(-\lambda(t-\tau)\right) f(\tau)\, d\tau, \tag{2.476}$$

and the right-sided tempered fractional integral of Riemann-Liouville type as

$$\left({}_{Cp}^{RL}I_{b-}^{\alpha,\lambda} f\right)(t) = \int_t^b \frac{1}{(\tau-t)^{\alpha+1}} E_{1,-\alpha}^{\kappa-\alpha}\left(-\lambda(\tau-t)\right) f(\tau)\, d\tau, \tag{2.477}$$

where $E_{1,v}^{\alpha-1}(t)$ is the Prabhakar function.

2.13.6 Tempered fractional integrals of a purely imaginary order

In particular, when $\alpha = i\chi$, $\chi > 0$, $-\infty < a < b < \infty$ and $\lambda \in \mathbb{C}$, the left-sided and right-sided tempered fractional integrals of a purely imaginary order in the sense of Riemann-Liouville type are defined as

$$\left({}_{Cp}^{RL}I_{a+}^{i\theta,\lambda} f\right)(t) = \int_a^t \frac{1}{(t-\tau)^{i\theta+1}} E_{1,-i\theta}^{1-i\theta}\left(-\lambda(t-\tau)\right) f(\tau)\, d\tau, \tag{2.478}$$

$$\left({}_{Cp}^{RL}I_{b-}^{i\theta,\lambda} f\right)(t) = \int_t^b \frac{1}{(\tau-t)^{i\theta+1}} E_{1,-i\theta}^{1-i\theta}\left(-\lambda(\tau-t)\right) f(\tau)\, d\tau, \tag{2.479}$$

$$\left({}_{Cp}I_{+}^{i\theta,\lambda} f\right)(t) = \int_{-\infty}^t \frac{1}{(t-\tau)^{i\theta+1}} E_{1,-i\theta}^{1-i\theta}\left(-\lambda(t-\tau)\right) f(\tau)\, d\tau, \tag{2.480}$$

and

$$\left({}_{Cp}I_{-}^{i\theta,\lambda} f\right)(t) = \int_t^{\infty} \frac{1}{(\tau-t)^{i\theta+1}} E_{1,-i\theta}^{1-i\theta}\left(-\lambda(\tau-t)\right) f(\tau)\, d\tau. \tag{2.481}$$

In a similar manner, when $\alpha \in \mathbb{R}$, $\kappa < \alpha < \kappa + 1$, $a = -\infty$, $b = \infty$ and $\lambda \in \mathbb{C}$, the left-sided tempered fractional derivative in the sense of Riemann-Liouville type is defined as

$$\left({}_{Cp}^{RL}I_{a+}^{\alpha,\lambda} f\right)(t) = \int_a^t \frac{1}{(t-\tau)^{\alpha+1}} E_{1,-\alpha}^{\kappa-\alpha}\left(-\lambda(t-\tau)\right) f(\tau)\, d\tau, \tag{2.482}$$

and the right-sided tempered fractional derivative in the sense of Riemann-Liouville type as

$$\left({}_{Cp}^{RL}I_{b-}^{\alpha,\lambda}f\right)(t) = \int_{t}^{b} \frac{1}{(\tau - t)^{\alpha+1}} E_{1,-\alpha}^{\kappa-\alpha}\left(-\lambda\left(\tau - t\right)\right) f\left(\tau\right) d\tau. \tag{2.483}$$

In particular, when $\alpha \in \mathbb{R}$, $0 < \alpha < 1$, $a = -\infty$, $b = \infty$ and $\lambda \in \mathbb{C}$, the left-sided tempered fractional derivative in the sense of Riemann-Liouville type is defined as

$$\left({}_{Cp}^{RL}I_{a+}^{\alpha,\lambda}f\right)(t) = \int_{a}^{t} \frac{1}{(t - \tau)^{\alpha+1}} E_{1,-\alpha}^{1-\alpha}\left(-\lambda\left(t - \tau\right)\right) f\left(\tau\right) d\tau, \tag{2.484}$$

and the right-sided tempered fractional derivative in the sense of Riemann-Liouville type as

$$\left({}_{Cp}^{RL}I_{b-}^{\alpha,\lambda}f\right)(t) = \int_{t}^{b} \frac{1}{(\tau - t)^{\alpha+1}} E_{1,-\alpha}^{1-\alpha}\left(-\lambda\left(\tau - t\right)\right) f\left(\tau\right) d\tau. \tag{2.485}$$

Analogous to the expressions of Eq.(2.484) and Eq.(2.485), when $\alpha \in \mathbb{C}$,$\kappa + 1 > Re\left(\alpha\right) > \kappa$ and $\lambda \in \mathbb{C}$, the left-sided tempered fractional derivative in the sense of Riemann-Liouville type is defined as

$$\left({}_{Cp}^{RL}I_{+}^{\alpha,\lambda}f\right)(t) = \int_{-\infty}^{t} \frac{1}{(t - \tau)^{\alpha+1}} E_{1,-\alpha}^{\kappa-\alpha}\left(-\lambda\left(t - \tau\right)\right) f\left(\tau\right) d\tau, \tag{2.486}$$

and the right-sided tempered fractional derivative in the sense of Riemann-Liouville type as

$$\left({}_{Cp}^{RL}I_{-}^{\alpha,\lambda}f\right)(t) = \int_{t}^{b} \frac{1}{(\tau - t)^{\alpha+1}} E_{1,-\alpha}^{\kappa-\alpha}\left(-\lambda\left(\tau - t\right)\right) f\left(\tau\right) d\tau, \tag{2.487}$$

respectively.

2.13.7 Tempered fractional derivatives in the sense of Liouville-Sonine and Liouville-Sonine-Caputo types

We now consider the left-sided and right-sided tempered fractional derivatives in the sense of Liouville-Sonine and Liouville-Sonine-Caputo types.

Definition 2.60 *Let* $\alpha \in \mathbb{C}$, $1 > Re\left(\alpha\right) > 0$, $-\infty < a < b < \infty$ *and* $\lambda \in \mathbb{C}$.

The left-sided tempered fractional derivative in the sense of Liouville-Sonine type is defined as

$$\left({}_{Cp}^{LS}D_{a+}^{\alpha,\lambda}f\right)(t)=\frac{1}{\Gamma(1-\alpha)}\int_a^t\frac{e^{-\lambda(t-\tau)}}{(t-\tau)^{\alpha}}f^{(1)}(\tau)\,d\tau, \tag{2.488}$$

and the right-sided tempered fractional derivative on the interval $[a,b]$ *in the sense of Liouville-Sonine type as*

$$\left({}_{Cp}^{LS}D_{b-}^{\alpha,\lambda}f\right)(t)=\frac{-1}{\Gamma(1-\alpha)}\int_t^b\frac{e^{-\lambda(\tau-t)}}{(\tau-t)^{\alpha}}f^{(1)}(\tau)\,d\tau, \tag{2.489}$$

respectively.

Definition 2.61 *Let* $\alpha\in\mathbb{C}$, $Re(\alpha)>0$, $\kappa=[Re(\alpha)]+1$, $-\infty<a<b<\infty$ *and* $\lambda\in\mathbb{C}$. *The left-sided tempered fractional derivative in the sense of Liouville-Sonine-Caputo type is defined as*

$$\left({}_{Cp}^{LSC}D_{a+}^{\alpha,\lambda}f\right)(t)=\frac{1}{\Gamma(\kappa-\alpha)}\int_a^t\frac{e^{-\lambda(t-\tau)}}{(t-\tau)^{\alpha-\kappa+1}}f^{(\kappa)}(\tau)\,d\tau, \tag{2.490}$$

and the right-sided tempered fractional derivative in the sense of Liouville-Sonine-Caputo type as

$$\left({}_{Cp}^{LSC}D_{b-}^{\alpha,\lambda}f\right)(t)=\frac{(-1)^{\kappa}}{\Gamma(\kappa-\alpha)}\int_t^b\frac{e^{-\lambda(\tau-t)}}{(\tau-t)^{\alpha-\kappa+1}}f^{(\kappa)}(\tau)\,d\tau, \tag{2.491}$$

respectively.

Definition 2.62 *Let* $\alpha\in\mathbb{R}$, $\kappa<\alpha<\kappa+1$, $a=-\infty$, $b=\infty$ *and* $\lambda\in\mathbb{C}$. *The left-sided tempered fractional derivative in the sense of Liouville-Sonine-Caputo type on the real line* $\mathbb{R}$ *is defined as*

$$\left({}_{Cp}^{LSC}D_{+}^{\alpha,\lambda}f\right)(t)=\frac{1}{\Gamma(\kappa-\alpha)}\int_{-\infty}^t\frac{e^{-\lambda(t-\tau)}}{(t-\tau)^{\alpha-\kappa+1}}f^{(1)}(\tau)\,d\tau, \tag{2.492}$$

and the right-sided tempered fractional derivative in the sense of Liouville-Sonine-Caputo type on the real line $\mathbb{R}$ *as*

$$\left({}_{Cp}^{LSC}D_{-}^{\alpha,\lambda}f\right)(t)=\frac{(-1)^{\kappa}}{\Gamma(\kappa-\alpha)}\int_t^{\infty}\frac{e^{-\lambda(\tau-t)}}{(\tau-t)^{\alpha-\kappa+1}}f^{(\kappa)}(\tau)\,d\tau, \tag{2.493}$$

respectively.

The properties of the tempered fractional derivatives and tempered fractional integrals are as follows:

Property 2.37 *Let $\alpha \in \mathbb{C}$, $1 > Re(\alpha) > 0$, $-\infty < a < b < \infty$, $\lambda \in \mathbb{C}$ and $f(t) \in AC[a,b]$. Then we have*

$$\left({}_{Cp}^{LS}D_{a+}^{\alpha,\lambda} f\right)(t) = \left({}_{Cp}^{RL}D_{a+}^{\alpha,\lambda} f\right)(t) + \frac{1}{\Gamma(1-\alpha)} \frac{e^{-\lambda(t-a)}}{(t-a)^{\alpha}} f(a), \tag{2.494}$$

and

$$\left({}_{Cp}^{LS}D_{b-}^{\alpha,\lambda} f\right)(t) = \left({}_{Cp}^{RL}D_{b-}^{\alpha,\lambda} f\right)(t) - \frac{1}{\Gamma(1-\alpha)} \frac{e^{-\lambda(b-t)}}{(b-t)^{\alpha}} f(b). \tag{2.495}$$

Property 2.38 *Let $\alpha \in \mathbb{C}$, $\kappa + 1 > Re(\alpha) > \kappa$, $-\infty < a < b < \infty$ and $\lambda \in \mathbb{C}$. The we have*

$$\left({}_{Cp}^{LSC}D_{a+}^{\alpha,\lambda} f\right)(t) = \left({}_{Cp}^{RL}D_{a+}^{\alpha,\lambda} f\right)(t) + \sum_{j=0}^{\kappa-1} \frac{e^{-\lambda(t-a)}(t-a)^{j-\alpha}}{\Gamma(1+j-\alpha)} f^{(j)}(a), \tag{2.496}$$

and

$$\left({}_{Cp}^{LSC}D_{b-}^{\alpha,\lambda} f\right)(t) = \left({}_{Cp}^{RL}D_{b-}^{\alpha,\lambda} f\right)(t) + \sum_{j=0}^{\kappa-1} \frac{(-1)^j e^{-\lambda(t-a)}(b-t)^{j-\alpha}}{\Gamma(1+j-\alpha)} f^{(j)}(b). \tag{2.497}$$

Property 2.39 *If $Re(\alpha) > 0$ and $Re(\beta) > 0$, then we have*

$$\left({}_{Cp}^{RL}I_{a+}^{\alpha,\lambda}\,{}_{Cp}^{RL}I_{a+}^{\beta,\lambda} f\right)(t) = \left({}_{Cp}^{RL}D_{a+}^{\alpha+\beta,\lambda} f\right)(t) \tag{2.498}$$

$$\left({}_{Cp}^{RL}I_{b-}^{\alpha,\lambda}\,{}_{Cp}^{RL}I_{b-}^{\beta,\lambda} f\right)(t) = \left({}_{Cp}^{RL}D_{b-}^{\alpha+\beta,\lambda} f\right)(t), \tag{2.499}$$

$$\left({}_{Cp}^{RL}I_{+}^{\alpha,\lambda}\,{}_{Cp}^{RL}I_{+}^{\beta,\lambda} f\right)(t) = \left({}_{Cp}^{RL}D_{+}^{\alpha+\beta,\lambda} f\right)(t), \tag{2.500}$$

and

$$\left({}_{Cp}^{RL}I_{-}^{\alpha,\lambda}\,{}_{Cp}^{RL}I_{-}^{\beta,\lambda} f\right)(t) = \left({}_{Cp}^{RL}D_{-}^{\alpha+\beta,\lambda} f\right)(t), \tag{2.501}$$

where $f(t) \in L_{\kappa}[a,b]$ $(1 \leq \kappa \leq \infty)$.

Property 2.40 *If $Re(\alpha) > 0$, then we have*

$$\left({}_{Cp}^{RL}D_{a+}^{\alpha,\lambda}\,{}_{Cp}^{RL}I_{a+}^{\alpha,\lambda} f\right)(t) = f(t) \tag{2.502}$$

$$\left({}_{Cp}^{RL}D_{b-}^{\alpha,\lambda}\,{}_{Cp}^{RL}I_{b-}^{\alpha,\lambda} f\right)(t) = f(t), \tag{2.503}$$

$$\left({}_{Cp}^{RL}D_{+}^{\alpha,\lambda}\,{}_{Cp}^{RL}I_{+}^{\alpha,\lambda} f\right)(t) = f(t), \tag{2.504}$$

and

$$\left({}_{Cp}^{RL}D_{-}^{\alpha,\lambda}\,{}_{Cp}^{RL}I_{-}^{\alpha,\lambda} f\right)(t) = f(t), \tag{2.505}$$

where $f(t) \in L_{\kappa}[a,b]$ $(1 \leq \kappa \leq \infty)$.

Property 2.41 *If $Re(\alpha) > 0$, then we have*

$$L\left\{\left({}_{Cp}^{RL}D_{0+}^{\alpha,\lambda}f\right)(t)\right\} = \left(s^{-1}\lambda + s\right)^{\alpha-\kappa} s^{\alpha} f(s), \tag{2.506}$$

$$L\left\{{}_{Cp}^{RL}I_{0+}^{\alpha,\lambda}f(t)\right\} = \left(s^{-1}\lambda + s\right)^{\alpha-\kappa} s^{\alpha} f(s), \tag{2.507}$$

$$L\left\{{}_{Cp}^{LSC}D_{0+}^{\alpha,\lambda}f(t)\right\} = \left(s^{-1}\lambda + s\right)^{\alpha-\kappa} s^{\alpha-\kappa} \left(s^{\kappa} f(s) - \sum_{j=0}^{\kappa-1} f^{(\kappa-j)}(0)\right), \tag{2.508}$$

and

$$L\left\{{}_{Cp}^{LS}D_{0+}^{\alpha,\lambda}f(t)\right\} = \left(s^{-1}\lambda + 1\right)^{\alpha-1} s^{\alpha-1} \left(sf(s) - f(0)\right). \tag{2.509}$$

Find that, when $D^{\kappa}\left(\left({}_{Cp}^{RL}I_{0+}^{\alpha,\lambda}f\right)(t)\right)|_{t=0} = 0$, we have

$$L\left\{\left({}_{Cp}^{RL}D_{0+}^{\alpha,\lambda}f\right)(t)\right\} = \left(1 + s^{-1}\lambda\right)^{\alpha-\kappa} s^{\alpha} f(s), \tag{2.510}$$

and when $D^{\kappa}\left(\left({}_{Cp}^{RL}I_{0+}^{\alpha,\lambda}f\right)(t)\right)|_{t=0} \neq 0$,

$$L\left\{\left({}_{Cp}^{RL}D_{0+}^{\alpha,\lambda}f\right)(t)\right\} = \left(1 + s^{-1}\lambda\right)^{\alpha-\kappa} s^{\alpha} f(s) - \sum_{\mu=0}^{\kappa-1} s^{\kappa-\mu-1} \left(D^{\mu}\left[{}_{Cp}^{RL}I_{0+}^{\alpha,\lambda}f(+0)\right]\right). \tag{2.511}$$

In particular, when $\kappa = 1$, there is

$$L\left\{{}_{Cp}^{RL}D_{0+}^{\alpha,\lambda}f(t)\right\} = \left(1 + s^{-1}\lambda\right)^{\alpha-1} s^{\alpha} f(s) - {}_{Cp}^{RL}I_{0+}^{\alpha,\lambda}f(+0), \tag{2.512}$$

and when $\kappa = 1$ and $D\left(\left({}_{Cp}^{RL}I_{0+}^{\alpha,\lambda}f\right)(t)\right)|_{t=0} = 0$,

$$L\left\{\left({}_{Cp}^{RL}D_{0+}^{\alpha,\lambda}f\right)(t)\right\} = \left(1 + s^{-1}\lambda\right)^{\alpha-1} s^{\alpha} f(s). \tag{2.513}$$

2.13.8 Tempered fractional derivatives involving power-sine and power-cosine functions

Definition 2.63 *Let $\alpha \in \mathbb{R}$, $-\infty < a < b < \infty$ and $\gamma > 0$. The left-sided tempered fractional derivative in the Liouville-Sonine-Caputo type involving the kernel of the cosine function is defined as*

$$\left({}_{Cb}^{LSC}D_{a+}^{\alpha,\gamma}f\right)(t) = \frac{1}{\Gamma(\kappa-\alpha)} \int_{a}^{t} \frac{\cos(-\gamma(t-\tau))}{(t-\tau)^{\alpha-\kappa+1}} f^{(\kappa)}(\tau)\, d\tau, \tag{2.514}$$

the right-sided tempered fractional derivative in the sense of Liouville-Sonine-Caputo type involving the kernel of the cosine function as

$$\left({}_{Cb}^{LSC}D_{b-}^{\alpha,\gamma}f\right)(t)=\frac{(-1)^{\kappa}}{\Gamma(\kappa-\alpha)}\int_{t}^{b}\frac{\cos\left(-\gamma\left(\tau-t\right)\right)}{\left(\tau-t\right)^{\alpha-\kappa+1}}f^{(\kappa)}(\tau)\,d\tau, \tag{2.515}$$

the left-sided tempered fractional derivative in the sense of Liouville-Sonine-Caputo type involving the kernel of the sine-power function as

$$\left({}_{Cc}^{LSC}D_{a+}^{\alpha,\gamma}f\right)(t)=\frac{1}{\Gamma(\kappa-\alpha)}\int_{a}^{t}\frac{\sin\left(-\gamma\left(t-\tau\right)\right)}{\left(t-\tau\right)^{\alpha-\kappa+1}}f^{(\kappa)}(\tau)\,d\tau, \tag{2.516}$$

the right-sided tempered fractional derivative in the sense of Liouville-Sonine-Caputo type involving the kernel of the sine-power function as

$$\left({}_{Cc}^{LSC}D_{b-}^{\alpha,\gamma}f\right)(t)=\frac{(-1)^{\kappa}}{\Gamma(\kappa-\alpha)}\int_{t}^{b}\frac{\sin\left(-\gamma\left(\tau-t\right)\right)}{\left(\tau-t\right)^{\alpha-\kappa+1}}f^{(\kappa)}(\tau)\,d\tau, \tag{2.517}$$

the left-sided tempered fractional derivative in the sense of Riemann-Liouville type involving the kernel of the cosine function is defined as

$$\left({}_{Cb}^{RL}D_{a+}^{\alpha,\gamma}f\right)(t)=\frac{1}{\Gamma(\kappa-\alpha)}\frac{d^{\kappa}}{dt^{\kappa}}\int_{a}^{t}\frac{\cos\left(-\gamma\left(t-\tau\right)\right)}{\left(t-\tau\right)^{\alpha-\kappa+1}}f(\tau)\,d\tau, \tag{2.518}$$

the right-sided tempered fractional derivative in the sense of Riemann-Liouville type involving the kernel of the cosine function as

$$\left({}_{Cb}^{RL}D_{b-}^{\alpha,\gamma}f\right)(t)=\frac{(-1)^{\kappa}}{\Gamma(\kappa-\alpha)}\frac{d^{\kappa}}{dt^{\kappa}}\int_{t}^{b}\frac{\cos\left(-\gamma\left(\tau-t\right)\right)}{\left(\tau-t\right)^{\alpha-\kappa+1}}f(\tau)\,d\tau, \tag{2.519}$$

the left-sided tempered fractional derivative in the sense of Riemann-Liouville type involving the kernel of the sine-power function as

$$\left({}_{Cc}^{RL}D_{a+}^{\alpha,\gamma}f\right)(t)=\frac{1}{\Gamma(\kappa-\alpha)}\frac{d^{\kappa}}{dt^{\kappa}}\int_{a}^{t}\frac{\sin\left(-\gamma\left(t-\tau\right)\right)}{\left(t-\tau\right)^{\alpha-\kappa+1}}f(\tau)\,d\tau, \tag{2.520}$$

the right-sided tempered fractional derivative in the sense of Riemann-Liouville type involving the kernel of the sine-power function as

$$\left({}_{Cc}^{RL}D_{b-}^{\alpha,\gamma}f\right)(t)=\frac{(-1)^{\kappa}}{\Gamma(\kappa-\alpha)}\frac{d^{\kappa}}{dt^{\kappa}}\int_{t}^{b}\frac{\sin\left(-\gamma\left(\tau-t\right)\right)}{\left(\tau-t\right)^{\alpha-\kappa+1}}f(\tau)\,d\tau, \tag{2.521}$$

the left-sided tempered fractional derivative in the sense of Riemann-Liouville type involving the kernel of the cosine function is defined as

$$\left({}_{Cb}^{RL}D_{a+}^{\alpha,\gamma}f\right)(t)=\frac{1}{\Gamma(1-\alpha)}\frac{d}{dt}\int_a^t\frac{\cos\left(-\gamma\left(t-\tau\right)\right)}{\left(t-\tau\right)^{\alpha}}f(\tau)\,d\tau, \tag{2.522}$$

the right-sided tempered fractional derivative in the sense of Riemann-Liouville type involving the kernel of the cosine function as

$$\left({}_{Cb}^{RL}D_{b-}^{\alpha,\gamma}f\right)(t)=\frac{-1}{\Gamma(1-\alpha)}\frac{d}{dt}\int_t^b\frac{\cos\left(-\gamma\left(\tau-t\right)\right)}{\left(\tau-t\right)^{\alpha}}f(\tau)\,d\tau, \tag{2.523}$$

the left-sided tempered fractional derivative in the sense of Riemann-Liouville type involving the kernel of the sine-power function as

$$\left({}_{Cc}^{RL}D_{a+}^{\alpha,\gamma}f\right)(t)=\frac{1}{\Gamma(1-\alpha)}\frac{d}{dt}\int_a^t\frac{\sin\left(-\gamma\left(t-\tau\right)\right)}{\left(t-\tau\right)^{\alpha}}f(\tau)\,d\tau, \tag{2.524}$$

and the right-sided tempered fractional derivative in the sense of Riemann-Liouville type involving the kernel of the sine-power function as

$$\left({}_{Cc}^{RL}D_{b-}^{\alpha,\gamma}f\right)(t)=\frac{-1}{\Gamma(1-\alpha)}\frac{d}{dt}\int_t^b\frac{\sin\left(-\gamma\left(\tau-t\right)\right)}{\left(\tau-t\right)^{\alpha}}f(\tau)\,d\tau. \tag{2.525}$$

2.13.9 Tempered fractional calculus involving power-Kohlrausch-Williams-Watts function

The conceptions of the fractional derivatives and fractional integrals with the weak singular power-Kohlrausch-Williams-Watts function kernel can be given as follows.

2.13.9.1 Tempered fractional derivative in the Liouville-Sonine-Caputo type involving the kernel of the power-Kohlrausch-Williams-Watts function

Definition 2.64 *The left-sided tempered fractional derivative in the Liouville-Sonine-Caputo type involving the kernel of the power-Kohlrausch-Williams-Watts function is defined as*

$$\left({}_{Vv}^{LSC}D_{a+}^{\alpha,\lambda,\gamma}f\right)(t)=\frac{1}{\Gamma(\kappa-\alpha)}\int_a^t\frac{e^{-\lambda(t-\tau)^{\gamma}}}{\left(t-\tau\right)^{\alpha-\kappa+1}}f^{(\kappa)}(\tau)\,d\tau, \tag{2.526}$$

the right-sided tempered fractional derivative involving the kernel of the power-Kohlrausch-Williams-Watts function in the sense of Liouville-Sonine-Caputo type as

$$\left({}_{Vv}^{LSC}D_{b-}^{\alpha,\lambda,\gamma}f\right)(t)=\frac{(-1)^{\kappa}}{\Gamma(\kappa-\alpha)}\int_{t}^{b}\frac{e^{-\lambda(\tau-t)^{\gamma}}}{(\tau-t)^{\alpha-\kappa+1}}f^{(\kappa)}(\tau)\,d\tau, \tag{2.527}$$

the left-sided tempered fractional derivative in the Liouville-Sonine type involving the kernel of the power-Kohlrausch-Williams-Watts function is defined as

$$\left({}_{Vv}^{LSC}D_{a+}^{\alpha,\lambda,\gamma}f\right)(t)=\frac{1}{\Gamma(1-\alpha)}\int_{a}^{t}\frac{e^{-\lambda(t-\tau)^{\gamma}}}{(t-\tau)^{\alpha}}f^{(1)}(\tau)\,d\tau, \tag{2.528}$$

the right-sided tempered fractional derivative in the sense of Liouville-Sonine type involving the kernel of the power-Kohlrausch-Williams-Watts function as

$$\left({}_{Vv}^{LSC}D_{b-}^{\alpha,\lambda,\gamma}f\right)(t)=\frac{-1}{\Gamma(1-\alpha)}\int_{t}^{b}\frac{e^{-\lambda(\tau-t)^{\gamma}}}{(\tau-t)^{\alpha}}f^{(1)}(\tau)\,d\tau, \tag{2.529}$$

the left-sided tempered fractional derivative in the sense of Riemann-Liouville type involving the kernel of the power-Kohlrausch-Williams-Watts function as

$$\left({}_{Vv}^{RL}D_{a+}^{\alpha,\lambda,\gamma}f\right)(t)=\frac{1}{\Gamma(\kappa-\alpha)}\frac{d^{\kappa}}{dt^{\kappa}}\int_{a}^{t}\frac{e^{-\lambda(t-\tau)^{\gamma}}}{(t-\tau)^{\alpha-\kappa+1}}f(\tau)\,d\tau, \tag{2.530}$$

and the right-sided tempered fractional derivative in the sense of Riemann-Liouville type involving the kernel of the power-Kohlrausch-Williams-Watts function as

$$\left({}_{Vv}^{RL}D_{b-}^{\alpha,\lambda,\gamma}f\right)(t)=\frac{(-1)^{\kappa}}{\Gamma(\kappa-\alpha)}\frac{d^{\kappa}}{dt^{\kappa}}\int_{t}^{b}\frac{e^{-\lambda(\tau-t)^{\gamma}}}{(\tau-t)^{\alpha-\kappa+1}}f(\tau)\,d\tau, \tag{2.531}$$

where $\alpha\in\mathbb{R}, -\infty<a<b<\infty$, $\gamma\geq 0$ *and* $\lambda\geq 0$.

2.13.9.2 Tempered fractional integral involving the kernel of the power-Kohlrausch-Williams-Watts function

Definition 2.65 *The left-sided tempered fractional derivative in the sense of Riemann-Liouville type involving the kernel of the power-Kohlrausch-Williams-Watts function is defined as*

$$\left({}_{Vv}^{LS}I_{a+}^{\alpha,\lambda,\gamma}f\right)(t)=\frac{1}{\Gamma(\alpha)}\int_{a}^{t}\frac{e^{-\lambda(t-\tau)^{\gamma}}}{(t-\tau)^{\alpha}}f(\tau)\,d\tau, \tag{2.532}$$

and the right-sided tempered fractional derivative in the sense of Riemann-Liouville type involving the kernel of the power-Kohlrausch-Williams-Watts function as

$$\left({}_{Vv}^{LS}I_{b-}^{\alpha,\lambda,\gamma}f\right)(t)=\frac{1}{\Gamma(\alpha)}\int_{t}^{b}\frac{e^{-\lambda(\tau-t)^{\gamma}}}{(\tau-t)^{\alpha}}f(\tau)\,d\tau, \tag{2.533}$$

where $\alpha\in\mathbb{R}$, $-\infty<a<b<\infty$, $\gamma\geq0$ *and* $\lambda\geq0$.

Definition 2.66 *The left-sided tempered fractional derivative in the sense of Riesz type, introduced by Chen and Deng in 2017, is defined as (see [287])*

$$\left({}_{Vcd}^{LSC}D_{a+}^{\alpha,\lambda,\gamma}f\right)(t)=\frac{1}{\Gamma(\alpha-1)}\int_{0}^{t}\frac{e^{-\lambda(t-\tau)^{\gamma}}}{(t-\tau)^{2-\alpha}}\,{}_{Rzl}D_{[a,b]}^{\beta}f(t)\,d\tau\;(t>a) \tag{2.534}$$

where the fractional derivative, being coupled with Riesz fractional derivative in the sense of Riemann-Liouville type, is defined as (see [5])

$${}_{Rzl}D_{[a,b]}^{\beta}f(t)=-\frac{\left({}_{L}D_{a+}^{\beta}f+{}_{L}D_{b-}^{\beta}f\right)(t)}{2\cos(\pi\beta/2)} \tag{2.535}$$

with the conditions

$${}_{L}D_{a+}^{\beta}f(t)=\left(D_{a+}^{\beta}f\right)(t)=\left(\frac{d}{dt}\right)^{2}\left(I_{a+}^{\beta}f\right)(t)\quad(1<\beta<2)$$

and

$${}_{L}D_{b-}^{\beta}f(t)=\left(D_{b-}^{\beta}f\right)(t)=\left(-\frac{d}{dt}\right)^{2}\left(I_{b-}^{\beta}f\right)(t)\quad(1<\beta<2).$$

This is the left-sided tempered fractional derivative, being coupled with Riesz fractional derivative in the sense of Liouville-Sonine-Caputo type.

In this case, we also write a generalization of Eq. (2.534) as

$$\left({}_{Vcd}^{LSC}D_{a+}^{\alpha,\lambda,\gamma}f\right)(t)=\frac{1}{\Gamma(\kappa-\alpha)}\int_{a}^{t}\frac{e^{-\lambda(t-\tau)^{\gamma}}}{(t-\tau)^{\alpha-\kappa+1}}\,{}_{Rzh}D_{[a,b]}^{\beta}f(t)\,d\tau, \tag{2.536}$$

where

$${}_{Rzh}D_{[a,b]}^{\beta}f(t)=-\frac{\left({}_{L}D_{a+}^{\beta}f+{}_{L}D_{b-}^{\beta}f\right)(t)}{2\cos(\pi\beta/2)} \tag{2.537}$$

with the conditions

$${}_{L}D_{a+}^{\beta}f(t)=\left(D_{a+}^{\beta}f\right)(t)=\left(\frac{d}{dt}\right)^{\kappa}\left(I_{a+}^{\beta}f\right)(t)\quad(\kappa<\beta<\kappa+1)$$

and

$$ {}_{L}D_{b-}^{\beta}f(t) = \left(D_{b-}^{\beta}f\right)(t) = \left(-\frac{d}{dt}\right)^{\kappa}\left(I_{b-}^{\beta}f\right)(t) \quad (\kappa < \beta < \kappa + 1). $$

This is the generalized version of Eq. (2.534) in the sense of Riemann-Liouville type.

As the generalized version, we also define the left-sided tempered fractional derivative, being coupled with Riesz fractional derivative in the sense of Liouville-Sonine-Caputo type, as

$$ \left({}_{Vcd}^{LSC}D_{a+}^{\alpha,\lambda,\gamma}f\right)(t) = \frac{1}{\Gamma(\kappa-\alpha)}\int_{a}^{t}\frac{e^{-\lambda(t-\tau)^{\gamma}}}{(t-\tau)^{\alpha-\kappa+1}}\,{}_{Rzd}D_{[a,b]}^{\beta}f(t)\,d\tau, \tag{2.538} $$

where

$$ {}_{Rzd}D_{[a,b]}^{\beta}f(t) = -\frac{\left({}_{Lr}D_{a+}^{\beta}f + {}_{Lr}D_{b-}^{\beta}f\right)(t)}{2\cos(\pi\beta/2)} \tag{2.539} $$

with the conditions

$$ {}_{Lr}D_{a+}^{\beta}f(t) = {}_{LSC}D_{a+}^{\beta}f(x) = I_{a+}^{\beta}\left[f^{(\kappa)}(x)\right] \quad (\kappa < \beta < \kappa+1) $$

and

$$ {}_{Lr}D_{b-}^{\beta}f(t) = {}_{LSC}D_{b-}^{\beta}f(x) = I_{b-}^{\beta}\left[f^{(\kappa)}(x)\right] \quad (\kappa < \beta < \kappa+1). $$

In fact, we also define the left-sided and right-sided tempered fractional derivative, being coupled with Liouville-Sonine-Caputo fractional derivative, as

$$ \left({}_{Hh}^{LSC}D_{a+}^{\alpha,\lambda,\gamma}f\right)(t) = \frac{1}{\Gamma(\kappa-\alpha)}\int_{a}^{t}\frac{e^{-\lambda(t-\tau)^{\gamma}}}{(t-\tau)^{\alpha-\kappa+1}}D_{a+}^{\beta}\left[f^{(\kappa)}(x)\right]d\tau, \tag{2.540} $$

the right-sided tempered fractional derivative, being coupled with Liouville-Sonine-Caputo fractional derivative, as

$$ \left({}_{Hh}^{LSC}D_{b-}^{\alpha,\lambda,\gamma}f\right)(t) = \frac{(-1)^{\kappa}}{\Gamma(\kappa-\alpha)}\int_{t}^{b}\frac{e^{-\lambda(t-\tau)^{\gamma}}}{(t-\tau)^{\alpha-\kappa+1}}D_{b-}^{\beta}\left[f^{(\kappa)}(x)\right]d\tau, \tag{2.541} $$

the left-sided tempered fractional derivative, being coupled with the left-sided Riemann-Liouville fractional derivative, as

$$ \left({}_{Hv}^{RL}D_{a+}^{\alpha,\lambda}f\right)(t) = \frac{1}{\Gamma(\kappa-\alpha)}\int_{a}^{t}\frac{e^{-\lambda(t-\tau)}}{(t-\tau)^{\alpha-\kappa+1}}\left[\left(\frac{d}{dt}\right)^{\kappa}\left(I_{a+}^{\beta}f\right)(t)\right]d\tau, \tag{2.542} $$

and the right-sided tempered fractional derivative, being coupled with the right-sided Riemann-Liouville fractional derivative, as

$$\left({}_{Hv}^{RL}D_{b-}^{\alpha,\lambda}f\right)(t)=\frac{(-1)^{\kappa}}{\Gamma(\kappa-\alpha)}\int\limits_{t}^{b}\frac{e^{-\lambda(\tau-t)}}{(\tau-t)^{\alpha-\kappa+1}}\left[\left(-\frac{d}{dt}\right)^{\kappa}\left(I_{b-}^{\beta}f\right)(t)\right]d\tau, \tag{2.543}$$

where $-\infty<a<b<\infty$, $\kappa+1>\alpha>\kappa$ and $\kappa<\beta<\kappa+1$.

Being motivated by the idea, the left-sided tempered fractional derivative, being coupled with the right-sided Liouville-Sonine-Caputo fractional derivative, is defined as

$$\left({}_{YL}^{LSC}D_{a+}^{\alpha,\lambda,\gamma}f\right)(t)=\frac{1}{\Gamma(\kappa-\alpha)}\int\limits_{a}^{t}\frac{e^{-\lambda(t-\tau)^{\gamma}}}{(t-\tau)^{\alpha-\kappa+1}}D_{a+}^{\beta}\left[f^{(\kappa)}(x)\right]d\tau, \tag{2.544}$$

and the right-sided tempered fractional derivative, being coupled with the right-sided Liouville-Sonine-Caputo fractional derivative, as

$$\left({}_{YL}^{LSC}D_{b-}^{\alpha,\lambda,\gamma}f\right)(t)=\frac{(-1)^{\kappa}}{\Gamma(\kappa-\alpha)}\int\limits_{t}^{b}\frac{e^{-\lambda(\tau-t)^{\gamma}}}{(\tau-t)^{\alpha-\kappa+1}}D_{b-}^{\beta}\left[f^{(\kappa)}(x)\right]d\tau, \tag{2.545}$$

where $-\infty<a<b<\infty$, $\kappa+1>\alpha>\kappa$ and $\kappa<\beta<\kappa+1$.

In particular, when $-\infty<a<b<\infty$, $1>\alpha>0$ and $0<\beta<1$, the left-sided tempered fractional derivative, being coupled with the left-sided Liouville-Sonine fractional derivative, is defined as

$$\left({}_{YL}^{LSC}D_{a+}^{\alpha,\lambda,\gamma}f\right)(t)=\frac{1}{\Gamma(1-\alpha)}\int\limits_{a}^{t}\frac{e^{-\lambda(t-\tau)^{\gamma}}}{(t-\tau)^{\alpha}}D_{a+}^{\beta}\left[f^{(1)}(x)\right]d\tau, \tag{2.546}$$

and the left-sided tempered fractional derivative, being coupled with the right-sided Liouville-Sonine fractional derivative, as

$$\left({}_{YL}^{LSC}D_{b-}^{\alpha,\lambda,\gamma}f\right)(t)=\frac{-1}{\Gamma(1-\alpha)}\int\limits_{t}^{b}\frac{e^{-\lambda(\tau-t)^{\gamma}}}{(\tau-t)^{\alpha}}D_{b-}^{\beta}\left[f^{(1)}(x)\right]d\tau. \tag{2.547}$$

Similarly, the left-sided tempered fractional derivative, being coupled with the left-sided Riemann-Liouville fractional derivative, is defined as

$$\left({}_{YV}^{RL}D_{a+}^{\alpha,\lambda,\gamma}f\right)(t)=\frac{1}{\Gamma(\kappa-\alpha)}\int\limits_{a}^{t}\frac{e^{-\lambda(t-\tau)^{\gamma}}}{(t-\tau)^{\alpha-\kappa+1}}\left[\left(\frac{d}{dt}\right)^{\kappa}\left(I_{a+}^{\beta}f\right)(t)\right]d\tau, \tag{2.548}$$

and the left-sided tempered fractional derivative, being coupled with the right-sided Riemann-Liouville fractional derivative, as

$$\left({}_{YV}^{RL}D_{b-}^{\alpha,\lambda,\gamma}f\right)(t)=\frac{(-1)^{\kappa}}{\Gamma(\kappa-\alpha)}\int_{t}^{b}\frac{e^{-\lambda(\tau-t)^{\gamma}}}{(\tau-t)^{\alpha-\kappa+1}}\left[\left(-\frac{d}{dt}\right)^{\kappa}\left(I_{b-}^{\beta}f\right)(t)\right]d\tau, \tag{2.549}$$

where $-\infty<a<b<\infty$ and $\kappa<\beta<\kappa+1$.

In particular, when $-\infty<a<b<\infty$, $1>\alpha>0$ and $0<\beta<1$, the left-sided tempered fractional derivative, being coupled with the left-sided Riemann-Liouville fractional derivative, is defined as

$$\left({}_{YV}^{RL}D_{a+}^{\alpha,\lambda,\gamma}f\right)(t)=\frac{1}{\Gamma(1-\alpha)}\int_{a}^{t}\frac{e^{-\lambda(t-\tau)^{\gamma}}}{(t-\tau)^{\alpha}}\left[\frac{d}{dt}\left(I_{a+}^{\beta}f\right)(t)\right]d\tau, \tag{2.550}$$

and the left-sided tempered fractional derivative, being coupled with the right-sided Riemann-Liouville fractional derivative, as

$$\left({}_{YV}^{RL}D_{b-}^{\alpha,\lambda,\gamma}f\right)(t)=\frac{(-1)^{\kappa}}{\Gamma(1-\alpha)}\int_{t}^{b}\frac{e^{-\lambda(\tau-t)^{\gamma}}}{(\tau-t)^{\alpha}}\left[-\frac{d}{dt}\left(I_{b-}^{\beta}f\right)(t)\right]d\tau, \tag{2.551}$$

where $-\infty<a<b<\infty$ and $\kappa<\beta<\kappa+1$.

We notice that Pollard and Widder considered the solution of the heat equation given as follows (see [288]):

$$g(t)=\frac{1}{4\pi}\int_{t}^{b}\frac{e^{-\frac{1}{4}(\tau-t)^{-1}}}{(\tau-t)^{\frac{3}{2}}}f(\tau)\,d\tau.$$

The left-sided tempered fractional integral on the interval $[a,b]$ is defined as (see [3, 75]; also see [289]):

$$\left({}_{Dv}^{LS}I_{a+}^{\alpha,\lambda}f\right)(t)=\frac{1}{\Gamma(\alpha)}\int_{a}^{t}\frac{e^{-\lambda(t-\tau)}}{(t-\tau)^{\alpha}}f(\tau)\,d\tau, \tag{2.552}$$

and the right-sided tempered fractional integral on the interval $[a,b]$ as (see [3, 75]; also see [289])

$$\left({}_{Dv}^{LS}I_{b-}^{\alpha,\lambda}f\right)(t)=\frac{1}{\Gamma(\alpha)}\int_{t}^{b}\frac{e^{-\lambda(\tau-t)}}{(\tau-t)^{\alpha}}f(\tau)\,d\tau, \tag{2.553}$$

where $\lambda\in\mathbb{C}$.

From the expressions Eq.(2.552) and Eq.(2.553), the left-sided tempered fractional integral involving the cosine function on the interval $[a, b]$ is defined as

$$\left({}_{Dvc}^{LS}I_{a+}^{\alpha,\lambda} f\right)(t) = \frac{1}{\Gamma(\alpha)} \int_a^t \frac{\cos(-\gamma(t-\tau))}{(t-\tau)^\alpha} f(\tau)\, d\tau, \tag{2.554}$$

the right-sided tempered fractional integral involving the cosine function as

$$\left({}_{Dvc}^{LS}I_{b-}^{\alpha,\lambda} f\right)(t) = \frac{1}{\Gamma(\alpha)} \int_t^b \frac{\cos(-\gamma(\tau-t))}{(t-\tau)^\alpha} f(\tau)\, d\tau, \tag{2.555}$$

the left-sided tempered fractional integral involving the sine function as

$$\left({}_{Dvs}^{LS}I_{a+}^{\alpha,\lambda} f\right)(t) = \frac{1}{\Gamma(\alpha)} \int_a^t \frac{\sin(-\gamma(t-\tau))}{(\tau-t)^\alpha} f(\tau)\, d\tau, \tag{2.556}$$

and the right-sided tempered fractional integral involving the sine function as

$$\left({}_{Dvs}^{LS}I_{b-}^{\alpha,\lambda} f\right)(t) = \frac{1}{\Gamma(\alpha)} \int_t^b \frac{\sin(-\gamma(\tau-t))}{(\tau-t)^\alpha} f(\tau)\, d\tau, \tag{2.557}$$

where $\gamma \in \mathbb{C}$.

The left-sided tempered fractional derivative on the interval $[a, b]$ is represented as (for $\lambda = 1$, see [3, 75]; also see [289])

$$\left({}_{Dv}^{LS}D_{a+}^{\alpha,\lambda} f\right)(t) = \frac{e^{-\lambda t}}{\Gamma(\kappa-\alpha)} \frac{d^\kappa}{dt^\kappa} \int_a^t \frac{e^{-\lambda\tau}}{(t-\tau)^{\alpha-\kappa+1}} f(\tau)\, d\tau, \tag{2.558}$$

and the right-sided tempered derivative on the interval $[a, b]$ as (for $\lambda = 1$, see [3, 75]; also see [289])

$$\left({}_{Dv}^{LS}D_{b-}^{\alpha,\lambda} f\right)(t) = \frac{(-1)^\kappa e^{-\lambda t}}{\Gamma(\kappa-\alpha)} \frac{d^\kappa}{dt^\kappa} \int_t^b \frac{e^{-\lambda\tau}}{(\tau-t)^{\alpha-\kappa+1}} f(\tau)\, d\tau, \tag{2.559}$$

where $\lambda \in \mathbb{C}$.

For $\lambda \in \mathbb{C}$, the left-sided tempered fractional derivative on the interval $[a, b]$ is represented as (for $\lambda = 1$, see [3, 75]; also see [289])

$$\left({}_{Dv}^{LS}D_{a+}^{\alpha,\lambda} f\right)(t) = \frac{e^{-\lambda t}}{\Gamma(\kappa-\alpha)} \int_a^t \frac{e^{\lambda\tau}}{(t-\tau)^{\alpha-\kappa+1}} \left(\frac{d}{d\tau}+\lambda\right)^\kappa f(\tau)\, d\tau, \tag{2.560}$$

which yields

$$\left({}^{LS}_{Dv}D^{\alpha,\lambda}_{a+}f\right)(t) = \frac{1}{\Gamma(\kappa-\alpha)} \int_a^t \frac{e^{-\lambda(t-\tau)}}{(t-\tau)^{\alpha-\kappa+1}} \left(\frac{d}{d\tau}+\lambda\right)^{\kappa} f(\tau)\, d\tau, \quad (2.561)$$

and the right-sided tempered derivative on the interval $[a,b]$ as (for $\lambda = 1$, see [3, 75]; also see [289])

$$\left({}^{LS}_{Dv}D^{\alpha,\lambda}_{b-}f\right)(t) = \frac{(-1)^{\kappa} e^{\lambda t}}{\Gamma(\kappa-\alpha)} \int_t^b \frac{e^{-\lambda\tau}}{(\tau-t)^{\alpha-\kappa+1}} \left(\frac{d}{d\tau}-\lambda\right)^{\kappa} f(\tau)\, d\tau, \quad (2.562)$$

which leads to

$$\left({}^{LS}_{Dv}D^{\alpha,\lambda}_{b-}f\right)(t) = \frac{(-1)^{\kappa}}{\Gamma(\kappa-\alpha)} \int_t^b \frac{e^{-\lambda(\tau-t)}}{(\tau-t)^{\alpha-\kappa+1}} (-1)^{\kappa} \left(\frac{d}{d\tau}-\lambda\right)^{\kappa} f(\tau)\, d\tau, \quad (2.563)$$

where

$$\left(\frac{d}{dt}+\lambda\right)^{\kappa} f(t) = e^{-\lambda t}\frac{d^{\kappa}}{dt^{\kappa}}\left(e^{\lambda t}f(t)\right)$$

and

$$\left(\frac{d}{dt}-\lambda\right)^{\kappa} f(t) = e^{\lambda t}\frac{d^{\kappa}}{dt^{\kappa}}\left(e^{-\lambda t}f(t)\right).$$

For $\lambda \in \mathbb{C}$, the left-sided tempered fractional derivative within the power-exponential function on the interval $[a,b]$ is represented as (for $\lambda = 1$, see [3, 75]; also see [289])

$$\left({}^{LS}_{Dv}D^{\alpha,\lambda}_{a+}f\right)(t) = \frac{e^{-\lambda t}}{\Gamma(1-\alpha)} \int_a^t \frac{e^{\lambda\tau}}{(t-\tau)^{\alpha}} \frac{d\left(e^{\lambda\tau}f(\tau)\right)}{d\tau} d\tau, \quad (2.564)$$

which reduces to

$$\left({}^{LS}_{Dv}D^{\alpha,\lambda}_{a+}f\right)(t) = \frac{1}{\Gamma(1-\alpha)} \int_a^t \frac{e^{\lambda(t-\tau)}}{(t-\tau)^{\alpha}} \frac{d\left(e^{\lambda\tau}f(\tau)\right)}{d\tau} d\tau, \quad (2.565)$$

and the right-sided tempered derivative on the interval $[a,b]$ as (for $\lambda = 1$, see [3, 75]; also see [289])

$$\left({}^{LS}_{Dv}D^{\alpha,\lambda}_{b-}f\right)(t) = \frac{-e^{-\lambda t}}{\Gamma(1-\alpha)} \int_t^b \frac{e^{\lambda\tau}}{(\tau-t)^{\alpha}} \frac{d\left(e^{-\lambda\tau}f(\tau)\right)}{d\tau} d\tau, \quad (2.566)$$

which can be written as

$$\left({}^{LS}_{Dv}D^{\alpha,\lambda}_{b-}f\right)(t) = \frac{-1}{\Gamma(1-\alpha)} \int_t^b \frac{e^{-\lambda(\tau-t)}}{(\tau-t)^{\alpha}} \frac{d\left(e^{-\lambda\tau}f(\tau)\right)}{d\tau} d\tau. \quad (2.567)$$

2.13.10 Sabzikar-Meerschaert-Chen tempered fractional calculus

Definition 2.67 *The left-sided tempered fractional derivative within the power-exponential function, introduced by Sabzikar, Meerschaert and Chen in 2015, is defined as (see [290])*

$$\left({}^{LS}_{Sme}D^{\alpha,\lambda}_{+}f\right)(t)=\frac{\alpha}{\Gamma(1-\alpha)}\int\limits_{0}^{\infty}\frac{e^{-\lambda\tau}\left(f(t)-f(t-\tau)\right)}{\tau^{1+\alpha}}d\tau, \qquad (2.568)$$

and the right-sided tempered fractional derivative within the power-exponential function as

$$\left({}^{LS}_{Sme}D^{\alpha,\lambda}_{-}f\right)(t)=\frac{\alpha}{\Gamma(1-\alpha)}\int\limits_{0}^{\infty}\frac{e^{-\lambda\tau}\left(f(t)-f(t+\tau)\right)}{\tau^{1+\alpha}}d\tau, \qquad (2.569)$$

where $\lambda\in\mathbb{C}$, which are called the left-sided and right-sided Sabzikar-Meerschaert-Chen tempered fractional derivatives (in honor of Sabzikar, Meerschaert and Chen) (see [290]).

In particular, with the use of the same idea in Eq. (2.444) and Eq. (2.445), we have

$$e^{-i\gamma(t-\tau)}=\cos\left(-\gamma(t-\tau)\right)+i\sin\left(-\gamma(t-\tau)\right) \qquad (2.570)$$

such that there are from Eq. (2.568) and Eq. (2.569)

$$\left({}^{LS}_{Smec}D^{\alpha,\gamma}_{+}f\right)(t)=\frac{\alpha}{\Gamma(1-\alpha)}\int\limits_{0}^{\infty}\frac{\cos(-\gamma\tau)\left(f(t)-f(t-\tau)\right)}{\tau^{1+\alpha}}d\tau, \qquad (2.571)$$

$$\left({}^{LS}_{Smec}D^{\alpha,\gamma}_{-}f\right)(t)=\frac{\alpha}{\Gamma(1-\alpha)}\int\limits_{0}^{\infty}\frac{\cos(-\gamma\tau)\left(f(t)-f(t+\tau)\right)}{\tau^{1+\alpha}}d\tau, \qquad (2.572)$$

$$\left({}^{LS}_{Smes}D^{\alpha,\gamma}_{+}f\right)(t)=\frac{\alpha}{\Gamma(1-\alpha)}\int\limits_{0}^{\infty}\frac{\sin(-\gamma\tau)\left(f(t)-f(t-\tau)\right)}{\tau^{1+\alpha}}d\tau, \qquad (2.573)$$

and

$$\left({}^{LS}_{Smec}D^{\alpha,\gamma}_{-}f\right)(t)=\frac{\alpha}{\Gamma(1-\alpha)}\int\limits_{0}^{\infty}\frac{\sin(-\gamma\tau)\left(f(t)-f(t+\tau)\right)}{\tau^{1+\alpha}}d\tau, \qquad (2.574)$$

where $\gamma\in\mathbb{C}$.

It is to say that Eq. (2.571) and Eq. (2.572) are called the left-sided and right-sided tempered fractional derivatives involving the cosine function, and Eq.(2.573) and Eq. (2.574) are called the left-sided and right-sided tempered fractional derivatives involving the sine function, respectively.

2.13.11 Torres tempered fractional derivatives

Definition 2.68 *The left-sided tempered fractional derivative within the kernel of the power-exponential function in the real line $\mathbb{R}$, introduced by Torres in 2017, is defined as (see [291]):*

$$\left({}_{Ct}D_{+}^{\alpha,\lambda} f\right)(t) = \lambda^{\alpha} f(t) + \frac{\alpha}{\Gamma(1-\alpha)} \int_{-\infty}^{t} \frac{f(t) - f(\tau)}{(t-\tau)^{\alpha+1}} e^{-\lambda(t-\tau)} d\tau, \quad (2.575)$$

and the right-sided tempered fractional derivative within the kernel of the power-exponential function on the real line $\mathbb{R}$ as

$$\left({}_{Ct}D_{-}^{\alpha,\lambda} f\right)(t) = \lambda^{\alpha} f(t) + \frac{\alpha}{\Gamma(1-\alpha)} \int_{t}^{\infty} \frac{f(t) - f(\tau)}{(\tau-t)^{\alpha+1}} e^{-\lambda(t-\tau)} d\tau, \quad (2.576)$$

where $1 > \alpha > 0$ and $\lambda \in \mathbb{C}$, which are called the left-sided and right-sided Torres tempered fractional derivatives (in honor of Torres), respectively.

When the Torres idea can be extended to that in the interval $[a, b]$, we have the following results.

Definition 2.69 *The left-sided tempered fractional derivative in the kernel of the power-exponential function on the interval $[a, b]$ is defined as:*

$$\left({}_{Ct}D_{a+}^{\alpha,\lambda} f\right)(t) = \lambda^{\alpha} f(t) + \frac{\alpha}{\Gamma(1-\alpha)} \int_{a}^{t} \frac{f(t) - f(\tau)}{(t-\tau)^{\alpha+1}} e^{-\lambda(t-\tau)} d\tau, \quad (2.577)$$

and the right-sided tempered fractional derivative in the kernel of the power-exponential function on the interval $[a, b]$ as

$$\left({}_{Ct}D_{b-}^{\alpha,\lambda} f\right)(t) = \lambda^{\alpha} f(t) + \frac{\alpha}{\Gamma(1-\alpha)} \int_{t}^{b} \frac{f(t) - f(\tau)}{(\tau-t)^{\alpha+1}} e^{-\lambda(t-\tau)} d\tau, \quad (2.578)$$

where $1 > \alpha > 0$, $-\infty < a < b < \infty$ and $\lambda \in \mathbb{C}$.

Definition 2.70 *The left-sided Torres tempered fractional derivatives involving the kernel of the power-cosine function are defined as*

$$\left({}_{Ct}D_{+}^{\alpha,\gamma} f\right)(t) = \lambda^{\alpha} f(t) + \frac{\alpha}{\Gamma(1-\alpha)} \int_{-\infty}^{t} \frac{f(t) - f(\tau)}{(t-\tau)^{\alpha+1}} \cos(-\gamma(t-\tau))\, d\tau, \quad (2.579)$$

the right-sided Torres tempered fractional derivative involving the kernel of the power-cosine function as

$$\left({}_{Ct}D_{-}^{\alpha,\lambda}f\right)(t)=\lambda^{\alpha}f(t)+\frac{\alpha}{\Gamma(1-\alpha)}\int_{t}^{\infty}\frac{f(t)-f(\tau)}{(\tau-t)^{\alpha+1}}\cos(-\gamma(\tau-t))\,d\tau, \tag{2.580}$$

the left-sided Torres tempered fractional derivative involving the kernel of the power-sine function as

$$\left({}_{Ct}D_{+}^{\alpha,\gamma}f\right)(t)=\lambda^{\alpha}f(t)+\frac{\alpha}{\Gamma(1-\alpha)}\int_{-\infty}^{t}\frac{f(t)-f(\tau)}{(t-\tau)^{\alpha+1}}\sin(-\gamma(t-\tau))\,d\tau, \tag{2.581}$$

and the right-sided Torres tempered fractional derivative involving the kernel of the power-sine function as

$$\left({}_{Ct}D_{-}^{\alpha,\lambda}f\right)(t)=\lambda^{\alpha}f(t)+\frac{\alpha}{\Gamma(1-\alpha)}\int_{t}^{\infty}\frac{f(t)-f(\tau)}{(\tau-t)^{\alpha+1}}\sin(-\gamma(\tau-t))\,d\tau, \tag{2.582}$$

where $1>\alpha>0$ *and* $\lambda\in\mathbb{C}$.

Similarly, the left-sided tempered fractional derivative within the kernel of the power-cosine function is defined as

$$\left({}_{Ctc}D_{+}^{\alpha,\lambda}f\right)(t)=\lambda^{\alpha}f(t)+\frac{\alpha}{\Gamma(1-\alpha)}\int_{-\infty}^{t}\frac{f(t)-f(\tau)}{(t-\tau)^{\alpha+1}}\cos(-\gamma(t-\tau))\,d\tau, \tag{2.583}$$

the right-sided tempered fractional derivative within the kernel of the power-cosine function as

$$\left({}_{Ctc}D_{-}^{\alpha,\lambda}f\right)(t)=\lambda^{\alpha}f(t)+\frac{\alpha}{\Gamma(1-\alpha)}\int_{t}^{\infty}\frac{f(t)-f(\tau)}{(\tau-t)^{\alpha+1}}\cos(-\gamma(\tau-t))\,d\tau, \tag{2.584}$$

the left-sided tempered fractional derivative within the kernel of the power-sine function as

$$\left({}_{Cts}D_{+}^{\alpha,\lambda}f\right)(t)=\lambda^{\alpha}f(t)+\frac{\alpha}{\Gamma(1-\alpha)}\int_{-\infty}^{t}\frac{f(t)-f(\tau)}{(t-\tau)^{\alpha+1}}\sin(-\gamma(t-\tau))\,d\tau, \tag{2.585}$$

the right-sided tempered fractional derivative within the kernel of the power-sine function as

$$\left({}_{Cts}D_{-}^{\alpha,\lambda}f\right)(t)=\lambda^{\alpha}f(t)+\frac{\alpha}{\Gamma(1-\alpha)}\int_{t}^{\infty}\frac{f(t)-f(\tau)}{(\tau-t)^{\alpha+1}}\sin(-\gamma(\tau-t))\,d\tau, \tag{2.586}$$

the left-sided tempered fractional derivative within the kernel of the power-cosine function is defined as

$$\left({}_{Ctc}D_{a+}^{\alpha,\lambda}f\right)(t)=\lambda^{\alpha}f(t)+\frac{\alpha}{\Gamma(1-\alpha)}\int_{a}^{t}\frac{f(t)-f(\tau)}{(t-\tau)^{\alpha+1}}\cos(-\gamma(t-\tau))\,d\tau, \tag{2.587}$$

the right-sided tempered fractional derivative within the kernel of the power-cosine function as

$$\left({}_{Ctc}D_{b-}^{\alpha,\lambda}f\right)(t)=\lambda^{\alpha}f(t)+\frac{\alpha}{\Gamma(1-\alpha)}\int_{t}^{b}\frac{f(t)-f(\tau)}{(\tau-t)^{\alpha+1}}\cos(-\gamma(\tau-t))\,d\tau, \tag{2.588}$$

the left-sided tempered fractional derivative within the kernel of the power-sine function as

$$\left({}_{Cts}D_{a+}^{\alpha,\lambda}f\right)(t)=\lambda^{\alpha}f(t)+\frac{\alpha}{\Gamma(1-\alpha)}\int_{a}^{t}\frac{f(t)-f(\tau)}{(t-\tau)^{\alpha+1}}\sin(-\gamma(t-\tau))\,d\tau, \tag{2.589}$$

and the right-sided tempered fractional derivative within the kernel of the power-sine function as

$$\left({}_{Cts}D_{b-}^{\alpha,\lambda}f\right)(t)=\lambda^{\alpha}f(t)+\frac{\alpha}{\Gamma(1-\alpha)}\int_{t}^{b}\frac{f(t)-f(\tau)}{(\tau-t)^{\alpha+1}}\sin(-\gamma(\tau-t))\,d\tau, \tag{2.590}$$

respectively, where $1>\alpha>0$ and $\lambda\in\mathbb{C}$.

Definition 2.71 *In this case, we define the left-sided tempered fractional derivative within the kernel of the power-exponential function in the real number $\mathbb{R}$ as*

$$\left({}_{Cy}D_{+}^{\alpha,\lambda}f\right)(t)=\frac{\alpha}{\Gamma(1-\alpha)}\int_{-\infty}^{t}\frac{f(t)-f(\tau)}{(t-\tau)^{\alpha+1}}e^{-\lambda(t-\tau)}\,d\tau, \tag{2.591}$$

and the right-sided tempered fractional derivative within the kernel of the power-exponential function in the real number $\mathbb{R}$ as

$$\left({}_{Cy}D_{-}^{\alpha,\lambda}f\right)(t)=\frac{\alpha}{\Gamma(1-\alpha)}\int_{t}^{\infty}\frac{f(t)-f(\tau)}{(\tau-t)^{\alpha+1}}e^{-\lambda(t-\tau)}\,d\tau, \tag{2.592}$$

where $1 > \alpha > 0$ *and* $\lambda \in \mathbb{C}$.

In particular, we have the following results.

Definition 2.72 *The left-sided tempered fractional derivative within the kernel of the power-exponential function on the interval* $[a, b]$ *is defined as*

$$\left({}_{Cy}D_{a+}^{\alpha,\lambda} f\right)(t) = \frac{\alpha}{\Gamma(1-\alpha)} \int\limits_{a}^{t} \frac{f(t) - f(\tau)}{(t-\tau)^{\alpha+1}} e^{-\lambda(t-\tau)} d\tau, \tag{2.593}$$

and the right-sided tempered fractional derivative within the kernel of the power-exponential function on the interval $[a, b]$ *as*

$$\left({}_{Cy}D_{b-}^{\alpha,\lambda} f\right)(t) = \frac{\alpha}{\Gamma(1-\alpha)} \int\limits_{t}^{b} \frac{f(t) - f(\tau)}{(\tau-t)^{\alpha+1}} e^{-\lambda(t-\tau)} d\tau, \tag{2.594}$$

where $1 > \alpha > 0$ *and* $\lambda \in \mathbb{C}$.

Meanwhile, the left-sided tempered fractional derivative within the kernel of the power-cosine function is defined as

$$\left({}_{Cyc}D_{a+}^{\alpha,\lambda} f\right)(t) = \frac{\alpha}{\Gamma(1-\alpha)} \int\limits_{a}^{t} \frac{f(t) - f(\tau)}{(t-\tau)^{\alpha+1}} \cos\left(-\gamma(t-\tau)\right) d\tau, \tag{2.595}$$

the left-sided tempered fractional derivative within the kernel of the power-cosine function as

$$\left({}_{Cyc}D_{b-}^{\alpha,\lambda} f\right)(t) = \frac{\alpha}{\Gamma(1-\alpha)} \int\limits_{t}^{b} \frac{f(t) - f(\tau)}{(\tau-t)^{\alpha+1}} \cos\left(-\gamma(\tau-t)\right) d\tau, \tag{2.596}$$

the left-sided tempered fractional derivative within the kernel of the power-sine function as

$$\left({}_{Cys}D_{a+}^{\alpha,\lambda} f\right)(t) = \frac{\alpha}{\Gamma(1-\alpha)} \int\limits_{a}^{t} \frac{f(t) - f(\tau)}{(t-\tau)^{\alpha+1}} \sin\left(-\gamma(t-\tau)\right) d\tau, \tag{2.597}$$

and the left-sided tempered fractional derivative within the kernel of the power-sine function as

$$\left({}_{Cys}D_{b-}^{\alpha,\lambda} f\right)(t) = \frac{\alpha}{\Gamma(1-\alpha)} \int\limits_{t}^{b} \frac{f(t) - f(\tau)}{(\tau-t)^{\alpha+1}} \sin\left(-\gamma(\tau-t)\right) d\tau, \tag{2.598}$$

where $1 > \alpha > 0$, $-\infty < a < b < \infty$ and $\gamma \in \mathbb{R}$.

In particular, when $\lambda = 0$, we have the following relations:

$$\left({}_{Cy}D_{+}^{\alpha,0} f\right)(t) = \left({}_{M}D_{+}^{\alpha} f\right)(t), \tag{2.599}$$

and

$$\left({}_{Cy}D_{-}^{\alpha,0} f\right)(t) = \left({}_{M}D_{-}^{\alpha} f\right)(t), \tag{2.600}$$

where $1 > \alpha > 0$.

Definition 2.73 *The left-sided tempered fractional derivative within the kernel of the power-exponential function in the interval $[a, b]$ is defined as*

$$\left({}_{Cy}D_{a+}^{\alpha,\lambda} f\right)(t) = \frac{\alpha}{\Gamma(1-\alpha)} \int_a^t \frac{f(t) - f(\tau)}{(t-\tau)^{\alpha+1}} e^{-\lambda(t-\tau)} d\tau, \tag{2.601}$$

and the right-sided tempered fractional derivative within the kernel of the power-exponential function as

$$\left({}_{Cy}D_{-}^{\alpha,\lambda} f\right)(t) = \frac{\alpha}{\Gamma(1-\alpha)} \int_t^b \frac{f(t) - f(\tau)}{(t-\tau)^{\alpha+1}} e^{-\lambda(t-\tau)} d\tau, \tag{2.602}$$

where $1 > \alpha > 0$ and $\lambda \in \mathbb{C}$, respectively.

In particular, when $a = 0$ and $b = \infty$, the left-sided tempered fractional derivative within the kernel of the power-exponential function is represented as

$$\left({}_{Cy}D_{0+}^{\alpha,\lambda} f\right)(t) := \frac{\alpha}{\Gamma(1-\alpha)} \int_0^t \frac{f(t) - f(\tau)}{(t-\tau)^{\alpha+1}} e^{-\lambda(t-\tau)} d\tau, \tag{2.603}$$

and the right-sided tempered fractional derivative within the kernel of the power-exponential function as

$$\left({}_{Cy}D_{-}^{\alpha,\lambda} f\right)(t) = \frac{\alpha}{\Gamma(1-\alpha)} \int_t^\infty \frac{f(t) - f(\tau)}{(t-\tau)^{\alpha+1}} e^{-\lambda(t-\tau)} d\tau, \tag{2.604}$$

where $1 > \alpha > 0$ and $\lambda \in \mathbb{C}$.

In fact, we also define the left-sided tempered fractional derivative within the kernel of the power-exponential function in sense of the Riesz type on the infinite interval $[a, b]$ as

$$\left({}_{Cs}D_{a+}^{\alpha,\lambda} f\right)(t) = \frac{\alpha}{\Gamma(1-\alpha)} \int_a^t \frac{f(t) - f(\tau)}{|t-\tau|^{\alpha+1}} e^{-\lambda(t-\tau)} d\tau, \tag{2.605}$$

and the right-sided tempered fractional derivative within the kernel of the power-exponential function in the sense of the Riesz type on the infinite interval $[a, b]$ as

$$\left({}_{Cs}D_{-}^{\alpha,\lambda} f\right)(t) = \frac{\alpha}{\Gamma(1-\alpha)} \int_{t}^{b} \frac{f(t) - f(\tau)}{|t-\tau|^{\alpha+1}} e^{-\lambda(t-\tau)} d\tau, \tag{2.606}$$

which yields that

$$\left({}_{Cs}D_{a+}^{\alpha,\lambda} f\right)(t) = \frac{\alpha}{\Gamma(1-\alpha)} \int_{-\infty}^{t} \frac{f(t) - f(\tau)}{|t-\tau|^{\alpha+1}} e^{-\lambda(t-\tau)} d\tau, \tag{2.607}$$

and

$$\left({}_{Cs}D_{-}^{\alpha,\lambda} f\right)(t) = \frac{\alpha}{\Gamma(1-\alpha)} \int_{t}^{-\infty} \frac{f(t) - f(\tau)}{|t-\tau|^{\alpha+1}} e^{-\lambda(t-\tau)} d\tau, \tag{2.608}$$

respectively, where $1 > \alpha > 0$ and $\lambda \in \mathbb{C}$.

In a similar way, we also define the left-sided tempered fractional derivative within the kernel of the power-exponential function in the sense of the Riesz type on the infinite interval $[a, b]$ by

$$\left({}_{Ch}D_{a+}^{\alpha,\lambda} f\right)(t) = \lambda^{\alpha} f(t) + \frac{\alpha}{\Gamma(1-\alpha)} \int_{a}^{t} \frac{f(t) - f(\tau)}{|t-\tau|^{\alpha+1}} e^{-\lambda(t-\tau)} d\tau, \tag{2.609}$$

and the right-sided tempered fractional derivative within the kernel of the power-exponential function in the sense of the Riesz type on the infinite interval $[a, b]$ by

$$\left({}_{Ct}D_{b-}^{\alpha,\lambda} f\right)(t) = \lambda^{\alpha} f(t) + \frac{\alpha}{\Gamma(1-\alpha)} \int_{t}^{b} \frac{f(t) - f(\tau)}{|t-\tau|^{\alpha+1}} e^{-\lambda(t-\tau)} d\tau, \tag{2.610}$$

which deduce that the left-sided tempered fractional derivative within the kernel of the power-exponential function in the sense of the Riesz type on the real line $\mathbb{R}$ is defined as

$$\left({}_{Ch}D_{+}^{\alpha,\lambda} f\right)(t) = \lambda^{\alpha} f(t) + \frac{\alpha}{\Gamma(1-\alpha)} \int_{-\infty}^{t} \frac{f(t) - f(\tau)}{|t-\tau|^{\alpha+1}} e^{-\lambda(t-\tau)} d\tau, \tag{2.611}$$

and the right-sided tempered fractional derivative within the kernel of the power-exponential function in the sense of the Riesz type on the real line $\mathbb{R}$ as

$$\left({}_{Ct}D_{-}^{\alpha,\lambda} f\right)(t) = \lambda^{\alpha} f(t) + \frac{\alpha}{\Gamma(1-\alpha)} \int_{t}^{\infty} \frac{f(t) - f(\tau)}{|t-\tau|^{\alpha+1}} e^{-\lambda(t-\tau)} d\tau, \tag{2.612}$$

where $1 > \alpha > 0$ and $\lambda \in \mathbb{C}$.

2.14 Tempered fractional derivatives involving Mittag-Leffler function

In this section, we consider the tempered fractional derivatives within the weak singularity and the features of the Mittag-Leffler kernel (see [235]).

Definition 2.74 *For $\lambda \in \mathbb{C}$, the left-sided tempered fractional derivative involving the kernel of the Mittag-Leffler function on the interval $[a, b]$ is defined as*

$$\left({}_{Ev}^{LS}D_{a+}^{\alpha,\beta,\lambda} f\right)(t) = \frac{1}{\Gamma(1-\alpha)} \int_a^t \frac{E_\alpha\left(\lambda(t-\tau)^\beta\right)}{(t-\tau)^\alpha} f^{(1)}(\tau)\, d\tau, \tag{2.613}$$

and the right-sided tempered derivative involving the kernel of the Mittag-Leffler function on the interval $[a, b]$ as

$$\left({}_{Ev}^{LS}D_{b-}^{\alpha,\beta,\lambda} f\right)(t) = -\frac{1}{\Gamma(1-\alpha)} \int_t^b \frac{E_\alpha\left(-\lambda(\tau-t)^\beta\right)}{(\tau-t)^\alpha} f^{(1)}(\tau)\, d\tau, \tag{2.614}$$

where $1 > \alpha > 0$, $1 > \beta > 0$ and $\lambda \in \mathbb{C}$.

Definition 2.75 *For $\lambda \in \mathbb{C}$, the left-sided tempered fractional derivative involving the kernel of the Mittag-Leffler function on the interval $[a, b]$ is defined as*

$$\left({}_{Ev}^{LS}D_{a+}^{\alpha,\beta,\lambda} f\right)(t) = \frac{1}{\Gamma(1-\alpha)} \frac{d}{dt} \int_a^t \frac{E_\alpha\left(\lambda(t-\tau)^\beta\right)}{(t-\tau)^\alpha} f(\tau)\, d\tau, \tag{2.615}$$

and the right-sided tempered derivative involving the kernel of the Mittag-Leffler function on the interval $[a, b]$ as

$$\left({}_{Ev}^{LS}D_{b-}^{\alpha,\beta,\lambda} f\right)(t) = -\frac{1}{\Gamma(1-\alpha)} \frac{d}{dt} \int_t^b \frac{E_\alpha\left(-\lambda(\tau-t)^\beta\right)}{(\tau-t)^\alpha} f(\tau)\, d\tau, \tag{2.616}$$

where $1 > \alpha > 0$, $1 > \beta > 0$ and $\lambda \in \mathbb{C}$.

Definition 2.76 *For $\lambda \in \mathbb{C}$, the left-sided tempered fractional derivative involving the kernel of the Mittag-Leffler function on the interval $[a, b]$ is defined as*

$$\left({}_{Ev}^{LS}D_{a+}^{\alpha,\alpha,\lambda} f\right)(t) = \frac{1}{\Gamma(1-\alpha)} \frac{d}{dt} \int_a^t \frac{E\left(\lambda(t-\tau)^\alpha\right)}{(t-\tau)^\alpha} f(\tau)\, d\tau, \tag{2.617}$$

and the right-sided tempered derivative involving the kernel of the Mittag-Leffler function on the interval $[a, b]$ *as*

$$\left({}_{Ev}^{LS}D_{b-}^{\alpha,\alpha,\lambda} f\right)(t) = -\frac{1}{\Gamma(1-\alpha)} \frac{d}{dt} \int_t^b \frac{E_\alpha\left(-\lambda(\tau-t)^\alpha\right)}{(\tau-t)^\alpha} f(\tau)\, d\tau, \tag{2.618}$$

where $1 > \alpha > 0$ *and* $\lambda \in \mathbb{C}$.

Similarly, the left-sided tempered fractional derivative involving the kernel of the Mittag-Leffler function is defined as

$$\left({}_{Ee}^{LS}D_{+}^{\alpha,\alpha,\lambda} f\right)(t) = \frac{\alpha}{\Gamma(1-\alpha)} \int_0^\infty \frac{E_\alpha(-\lambda\tau^\alpha)\left(f(t) - f(t-\tau)\right)}{\tau^{1+\alpha}} d\tau, \tag{2.619}$$

and the right-sided tempered fractional derivative involving the kernel of the Mittag-Leffler function as

$$\left({}_{Ee}^{LS}D_{-}^{\alpha,\alpha,\lambda} f\right)(t) = \frac{\alpha}{\Gamma(1-\alpha)} \int_0^\infty \frac{E_\alpha(-\lambda\tau^\alpha)\left(f(t) - f(t+\tau)\right)}{\tau^{1+\alpha}} d\tau, \tag{2.620}$$

where $1 > \alpha > 0$ and $\lambda \in \mathbb{C}$.

2.15 Liouville-Weyl fractional calculus

In this section, we introduce the Liouville-Weyl fractional integral and Liouville-Weyl fractional derivative (in honor of Liouville and Weyl) (see [15, 16, 33]; also see [236]).

Definition 2.77 *Let* $\alpha \in \mathbb{C}$, $Re(\alpha) > 0$ *and* $-\infty < x < b < \infty$. *The left-sided Liouville-Weyl fractional integral is defined as (see [3, 75])*

$$\left({}_{Cp}^{LW}I_{+}^{\alpha} f\right)(t) = \frac{1}{\Gamma(\alpha)} \int_{-\infty}^{t} \frac{1}{(t-\tau)^{1-\alpha}} f(\tau)\, d\tau. \tag{2.621}$$

Definition 2.78 *Let* $\alpha \in \mathbb{C}$, $Re(\alpha) > 0$ *and* $a < x < \infty$. *The right-sided Liouville-Weyl fractional integral is defined as (see [3, 75])*

$$\left({}_{Cp}^{LW}I_{-}^{\alpha} f\right)(t) = \frac{1}{\Gamma(\alpha)} \int_{t}^{\infty} \frac{1}{(\tau-t)^{1-\alpha}} f(\tau)\, d\tau. \tag{2.622}$$

Definition 2.79 *Let $\kappa+1 > \alpha > \kappa$ and $-\infty < x < b$. The left-sided Liouville-Weyl fractional derivative is defined as (see [236])*

$$\left({}_{Cp}^{LW}D_{+}^{\alpha}f\right)(t) = \frac{1}{\Gamma(\kappa-\alpha)}\frac{d^{\kappa}}{dt^{\kappa}}\int_{-\infty}^{t}\frac{1}{(t-\tau)^{\alpha-\kappa+1}}f(\tau)\,d\tau. \tag{2.623}$$

Definition 2.80 *Let $\kappa+1 > \alpha > \kappa$ and $a < x < \infty$. The right-sided Liouville-Weyl fractional derivative is defined as (see [236])*

$$\left({}_{Cp}^{LW}D_{-}^{\alpha}f\right)(t) = \frac{1}{\Gamma(\kappa-\alpha)}\frac{d^{\kappa}}{dt^{\kappa}}\int_{t}^{\infty}\frac{1}{(\tau-t)^{\alpha-\kappa+1}}f(\tau)\,d\tau. \tag{2.624}$$

Here, the Liouville-Weyl fractional derivatives and the Liouville-Weyl fractional integrals are called the Liouville-Weyl fractional calculus.

The properties of the Liouville-Weyl fractional derivatives and the Liouville-Weyl fractional integrals are represented as follows.

Property 2.42 *(see [236]) Let $Re(\alpha) > 0$ and $Re(\beta) > 0$. Then we have*

$$\left({}_{Cp}^{LW}I_{+}^{\alpha}\,{}_{Cp}^{LW}I_{+}^{\beta}f\right)(t) = \left({}_{Cp}^{LW}I_{+}^{\alpha+\beta}f\right)(t) \tag{2.625}$$

and

$$\left({}_{Cp}^{LW}I_{-}^{\alpha}\,{}_{Cp}^{LW}I_{-}^{\beta}f\right)(t) = \left({}_{Cp}^{LW}I_{-}^{\alpha+\beta}f\right)(t). \tag{2.626}$$

Property 2.43 *(see [236]) Let $Re(\alpha) > 0$. Then we have*

$$F\left\{\left({}_{Cp}^{LW}I_{+}^{\alpha}f\right)(t)\right\}(\omega) = \frac{f(\omega)}{(-i\omega)^{\alpha}} = \frac{e^{i(sgn\omega)\alpha\pi/2}}{|\omega|^{\alpha}}, \tag{2.627}$$

$$F\left\{\left({}_{Cp}^{LW}I_{-}^{\alpha}f\right)(t)\right\}(\omega) = \frac{f(\omega)}{(i\omega)^{\alpha}} = \frac{e^{-i(sgn\omega)\alpha\pi/2}}{|\omega|^{\alpha}}, \tag{2.628}$$

$$F\left\{\left({}_{Cp}^{LW}D_{+}^{\alpha}f\right)(t)\right\}(\omega) = (-i\omega)^{\alpha}f(\omega) = |\omega|^{\alpha}e^{-i(sgn\omega)\alpha\pi/2} \tag{2.629}$$

and

$$F\left\{\left({}_{Cp}^{LW}D_{-}^{\alpha}f\right)(t)\right\}(\omega) = (-i\omega)^{\alpha}f(\omega) = |\omega|^{\alpha}e^{i(sgn\omega)\alpha\pi/2}. \tag{2.630}$$

2.16 Kilbas-Srivastava-Trujillo fractional calculus with respect to another function

The concepts of Liouville-Weyl fractional derivatives and Liouville-Weyl fractional integrals with respect to another function, introduced by Kilbas, Srivastava and Trujillo in 2004 (see [152]), which are called the Kilbas-Srivastava-Trujillo fractional calculus with respect to another function in honor of Kilbas, Srivastava and Trujillo, are given as follows.

Definition 2.81 *Let $\alpha \in \mathbb{C}$, $Re(\alpha) > 0$, $-\infty < x < b < \infty$ and $h^{(1)}(t) > 0$. The left-sided Kilbas-Srivastava-Trujillo fractional integral with respect to another function is defined as (see [152])*

$$\left({}_{Cp}^{LW}I_{+,h}^{\alpha}f\right)(t) = \frac{1}{\Gamma(\alpha)}\int_{-\infty}^{t}\frac{h^{(1)}(\tau)}{(h(t)-h(\tau))^{1-\alpha}}f(\tau)\,d\tau, \tag{2.631}$$

and the right-sided Kilbas-Srivastava-Trujillo fractional integral with respect to another function is defined as (see [152])

$$\left({}_{Cp}^{LW}I_{-,h}^{\alpha}f\right)(t) = \frac{1}{\Gamma(\alpha)}\int_{t}^{\infty}\frac{h^{(1)}(\tau)}{(h(\tau)-h(t))^{1-\alpha}}f(\tau)\,d\tau. \tag{2.632}$$

Definition 2.82 *Let $\alpha \in \mathbb{C}$, $\kappa+1 > Re(\alpha) > \kappa$, $-\infty < x < b < \infty$ and $h^{(1)}(t) > 0$. The left-sided Kilbas-Srivastava-Trujillo fractional derivative with respect to another function is defined as (see [152])*

$$\left({}_{Cp}^{LW}D_{+,h}^{\alpha}f\right)(t) = \frac{1}{\Gamma(\kappa-\alpha)}\left(\frac{1}{h^{(1)}(t)}\frac{d}{dt}\right)^{\kappa}\int_{-\infty}^{t}\frac{h^{(1)}(t)}{(h(t)-h(\tau))^{\alpha-\kappa+1}}f(\tau)\,d\tau, \;. \tag{2.633}$$

and the right-sided Kilbas-Srivastava-Trujillo fractional derivative with respect to another function is defined as (see [152])

$$\left({}_{Cp}^{LW}D_{-,h}^{\alpha}f\right)(t) = \frac{1}{\Gamma(\kappa-\alpha)}\left(-\frac{1}{h^{(1)}(t)}\frac{d}{dt}\right)^{\kappa}\int_{t}^{\infty}\frac{h^{(1)}(\tau)}{(h(\tau)-h(t))^{\alpha-\kappa+1}}f(\tau)\,d\tau. \tag{2.634}$$

Property 2.44 *(see [152])*
Let $Re(\alpha) > 0$, $Re(\beta) > 0$ and $h^{(1)}(t) > 0$. Then we have (see [152])

$$\left({}_{Cp}^{LW}I_{+,h}^{\alpha}\,{}_{Cp}^{LW}I_{+,h}^{\beta}f\right)(t) = \left({}_{Cp}^{LW}I_{+,h}^{\alpha+\beta}f\right)(t) \tag{2.635}$$

and

$$\left({}_{Cp}^{LW}I_{-,h}^{\alpha}\,{}_{Cp}^{LW}I_{-,h}^{\beta}f\right)(t) = \left({}_{Cp}^{LW}I_{-,h}^{\alpha+\beta}f\right)(t). \tag{2.636}$$

Property 2.45 *(see [152])*
Let $Re(\alpha) > 0$, $f(t) \in L_{\kappa}(-\infty,\infty)$ and $h^{(1)}(t) > 0$. Then we have (see [152])

$$\left({}_{Cp}^{LW}D_{+,h}^{\alpha}\,{}_{Cp}^{LW}I_{+,h}^{\beta}f\right)(t) = f(t) \tag{2.637}$$

and

$$\left({}_{Cp}^{LW}D_{-,h}^{\alpha}\,{}_{Cp}^{LW}I_{-,h}^{\beta}f\right)(t) = f(t). \tag{2.638}$$

2.17 Sousa-de Oliveira fractional derivative with respect to another function

The conceptions of the Sousa-de Oliveira fractional derivatives with respect to another function in honor of Sousa and de Oliveira are presented as follows [292].

Definition 2.83 *Let $\kappa+1>\alpha>\kappa$, $-\infty<x<b$, $f(t)\in AC_\kappa(-\infty,\infty)$ and $h^{(1)}(t)>0$. The left-sided Liouville fractional derivative with respect to another function is defined as (see [292])*

$$\left({}_{Cp}^{L}D_{+,h}^{\alpha}f\right)(t)=\frac{1}{\Gamma(\kappa-\alpha)}\int_{-\infty}^{t}\frac{h^{(1)}(\tau)}{(h(t)-h(\tau))^{\alpha-\kappa+1}}\left(\frac{1}{h^{(1)}(\tau)}\frac{d}{d\tau}\right)^{\kappa}f_h^{(\kappa)}(\tau)\,d\tau. \tag{2.639}$$

Definition 2.84 *Let $\kappa+1>\alpha>\kappa$, $a<x<\infty$, $f(t)\in AC_\kappa(-\infty,\infty)$ and $h^{(1)}(t)>0$. The right-sided Liouville fractional derivative with respect to another function is defined as (see [292])*

$$\left({}_{Cp}^{L}D_{-,h}^{\alpha}f\right)(t)=\frac{1}{\Gamma(\kappa-\alpha)}\int_{t}^{\infty}\frac{h^{(1)}(\tau)}{(h(\tau)-h(t))^{\alpha-\kappa+1}}\left(-\frac{1}{h^{(1)}(\tau)}\frac{d}{d\tau}\right)^{\kappa}f_h^{(\kappa)}(\tau)\,d\tau. \tag{2.640}$$

Definition 2.85 *Let $\alpha\in\mathbb{C}$, $1>Re(\alpha)>0$, $-\infty<x<b$, $f(t)\in AC_1(-\infty,\infty)$ and $h^{(1)}(t)>0$. The left-sided Liouville fractional derivative with respect to another function is defined as*

$$\left({}_{Cp}^{L}D_{+,h}^{\alpha}f\right)(t)=\frac{1}{\Gamma(1-\alpha)}\int_{-\infty}^{t}\frac{h^{(1)}(\tau)}{(h(t)-h(\tau))^{\alpha}}\left(\frac{1}{h^{(1)}(\tau)}\frac{d}{d\tau}\right)f_h^{(1)}(\tau)\,d\tau. \tag{2.641}$$

Definition 2.86 *Let $1>Re(\alpha)>0$,$a<x<\infty$, $f(t)\in AC_1(-\infty,\infty)$ and $h^{(1)}(t)>0$. The right-sided Liouville fractional derivative with respect to another function is defined as*

$$\left({}_{Cp}^{L}D_{-,h}^{\alpha}f\right)(t)=\frac{1}{\Gamma(1-\alpha)}\int_{t}^{\infty}\frac{h^{(1)}(\tau)}{(h(\tau)-h(t))^{\alpha}}\left(-\frac{1}{h^{(1)}(\tau)}\frac{d}{d\tau}\right)f_h^{(1)}(\tau)\,d\tau. \tag{2.642}$$

Property 2.46 *Let $Re(\alpha)>0$, $f(t)\in AC_\kappa(-\infty,\infty)$ and $h^{(1)}(t)>0$. Then we have*

$$\left({}_{Cp}^{L}D_{+,h}^{\alpha}\,{}_{Cp}^{LW}I_{+,h}^{\beta}f\right)(t)=f(t) \tag{2.643}$$

and

$$\left({}_{Cp}^{L}D_{-,h}^{\alpha}\,{}_{Cp}^{LW}I_{-,h}^{\beta}f\right)(t)=f(t). \tag{2.644}$$

2.18 Liouville-Weyl fractional calculus with respect to another function in the Sense of Riesz, Feller and Richard types

The conceptions of the Liouville-Weyl fractional integrals, Liouville-Weyl fractional derivatives and Liouville fractional derivatives with respect to another function in the sense of Riesz, Feller and Richard types are given as follows.

2.18.1 Liouville-Weyl fractional integrals with respect to another function in the sense of Riesz, Feller and Richard types

Definition 2.87 *Let* $0 < \alpha < 1$, $f \in L_1(-\infty, \infty)$, $a = -\infty$, $b = \infty$ *and* $h^{(1)}(t) > 0$. *The fractional integral with respect to another function in the sense of Liouville-Weyl and Riesz types is defined as*

$$ {}_{Rz}^{Hi}I_{[-\infty,\infty],h}^{\alpha} f(t) = \frac{\left(I_{+,h}^{\alpha} f + I_{-,h}^{\alpha} f\right)(t)}{2\cos(\pi\alpha/2)}, \tag{2.645}$$

the Liouville-Weyl fractional integral with respect to another function in the sense of Riemann-Liouville and Feller types with respect to another function as

$$ {}_{Rz}^{Fe}I_{[-\infty,\infty],h}^{\alpha} f(t) = H_{-}(\vartheta,\alpha)\, I_{+,h}^{\alpha} f(t) + H_{+}(\vartheta,\alpha)\, I_{-,h}^{\alpha} f(t), \tag{2.646}$$

and the fractional Liouville-Weyl integral with respect to another function in the sense of Riemann-Liouville and Richard types as

$$ {}_{Rz}^{Ri}I_{[-\infty,\infty],h}^{\alpha} f(t) = \frac{\left(I_{+,h}^{\alpha} f + I_{-,h}^{\alpha} f\right)(t)}{2\sin(\pi\alpha/2)}. \tag{2.647}$$

2.18.2 Liouville-Weyl fractional derivatives with respect to another function in the sense of Riesz, Feller and Richard types

Definition 2.88 *Let* $0 < \alpha < 1$, $f \in L_1(-\infty, \infty)$ $a = -\infty$ *and* $b = \infty$. *The fractional derivative with respect to another function in the sense of Liouville-Weyl and Riesz types is defined as*

$$ {}_{Rz}^{Hi}D_{[-\infty,\infty],h}^{\alpha} f(t) = -\frac{\left(D_{+,h}^{\alpha} f + D_{-,h}^{\alpha} f\right)(t)}{2\cos(\pi\alpha/2)}, \tag{2.648}$$

the fractional derivative with respect to another function in the sense of Liouville-Weyl and Feller types with respect to another function as

$$ {}_{Rz}^{Fe}D_{[-\infty,\infty],h}^{\alpha} f(t) = -\left(H_{+}(\vartheta,\alpha) D_{+,h}^{\alpha} f(t) + H_{-}(\vartheta,\alpha) D_{-,h}^{\alpha} f(t)\right), \tag{2.649} $$

and the fractional derivative with respect to another function in the sense of Liouville-Weyl and Richard types as

$$ {}_{Rz}^{Ri}D_{[-\infty,\infty],h}^{\alpha} f(t) = -\frac{\left(D_{+,h}^{\alpha} f + D_{-,h}^{\alpha} f\right)(t)}{2\sin(\pi\alpha/2)}. \tag{2.650} $$

Definition 2.89 *Let* $0 < \alpha < 1$, $a = -\infty$, $b = \infty$ *and* $f \in AC_{\kappa}(-\infty,\infty)$. *The fractional derivative with respect to another function in the sense of Liouville and Riesz types is defined as*

$$ {}_{Rzv}^{Ri}D_{[-\infty,\infty],h}^{\alpha} f(t) = -\frac{\left({}_{LCS}D_{+,h}^{\alpha} f + {}_{LCS}D_{-,h}^{\alpha} f\right)(t)}{2\cos(\pi\alpha/2)}, \tag{2.651} $$

the fractional derivative with respect to another function in the sense of Liouville and Feller types as

$$ {}_{Rzv}^{Fe}D_{[-\infty,\infty],h}^{\alpha} f(t) = -\left(H_{+}(\vartheta,\alpha)\,{}_{LCS}D_{+,h}^{\alpha} f(t) + H_{-}(\vartheta,\alpha)\,{}_{LCS}D_{-,h}^{\alpha} f(t)\right), \tag{2.652} $$

and the fractional derivative with respect to another function in the sense of Liouville and Richard types as

$$ {}_{Rzv}^{Ri}D_{[-\infty,\infty],h}^{\alpha} f(t) = -\frac{\left({}_{LCS}D_{+,h}^{\alpha} f + {}_{LCS}D_{-,h}^{\alpha} f\right)(t)}{2\sin(\pi\alpha/2)}. \tag{2.653} $$

Definition 2.90 *Let* $0 < \alpha < 1$, $a = -\infty$, $b = \infty$ *and* $f \in AC_{\kappa}(-\infty,\infty)$. *The fractional derivative with respect to another function in the sense of Liouville and Riesz types is defined as*

$$ {}_{Rzv}^{Riv}D_{[-\infty,\infty],h}^{\alpha} f(t) = -\frac{\left({}_{LC}D_{+,h}^{\alpha} f + {}_{LC}D_{-,h}^{\alpha} f\right)(t)}{2\cos(\pi\alpha/2)}, \tag{2.654} $$

the fractional derivative with respect to another function in the sense of Liouville and Feller types as

$$ {}_{Rzv}^{Fev}D_{[-\infty,\infty],h}^{\alpha} f(t) = -\left(H_{+}(\vartheta,\alpha)\,{}_{LC}D_{+,h}^{\alpha} f(t) + H_{-}(\vartheta,\alpha)\,{}_{LC}D_{-,h}^{\alpha} f(t)\right), \tag{2.655} $$

and the fractional derivative with respect to another function in the sense of Liouville and Richard types as

$$ {}_{Rzv}^{Riv}D_{[-\infty,\infty],h}^{\alpha} f(t) = \frac{\left({}_{LC}D_{+,h}^{\alpha} f + {}_{LC}D_{-,h}^{\alpha} f\right)(t)}{2\sin(\pi\alpha/2)}. \tag{2.656} $$

Definition 2.91 *Let* $0 < Re(\alpha) < 1$, $f \in L_1(-\infty,\infty)$, $a = -\infty$, $b = \infty$ *and* $h^{(1)}(t) > 0$. *The fractional integral with respect to another function in the sense of Liouville-Weyl and Riesz types is defined as*

$$ {}^{Hi}_{Rz}I^{\alpha}_{[-\infty,\infty],h}f(t) = \frac{\left(I^{\alpha}_{+,h}f + I^{\alpha}_{-,h}f\right)(t)}{2\cos(\pi\alpha/2)}, \tag{2.657} $$

the Liouville-Weyl fractional integral with respect to another function in the sense of Riemann-Liouville and Feller types with respect to another function as

$$ {}^{Fe}_{Rz}I^{\alpha}_{[-\infty,\infty],h}f(t) = H_-(\vartheta,\alpha)\,I^{\alpha}_{+,h}f(t) + H_+(\vartheta,\alpha)\,I^{\alpha}_{-,h}f(t), \tag{2.658} $$

and the fractional Liouville-Weyl integral with respect to another function in the sense of Riemann-Liouville and Richard types as

$$ {}^{Ri}_{Rz}I^{\alpha}_{[-\infty,\infty],h}f(t) = \frac{\left(I^{\alpha}_{+,h}f + I^{\alpha}_{-,h}f\right)(t)}{2\sin(\pi\alpha/2)}. \tag{2.659} $$

Definition 2.92 *Let* $0 < Re(\alpha) < 1$, $f \in L_1(-\infty,\infty)$ $a = -\infty$ *and* $b = \infty$. *The fractional derivative with respect to another function in the sense of Liouville-Weyl and Riesz types is defined as*

$$ {}^{Hi}_{Rz}D^{\alpha}_{[-\infty,\infty],h}f(t) = -\frac{\left(D^{\alpha}_{+,h}f + D^{\alpha}_{-,h}f\right)(t)}{2\cos(\pi\alpha/2)}, \tag{2.660} $$

the fractional derivative with respect to another function in the sense of Liouville-Weyl and Feller types with respect to another function as

$$ {}^{Fe}_{Rz}D^{\alpha}_{[-\infty,\infty],h}f(t) = -\left(H_+(\vartheta,\alpha)\,D^{\alpha}_{+,h}f(t) + H_-(\vartheta,\alpha)\,D^{\alpha}_{-,h}f(t)\right), \tag{2.661} $$

and the fractional derivative with respect to another function in the sense of Liouville-Weyl and Richard types as

$$ {}^{Ri}_{Rz}D^{\alpha}_{[-\infty,\infty],h}f(t) = -\frac{\left(D^{\alpha}_{+,h}f + D^{\alpha}_{-,h}f\right)(t)}{2\sin(\pi\alpha/2)}. \tag{2.662} $$

Definition 2.93 *Let* $0 < Re(\alpha) < 1$, $a = -\infty$, $b = \infty$ *and* $f \in AC_{\kappa}(-\infty,\infty)$. *The fractional derivative with respect to another function in the sense of Liouville and Riesz types is defined as*

$$ {}^{Ri}_{Rzv}D^{\alpha}_{[-\infty,\infty],h}f(t) = -\frac{\left({}_{LCS}D^{\alpha}_{+,h}f + {}_{LCS}\,D^{\alpha}_{-,h}f\right)(t)}{2\cos(\pi\alpha/2)}, \tag{2.663} $$

the fractional derivative with respect to another function in the sense of Liouville and Feller types as

$$ {}_{Rzv}^{Fe}D^{\alpha}_{[-\infty,\infty],h} f(t) = -\left(H_{+}(\vartheta,\alpha)_{LCS}\, D^{\alpha}_{+,h} f(t) + H_{-}(\vartheta,\alpha)\, {}_{LCS}D^{\alpha}_{-,h} f(t)\right), \tag{2.664} $$

and the fractional derivative with respect to another function in the sense of Liouville and Richard types as

$$ {}_{Rzv}^{Ri}D^{\alpha}_{[-\infty,\infty],h} f(t) = -\frac{\left({}_{LCS}D^{\alpha}_{+,h} f +_{LCS} D^{\alpha}_{-,h} f\right)(t)}{2\sin(\pi\alpha/2)}. \tag{2.665} $$

Definition 2.94 *Let* $0 < Re(\alpha) < 1$, $a = -\infty$, $b = \infty$ *and* $f \in AC_{\kappa}(-\infty,\infty)$. *The fractional derivative with respect to another function in the sense of Liouville and Riesz types is defined as*

$$ {}_{Rzv}^{Riv}D^{\alpha}_{[-\infty,\infty],h} f(t) = -\frac{\left({}_{LC}D^{\alpha}_{+,h} f +_{LC} D^{\alpha}_{-,h} f\right)(t)}{2\cos(\pi\alpha/2)}, \tag{2.666} $$

the fractional derivative with respect to another function in the sense of Liouville and Feller types as

$$ {}_{Rzv}^{Fev}D^{\alpha}_{[-\infty,\infty],h} f(t) = -\left(H_{+}(\vartheta,\alpha)_{LC}\, D^{\alpha}_{+,h} f(t) + H_{-}(\vartheta,\alpha)_{LC}\, D^{\alpha}_{-,h} f(t)\right), \tag{2.667} $$

and the fractional derivative with respect to another function in the sense of Liouville and Richard types as

$$ {}_{Rzv}^{Riv}D^{\alpha}_{[-\infty,\infty],h} f(t) = \frac{\left({}_{LC}D^{\alpha}_{+,h} f +_{LC} D^{\alpha}_{-,h} f\right)(t)}{2\sin(\pi\alpha/2)}. \tag{2.668} $$

Definition 2.95 *Let* $0 < Re(\alpha) < 1$, $f \in L_1(-\infty,\infty)$, $a = -\infty$, $b = \infty$ *and* $h^{(1)}(t) > 0$. *The Tarasov type fractional derivative with respect to another function in the sense of Grünwald–Letnikov–Riesz type is defined as*

$$ {}_{Ta}D^{\alpha}_{t;[-\infty,\infty],h} f(t) = \frac{1}{2cos(\alpha\pi/2)}\left[{}_{M}^{V}D^{\alpha}_{+,h} f(t) +_{M}^{V} D^{\alpha}_{-,h} f(t)\right], \tag{2.669} $$

the Tarasov type fractional derivative with respect to another function in the sense of Grünwald–Letnikov–Feller type as

$$ {}_{Ta}D^{\alpha}_{t;[-\infty,\infty],h} f(t) = H_{+}(\vartheta,\alpha) \times {}_{M}^{V}D^{\alpha}_{+,h} f(t) + H_{-}(\vartheta,\alpha) \times {}_{M}^{V}D^{\alpha}_{-,h} f(t), \tag{2.670} $$

and the Tarasov type fractional derivative with respect to another function in the sense of Grünwald–Letnikov–Richard type as

$$ {}_{Ta}D^{\alpha}_{t;[-\infty,\infty],h} f(t) = -\frac{1}{2sin(\alpha\pi/2)}\left[{}_{M}^{V}D^{\alpha}_{+,h} f(t) + {}_{M}^{V}D^{\alpha}_{-,h} f(t)\right]. \tag{2.671} $$

Definition 2.96 *Let $0 < \alpha < 1$, $f \in L_1(-\infty, \infty)$, $a = -\infty$, $b = \infty$ and $h^{(1)}(t) > 0$. The Tarasov type fractional derivative with respect to another function in the sense of Grünwald–Letnikov–Riesz type is defined as*

$$ {}_{Ta}D^{\alpha}_{t;[-\infty,\infty],h} f(t) = \frac{1}{2cos(\alpha\pi/2)} \left[{}^{V}_{M}D^{\alpha}_{+,h} f(t) + {}^{V}_{M}D^{\alpha}_{-,h} f(t) \right], \tag{2.672} $$

the Tarasov type fractional derivative with respect to another function in the sense of Grünwald–Letnikov–Feller type as

$$ {}_{Ta}D^{\alpha}_{t;[-\infty,\infty],h} f(t) = H_+(\vartheta, \alpha) \times {}^{V}_{M} D^{\alpha}_{+,h} f(t) + H_-(\vartheta, \alpha) \times {}^{V}_{M} D^{\alpha}_{-,h} f(t), \tag{2.673} $$

and the Tarasov type fractional derivative with respect to another function in the sense of Grünwald–Letnikov–Richard type as

$$ {}_{Ta}D^{\alpha}_{t;[-\infty,\infty],h} f(t) = -\frac{1}{2sin(\alpha\pi/2)} \left[{}^{V}_{M}D^{\alpha}_{+,h} f(t) + {}^{V}_{M} D^{\alpha}_{-,h} f(t) \right]. \tag{2.674} $$

2.19 Hilfer derivatives

The conceptions of the fractional derivatives [292], corned in 2000 by Hilfer, which have the Riemann-Liouville and the Caputo derivatives as specific cases [293, 294] and the related topics are presented as follows.

2.19.1 Hilfer derivatives

Definition 2.97 *Let $0 < \alpha < 1$, $0 \le \beta \le 1$, $f \in L_1(a, b)$ and $-\infty < a < b < \infty$. The left-sided fractional derivative is defined as (see [292])*

$$ D^{\alpha,\beta}_{a+} f(t) = \left(I^{\beta(1-\alpha)}_{a+} \frac{d}{dt} \left(I^{(1-\beta)(1-\alpha)}_{a+} f \right) \right)(t), \tag{2.675} $$

where $I^{(1-\beta)(1-\alpha)}_{a+} f(t) \in AC_1(a, b)$, and the right-sided fractional derivative as

$$ D^{\alpha,\beta}_{b-} f(t) = \left(-I^{\beta(1-\alpha)}_{b-} \frac{d}{dt} \left(I^{(1-\beta)(1-\alpha)}_{b-} f \right) \right)(t), \tag{2.676} $$

where $I^{(1-\beta)(1-\alpha)}_{b-} f(t) \in AC_1(-\infty, \infty)$, which are called the left-sided and right-sided Hilfer derivatives (in honor of Hilfer), respectively.

Definition 2.98 *Let $\kappa < \alpha < 1 + \kappa$, $\kappa = [\alpha] + 1$, $0 \le \beta \le 1$, $f \in L_\kappa(a, b)$ and $-\infty < a < b < \infty$. The left-sided Hilfer fractional derivative is defined as*

$$ D^{\alpha,\beta,\kappa}_{a+} f(t) = \left(I^{\beta(\kappa-\alpha)}_{a+} \frac{d^\kappa}{dt^\kappa} \left(I^{(1-\beta)(\kappa-\alpha)}_{a+} f \right) \right)(t), \tag{2.677} $$

where $I_{a+}^{(1-\beta)(1-\alpha)} f(t) \in AC_\kappa(a,b)$, *and the right-sided Hilfer fractional derivative as*

$$D_{b-}^{\alpha,\beta,\kappa} f(t) = \left((-1)^\kappa I_{b-}^{\beta(\kappa-\alpha)} \frac{d^\kappa}{dt^\kappa} \left(I_{b-}^{(1-\beta)(\kappa-\alpha)} f \right) \right)(t), \tag{2.678}$$

where $I_{b-}^{(1-\beta)(1-\alpha)} f(t) \in AC_\kappa(a,b)$, *respectively.*

In this case, we set up the following relations:

Property 2.47 *Let* $0 < \alpha < 1$, $0 \le \beta \le 1$ *and* $f \in L_1(a,b)$. *Then we have*

$$D_{a+}^{\alpha,\beta} f(t) = D_{a+}^{\alpha,\beta,1} f(t) \tag{2.679}$$

and

$$D_{b-}^{\alpha,\beta} f(t) = D_{b-}^{\alpha,\beta,1} f(t). \tag{2.680}$$

Property 2.48 *(1) Suppose that* $0 < \alpha < 1$, $0 \le \beta \le 1$, $f \in L_1(a,b)$ *and* $I_{a+}^{(1-\beta)(1-\alpha)} f(t) \in AC_1(a,b)$, *then we have (see [293])*

$$L\left\{ D_{0+}^{\alpha,\beta} f(t) \right\}(s) = s^\alpha f(s) - s^{\beta(1-\alpha)} I_{0+}^{(1-\beta)(1-\alpha)} f(0). \tag{2.681}$$

(2) Suppose that $\kappa < \alpha < 1+\kappa$, $\kappa = [\alpha]+1$, $0 \le \beta \le 1$, $f \in L_\kappa(a,b)$, $-\infty < a < b < \infty$ *and* $I_{0+}^{(1-\beta)(1-\alpha)} f(t) \in AC_\kappa(a,b)$, *then we have*

$$L\left\{ D_{0+}^{\alpha,\beta,\kappa} f(t) \right\}(s) = s^\alpha f(s) - s^{\beta(\kappa-\alpha)} I_{0+}^{(1-\beta)(\kappa-\alpha)} f(0). \tag{2.682}$$

Definition 2.99 *Let* $0 < \alpha < 1$, $0 \le \beta \le 1$, $f \in L_1(-\infty,\infty)$, $a = -\infty$ *and* $b = \infty$. *The left-sided fractional derivative in the sense of Hilfer type can be represented as*

$$D_{+}^{\alpha,\beta} f(t) = \left(I_{+}^{\beta(1-\alpha)} \frac{d}{dt} \left(I_{+}^{(1-\beta)(1-\alpha)} f \right) \right)(t), \tag{2.683}$$

where $I_{+}^{(1-\beta)(1-\alpha)} f(t) \in AC_1(-\infty,\infty)$, *and the right-sided Hilfer fractional derivative in the sense of Hilfer type as*

$$D_{-}^{\alpha,\beta} f(t) = \left(-I_{-}^{\beta(1-\alpha)} \frac{d}{dt} \left(I_{-}^{(1-\beta)(1-\alpha)} f \right) \right)(t), \tag{2.684}$$

where $I_{-}^{(1-\beta)(1-\alpha)} f(t) \in AC_1(-\infty,\infty)$.

Definition 2.100 *Let* $\kappa < \alpha < 1+\kappa$, $\kappa = [\alpha]+1$, $0 \le \beta \le 1$, $f \in L_\kappa(-\infty,\infty)$, $a = -\infty$ *and* $b = \infty$. *The left-sided Hilfer fractional derivative can be represented as*

$$D_{+}^{\alpha,\beta,\kappa} f(t) = \left(I_{+}^{\beta(1-\alpha)} \frac{d^\kappa}{dt^\kappa} \left(I_{+}^{(1-\beta)(\kappa-\alpha)} f \right) \right)(t), \tag{2.685}$$

where $I_{+}^{(1-\beta)(\kappa-\alpha)} f(t) \in AC_{\kappa}(-\infty,\infty)$, *and the right-sided fractional derivative as*

$$D_{-}^{\alpha,\beta,\kappa} f(t) = \left((-1)^{\kappa} I_{-}^{\beta(1-\alpha)} \frac{d^{\kappa}}{dt^{\kappa}} \left(I_{-}^{(1-\beta)(\kappa-\alpha)} f \right) \right)(t), \tag{2.686}$$

where $I_{-}^{(1-\beta)(\kappa-\alpha)} f(t) \in AC_{\kappa}(-\infty,\infty)$.

Definition 2.101 *Let* $0 < \alpha < 1$, $0 \leq \beta \leq 1$, $f \in L_1(a,b)$, $-\infty < a < b < \infty$, $I_{a+}^{(1-\beta)(1-\alpha)} f(t) \in AC_1(a,b)$ *and* $I_{b-}^{(1-\beta)(1-\alpha)} f(t) \in AC_1(a,b)$. *The fractional derivative in the sense of Riesz and Hilfer types is defined as*

$${}_{Rz}^{Hi}D_{[a,b]}^{\alpha,\beta} f(t) = -\frac{\left(D_{a+}^{\alpha,\beta} f + D_{b-}^{\alpha,\beta} f \right)(t)}{2\cos(\pi\alpha/2)}, \tag{2.687}$$

the fractional derivative in the sense of Feller and Hilfer types as

$${}_{Rz}^{Fe}D_{[a,b]}^{\alpha,\beta} f(t) = -\frac{\sin((\alpha+\vartheta)\pi/2)}{\sin(\pi\vartheta)} \times D_{a+}^{\alpha,\beta} f(t) - \frac{\sin((\alpha-\vartheta)\pi/2)}{\sin(\pi\vartheta)} \times D_{b-}^{\alpha,\beta} f(t), \tag{2.688}$$

and the fractional derivative in the sense of Richard and Hilfer types as

$${}_{Rz}^{Ri}D_{[a,b]}^{\alpha,\beta} f(t) = -\frac{\left(D_{a+}^{\alpha,\beta} f + D_{b-}^{\alpha,\beta} f \right)(t)}{2\sin(\pi\alpha/2)}. \tag{2.689}$$

Definition 2.102 *Let* $0 < Re(\alpha) < 1$, $0 \leq \beta \leq 1$, $f \in L_1(a,b)$ *and* $-\infty < a < b < \infty$. *The left-sided Hilfer fractional derivative is defined as*

$$D_{a+}^{\alpha,\beta} f(t) = \left(I_{a+}^{\beta(1-\alpha)} \frac{d}{dt} \left(I_{a+}^{(1-\beta)(1-\alpha)} f \right) \right)(t), \tag{2.690}$$

where $I_{a+}^{(1-\beta)(1-\alpha)} f(t) \in AC_1(a,b)$, *and the right-sided Hilfer fractional derivative as*

$$D_{b-}^{\alpha,\beta} f(t) = \left(-I_{b-}^{\beta(1-\alpha)} \frac{d}{dt} \left(I_{b-}^{(1-\beta)(1-\alpha)} f \right) \right)(t), \tag{2.691}$$

where $I_{b-}^{(1-\beta)(1-\alpha)} f(t) \in AC_1(-\infty,\infty)$, *respectively.*

Definition 2.103 *Let* $\kappa < Re(\alpha) < 1+\kappa$, $\kappa = [\alpha]+1$, $0 \leq \beta \leq 1$, $f \in L_{\kappa}(a,b)$ *and* $-\infty < a < b < \infty$. *The left-sided Hilfer fractional derivative is defined as*

$$D_{a+}^{\alpha,\beta,\kappa} f(t) = \left(I_{a+}^{\beta(\kappa-\alpha)} \frac{d^{\kappa}}{dt^{\kappa}} \left(I_{a+}^{(1-\beta)(\kappa-\alpha)} \right) \right), \tag{2.692}$$

where $I_{a+}^{(1-\beta)(1-\alpha)} f(t) \in AC_{\kappa}(a,b)$, *and the right-sided Hilfer fractional derivative as*

$$D_{b-}^{\alpha,\beta,\kappa} f(t) = \left((-1)^{\kappa} I_{b-}^{\beta(\kappa-\alpha)} \frac{d^{\kappa}}{dt^{\kappa}} \left(I_{b-}^{(1-\beta)(\kappa-\alpha)} f \right) \right)(t), \tag{2.693}$$

where $I_{b-}^{(1-\beta)(1-\alpha)} f(t) \in AC_{\kappa}(a,b)$, *respectively.*

Definition 2.104 *Let $\kappa < Re(\alpha) < 1+\kappa$, $\kappa = [\alpha]+1$, $0 \le \beta \le 1$, $f \in L_\kappa(a,b)$, $a = -\infty$ and $b = \infty$. The left-sided Hilfer fractional derivative is defined as*

$$D_{+}^{\alpha,\beta,\kappa} f(t) = \left(I_{+}^{\beta(\kappa-\alpha)} \frac{d^\kappa}{dt^\kappa} \left(I_{+}^{(1-\beta)(\kappa-\alpha)} \right) \right), \tag{2.694}$$

where $I_{+}^{(1-\beta)(1-\alpha)} f(t) \in AC_\kappa(a,b)$, and the right-sided Hilfer fractional derivative as

$$D_{-}^{\alpha,\beta,\kappa} f(t) = \left((-1)^\kappa I_{-}^{\beta(\kappa-\alpha)} \frac{d^\kappa}{dt^\kappa} \left(I_{-}^{(1-\beta)(\kappa-\alpha)} f \right) \right)(t), \tag{2.695}$$

where $I_{-}^{(1-\beta)(1-\alpha)} f(t) \in AC_\kappa(a,b)$, respectively.

2.19.2 Sousa-de Oliveira fractional derivatives with respect to another function in the sense of Hilfer type

The conceptions of the Hilfer fractional derivative with respect to another function, which are called the Sousa-de Oliveira fractional derivatives with respect to another function, are presented as follows [295].

Definition 2.105 *Let $0 < \alpha < 1$, $0 \le \beta \le 1$, $f \in L_1(a,b)$, $-\infty < a < b < \infty$ and $h^{(1)}(t) > 0$. The left-sided Hilfer fractional derivative with respect to another function is defined as*

$$D_{a+,h}^{\alpha,\beta} f(t) = \left(I_{a+,h}^{\beta(1-\alpha)} \left(\frac{1}{h^{(1)}(t)} \frac{d}{dt} \right) \left(I_{a+,h}^{(1-\beta)(1-\alpha)} f \right) \right)(t), \tag{2.696}$$

where $I_{a+,h}^{(1-\beta)(1-\alpha)} f(t) \in AC_1(a,b)$, and the right-sided Hilfer fractional derivative with respect to another function as

$$D_{b-,h}^{\alpha,\beta} f(t) = \left(-I_{b-,h}^{\beta(1-\alpha)} \left(\frac{1}{h^{(1)}(t)} \frac{d}{dt} \right) \left(I_{b-,h}^{(1-\beta)(1-\alpha)} f \right) \right)(t), \tag{2.697}$$

where $I_{b-,h}^{(1-\beta)(1-\alpha)} f(t) \in AC_1(-\infty,\infty)$, respectively.

Definition 2.106 *Let $0 < Re(\alpha) < 1$, $0 \le Re(\beta) \le 1$, $f \in L_1(a,b)$, $-\infty < a < b < \infty$ and $h^{(1)}(t) > 0$. The left-sided Hilfer fractional derivative with respect to another function is defined as*

$$D_{a+,h}^{\alpha,\beta} f(t) = \left(I_{a+,h}^{\beta(1-\alpha)} \left(\frac{1}{h^{(1)}(t)} \frac{d}{dt} \right) \left(I_{a+,h}^{(1-\beta)(1-\alpha)} f \right) \right)(t), \tag{2.698}$$

where $I_{a+,h}^{(1-\beta)(1-\alpha)} f(t) \in AC_1(a,b)$, and the right-sided Hilfer fractional derivative with respect to another function as

$$D_{b-,h}^{\alpha,\beta} f(t) = \left(-I_{b-,h}^{\beta(1-\alpha)} \left(\frac{1}{h^{(1)}(t)} \frac{d}{dt} \right) \left(I_{b-,h}^{(1-\beta)(1-\alpha)} f \right) \right)(t), \tag{2.699}$$

where $I_{b-,h}^{(1-\beta)(1-\alpha)} f(t) \in AC_1(-\infty,\infty)$, respectively.

2.19.3 Riesz, Feller and Richard fractional derivatives with respect to another function via Hilfer fractional derivative

The conceptions of the Riesz, Feller and Richard fractional derivatives with respect to another function coupling with Hilfer fractional derivative are given as follows.

Definition 2.107 *Let* $0 < \alpha < 1$, $0 \le \beta \le 1$, $f \in L_1(a,b)$, $-\infty < a < b < \infty$, $I_{a+,h}^{(1-\beta)(1-\alpha)} f(t) \in AC_1(a,b)$ *and* $I_{b-,h}^{(1-\beta)(1-\alpha)} f(t) \in AC_1(a,b)$. *The fractional derivative in the sense of Riesz and Hilfer types with respect to another function is defined as*

$$ {}_{Rz}^{Hi}D_{[a,b],h}^{\alpha,\beta} f(t) = -\frac{\left(D_{a+,h}^{\alpha,\beta} f + D_{b-,h}^{\alpha,\beta} f\right)(t)}{2\cos(\pi\alpha/2)}, \tag{2.700} $$

the fractional derivative with respect to another function in the sense of Feller and Hilfer types as

$$ {}_{Rz}^{Fe}D_{[a,b],h}^{\alpha,\beta} f(t) = -\left(H_{+}(\vartheta,\alpha) D_{a+,h}^{\alpha,\beta} f(t) + H_{-}(\vartheta,\alpha) D_{b-,h}^{\alpha,\beta} f(t)\right), \tag{2.701} $$

and the fractional derivative with respect to another function in the sense of Richard and Hilfer types as

$$ {}_{Rz}^{Ri}D_{[a,b],h}^{\alpha,\beta} f(t) = -\frac{\left(D_{a+,h}^{\alpha,\beta} f + D_{b-,h}^{\alpha,\beta} f\right)(t)}{2\sin(\pi\alpha/2)}. \tag{2.702} $$

Definition 2.108 *Let* $0 < \alpha < 1$, $0 \le \beta \le 1$, $f \in L_1(-\infty,\infty)$, $a = -\infty$, $b = \infty$, $I_{+,h}^{(1-\beta)(1-\alpha)} f(t) \in AC_1(-\infty,\infty)$ *and* $I_{-,h}^{(1-\beta)(1-\alpha)} f(t) \in AC_1(-\infty,\infty)$. *The fractional derivative in the sense of Riesz and Hilfer types with respect to another function is defined as*

$$ {}_{Rz}^{Hi}D_{[-\infty,\infty],h}^{\alpha,\beta} f(t) = -\frac{\left(D_{+,h}^{\alpha,\beta} f + D_{-,h}^{\alpha,\beta} f\right)(t)}{2\cos(\pi\alpha/2)}, \tag{2.703} $$

the fractional derivative with respect to another function in the sense of Feller and Hilfer types as

$$ {}_{Rz}^{Fe}D_{[-\infty,\infty],h}^{\alpha,\beta} f(t) = -\left(H_{+}(\vartheta,\alpha) D_{+,h}^{\alpha,\beta} f(t) + H_{-}(\vartheta,\alpha) D_{-,h}^{\alpha,\beta} f(t)\right), \tag{2.704} $$

and the fractional derivative with respect to another function in the sense of Richard and Hilfer types as

$$ {}_{Rz}^{Ri}D_{[-\infty,\infty],h}^{\alpha,\beta} f(t) = -\frac{\left(D_{+,h}^{\alpha,\beta} f + D_{-,h}^{\alpha,\beta} f\right)(t)}{2\sin(\pi\alpha/2)}. \tag{2.705} $$

2.20 Applications

2.20.1 Relaxation equations within fractional derivatives

In view of the anomalous relaxation behaviors in nature, one may utilize the fractional derivatives to propose the mathematical models. Now we give the examples and the open problems for the anomalous relaxation in complex phenomenon.

Example 2.1 *The mathematical model of the relaxation within the Riemann-Liouville fractional derivative can be given as (see [152]; also see [301])*

$$\left(D_{0+}^{\alpha} f\right)(t) + \gamma f(t) = 0 \, (0 < \alpha < 1), \tag{2.706}$$

subjected to the initial condition

$$\left(I_{0+}^{1-\alpha} f\right)(0+) = h, \tag{2.707}$$

and Fourier-transformation solution is (for example, K. S. Cole and R. H. Cole introduced to the Cole-Cole model in 1941, see [302])

$$f(\omega) = \frac{h}{(i\omega)^{\alpha} + \gamma}, \tag{2.708}$$

which reduces to (see [152])

$$f(t) = ht^{\alpha-1} E_{\alpha,\alpha}\left(-\gamma t^{\alpha}\right), \tag{2.709}$$

where h and γ are the constants and the Hille-Tamarkin function (also called the $\alpha-$exponential function (see [152]), structured in 1930, can be written as (see [37])

$$e_{\alpha}^{\gamma t} = t^{\alpha-1} E_{\alpha,\alpha}\left(\gamma t^{\alpha}\right) = \frac{dE_{\alpha}\left(\gamma t^{\alpha}\right)}{dt}. \tag{2.710}$$

Example 2.2 *The mathematical model of the relaxation within the Liouville-Sonine fractional derivative can be written as (see [304])*

$$\left({}_{LS}D_{0+}^{\alpha} f\right)(t) + \gamma f(t) = 0 \, (0 < \alpha < 1), \tag{2.711}$$

subjected to the initial condition

$$f(0+) = h, \tag{2.712}$$

and the Laplace-transformation solution can be represented in the form [152]:

$$f(s) = \frac{h}{s^{\alpha} + \gamma}, \tag{2.713}$$

which reduces to [152]

$$f(t) = ht^{\alpha-1} E_{\alpha,\alpha}\left(-\gamma t^{\alpha}\right), \tag{2.714}$$

where h and γ are the constants.

Example 2.3 *The mathematical model of the relaxation within the Liouville-Grünwald-Letnikov fractional derivative can be written as*

$$ {}_{d}D_{0+}^{\alpha} f(t) + \gamma f(t) = h\delta(t) \quad (0 < \alpha < 1), \tag{2.715} $$

with the Laplace-transformation solution represented as

$$ f(s) = \frac{h}{s^{\alpha} + \gamma}, \tag{2.716} $$

which reduces to [152]

$$ f(t) = ht^{\alpha-1} E_{\alpha,\alpha}(-\gamma t^{\alpha}), \tag{2.717} $$

where h and γ are the constants.

Example 2.4 *The mathematical model of the relaxation within the Hilfer fractional derivative can be written as [292]*

$$ D_{0+}^{\alpha,\beta} f(t) + \gamma f(t) = 0 \quad (0 < \alpha < 1), \tag{2.718} $$

subjected to the condition

$$ I_{0+}^{(1-\alpha)(1-\beta)} = f(0), \tag{2.719} $$

and the Laplace transformation solution of the model is represented in the form

$$ f(s) = \frac{s^{\beta(1-\alpha)} f(0)}{s^{\alpha} + \gamma} = \frac{s^{\beta(1-\alpha)-\alpha} f(0)}{1 + \gamma s^{-\alpha}} \tag{2.720} $$

with the solution

$$ f(t) = f(0)\, t^{\alpha-\beta(1-\alpha)-1} E_{\alpha,\alpha-\beta(1-\alpha)}(-\gamma t^{\alpha}), $$

where $f(0)$ and γ are the constants.

There are some open problems for the mathematical model of the relaxation as follows.

Example 2.5 *The mathematical model of the relaxation within the Tarasov fractional derivative can be written as*

$$ {}_{Ta}D_{t;[a,b]}^{\alpha} f(t) + \gamma f(t) = 0 \quad (0 < \alpha < 1), \tag{2.721} $$

which reduces to

$$ \frac{\alpha}{2\Gamma(1-\alpha)\cos(\alpha\pi/2)} \left({}_{d}D_{a+}^{\alpha} f(t) + {}_{d}D_{b-}^{\alpha} f(t) \right) + \gamma f(t) = 0, \tag{2.722} $$

where γ is the constant.

Example 2.6 *The mathematical model of the relaxation within the Tarasov type fractional derivative of the Grünwald–Letnikov–Feller type can be written as*

$$ {}_{FV}D^{\alpha}_{t;[a,b];\vartheta}f(t) + \gamma f(t) = 0 \quad (0 < \alpha < 1), \tag{2.723}$$

which reduces to

$$ {}_{Ta}D^{\alpha}_{t;[a,b]}f(t) + \gamma f(t) = 0 \quad (0 < \alpha < 1), \tag{2.724}$$

where ${}_{FV}D^{\alpha}_{t;[a,b];0}f(t) ={}_{Ta} D^{\alpha}_{t;[a,b]}f(t)$, *and the mathematical model of the relaxation within the Tarasov type fractional derivative of the Grünwald–Letnikov–Richard type can be represented in the form:*

$$ {}_{L}D^{\alpha}_{t;[a,b]}f(t) + \gamma f(t) = 0 \quad (0 < \alpha < 1), \tag{2.725}$$

where ${}_{FV}D^{\alpha}_{t;[a,b];1}f(t) ={}_{L} D^{\alpha}_{t;[a,b]}f(t)$ *and* γ *is the constant.*

Example 2.7 *The mathematical model of the relaxation within the Erdélyi-Kober type fractional derivative is represented in the form:*

$$ \left(D^{\alpha}_{a+;\sigma,\eta}f\right)(t) + \gamma f(t) = 0 \quad (0 < \alpha < 1), \tag{2.726}$$

which yields that the mathematical model of the relaxation within the fractional derivative of Erdélyi-Kober-Riesz type is given as

$$ {}_{Rw}D^{\alpha}_{\mathbb{R}}f(t) + \gamma f(t) = 0 \quad (0 < \alpha < 1), \tag{2.727}$$

the mathematical model of the relaxation within the fractional derivative of Erdélyi-Kober-Feller type as

$$ {}_{Rv}I^{\alpha}_{\mathbb{R}}f(t) + \gamma f(t) = 0 \quad (0 < \alpha < 1), \tag{2.728}$$

and the mathematical model of the relaxation within the fractional derivative of Erdélyi-Kober-Richard type as

$$ {}_{Ro}D^{\alpha}_{\mathbb{R}}f(t) + \gamma f(t) = 0 \quad (0 < \alpha < 1), \tag{2.729}$$

where γ *is the constant.*

Example 2.8 *The mathematical model of the relaxation within the Katugampola fractional derivative can be presented as*

$$ \left({}_{K}D^{\alpha}_{a+;\sigma}f\right)(t) + \gamma f(t) = 0 \quad (0 < \alpha < 1), \tag{2.730}$$

where γ *is the constant.*

Example 2.9 *The mathematical model of the relaxation within the Hadamard type fractional derivative can be represented as*

$$ \left({}_{Hv}D^{\alpha}_{a+}f\right)(t) + \gamma f(t) = 0 \quad (0 < \alpha < 1), \tag{2.731}$$

where γ *is the constant.*

Example 2.10 *The mathematical models of the relaxation within the tempered fractional derivative are presented as*

$$\left({}_{Cp}^{RL}D_{a+}^{\alpha,\lambda}f\right)(t)+\gamma f(t)=0, \tag{2.732}$$

$$\left({}_{Cp}^{RL}D_{a+;h}^{\alpha,\lambda}f\right)(t)+\gamma f(t)=0, \tag{2.733}$$

$$\left({}_{Cp}^{LS}D_{a+}^{\alpha,\lambda}f\right)(t)+\gamma f(t)=0, \tag{2.734}$$

$$\left({}_{Cb}^{RL}D_{a+}^{\alpha,\gamma}f\right)(t)+\gamma f(t)=0, \tag{2.735}$$

$$\left({}_{Cc}^{RL}D_{a+}^{\alpha,\gamma}f\right)(t)+\gamma f(t)=0, \tag{2.736}$$

$$\left({}_{Vv}^{LS}D_{a+}^{\alpha,\lambda,\gamma}f\right)(t)+\gamma f(t)=0, \tag{2.737}$$

$$\left({}_{Vcd}^{LS}D_{a+}^{\alpha,\lambda,\gamma}f\right)(t)+\gamma f(t)=0, \tag{2.738}$$

$$\left({}_{YV}^{RL}D_{a+}^{\alpha,\lambda,\gamma}f\right)(t)+\gamma f(t)=0, \tag{2.739}$$

$$\left({}_{Cs}D_{a+}^{\alpha,\lambda}f\right)(t)+\gamma f(t)=0, \tag{2.740}$$

$$\left({}_{Ch}D_{a+}^{\alpha,\lambda}f\right)(t)+\gamma f(t)=0 \tag{2.741}$$

and

$$\left({}_{Ev}^{LS}D_{a+}^{\alpha,\beta,\lambda}f\right)(t)+\gamma f(t)=0, \tag{2.742}$$

where γ is the constant and α $(0<\alpha<1)$.

Example 2.11 *The mathematical model of the relaxation within the Sousa-de Oliveira fractional derivative with respect to another function is represented as*

$$D_{a+,h}^{\alpha,\beta}f(t)+\gamma f(t)=0\ \ (0<\alpha<1), \tag{2.743}$$

where γ is the constant.

Example 2.12 *The mathematical model of the relaxation within the Riesz fractional derivative with respect to another function is written as*

$${}_{Rz}^{Hi}D_{[a,b],h}^{\alpha,\beta}f(t)+\gamma f(t)=0\ \ (0<\alpha<1), \tag{2.744}$$

the mathematical model of the relaxation within the Feller fractional derivative with respect to another function as

$${}_{Rz}^{Fe}D_{[a,b],h}^{\alpha,\beta}f(t)+\gamma f(t)=0\ \ (0<\alpha<1), \tag{2.745}$$

the mathematical model of the relaxation within the Richard fractional derivative with respect to another function as

$${}_{Rz}^{Ri}D_{[a,b],h}^{\alpha,\beta}f(t)+\gamma f(t)=0\ \ (0<\alpha<1), \tag{2.746}$$

the mathematical model of the relaxation within the Riemann-Liouville fractional derivative with respect to another function as

$$\left(D_{a+,h}^{\alpha} f\right)(t) + \gamma f(t) = 0 \ \ (0 < \alpha < 1), \tag{2.747}$$

and the mathematical model of the relaxation within the Almeida fractional derivatives with respect to another function as

$$\left({}_{LS}D_{a+,h}^{\alpha} f\right)(t) + \gamma f(t) = 0 \ \ (0 < \alpha < 1), \tag{2.748}$$

where γ is the constant.

2.20.2 Rheological models within fractional derivative

The Nutting phenomenon, first proposed in 1921 by Nutting, was considered model to the elastic viscous deformation (see [339, 340]). In order to solve the above problems, the mathematical model for the deformation (see [338]), proposed by Blair in 1947, which is called the Scott Blair model in honor of the G.W. Scott Blair, was developed by many scientists, for example, Bagley and Torvik (see [139]), Mainardi and Spada (see [341]), Stiassnie (see [342]) and so on. Gerasimov reported the general law of the deformation based on the Liouville fractional derivative (see [269]) in 1948. The mathematical model via Liouville-Sonine fractional derivative was proposed by Smit and De Vries in 1970 and by Caputo and Mainardi in 1971. However, the Liouville and Sonine work of the study in the operator is not noticed in the results (see [36, 68, 270, 271]).

We now present the proposed models for the law of deformation and some of the open problems of the mathematical models for the law of deformation as follows.

Example 2.13 *The Scott Blair model for the law of deformation via Riemann-Liouville (or Riemann) fractional derivative is represented in the form [338]:*

$$\sigma(t) = \xi\left(D_{0+}^{\alpha}\varepsilon\right)(t) \ \ (0 < \alpha < 1), \tag{2.749}$$

which can be written as

$$\sigma(t) = \xi \frac{1}{\Gamma(1-\alpha)} \frac{d}{dt} \int_{0}^{t} \frac{\varepsilon(\tau)}{(\tau - t)^{\alpha}} d\tau \ \ (0 < \alpha < 1), \tag{2.750}$$

where α $(0 < \alpha < 1)$ and ξ are the material constants, $\sigma(t)$ is the stress, and $\varepsilon(t)$ is the strain.

Example 2.14 *The Gerasimov model for the law of deformation via Liouville fractional derivative can be written as (also see [269])*

$$\sigma(t) = \xi_L D_{-\infty}^{\alpha}\varepsilon(t) \ \ (0 < \alpha < 1), \tag{2.751}$$

which becomes

$$\sigma(t) = \frac{\xi}{\Gamma(1-\alpha)} \int\limits_{-\infty}^{t} \frac{\varepsilon^{(1)}(\tau)}{(t-\tau)^{\alpha}} d\tau \quad (0 < \alpha < 1), \tag{2.752}$$

and the Gerasimov model for the law of deformation via Liouville-Weyl fractional derivative as (also see [269]; also see [150])

$$\sigma(t) = \xi \left({}^{LW}_{Cp} D^{\alpha}_{-} \varepsilon \right)(t) \quad (0 < \alpha < 1), \tag{2.753}$$

which is

$$\sigma(t) = \frac{\xi}{\Gamma(1-\alpha)} \frac{d}{dt} \int\limits_{-\infty}^{t} \frac{1}{(\tau - t)^{\alpha}} \varepsilon(\tau)\, d\tau, \tag{2.754}$$

where α $(0 < \alpha < 1)$ *and* ξ *are the material constants,* $\sigma(t)$ *is the stress, and* $\varepsilon(t)$ *is the strain.*

Example 2.15 *The mathematical model for the law of deformation via Liouville-Sonine fractional derivative is presented as (see [36, 68, 270, 271])*

$$\sigma(t) = \frac{\xi}{\Gamma(1-\alpha)} \int\limits_{0}^{t} \frac{\varepsilon^{(1)}(\tau)}{(t-\tau)^{\alpha}} d\tau, \tag{2.755}$$

where α $(0 < \alpha < 1)$ *and* ξ *are the material constants,* $\sigma(t)$ *is the stress, and* $\varepsilon(t)$ *is the strain.*

Example 2.16 *The mathematical model for the law of deformation via Hilfer fractional derivative is represented as*

$$\sigma(t) = \xi \left(I_{0+}^{\beta(1-\alpha)} \frac{d}{dt} \left(I_{0+}^{(1-\beta)(1-\alpha)} \varepsilon \right) \right)(t) \quad (0 < \alpha < 1), \tag{2.756}$$

the mathematical model for the law of deformation via Sousa-de Oliveira fractional derivative with respect to another function as

$$\sigma(t) = \xi \left(I_{0+,h}^{\beta(1-\alpha)} \frac{d}{dt} \left(I_{0+,h}^{(1-\beta)(1-\alpha)} \varepsilon \right) \right)(t) \quad (0 < \alpha < 1), \tag{2.757}$$

and the mathematical model for the law of deformation via Hilfer fractional derivative with respect to another function in the sense of Liouville-Weyl type as

$$\sigma(t) = \xi \left(I_{+,h}^{\beta(1-\alpha)} \frac{d}{dt} \left(I_{+,h}^{(1-\beta)(1-\alpha)} \varepsilon \right) \right)(t) \quad (0 < \alpha < 1), \tag{2.758}$$

where α $(0 < \alpha < 1)$ *and* ξ *are the material constants,* $h(t)$ *is the basic function,* $\sigma(t)$ *is the stress, and* $\varepsilon(t)$ *is the strain.*

Example 2.17 *The Colombaro-Garra-Giusti-Mainardi model for the law of deformation via Almeida fractional derivative with respect to another function in the sense of Liouville-Sonine type is presented as (see [343])*

$$\sigma(t) = \frac{\xi}{\Gamma(1-\alpha)} \int_0^t \frac{h^{(1)}(t)}{(h(t)-h(\tau))^{\alpha}} \left(\frac{1}{h^{(1)}(\tau)} \varepsilon_h^{(1)}(\tau) \right) d\tau, \tag{2.759}$$

where α $(0<\alpha<1)$*and* ξ *are the material constants,* $h(t)$ *is the basic function,* $\sigma(t)$ *is the stress, and* $\varepsilon(t)$ *is the strain.*

Example 2.18 *The Garra-Mainardi model for the law of deformation via Riemann-Liouville fractional derivative with respect to another function is presented as (see [344])*

$$\sigma(t) = \frac{\xi}{\Gamma(1-\alpha) h^{(1)}(t)} \frac{d}{dt} \int_0^t \frac{h^{(1)}(t)}{(h(t)-h(\tau))^{\alpha}} \varepsilon(\tau) d\tau, \tag{2.760}$$

where α $(0<\alpha<1)$ *and* ξ *are the material constants,* $h(t)$ *is the basic function,* $\sigma(t)$ *is the stress, and* $\varepsilon(t)$ *is the strain.*

Example 2.19 *The mathematical model for the law of deformation via Liouville fractional derivative with respect to another function is represented as*

$$\sigma(t) = \frac{\xi}{\Gamma(1-\alpha)} \int_{-\infty}^t \frac{h^{(1)}(t)}{(h(t)-h(\tau))^{\alpha}} \left(\frac{1}{h^{(1)}(\tau)} \varepsilon_h^{(1)}(\tau) \right) d\tau, \tag{2.761}$$

and the mathematical model for the law of deformation via Kilbas-Srivastava-Trujillo fractional derivative with respect to another function as

$$\sigma(t) = \frac{\xi}{\Gamma(1-\alpha) h^{(1)}(t)} \frac{d}{dt} \int_{-\infty}^t \frac{h^{(1)}(t)}{(h(t)-h(\tau))^{\alpha}} \varepsilon(\tau) d\tau, \tag{2.762}$$

where α $(0<\alpha<1)$ *and* ξ *are the material constants,* $h(t)$ *is the basic function,* $\sigma(t)$ *is the stress, and* $\varepsilon(t)$ *is the strain.*

Example 2.20 *The mathematical model for the law of deformation via tempered fractional derivative in the sense of Riemann-Liouville type is represented as*

$$\sigma(t) = \frac{\xi}{\Gamma(1-\alpha)} \frac{d}{dt} \int_0^t \frac{\varepsilon(\tau)}{(t-\tau)^{\alpha}} e^{-\lambda(t-\tau)} d\tau, \tag{2.763}$$

the mathematical model for the law of deformation via tempered fractional derivative in the sense of Liouville-Sonine type as

$$\sigma(t) = \frac{\xi}{\Gamma(1-\alpha)} \int_0^t \frac{e^{-\lambda(t-\tau)}}{(t-\tau)^{\alpha}} \varepsilon^{(1)}(\tau) d\tau, \tag{2.764}$$

and the mathematical model for the law of deformation via tempered fractional derivative with respect to another function in the sense of Riemann-Liouville type as

$$\sigma(t) = \frac{\xi}{\Gamma(1-\alpha)} \left(\frac{1}{h^{(1)}(t)} \frac{d}{dt} \right) \int_{-\infty}^{t} \frac{e^{-\lambda(h(t)-h(\tau))} h^{(1)}(\tau)}{(h(t)-h(\tau))^{\alpha}} \varepsilon(\tau)\, d\tau, \qquad (2.765)$$

where α $(0 < \alpha < 1)$ *and* ξ *are the material constants,* $h(t)$ *is the basic function,* $\sigma(t)$ *is the stress, and* $\varepsilon(t)$ *is the strain.*

Chapter 3

General Fractional Derivatives of Constant Order and Applications

In this chapter, we present the general fractional derivatives and integrals of constant order involving the kernel of the special functions, such as the Mittag-Leffler function, Kohlrausch-Williams-Watts function, Wiman function, Prabhakar function, Gorenflo-Mainardi function, Miller-Ross function, Lorenzo-Hartley function, subcosine function and subsine function. The general fractional derivatives and the general fractional integrals with respect to another function are also presented. The general fractional derivatives are also considered to describe the relaxation and rheological models in complex phenomena.

3.1 General fractional derivatives involving the kernel of Mittag-Leffler function

In this section, the family of the general fractional derivatives involving the kernel of Mittag-Leffler function is introduced in detail.

3.1.1 General fractional derivatives involving the kernel of Mittag-Leffler function without power law

Definition 3.1 *Let $\kappa \geq 0$, $\nu \in \mathbb{C}$, $\lambda \in \mathbb{C}$ and $Re(\nu) > 0$. The left-sided generalized derivative involving the kernel of Mittag-Leffler function in the sense of Liouville-Sonine-Caputo type is defined as*

$$ {}_{Ml}^{LSC}D_{a+}^{\kappa,\lambda;\nu} f(t) = \int_{a}^{t} E_{\nu}\left(-\lambda\left(t-\tau\right)\right) f^{(\kappa)}\left(\tau\right) d\tau, \tag{3.1}$$

and the right-sided generalized derivative involving the kernel of Mittag-Leffler function in the sense of Liouville-Sonine-Caputo type as

$$ {}_{Ml}^{LSC}D_{b-}^{\kappa,\lambda;\nu} f(t) := (-1)^{\kappa} \int_t^b E_\nu\left(-\lambda(\tau - t)\right) f^{(\kappa)}(\tau)\, d\tau. \tag{3.2} $$

Definition 3.2 *The left-sided generalized derivative involving the kernel of Mittag-Leffler function in the sense of Riemann-Liouville type is defined as*

$$ {}_{Ml}^{RL}D_{a+}^{\kappa,\lambda;\nu} f(t) = \frac{d^{\kappa}}{dt^{\kappa}} \int_a^t E_\nu\left(-\lambda(t - \tau)\right) f(\tau)\, d\tau, \tag{3.3} $$

and the right-sided generalized derivative involving the kernel of Mittag-Leffler function in the sense of Riemann-Liouville type is defined as

$$ {}_{Ml}^{RL}D_{b-}^{\kappa,\lambda;\nu} f(t) = (-1)^{\kappa} \frac{d^{\kappa}}{dt^{\kappa}} \int_t^b E_\nu\left(-\lambda(\tau - t)\right) f(\tau)\, d\tau. \tag{3.4} $$

In particular, when $\kappa = 1$, $\nu \in \mathbb{C}$, $\lambda \in \mathbb{C}$ and $Re(\nu) > 0$, the left-sided generalized derivative involving the kernel of Mittag-Leffler function in the sense of Liouville-Sonine type is defined as (see [299])

$$ {}_{Ml}^{LSC}D_{a+}^{1,\lambda;\nu} f(t) = \int_a^t E_\nu\left(-\lambda(t - \tau)\right) f^{(1)}(\tau)\, d\tau, \tag{3.5} $$

and the right-sided generalized derivative involving the kernel of Mittag-Leffler function in the sense of Liouville-Sonine type as

$$ {}_{Ml}^{LSC}D_{b-}^{1,\lambda;\nu} f(t) = -\int_t^b E_\nu\left(-\lambda(\tau - t)\right) f^{(1)}(\tau)\, d\tau. \tag{3.6} $$

On the other hand, the left-sided generalized derivative involving the kernel of Mittag-Leffler function in the sense of Riemann-Liouville type is defined as

$$ {}_{Ml}^{RL}D_{a+}^{1,\lambda;\nu} f(t) = \frac{d}{dt} \int_a^t E_\nu\left(-\lambda(t - \tau)\right) f(\tau)\, d\tau, \tag{3.7} $$

and the right-sided generalized derivative involving the kernel of Mittag-Leffler function in the sense of Riemann-Liouville type is defined as

$$ {}_{Ml}^{RL}D_{b-}^{1,\lambda;\nu} f(t) = -\frac{d}{dt} \int_t^b E_\nu\left(-\lambda(\tau - t)\right) f(\tau)\, d\tau. \tag{3.8} $$

3.1.2 General fractional derivatives involving the kernel of Mittag-Leffler-Gauss function with power law

For $J(\alpha) = \frac{1+\alpha^2}{\sqrt{\pi^\alpha(1-\alpha)}}$ and $\lambda = \frac{\alpha}{1-\alpha}$, the left-sided general fractional derivative involving the kernel of Mittag-Leffler-Gauss function with power law in the sense of Liouville-Sonine-Caputo type is defined as (see [299])

$$\begin{aligned} & {}^{LSC}_{Mle}D^{\alpha,\kappa,\lambda}_{a+} f(t) \\ & = J(\beta) \int_a^t E_\alpha\left(-\lambda (t-\tau)^{2\alpha}\right) f^{(\kappa)}(\tau)\, d\tau \\ & = \frac{1+\alpha^2}{\sqrt{\pi^\alpha(1-\alpha)}} \int_a^t E_\alpha\left(-\frac{\alpha}{1-\alpha}(t-\tau)^{2\alpha}\right) f^{(\kappa)}(\tau)\, d\tau \end{aligned} . \tag{3.9}$$

Definition 3.3 *Let $\alpha \in \mathbb{R}$, $\kappa = [\alpha] + 1$, $\kappa < \alpha < \kappa + 1$ and $\lambda \in \mathbb{C}$. The left-sided general fractional derivative involving the kernel of Mittag-Leffler-Gauss function with power law in the sense of Liouville-Sonine-Caputo type is defined as*

$$ {}^{LSC}_{Mlt}D^{\alpha,\kappa,\lambda}_{a+} f(t) = \int_a^t E_{2\alpha}\left(-\lambda (t-\tau)^{2\alpha}\right) f^{(\kappa)}(\tau)\, d\tau, \tag{3.10}$$

and the right-sided general fractional derivative involving the kernel of Mittag-Leffler-Gauss function with power law in the sense of Liouville-Sonine-Caputo type as

$$ {}^{LSC}_{Mlt}D^{\alpha,\kappa,\lambda}_{b-} f(t) = (-1)^\kappa \int_t^b E_{2\alpha}\left(-\lambda (\tau-t)^{2\alpha}\right) f^{(\kappa)}(\tau)\, d\tau. \tag{3.11}$$

Definition 3.4 *The left-sided general fractional derivative involving the kernel of Mittag-Leffler-Gauss function with power law in the sense of Riemann-Liouville type is defined as*

$$ {}^{RL}_{Mlt}D^{\alpha,\kappa,\lambda}_{a+} f(t) = \frac{d^\kappa}{dt^\kappa} \int_a^t E_{2\alpha}\left(-\lambda (t-\tau)^{2\alpha}\right) f(\tau)\, d\tau, \tag{3.12}$$

and the right-sided general fractional derivative involving the kernel of Mittag-Leffler-Gauss function with power law in the sense of Riemann-Liouville type is defined as

$$ {}^{RL}_{Mlt}D^{\alpha,\kappa,\lambda}_{b-} f(t) = (-1)^\kappa \frac{d^\kappa}{dt^\kappa} \int_t^b E_{2\alpha}\left(-\lambda (\tau-t)^{2\alpha}\right) f(\tau)\, d\tau. \tag{3.13}$$

For $\kappa = 1$ we have:

$$ {}^{LSC}_{Mlt}D^{\alpha,\kappa,\lambda}_{a+} f(t) = {}^{RL}_{Mlt}D^{\alpha,\kappa,\lambda}_{a+} f(t) + E_{2\alpha}\left(-\lambda (t-a)^{2\alpha}\right) f(a) \tag{3.14}$$

and

$$ {}^{LSC}_{Mlt}D^{\alpha,\kappa,\lambda}_{b-}f(t) = {}^{RL}_{Mlt}D^{\alpha,\kappa,\lambda}_{b-}f(t) + E_{2\alpha}\left(-\lambda (b-t)^{2\alpha}\right) f(b). \tag{3.15} $$

Definition 3.5 *Let $\alpha > 0$, $1+\kappa > \alpha > \kappa$, $\kappa = [\alpha]+1$, $-\infty < a < b < \infty$ and $\lambda \in \mathbb{C}$. The left-sided general fractional integral, as the inverse operator of the general fractional derivative involving the kernel of Mittag-Leffler-Gauss function with power law, is defined as*

$$ {}^{RL}_{Mlt}I^{\alpha,\kappa,\lambda}_{a+}f(t) = \int_a^t \frac{1}{(t-\tau)^{2\alpha+1-\kappa}} E^{-1}_{2\alpha,\kappa-2\alpha}\left(-\lambda (t-\tau)^{2\alpha}\right) f(\tau)\, d\tau, \tag{3.16} $$

and the right-sided general fractional integral as

$$ {}^{RL}_{Mlt}I^{\alpha,\kappa,\lambda}_{b-}f(t) = (-1)^{\kappa} \int_t^b \frac{1}{(\tau-t)^{2\alpha+1-\kappa}} E^{-1}_{2\alpha,\kappa-2\alpha}\left(-\lambda (\tau-t)^{2\alpha}\right) f(\tau)\, d\tau. \tag{3.17} $$

Property 3.1 *Let $\alpha > 0$, $1+\kappa > \alpha > \kappa$, $\kappa = [\alpha]+1$, $-\infty < a < b < \infty$ and $\lambda \in \mathbb{C}$. Then we have*

$$ L\left\{ {}^{RL}_{Mlt}D^{\alpha,\kappa,\lambda}_{0+}f(t)\right\} = s^{\kappa-1}\left(1+\lambda s^{-2\alpha}\right)^{-1} f(s), \tag{3.18} $$

$$ L\left\{ {}^{LSC}_{Mlt}D^{\alpha,\kappa,\lambda}_{0+}f(t)\right\} = s^{-1}\left(1+\lambda s^{-2\alpha}\right)^{-1}\left(s^{\kappa} f(s) - \sum_{j=1}^{\kappa} s^{\kappa-j} f^{(j-1)}(0)\right) \tag{3.19} $$

and

$$ L\left\{ {}^{RL}_{Mlt}I^{\alpha,\kappa,\lambda}_{0+}f(t)\right\} = s^{1-\kappa}\left(1+\lambda s^{-2\alpha}\right) f(s). \tag{3.20} $$

Note that, when $D^{\kappa}\left({}^{RL}_{Mlt}I^{\alpha,\kappa,\lambda}_{0+}f(t)\right)|_{t=0} = 0$, there is

$$ L\left\{ {}^{RL}_{Mlt}D^{\alpha,\kappa,\lambda}_{0+}f(t)\right\} = s^{\kappa-1}\left(1+\lambda s^{-2\alpha}\right)^{-1} f(s). \tag{3.21} $$

and when $D^{\kappa}\left({}^{RL}_{Mlt}I^{\alpha,\kappa,\lambda}_{0+}f(t)\right)|_{t=0} \neq 0$, there is

$$ \begin{aligned} L\left\{ {}^{RL}_{Mlt}D^{\alpha,\kappa,\lambda}_{0+}f(t)\right\} &= s^{\kappa-1}\left(1+\lambda s^{-2\alpha}\right)^{-1} f(s) \\ &\quad - \sum_{\kappa=0}^{\kappa-1} s^{\kappa-\mu-1}\left(D^{\mu}\left[{}^{RL}_{Mlt}I^{\alpha,\kappa,\lambda}_{0+}f(0+)\right]\right). \end{aligned} \tag{3.22} $$

In particular, when $\kappa = 1$, we have

$$ L\left\{ {}^{RL}_{Mlt}D^{\alpha,1,\lambda}_{0+}f(t)\right\} = \left(1+\lambda s^{-2\alpha}\right)^{-1} f(s) - {}^{RL}_{Mlt}I^{\alpha,1,\lambda}_{0+}f(0+). \tag{3.23} $$

Property 3.2 *Let $\alpha > 0$, $1+\kappa > \alpha > \kappa$, $\kappa = [\alpha]+1$, $-\infty < a < b < \infty$, $\lambda \in \mathbb{C}$ and $f(t) \in L_l(a,b)$ $(1 \le l < \infty)$. Then we have*

$$\left({}^{RL}_{Mlt}D^{\alpha,\kappa,\lambda}_{a+}\,{}^{RL}_{Mlt}I^{\alpha,\kappa,\lambda}_{a+}f\right)(t) = f(t), \tag{3.24}$$

and

$$\left({}^{RL}_{Mlt}D^{\alpha,\kappa,\lambda}_{+}\,{}^{RL}_{Mlt}I^{\alpha,\kappa,\lambda}_{+}f\right)(t) = f(t). \tag{3.25}$$

Property 3.3 *Let $\alpha > 0$, $1+\kappa > \alpha > \kappa$, $\kappa = [\alpha]+1$, $-\infty < a < b < \infty$, $\lambda \in \mathbb{C}$ and $f(t) \in AC^{\kappa}(a,b)$. Then we have*

$$\left({}^{LCS}_{Mlt}D^{\alpha,\kappa,\lambda}_{a+}\,{}^{RL}_{Mlt}I^{\alpha,\kappa,\lambda}_{a+}f\right)(t) = f(t), \tag{3.26}$$

and

$$\left({}^{LCS}_{Mlt}D^{\alpha,\kappa,\lambda}_{b-}\,{}^{RL}_{Mlt}I^{\alpha,\kappa,\lambda}_{b-}f\right)(t) = f(t). \tag{3.27}$$

3.1.3 General fractional derivatives involving the kernel of Mittag-Leffler function with power law

The definitions and properties of the general fractional derivatives involving the kernel of Mittag-Leffler function with power law are given as follows.

Definition 3.6 *Let $\alpha \in \mathbb{R}$, $\kappa = [\alpha]+1$, $\kappa < \alpha < \kappa+1$, $\upsilon \in \mathbb{C}$, $\lambda \in \mathbb{C}$ and $Re(\upsilon) > 0$. The left-sided general fractional derivative involving the kernel of Mittag-Leffler function with power law in the sense of Liouville-Sonine-Caputo type is defined as (for $\lambda = \pm 1$, see [116])*

$${}^{LSC}_{Ml}D^{\alpha,\kappa,\lambda}_{a+}f(t) = \int_a^t E_\alpha\left(-\lambda(t-\tau)^\alpha\right) f^{(\kappa)}(\tau)\,d\tau, \tag{3.28}$$

and the right-sided general fractional derivative involving the kernel of Mittag-Leffler function with power law in the sense of Liouville-Sonine-Caputo type as (for $\lambda = \pm 1$, see [116])

$${}^{LSC}_{Ml}D^{\alpha,\kappa,\lambda}_{b-}f(t) = (-1)^\kappa \int_t^b E_\alpha\left(-\lambda(\tau-t)^\alpha\right) f^{(\kappa)}(\tau)\,d\tau. \tag{3.29}$$

The left-sided general fractional derivative involving the kernel of Mittag-Leffler function with power law in the sense of Riemann-Liouville type is defined as (for $\lambda = \pm 1$, see [116, 301] and for $\kappa = 1$, see [37])

$${}^{RL}_{Ml}D^{\alpha,\kappa,\lambda}_{a+}f(t) = \frac{d^\kappa}{dt^\kappa}\int_a^t E_\alpha\left(-\lambda(t-\tau)^\alpha\right) f(\tau)\,d\tau, \tag{3.30}$$

and the right-sided general fractional derivative involving the kernel of Mittag-Leffler function with power law in the sense of Riemann-Liouville type is defined as (for $\lambda = \pm 1$, see [116])

$$ {}_{Ml}^{RL}D_{b-}^{\alpha,\kappa,\lambda} f(t) = (-1)^{\kappa} \frac{d^{\kappa}}{dt^{\kappa}} \int_{t}^{b} E_{\alpha}\left(-\lambda(\tau - t)^{\alpha}\right) f(\tau)\, d\tau. \tag{3.31} $$

When $\kappa = 1$, the left-sided general fractional derivative involving the kernel of Mittag-Leffler function with power law in the sense of Riemann-Liouville type is defined as (see [301, 302])

$$ {}_{Ml}^{RL}D_{a+}^{\alpha,\lambda} f(t) = \tfrac{d}{dt} \int_{a}^{t} E_{\alpha}\left(-\lambda(t - \tau)^{\alpha}\right) f(\tau)\, d\tau, \tag{3.32} $$

and the right-sided general fractional derivative involving the kernel of Mittag-Leffler function with power law in the sense of Riemann-Liouville type is defined as

$$ {}_{Ml}^{RL}D_{b-}^{\alpha,\lambda} f(t) = -\tfrac{d}{dt} \int_{t}^{b} E_{\alpha}\left(-\lambda(\tau - t)^{\alpha}\right) f(\tau)\, d\tau. \tag{3.33} $$

This is also called the Hille-Tamarkin general fractional derivative (see [301, 302]). Here, for $\kappa = 1$ we have (see [302])

$$ {}_{Ml}^{LSC}D_{a+}^{\alpha,1,\lambda} f(t) = {}_{Ml}^{RL}D_{a+}^{\alpha,1,\lambda} f(t) + E_{\alpha}\left(-\lambda(t - a)^{\alpha}\right) f(a) \tag{3.34} $$

and

$$ {}_{Ml}^{LSC}D_{b-}^{\alpha,\kappa,\lambda} f(t) = {}_{Ml}^{LSC}D_{b-}^{\alpha,\kappa,\lambda} f(t) + E_{\alpha}\left(-\lambda(b - t)^{\alpha}\right) f(b). \tag{3.35} $$

In particular, when $\kappa = 1$ and $a = 0$, we have

$$ \frac{d}{dt}\int_{0}^{t} E_{\alpha}\left(-\lambda(t-\tau)^{\alpha}\right) f(\tau)\, d\tau = \int_{0}^{t} E_{\alpha}\left(-\lambda(t-\tau)^{\alpha}\right) f^{(1)}(\tau) d\tau - E_{\alpha}\left(-\lambda t^{\alpha}\right) f(0), \tag{3.36} $$

such that, by using the expression

$$ \int_{0}^{t} E_{\alpha}\left(-\lambda\tau^{\alpha}\right) f^{(1)}(t-\tau)\, d\tau = \int_{0}^{t} E_{\alpha}\left(-\lambda(t-\tau)^{\alpha}\right) f^{(1)}(\tau)\, d\tau, \tag{3.37} $$

there is the Gorenflo-Mainardi formulation (see [303])

$$ \frac{d}{dt}\int_{0}^{t} E_{\alpha}\left(-\lambda(t-\tau)^{\alpha}\right) f(\tau)\, d\tau = \int_{0}^{t} E_{\alpha}\left(-\lambda\tau^{\alpha}\right) f^{(1)}(t-\tau)\, d\tau + E_{\alpha}\left(-\lambda t^{\alpha}\right) f(0). \tag{3.38} $$

In particular, when $\kappa = 1$, $J(\alpha) = \frac{\Im(\alpha)}{1-\alpha}$, where $\Im(\alpha)$ is the normalization constant with the conditions $\Im(0) = 1$ and $\Im(1) = 1$, and $\lambda = \frac{\alpha}{1-\alpha}$, the left-sided general fractional derivative involving the kernel of Mittag-Leffler function with power law in the sense of Liouville-Sonine-Caputo type is defined as (see [105])

$$ {}_{MlA}^{LSC}D_{a+}^{\alpha,\lambda} f(t) = \frac{\Im(\alpha)}{1-\alpha} \int_a^t E_\alpha \left(-\frac{\alpha}{1-\alpha} (t-\tau)^\alpha \right) f^{(1)}(\tau)\, d\tau, \tag{3.39} $$

and the right-sided general fractional derivative involving the kernel of Mittag-Leffler function with power law in the sense of Liouville-Sonine-Caputo type as

$$ {}_{MlA}^{LSC}D_{b-}^{\alpha,\lambda} f(t) = \frac{\Im(\alpha)}{1-\alpha} \int_t^b E_\alpha \left(-\frac{\alpha}{1-\alpha} (\tau-t)^\alpha \right) f^{(1)}(\tau)\, d\tau. \tag{3.40} $$

The left-sided general fractional derivative involving the kernel of Prabhakar function with power law in the sense of Riemann-Liouville type is defined as (see [105])

$$ {}_{MlA}^{RL}D_{a+}^{\alpha,\kappa,\lambda} f(t) = J(\alpha) \frac{d}{dt} \int_a^t E_\alpha \left(-\lambda (t-\tau)^\alpha \right) f(\tau)\, d\tau, \tag{3.41} $$

and the right-sided general fractional derivative involving the kernel of Mittag-Leffler function with power law in the sense of Riemann-Liouville type is defined as (see [105])

$$ {}_{MlA}^{RL}D_{b-}^{\alpha,\kappa,\lambda} f(t) = (-1)^\kappa J(\alpha) \frac{d}{dt} \int_t^b E_\alpha \left(-\lambda (\tau-t)^\alpha \right) f(\tau)\, d\tau. \tag{3.42} $$

In particular, when $\kappa > 1$, $J(\alpha) = \frac{\Im(\alpha)}{1-\alpha}$, where $\Im(\alpha)$ is the normalization constant with the conditions $\Im(0) = 1$ and $\Im(1) = 1$, and $\lambda = \frac{\alpha}{1-\alpha}$, the left-sided general fractional derivative involving the kernel of Mittag-Leffler function with power law in the sense of Liouville-Sonine-Caputo type is defined as (see [105])

$$ {}_{MlA}^{LSC}D_{a+}^{\alpha,\kappa,\lambda} f(t) = \frac{\Im(\alpha)}{1-\alpha} \int_a^t E_\alpha \left(-\frac{\alpha}{1-\alpha} (t-\tau)^\alpha \right) f^{(\kappa+1)}(\tau)\, d\tau \tag{3.43} $$

and the right-sided general fractional derivative involving the kernel of Mittag-Leffler function with power law in the sense of Liouville-Sonine-Caputo type as

$$ {}_{MlA}^{LSC}D_{b-}^{\alpha,\kappa,\lambda} f(t) = \frac{\Im(\alpha)}{1-\alpha} \int_t^b E_\alpha \left(-\frac{\alpha}{1-\alpha} (\tau-t)^\alpha \right) f^{(1+\kappa)}(\tau)\, d\tau. \tag{3.44} $$

The left-sided general fractional derivative involving the kernel of Mittag-Leffler function with power law in the sense of Riemann-Liouville type is defined as (see [105])

$$ {}_{MlA}^{RL}D_{a+}^{\alpha,\kappa,\lambda} f(t) = J(\alpha) \frac{d^{\kappa+1}}{dt^{\kappa+1}} \int_a^t E_\alpha\left(-\lambda (t-\tau)^\alpha\right) f(\tau)\, d\tau, \tag{3.45} $$

and the right-sided general fractional derivative involving the kernel of Mittag-Leffler function with power law in the sense of Riemann-Liouville type is defined as (see [105])

$$ {}_{MlA}^{RL}D_{b-}^{\alpha,\kappa,\lambda} f(t) = (-1)^\kappa J(\alpha) \frac{d^{\kappa+1}}{dt^{\kappa+1}} \int_t^b E_\alpha\left(-\lambda (\tau-t)^\alpha\right) f(\tau)\, d\tau. \tag{3.46} $$

In fact, for $\alpha > 0$, $1+\kappa > \alpha > \kappa$, $\kappa = [\alpha]+1$, $-\infty < a < b < \infty$ and $\lambda \in \mathbb{C}$, the left-sided general fractional integral, as the inverse operator of the general fractional derivative involving the kernel of Mittag-Leffler function with power law, is defined as

$$ {}_{Ml}^{RL}I_{a+}^{\alpha,\kappa,\lambda} f(t) = \int_a^t \frac{1}{(t-\tau)^{\alpha+1-\kappa}} E_{\alpha,\kappa-\alpha}^{-1}\left(-\lambda (t-\tau)^\alpha\right) f(\tau)\, d\tau, \tag{3.47} $$

and the right-sided general fractional integral as

$$ {}_{Ml}^{RL}I_{b-}^{\alpha,\kappa,\lambda} f(t) = (-1)^\kappa \int_t^b \frac{1}{(\tau-t)^{\alpha+1-\kappa}} E_{\alpha,\kappa-\alpha}^{-1}\left(-\lambda (\tau-t)^\alpha\right) f(\tau)\, d\tau. \tag{3.48} $$

Definition 3.7 *For $a = -\infty$ and $b = \infty$, the left-sided general fractional derivative involving the kernel of Mittag-Leffler function with power law in the sense of Riemann-Liouville type is defined as*

$$ {}_{Ml}^{RL}D_{+}^{\alpha,\kappa,\lambda} f(t) = \frac{d^\kappa}{dt^\kappa} \int_{-\infty}^t E_\alpha\left(-\lambda (t-\tau)^\alpha\right) f(\tau)\, d\tau, \tag{3.49} $$

and the right-sided general fractional derivative involving the kernel of Mittag-Leffler function with power law in the sense of Riemann-Liouville type is defined as

$$ {}_{Ml}^{RL}D_{-}^{\alpha,\kappa,\lambda} f(t) = (-1)^\kappa \frac{d^\kappa}{dt^\kappa} \int_t^\infty E_\alpha\left(-\lambda (\tau-t)^\alpha\right) f(\tau)\, d\tau. \tag{3.50} $$

Definition 3.8 *For $a = -\infty$ and $b = \infty$, the left-sided general fractional derivative involving the kernel of Mittag-Leffler function with power law in the sense of Liouville-Sonine-Caputo type is defined as*

$$ {}_{Ml}^{LSC}D_{+}^{\alpha,\kappa,\lambda} f(t) = \int_{-\infty}^{t} E_\alpha\left(-\lambda(t-\tau)^\alpha\right) f^{(\kappa)}(\tau)\, d\tau, \tag{3.51} $$

and the right-sided general fractional derivative involving the kernel of Mittag-Leffler function with power law in the sense of Liouville-Sonine-Caputo type as

$$ {}_{Ml}^{LSC}D_{-}^{\alpha,\kappa,\lambda} f(t) = (-1)^\kappa \int_{t}^{\infty} E_\alpha\left(-\lambda(\tau-t)^\alpha\right) f^{(\kappa)}(\tau)\, d\tau. \tag{3.52} $$

Definition 3.9 *For $a = -\infty$ and $b = \infty$, the left-sided general fractional integral, as the inverse operator of the general fractional derivative involving the kernel of Mittag-Leffler function with power law, is defined as*

$$ {}_{Ml}^{RL}I_{+}^{\alpha,\kappa,\lambda} f(t) = \int_{-\infty}^{t} \frac{1}{(t-\tau)^{\alpha+1-\kappa}} E_{\alpha,\kappa-\alpha}^{-1}\left(-\lambda(t-\tau)^\alpha\right) f(\tau)\, d\tau, \tag{3.53} $$

and the right-sided general fractional integral as

$$ {}_{Ml}^{RL}I_{-}^{\alpha,\kappa,\lambda} f(t) = (-1)^\kappa \int_{t}^{-\infty} \frac{1}{(\tau-t)^{\alpha+1-\kappa}} E_{\alpha,\kappa-\alpha}^{-1}\left(-\lambda(\tau-t)^\alpha\right) f(\tau)\, d\tau. \tag{3.54} $$

The properties of the general fractional derivative involving the kernel of Mittag-Leffler function with power law are given as follows.

Property 3.4 *Let $\alpha > 0$, $1+\kappa > \alpha > \kappa$, $\kappa = [\alpha]+1$, $-\infty < a < b < \infty$ and $\lambda \in \mathbb{C}$. Then we have*

$$ L\left\{{}_{Ml}^{RL}D_{0+}^{\alpha,\kappa,\lambda} f(t)\right\} = s^{\kappa-1}\left(1+\lambda s^{-\alpha}\right)^{-1} f(s), \tag{3.55} $$

$$ L\left\{{}_{Ml}^{LSC}D_{0+}^{\alpha,\kappa,\lambda} f(t)\right\} = s^{-1}\left(1+\lambda s^{-\alpha}\right)^{-1} \left(s^\kappa f(s) - \sum_{j=1}^{\kappa} s^{\kappa-j} f^{(j-1)}(0)\right) \tag{3.56} $$

and

$$ L\left\{{}_{Ml}^{RL}I_{0+}^{\alpha,\kappa,\lambda} f(t)\right\} = s^{1-\kappa}\left(1+\lambda s^{-\alpha}\right) f(s). \tag{3.57} $$

Note that, when $D^\kappa\left({}_{Ml}^{RL}I_{0+}^{\alpha,\kappa,\lambda} f(t)\right)|_{t=0} = 0$, there is

$$ L\left\{{}_{Ml}^{RL}D_{0+}^{\alpha,\kappa,\lambda} f(t)\right\} = s^{\kappa-1}\left(1+\lambda s^{-\alpha}\right)^{-1} f(s) \tag{3.58} $$

and when $D^{\kappa}\left({}^{RL}_{Ml}I^{\alpha,\kappa,\lambda}_{0+}f(t)\right)|_{t=0}\neq 0$, there is

$$L\left\{{}^{RL}_{Ml}D^{\alpha,\kappa,\lambda}_{0+}f(t)\right\}=s^{\kappa-1}\left(1+\lambda s^{-\alpha}\right)^{-1}f(s)$$
$$-\sum_{\kappa=0}^{\kappa-1}s^{\kappa-\mu-1}\left(D^{\mu}\left[{}^{RL}_{Ml}I^{\alpha,\kappa,\lambda}_{0+}f(+0)\right]\right). \tag{3.59}$$

In particular, when $\kappa=1$, we have

$$L\left\{{}^{RL}_{Ml}D^{\alpha,1,\lambda}_{0+}f(t)\right\}=\left(1+\lambda s^{-\alpha}\right)^{-1}f(s)-{}^{RL}_{Ml}I^{\alpha,1,\lambda}_{0+}f(+0). \tag{3.60}$$

Property 3.5 *Let $\alpha>0$, $1+\kappa>\alpha>\kappa$, $\kappa=[\alpha]+1$, $-\infty<a<b<\infty$, $\lambda\in\mathbb{C}$ and $f(t)\in L_l(a,b)$ $(1\leq l<\infty)$. Then we have*

$$\left({}^{RL}_{Ml}D^{\alpha,\kappa,\lambda}_{a+}\,{}^{RL}_{Ml}I^{\alpha,\kappa,\lambda}_{a+}f\right)(t)=f(t), \tag{3.61}$$

$$\left({}^{RL}_{Ml}D^{\alpha,\kappa,\lambda}_{+}\,{}^{RL}_{Ml}I^{\alpha,\kappa,\lambda}_{+}f\right)(t)=f(t), \tag{3.62}$$

$$\left({}^{RL}_{Ml}D^{\alpha,\kappa,\lambda}_{b-}\,{}^{RL}_{Ml}I^{\alpha,\kappa,\lambda}_{b-}f\right)(t)=f(t) \tag{3.63}$$

and

$$\left({}^{RL}_{Ml}D^{\alpha,\kappa,\lambda}_{-}\,{}^{RL}_{Ml}I^{\alpha,\kappa,\lambda}_{-}f\right)(t)=f(t). \tag{3.64}$$

Property 3.6 *Let $\alpha>0$, $1+\kappa>\alpha>\kappa$, $\kappa=[\alpha]+1$, $-\infty<a<b<\infty$, $\lambda\in\mathbb{C}$ and $f(t)\in AC^{\kappa}(a,b)$. Then we have*

$$\left({}^{LCS}_{Ml}D^{\alpha,\kappa,\lambda}_{a+}\,{}^{RL}_{Ml}I^{\alpha,\kappa,\lambda}_{a+}f\right)(t)=f(t), \tag{3.65}$$

$$\left({}^{LCS}_{Ml}D^{\alpha,\kappa,\lambda}_{b-}\,{}^{RL}_{Ml}I^{\alpha,\kappa,\lambda}_{b-}f\right)(t)=f(t). \tag{3.66}$$

$$\left({}^{LCS}_{Ml}D^{\alpha,\kappa,\lambda}_{+}\,{}^{RL}_{Ml}I^{\alpha,\kappa,\lambda}_{+}f\right)(t)=f(t) \tag{3.67}$$

and

$$\left({}^{LCS}_{Ml}D^{\alpha,\kappa,\lambda}_{-}\,{}^{RL}_{Ml}I^{\alpha,\kappa,\lambda}_{-}f\right)(t)=f(t). \tag{3.68}$$

For more details of the general fractional derivatives involving the kernel of Mittag-Leffler function, see [297, 298, 299, 300, 301, 302, 303, 304, 305].

3.1.4 General fractional derivatives involving the kernel of Mittag-Leffler function with the negative power law

Definition 3.10 *For $\alpha>0$, $1+\kappa>\alpha>\kappa$, $\kappa=[\alpha]+1$, $-\infty<a<b<\infty$ and $\lambda\in\mathbb{C}$, the left-sided general fractional integral is defined as*

$${}^{RL}_{Mlr}I^{\alpha,\kappa,\lambda}_{a+}f(t)=\int_a^t\frac{1}{|t-\tau|^{\alpha+1-\kappa}}E^{-1}_{\alpha,\kappa-\alpha}\left(-\lambda|t-\tau|^{\alpha}\right)f(\tau)\,d\tau, \tag{3.69}$$

and the right-sided general fractional integral with the negative power law as

$$ {}^{RL}_{Mlr}I^{\alpha,\kappa,\lambda}_{b-} f(t) = \int_t^b \frac{1}{|\tau - t|^{\alpha+1-\kappa}} E^{-1}_{\alpha,\kappa-\alpha}\left(-\lambda\,|\tau - t|^{\alpha}\right) f(\tau)\, d\tau. \tag{3.70} $$

Definition 3.11 *For $\alpha > 0$, $1 + \kappa > \alpha > \kappa$, $\kappa = [\alpha] + 1$, $-\infty < a < b < \infty$ and $\lambda \in \mathbb{C}$, the left-sided general fractional derivative involving the kernel of Mittag-Leffler function with the negative power law in the sense of Riemann-Liouville type is defined as*

$$ {}^{RL}_{Mlr}D^{\alpha,\kappa,\lambda}_{a+} f(t) = \frac{d^{\kappa}}{dt^{\kappa}} \int_a^t E_{\alpha}\left(-\lambda\,|t - \tau|^{\alpha}\right) f(\tau)\, d\tau, \tag{3.71} $$

and the right-sided general fractional derivative involving the kernel of Mittag-Leffler function with the negative power law in the sense of Riemann-Liouville type is defined as

$$ {}^{RL}_{Mlr}D^{\alpha,\kappa,\lambda}_{b-} f(t) = (-1)^{\kappa} \frac{d^{\kappa}}{dt^{\kappa}} \int_t^b E_{\alpha}\left(-\lambda\,|\tau - t|^{\alpha}\right) f(\tau)\, d\tau. \tag{3.72} $$

Definition 3.12 *For $\alpha > 0$, $1 + \kappa > \alpha > \kappa$, $\kappa = [\alpha] + 1$, $-\infty < a < b < \infty$ and $\lambda \in \mathbb{C}$. The left-sided general fractional derivative involving the kernel of Mittag-Leffler function with the negative power law in the sense of Liouville-Sonine-Caputo type is defined as*

$$ {}^{LSC}_{Mlr}D^{\alpha,\kappa,\lambda}_{a+} f(t) = \int_a^t E_{\alpha}\left(-\lambda\,|t - \tau|^{\alpha}\right) f^{(\kappa)}(\tau)\, d\tau, \tag{3.73} $$

and the right-sided general fractional derivative involving the kernel of Mittag-Leffler function with the negative power law in the sense of Liouville-Sonine-Caputo type as

$$ {}^{LSC}_{Mlr}D^{\alpha,\kappa,\lambda}_{b-} f(t) = (-1)^{\kappa} \int_t^b E_{\alpha}\left(-\lambda\,|\tau - t|^{\alpha}\right) f^{(\kappa)}(\tau)\, d\tau. \tag{3.74} $$

In particular, for $\kappa = 1$ we have:

$$ {}^{LSC}_{Mlr}D^{\alpha,1,\lambda}_{a+} f(t) = {}^{RL}_{Mlr}D^{\alpha,1,\lambda}_{a+} f(t) + E_{\alpha}\left(-\lambda\,|t - a|^{\alpha}\right) f(a) \tag{3.75} $$

and

$$ {}^{LSC}_{Mlr}D^{\alpha,1,\lambda}_{b-} f(t) = {}^{RL}_{Mlr}D^{\alpha,1,\lambda}_{b-} f(t) + E_{\alpha}\left(-\lambda\,|b - t|^{\alpha}\right) f(b). \tag{3.76} $$

For $\alpha > 0$, $1+\kappa > \alpha > \kappa$, $\kappa = [\alpha]+1$, $-\infty < a < b < \infty$ and $\lambda \in \mathbb{C}$, the left-sided general fractional integral with the negative power law is defined as

$$ {}_{Mlr}^{RL}I_{+}^{\alpha,\kappa,\lambda} f(t) = \int_{-\infty}^{t} \frac{1}{|t-\tau|^{\alpha+1-\kappa}} E_{\alpha,\kappa-\alpha}^{-1}\left(-\lambda |t-\tau|^{\alpha}\right) f(\tau)\, d\tau, \tag{3.77} $$

and the right-sided general fractional integral with power law as

$$ {}_{Mlr}^{RL}I_{-}^{\alpha,\kappa,\lambda} f(t) = \int_{t}^{\infty} \frac{1}{|\tau-t|^{\alpha+1-\kappa}} E_{\alpha,\kappa-\alpha}^{-1}\left(-\lambda |\tau-t|^{\alpha}\right) f(\tau)\, d\tau. \tag{3.78} $$

Definition 3.13 *Fr $a = -\infty$ and $b = \infty$, the left-sided general fractional derivative involving the kernel of Mittag-Leffler function with the negative power law in the sense of Riemann-Liouville type is defined as*

$$ {}_{Mlr}^{RL}D_{+}^{\alpha,\kappa,\lambda} f(t) = \frac{d^{\kappa}}{dt^{\kappa}} \int_{-\infty}^{t} E_{\alpha}\left(-\lambda |t-\tau|^{\alpha}\right) f(\tau)\, d\tau, \tag{3.79} $$

and the right-sided general fractional derivative involving the kernel of Mittag-Leffler function with the negative power law in the sense of Riemann-Liouville type is defined as

$$ {}_{Mlr}^{RL}D_{-}^{\alpha,\kappa,\lambda} f(t) = (-1)^{\kappa} \frac{d^{\kappa}}{dt^{\kappa}} \int_{t}^{\infty} E_{\alpha}\left(-\lambda |\tau-t|^{\alpha}\right) f(\tau)\, d\tau. \tag{3.80} $$

Definition 3.14 *For $a = -\infty$ and $b = \infty$, the left-sided general fractional derivative involving the kernel of Mittag-Leffler function with the negative power law in the sense of Liouville-Sonine-Caputo type is defined as*

$$ {}_{Mlr}^{LSC}D_{+}^{\alpha,\kappa,\lambda} f(t) = \int_{-\infty}^{t} E_{\alpha}\left(-\lambda |t-\tau|^{\alpha}\right) f^{(\kappa)}(\tau)\, d\tau, \tag{3.81} $$

and the right-sided general fractional derivative involving the kernel of Mittag-Leffler function with the negative power law in the sense of Liouville-Sonine-Caputo type as

$$ {}_{Mlr}^{LSC}D_{-}^{\alpha,\kappa,\lambda} f(t) = (-1)^{\kappa} \int_{t}^{\infty} E_{\alpha}\left(-\lambda |\tau-t|^{\alpha}\right) f^{(\kappa)}(\tau)\, d\tau. \tag{3.82} $$

3.2 General fractional derivatives involving the kernel of Mittag-Leffler function in the sense of Riesz, Feller and Richard types

The general fractional calculus involving the kernel of Mittag-Leffler function in the sense of Riesz, Feller and Richard types is given as follows.

In the above conditions, the general fractional integral in the sense of Riesz type is defined as

$$\begin{aligned}
{}_{Ml}^{Rze}I_{[a,b]}^{\alpha,\kappa,\lambda}f(t) &= \frac{\left({}_{Ml}^{RL}I_{a+}^{\alpha,\kappa,\lambda}f+{}_{Ml}^{RL}I_{b-}^{\alpha,\kappa,\lambda}f\right)(t)}{2\cos(\pi\alpha/2)} \\
&= \frac{1}{2\cos(\pi\alpha/2)}\left[\int_a^t \frac{1}{(t-\tau)^{\alpha+1-\kappa}}E_{\alpha,\kappa-\alpha}^{-1}\left(-\lambda(t-\tau)^{\alpha}\right)f(\tau)\,d\tau\right. \\
&\left.+\int_t^b \frac{1}{(\tau-t)^{\alpha+1-\kappa}}E_{\alpha,\kappa-\alpha}^{-1}\left(-\lambda(\tau-t)^{\alpha}\right)f(\tau)\,d\tau\right],
\end{aligned} \tag{3.83}$$

the general fractional integral in the sense of Feller type as

$$\begin{aligned}
{}_{Ml}^{Rzf}I_{[a,b]}^{\alpha,\kappa,\lambda}f(t) &= \left(H_{-}(\vartheta,\alpha)\,{}_{Ml}^{RL}I_{a+}^{\alpha,\kappa,\lambda}f+H_{-}(\vartheta,\alpha)\,{}_{Ml}^{RL}I_{b-}^{\alpha,\kappa,\lambda}f\right)(t) \\
&= \frac{\sin((\alpha-\vartheta)\pi/2)}{\sin(\pi\vartheta)}\times\int_a^t \frac{1}{(t-\tau)^{\alpha+1-\kappa}}E_{\alpha,\kappa-\alpha}^{-1}\left(-\lambda(t-\tau)^{\alpha}\right)f(\tau)\,d\tau \\
&+\frac{\sin((\alpha+\vartheta)\pi/2)}{\sin(\pi\vartheta)}\times\int_t^b \frac{1}{(\tau-t)^{\alpha+1-\kappa}}E_{\alpha,\kappa-\alpha}^{-1}\left(-\lambda(\tau-t)^{\alpha}\right)f(\tau)\,d\tau,
\end{aligned} \tag{3.84}$$

and the general fractional integral in the sense of Richard type as

$$\begin{aligned}
{}_{Ml}^{Rzg}I_{[a,b]}^{\alpha,\kappa,\lambda}f(t) &= \frac{\left({}_{Ml}^{RL}I_{a+}^{\alpha,\kappa,\lambda}f+{}_{Ml}^{RL}I_{b-}^{\alpha,\kappa,\lambda}f\right)(t)}{2\sin(\pi\alpha/2)} \\
&= \frac{1}{2\sin(\pi\alpha/2)}\left[\int_a^t \frac{1}{(t-\tau)^{\alpha+1-\kappa}}E_{\alpha,\kappa-\alpha}^{-1}\left(-\lambda(t-\tau)^{\alpha}\right)f(\tau)\,d\tau\right. \\
&\left.+\int_t^b \frac{1}{(\tau-t)^{\alpha+1-\kappa}}E_{\alpha,\kappa-\alpha}^{-1}\left(-\lambda(\tau-t)^{\alpha}\right)f(\tau)\,d\tau\right].
\end{aligned} \tag{3.85}$$

Definition 3.15 *The general fractional derivatives in the sense of Riesz type are defined as*

$$\begin{aligned}
{}_{Ml}^{Rzh}D_{[a,b]}^{\alpha,\kappa,\lambda}f(t) &= \frac{\left({}_{Ml}^{RL}D_{a+}^{\alpha,\kappa,\lambda}f+{}_{Ml}^{RL}D_{b-}^{\alpha,\kappa,\lambda}f\right)(t)}{2\cos(\pi\alpha/2)} \\
&= -\frac{1}{2\cos(\pi\alpha/2)}\left[\frac{d^{\kappa}}{dt^{\kappa}}\int_a^t E_{\nu}\left(-\lambda(t-\tau)^{\alpha}\right)f(\tau)\,d\tau\right. \\
&\left.+(-1)^{\kappa}\frac{d^{\kappa}}{dt^{\kappa}}\int_t^b E_{\alpha}\left(-\lambda(\tau-t)^{\alpha}\right)f(\tau)\,d\tau\right],
\end{aligned} \tag{3.86}$$

and

$$\begin{aligned}{}^{Rzh}_{Mlr}D^{\alpha,\kappa,\lambda}_{[a,b]}f(t) &= -\frac{\left({}^{LSC}_{Ml}D^{\alpha,\kappa,\lambda}_{a+}f + {}^{LSC}_{Ml}D^{\alpha,\kappa,\lambda}_{b-}f\right)(t)}{2\cos(\pi\alpha/2)} \\ &= -\frac{1}{2\cos(\pi\alpha/2)}\left[\int_a^t E_\nu\left(-\lambda(t-\tau)^\alpha\right) f^{(\kappa)}(\tau)\,d\tau \right. \\ &\left. + (-1)^\kappa \int_t^b E_\alpha\left(-\lambda(\tau-t)^\alpha\right) f^{(\kappa)}(\tau)\,d\tau\right],\end{aligned} \tag{3.87}$$

the general fractional derivative in the sense of Feller type as

$$\begin{aligned}{}^{Rzj}_{Ml}D^{\alpha,\kappa,\lambda}_{[a,b]}f(t) &= -\left(H_+(\vartheta,\alpha)\,{}^{RL}_{Ml}D^{\alpha,\kappa,\lambda}_{a+}f + H_-(\vartheta,\alpha)\,{}^{RL}_{Ml}D^{\alpha,\kappa,\lambda}_{b-}f\right)(t) \\ &= -\frac{\sin((\alpha+\vartheta)\pi/2)}{\sin(\pi\vartheta)} \times \left(\frac{d^\kappa}{dt^\kappa}\int_a^t E_\nu\left(-\lambda(t-\tau)^\alpha\right) f(\tau)\,d\tau\right) \\ &- \frac{\sin((\alpha-\vartheta)\pi/2)}{\sin(\pi\vartheta)} \times \left((-1)^\kappa \frac{d^\kappa}{dt^\kappa}\int_t^b E_\alpha\left(-\lambda(\tau-t)^\alpha\right) f(\tau)\,d\tau\right),\end{aligned} \tag{3.88}$$

and

$$\begin{aligned}{}^{Rzj}_{Mlr}D^{\alpha,\kappa,\lambda}_{[a,b]}f(t) &= -\left(H_+(\vartheta,\alpha)\,{}^{LSC}_{Ml}D^{\alpha,\kappa,\lambda}_{a+}f + H_-(\vartheta,\alpha)\,{}^{LSC}_{Ml}D^{\alpha,\kappa,\lambda}_{b-}f\right)(t) \\ &= -\frac{\sin((\alpha+\vartheta)\pi/2)}{\sin(\pi\vartheta)} \times \left(\int_a^t E_\nu\left(-\lambda(t-\tau)^\alpha\right) f^{(\kappa)}(\tau)\,d\tau\right) \\ &- \frac{\sin((\alpha-\vartheta)\pi/2)}{\sin(\pi\vartheta)} \times \left((-1)^\kappa \int_t^b E_\alpha\left(-\lambda(\tau-t)^\alpha\right) f^{(\kappa)}(\tau)\,d\tau\right),\end{aligned} \tag{3.89}$$

and the general fractional derivative in the sense of Richard type as

$$\begin{aligned}{}^{Rzl}_{Ml}D^{\alpha,\kappa,\lambda}_{[a,b]}f(t) &= \frac{\left({}^{RL}_{Ml}D^{\alpha,\kappa,\lambda}_{a+}f + {}^{RL}_{Ml}D^{\alpha,\kappa,\lambda}_{b-}f\right)(t)}{2\cos(\pi\alpha/2)} \\ &= -\frac{1}{2\sin(\pi\alpha/2)}\left[\frac{d^\kappa}{dt^\kappa}\int_a^t E_\nu\left(-\lambda(t-\tau)^\alpha\right) f(\tau)\,d\tau \right. \\ &\left. + (-1)^\kappa \frac{d^\kappa}{dt^\kappa}\int_t^b E_\alpha\left(-\lambda(\tau-t)^\alpha\right) f(\tau)\,d\tau\right],\end{aligned} \tag{3.90}$$

and

$$\begin{aligned}{}^{Rzl}_{Mlr}D^{\alpha,\kappa,\lambda}_{[a,b]}f(t) &= -\frac{\left({}^{LSC}_{Ml}D^{\alpha,\kappa,\lambda}_{a+}f + {}^{LSC}_{Ml}D^{\alpha,\kappa,\lambda}_{b-}f\right)(t)}{2\cos(\pi\alpha/2)} \\ &= -\frac{1}{2\sin(\pi\alpha/2)}\left[\int_a^t E_\nu\left(-\lambda(t-\tau)^\alpha\right) f^{(\kappa)}(\tau)\,d\tau \right. \\ &\left. + (-1)^\kappa \int_t^b E_\alpha\left(-\lambda(\tau-t)^\alpha\right) f^{(\kappa)}(\tau)\,d\tau\right].\end{aligned} \tag{3.91}$$

In another way, the general fractional integral in the sense of Riesz type is defined as

$$
\begin{aligned}
{}_{Mlu}^{Rze}I_{[a,b]}^{\alpha,\kappa,\lambda} f(t) &= \frac{\left({}_{Mlr}^{RL}I_{a+}^{\alpha,\kappa,\lambda} f + {}_{Mlr}^{RL}I_{b-}^{\alpha,\kappa,\lambda} f\right)(t)}{2\cos(\pi\alpha/2)} \\
&= \frac{1}{2\cos(\pi\alpha/2)}\left[\int_a^t \frac{1}{|t-\tau|^{\alpha+1-\kappa}} E_{\alpha,\kappa-\alpha}^{-1}\left(-\lambda|t-\tau|^{\alpha}\right) f(\tau)\,d\tau\right. \\
&\left.+\int_t^b \frac{1}{|t-\tau|^{\alpha+1-\kappa}} E_{\alpha,\kappa-\alpha}^{-1}\left(-\lambda|t-\tau|^{\alpha}\right) f(\tau)\,d\tau\right],
\end{aligned}
\tag{3.92}
$$

the general fractional integral in the sense of Feller type as

$$
\begin{aligned}
{}_{Mly}^{Rzf}I_{[a,b]}^{\alpha,\kappa,\lambda} f(t) &= \left(H_{-}(\vartheta,\alpha)\,{}_{Mlr}^{RL}I_{a+}^{\alpha,\kappa,\lambda} f + H_{-}(\vartheta,\alpha)\,{}_{Mlr}^{RL}I_{b-}^{\alpha,\kappa,\lambda} f\right)(t) \\
&= \frac{\sin((\alpha-\vartheta)\pi/2)}{\sin(\pi\vartheta)} \times \int_a^t \frac{1}{|t-\tau|^{\alpha+1-\kappa}} E_{\alpha,\kappa-\alpha}^{-1}\left(-\lambda|t-\tau|^{\alpha}\right) f(\tau)\,d\tau \\
&+ \frac{\sin((\alpha+\vartheta)\pi/2)}{\sin(\pi\vartheta)} \times \int_t^b \frac{1}{|t-\tau|^{\alpha+1-\kappa}} E_{\alpha,\kappa-\alpha}^{-1}\left(-\lambda|t-\tau|^{\alpha}\right) f(\tau)\,d\tau,
\end{aligned}
\tag{3.93}
$$

and the general fractional integral in the sense of Richard type as

$$
\begin{aligned}
{}_{Mlh}^{Rzg}I_{[a,b]}^{\alpha,\kappa,\lambda} f(t) &= \frac{\left({}_{Mlr}^{RL}I_{a+}^{\alpha,\kappa,\lambda} f + {}_{Mlr}^{RL}I_{b-}^{\alpha,\kappa,\lambda} f\right)(t)}{2\sin(\pi\alpha/2)} \\
&= \frac{1}{2\sin(\pi\alpha/2)}\left[\int_a^t \frac{1}{|t-\tau|^{\alpha+1-\kappa}} E_{\alpha,\kappa-\alpha}^{-1}\left(-\lambda|t-\tau|^{\alpha}\right) f(\tau)\,d\tau\right. \\
&\left.+\int_t^b \frac{1}{|t-\tau|^{\alpha+1-\kappa}} E_{\alpha,\kappa-\alpha}^{-1}\left(-\lambda|t-\tau|^{\alpha}\right) f(\tau)\,d\tau\right].
\end{aligned}
\tag{3.94}
$$

Definition 3.16 *The general fractional derivatives in the sense of Riesz type are defined as*

$$
\begin{aligned}
{}_{Mlu}^{Rzh}D_{[a,b]}^{\alpha,\kappa,\lambda} f(t) &= \frac{\left({}_{Mlr}^{RL}D_{a+}^{\alpha,\kappa,\lambda} f + {}_{Mlr}^{RL}D_{b-}^{\alpha,\kappa,\lambda} f\right)(t)}{2\cos(\pi\alpha/2)} \\
&= -\frac{1}{2\cos(\pi\alpha/2)}\left[\frac{d^{\kappa}}{dt^{\kappa}}\int_a^t E_{\nu}\left(-\lambda|t-\tau|^{\alpha}\right) f(\tau)\,d\tau\right. \\
&\left.+(-1)^{\kappa}\frac{d^{\kappa}}{dt^{\kappa}}\int_t^b E_{\alpha}\left(-\lambda|t-\tau|^{\alpha}\right) f(\tau)\,d\tau\right],
\end{aligned}
\tag{3.95}
$$

and

$$
\begin{aligned}
{}_{Mlv}^{Rzh}D_{[a,b]}^{\alpha,\kappa,\lambda} f(t) &= -\frac{\left({}_{Mlr}^{LSC}D_{a+}^{\alpha,\kappa,\lambda} f + {}_{Mlr}^{LSC}D_{b-}^{\alpha,\kappa,\lambda} f\right)(t)}{2\cos(\pi\alpha/2)} \\
&= -\frac{1}{2\cos(\pi\alpha/2)}\left[\int_a^t E_{\nu}\left(-\lambda(t-\tau)^{\alpha}\right) f^{(\kappa)}(\tau)\,d\tau\right. \\
&\left.+(-1)^{\kappa}\int_t^b E_{\alpha}\left(-\lambda(\tau-t)^{\alpha}\right) f^{(\kappa)}(\tau)\,d\tau\right],
\end{aligned}
\tag{3.96}
$$

the general fractional derivative in the sense of Feller type as

$$
\begin{aligned}
& {}^{Rzj}_{Mly}D^{\alpha,\kappa,\lambda}_{[a,b]} f(t) = -\left(H_+(\vartheta,\alpha)\,{}^{RL}_{Mlr}D^{\alpha,\kappa,\lambda}_{a+} f + H_-(\vartheta,\alpha)\,{}^{RL}_{Mlr}D^{\alpha,\kappa,\lambda}_{b-} f\right)(t) \\
& = -\frac{\sin((\alpha+\vartheta)\pi/2)}{\sin(\pi\vartheta)} \times \left(\frac{d^\kappa}{dt^\kappa}\int_a^t E_\nu\left(-\lambda\left|t-\tau\right|^\alpha\right) f(\tau)\, d\tau\right) \\
& -\frac{\sin((\alpha-\vartheta)\pi/2)}{\sin(\pi\vartheta)} \times \left((-1)^\kappa \frac{d^\kappa}{dt^\kappa}\int_t^b E_\alpha\left(-\lambda\left|t-\tau\right|^\alpha\right) f(\tau)\, d\tau\right),
\end{aligned}
\tag{3.97}
$$

and

$$
\begin{aligned}
& {}^{Rzj}_{Mlz}D^{\alpha,\kappa,\lambda}_{[a,b]} f(t) = -\left(H_+(\vartheta,\alpha)\,{}^{LSC}_{Mlr}D^{\alpha,\kappa,\lambda}_{a+} f + H_-(\vartheta,\alpha)\,{}^{LSC}_{Mlr}D^{\alpha,\kappa,\lambda}_{b-} f\right)(t) \\
& = -\frac{\sin((\alpha+\vartheta)\pi/2)}{\sin(\pi\vartheta)} \times \left(\int_a^t E_\nu\left(-\lambda\left|t-\tau\right|^\alpha\right) f^{(\kappa)}(\tau)\, d\tau\right) \\
& -\frac{\sin((\alpha-\vartheta)\pi/2)}{\sin(\pi\vartheta)} \times \left((-1)^\kappa \int_t^b E_\alpha\left(-\lambda\left|t-\tau\right|^\alpha\right) f^{(\kappa)}(\tau)\, d\tau\right),
\end{aligned}
\tag{3.98}
$$

and the general fractional derivative in the sense of Richard type as

$$
\begin{aligned}
& {}^{Rzl}_{Mlh}D^{\alpha,\kappa,\lambda}_{[a,b]} f(t) = \frac{\left({}^{RL}_{Mlr}D^{\alpha,\kappa,\lambda}_{a+} f + {}^{RL}_{Mlr}D^{\alpha,\kappa,\lambda}_{b-} f\right)(t)}{2\cos(\pi\alpha/2)} \\
& = -\frac{1}{2\sin(\pi\alpha/2)}\left[\frac{d^\kappa}{dt^\kappa}\int_a^t E_\nu\left(-\lambda\left|t-\tau\right|^\alpha\right) f(\tau)\, d\tau\right. \\
& \left. +(-1)^\kappa \frac{d^\kappa}{dt^\kappa}\int_t^b E_\alpha\left(-\lambda\left|t-\tau\right|^\alpha\right) f(\tau)\, d\tau\right],
\end{aligned}
\tag{3.99}
$$

$$
\begin{aligned}
& {}^{Rzl}_{Mlj}D^{\alpha,\kappa,\lambda}_{[a,b]} f(t) = -\frac{\left({}^{LSC}_{Mlr}D^{\alpha,\kappa,\lambda}_{a+} f + {}^{LSC}_{Mlr}D^{\alpha,\kappa,\lambda}_{b-} f\right)(t)}{2\cos(\pi\alpha/2)} \\
& = -\frac{1}{2\sin(\pi\alpha/2)}\left[\int_a^t E_\nu\left(-\lambda\left|t-\tau\right|^\alpha\right) f^{(\kappa)}(\tau)\, d\tau\right. \\
& \left. +(-1)^\kappa \int_t^b E_\alpha\left(-\lambda\left|t-\tau\right|^\alpha\right) f^{(\kappa)}(\tau)\, d\tau\right].
\end{aligned}
\tag{3.100}
$$

When $a = -\infty$ and $b = \infty$, the general fractional integral in the sense of Riesz type is defined as

$$
\begin{aligned}
& {}^{Rze}_{Mlr}I^{\alpha,\kappa,\lambda}_{[-\infty,\infty]} f(t) = \frac{\left({}^{RL}_{Mlr}I^{\alpha,\kappa,\lambda}_{+} f + {}^{RL}_{Mlr}I^{\alpha,\kappa,\lambda}_{-} f\right)(t)}{2\cos(\pi\alpha/2)} \\
& = \frac{1}{2\cos(\pi\alpha/2)}\left[\int_{-\infty}^t \frac{1}{\left|t-\tau\right|^{\alpha+1-\kappa}} E^{-1}_{\alpha,\kappa-\alpha}\left(-\lambda\left|t-\tau\right|^\alpha\right) f(\tau)\, d\tau\right. \\
& \left. +\int_t^{-\infty} \frac{1}{\left|t-\tau\right|^{\alpha+1-\kappa}} E^{-1}_{\alpha,\kappa-\alpha}\left(-\lambda\left|t-\tau\right|^\alpha\right) f(\tau)\, d\tau\right],
\end{aligned}
\tag{3.101}
$$

the general fractional integral in the sense of Feller type as

$$
\begin{aligned}
&{}^{Rzf}_{Mlr}I^{\alpha,\kappa,\lambda}_{[-\infty,\infty]}f(t) = \left(H_{-}(\vartheta,\alpha)\,{}^{RL}_{Mlr}I^{\alpha,\kappa,\lambda}_{+}f + H_{-}(\vartheta,\alpha)\,{}^{RL}_{Mlr}I^{\alpha,\kappa,\lambda}_{-}f\right)(t)\\
&= \frac{\sin((\alpha-\vartheta)\pi/2)}{\sin(\pi\vartheta)} \times \int_{-\infty}^{t} \frac{1}{|t-\tau|^{\alpha+1-\kappa}} E^{-1}_{\alpha,\kappa-\alpha}\left(-\lambda\,|t-\tau|^{\alpha}\right) f(\tau)\,d\tau\\
&+\frac{\sin((\alpha+\vartheta)\pi/2)}{\sin(\pi\vartheta)} \times \int_{t}^{\infty} \frac{1}{|t-\tau|^{\alpha+1-\kappa}} E^{-1}_{\alpha,\kappa-\alpha}\left(-\lambda\,|t-\tau|^{\alpha}\right) f(\tau)\,d\tau,
\end{aligned}
\tag{3.102}
$$

and the general fractional integral in the sense of Richard type as

$$
\begin{aligned}
&{}^{Rzg}_{Mlr}I^{\alpha,\kappa,\lambda}_{[-\infty,\infty]}f(t) = \frac{\left({}^{RL}_{Mlr}I^{\alpha,\kappa,\lambda}_{+}f + {}^{RL}_{Mlr}I^{\alpha,\kappa,\lambda}_{-}f\right)(t)}{2\sin(\pi\alpha/2)}\\
&= \frac{1}{2\sin(\pi\alpha/2)}\left[\int_{-\infty}^{t} \frac{1}{|t-\tau|^{\alpha+1-\kappa}} E^{-1}_{\alpha,\kappa-\alpha}\left(-\lambda\,|t-\tau|^{\alpha}\right) f(\tau)\,d\tau\right.\\
&\left.+\int_{t}^{\infty} \frac{1}{|t-\tau|^{\alpha+1-\kappa}} E^{-1}_{\alpha,\kappa-\alpha}\left(-\lambda\,|t-\tau|^{\alpha}\right) f(\tau)\,d\tau\right].
\end{aligned}
\tag{3.103}
$$

Definition 3.17 *The general fractional derivative in the sense of Riesz type are defined as*

$$
\begin{aligned}
&{}^{Rzh}_{Mlr}D^{\alpha,\kappa,\lambda}_{[-\infty,\infty]}f(t) = \frac{\left({}^{RL}_{Mlr}D^{\alpha,\kappa,\lambda}_{+}f + {}^{RL}_{Mlr}D^{\alpha,\kappa,\lambda}_{-}f\right)(t)}{2\cos(\pi\alpha/2)}\\
&= -\frac{1}{2\cos(\pi\alpha/2)}\left[\frac{d^{\kappa}}{dt^{\kappa}}\int_{-\infty}^{t} E_{\nu}\left(-\lambda\,|t-\tau|^{\alpha}\right) f(\tau)\,d\tau\right.\\
&\left.+(-1)^{\kappa}\frac{d^{\kappa}}{dt^{\kappa}}\int_{t}^{\infty} E_{\alpha}\left(-\lambda\,|t-\tau|^{\alpha}\right) f(\tau)\,d\tau\right],
\end{aligned}
\tag{3.104}
$$

and

$$
\begin{aligned}
&{}^{Rzh}_{Mlr}D^{\alpha,\kappa,\lambda}_{[-\infty,\infty]}f(t) = -\frac{\left({}^{LSC}_{Mlr}D^{\alpha,\kappa,\lambda}_{+}f + {}^{LSC}_{Mlr}D^{\alpha,\kappa,\lambda}_{-}f\right)(t)}{2\cos(\pi\alpha/2)}\\
&= -\frac{1}{2\cos(\pi\alpha/2)}\left[\int_{-\infty}^{t} E_{\nu}\left(-\lambda\,|t-\tau|^{\alpha}\right) f^{(\kappa)}(\tau)\,d\tau\right.\\
&\left.+(-1)^{\kappa}\int_{t}^{\infty} E_{\alpha}\left(-\lambda\,|t-\tau|^{\alpha}\right) f^{(\kappa)}(\tau)\,d\tau\right],
\end{aligned}
\tag{3.105}
$$

the general fractional derivative in the sense of Feller type as

$$
\begin{aligned}
&{}^{Rzj}_{Ml}D^{\alpha,\kappa,\lambda}_{[-\infty,\infty]}f(t) = -\left(H_{+}(\vartheta,\alpha)\,{}^{RL}_{Mlr}D^{\alpha,\kappa,\lambda}_{+}f + H_{-}(\vartheta,\alpha)\,{}^{RL}_{Mlr}D^{\alpha,\kappa,\lambda}_{-}f\right)(t)\\
&= -\frac{\sin((\alpha+\vartheta)\pi/2)}{\sin(\pi\vartheta)} \times \left(\frac{d^{\kappa}}{dt^{\kappa}}\int_{-\infty}^{t} E_{\nu}\left(-\lambda\,|t-\tau|^{\alpha}\right) f(\tau)\,d\tau\right)\\
&-\frac{\sin((\alpha-\vartheta)\pi/2)}{\sin(\pi\vartheta)} \times \left((-1)^{\kappa}\frac{d^{\kappa}}{dt^{\kappa}}\int_{t}^{\infty} E_{\alpha}\left(-\lambda\,|t-\tau|^{\alpha}\right) f(\tau)\,d\tau\right),
\end{aligned}
\tag{3.106}
$$

and

$$
\begin{aligned}
&{}_{Mlr}^{Rzj}D_{[-\infty,\infty]}^{\alpha,\kappa,\lambda}f(t) = -\left(H_{+}(\vartheta,\alpha)\,{}_{Mlr}^{LSC}D_{+}^{\alpha,\kappa,\lambda}f + H_{-}(\vartheta,\alpha)\,{}_{Mlr}^{LSC}D_{-}^{\alpha,\kappa,\lambda}f\right)(t) \\
&= -\frac{\sin((\alpha+\vartheta)\pi/2)}{\sin(\pi\vartheta)} \times \left(\int_{-\infty}^{t} E_{\nu}\left(-\lambda\left|t-\tau\right|^{\alpha}\right) f^{(\kappa)}(\tau)\, d\tau\right) \\
&-\frac{\sin((\alpha-\vartheta)\pi/2)}{\sin(\pi\vartheta)} \times \left((-1)^{\kappa}\int_{t}^{\infty} E_{\alpha}\left(-\lambda\left|t-\tau\right|^{\alpha}\right) f^{(\kappa)}(\tau)\, d\tau\right),
\end{aligned}
\tag{3.107}
$$

and the general fractional derivative in the sense of Richard type as

$$
\begin{aligned}
&{}_{Mlr}^{Rzl}D_{[-\infty,\infty]}^{\alpha,\kappa,\lambda}f(t) = \frac{\left({}_{Mlr}^{RL}D_{+}^{\alpha,\kappa,\lambda}f + {}_{Mlr}^{RL}D_{-}^{\alpha,\kappa,\lambda}f\right)(t)}{2\cos(\pi\alpha/2)} \\
&= -\frac{1}{2\sin(\pi\alpha/2)}\left[\frac{d^{\kappa}}{dt^{\kappa}}\int_{-\infty}^{t} E_{\nu}\left(-\lambda\left|t-\tau\right|^{\alpha}\right) f(\tau)\, d\tau\right. \\
&\left.+(-1)^{\kappa}\frac{d^{\kappa}}{dt^{\kappa}}\int_{t}^{\infty} E_{\alpha}\left(-\lambda\left|t-\tau\right|^{\alpha}\right) f(\tau)\, d\tau\right],
\end{aligned}
\tag{3.108}
$$

and

$$
\begin{aligned}
&{}_{Mlr}^{Rzl}D_{[-\infty,\infty]}^{\alpha,\kappa,\lambda} = -\frac{\left({}_{Mlr}^{LSC}D_{a+}^{\alpha,\kappa,\lambda}f + {}_{Mlr}^{LSC}D_{b-}^{\alpha,\kappa,\lambda}f\right)(t)}{2\cos(\pi\alpha/2)} \\
&= -\frac{1}{2\sin(\pi\alpha/2)}\left[\int_{-\infty}^{t} E_{\nu}\left(-\lambda\left|t-\tau\right|^{\alpha}\right) f^{(\kappa)}(\tau)\, d\tau\right. \\
&\left.+(-1)^{\kappa}\int_{t}^{\infty} E_{\alpha}\left(-\lambda\left|t-\tau\right|^{\alpha}\right) f^{(\kappa)}(\tau)\, d\tau\right].
\end{aligned}
\tag{3.109}
$$

3.3 General fractional derivatives involving the kernel of Mittag-Leffler function with respect to another function

The definitions and properties of the general fractional derivatives involving the kernel of Mittag-Leffler function with respect to another function are presented as follows.

Definition 3.18 *Let $\alpha > 0$, $1+\kappa > \alpha > \kappa$, $\kappa = [\alpha]+1$, $-\infty < a < b < \infty$, $\lambda \in \mathbb{C}$ and $h^{(\alpha)}(t) > 0$. The left-sided fractional integral with respect to another function, as the inverse operator of the general fractional derivative involving the kernel of Mittag-Leffler function with power law, is defined as*

$$
\begin{aligned}
&{}_{Ml}^{RL}I_{a+,h}^{\alpha,\kappa,\lambda}f(t) \\
&= \int_{a}^{t} \frac{1}{(h(t)-h(\tau))^{\alpha+1-\kappa}} E_{\alpha,\kappa-\alpha}^{-1}\left(-\lambda\left(h(t)-h(\tau)\right)^{\alpha}\right) f(\tau)\, h^{(1)}(\tau)\, d\tau,
\end{aligned}
\tag{3.110}
$$

and the right-sided fractional integral with respect to another function as

$$
\begin{aligned}
&{}^{RL}_{Ml}I^{\alpha,\kappa,\lambda}_{b-,h} f(t) \\
&= (-1)^{\kappa} \int_t^b \frac{1}{(h(\tau)-h(t))^{\alpha+1-\kappa}} E^{-1}_{\alpha,\kappa-\alpha}\left(-\lambda\left(h(\tau)-h(t)\right)^{\alpha}\right) f(\tau)\, h^{(1)}(\tau)\, d\tau,
\end{aligned}
\tag{3.111}
$$

respectively.

Definition 3.19 *The left-sided general fractional derivative involving the kernel of Mittag-Leffler function with power law in the sense of Riemann-Liouville type with respect to another function is defined as*

$$
{}^{RL}_{Ml}D^{\alpha,\kappa,\lambda}_{a+,h} f(t) = \left(\frac{1}{h^{(1)}(t)}\frac{d}{dt}\right)^{\kappa} \int_a^t E_{\alpha}\left(-\lambda\left(h(t)-h(\tau)\right)^{\alpha}\right) f(\tau)\, h^{(1)}(\tau)\, d\tau,
\tag{3.112}
$$

and the right-sided general fractional derivative involving the kernel of Mittag-Leffler function with power law in the sense of Riemann-Liouville type with respect to another function as

$$
{}^{RL}_{Ml}D^{\alpha,\kappa,\lambda}_{b-,h} f(t) = (-1)^{\kappa}\left(\frac{1}{h^{(1)}(t)}\frac{d}{dt}\right)^{\kappa} \int_t^b E_{\alpha}\left(-\lambda\left(h(\tau)-h(t)\right)^{\alpha}\right) f(\tau)\, h^{(1)}(\tau)\, d\tau.
\tag{3.113}
$$

Definition 3.20 *In a similar way, the left-sided general fractional derivative involving the kernel of Mittag-Leffler function with power law in the sense of Liouville-Sonine-Caputo type with respect to another function is defined as*

$$
{}^{LSC}_{Ml}D^{\alpha,\kappa,\lambda}_{a+,h} f(t) = \int_a^t E_{\alpha}\left(-\lambda\left(h(t)-h(\tau)\right)^{\alpha}\right) \left(\frac{1}{h^{(1)}(\tau)} f^{(1)}_{h}(\tau)\right)^{\kappa}(\tau)\, h^{(1)}(\tau)\, d\tau,
\tag{3.114}
$$

and the right-sided general fractional derivative involving the kernel of Mittag-Leffler function with power law in the sense of Liouville-Sonine-Caputo type with respect to another function as

$$
\begin{aligned}
&{}^{LSC}_{Ml}D^{\alpha,\kappa,\lambda}_{b-,h} f(t) \\
&= (-1)^{\kappa} \int_t^{\infty} E_{\alpha}\left(-\lambda\left(h(\tau)-h(t)\right)^{\alpha}\right) \left(\frac{1}{h^{(1)}(\tau)} f^{(1)}_{h}(\tau)\right)^{\kappa}(\tau)\, h^{(1)}(\tau)\, d\tau,
\end{aligned}
\tag{3.115}
$$

where $\left(\frac{1}{h^{(1)}(\tau)} f^{(1)}_{h}(\tau)\right)^{\kappa}(\tau) = \left(\frac{1}{h^{(1)}(\tau)}\frac{df(h(\tau))}{dh(\tau)}\right)^{\kappa}$, *respectively.*

Property 3.7 *Let* $\alpha > 0, 1+\kappa > \alpha > \kappa$, $\kappa = [\alpha]+1$, $-\infty < a < b < \infty$, $\lambda \in \mathbb{C}$ *and* $f(t) \in L_l(a,b)$ $(1 \le l < \infty)$. *Then we have*

$$
\left({}^{RL}_{Ml}D^{\alpha,\kappa,\lambda}_{a+,h}\, {}^{RL}_{Ml}I^{\alpha,\kappa,\lambda}_{a+,h} f\right)(t) = f(t),
\tag{3.116}
$$

and

$$\left({}^{RL}_{Ml}D^{\alpha,\kappa,\lambda}_{b-,h}\,{}^{RL}_{Ml}I^{\alpha,\kappa,\lambda}_{b-,h}f\right)(t) = f(t). \tag{3.117}$$

Property 3.8 *Let* $\alpha > 0,\ 1+\kappa > \alpha > \kappa,\ \kappa = [\alpha]+1,\ -\infty < a < b < \infty,$ $\lambda \in \mathbb{C}$ *and* $f(t) \in AC^{\kappa}(a,b)$*. Then we have*

$$\left({}^{LCS}_{Ml}D^{\alpha,\kappa,\lambda}_{a+,h}\,{}^{RL}_{Ml}I^{\alpha,\kappa,\lambda}_{a+,h}f\right)(t) = f(t), \tag{3.118}$$

and

$$\left({}^{LCS}_{Ml}D^{\alpha,\kappa,\lambda}_{b-,h}\,{}^{RL}_{Ml}I^{\alpha,\kappa,\lambda}_{b-,h}f\right)(t) = f(t). \tag{3.119}$$

For more details of the general fractional derivative involving the kernel of Mittag-Leffler function, see [297, 298, 299, 301, 302, 304].

3.4 General fractional derivatives involving the kernel of Kohlrausch-Williams-Watts function

The general fractional derivatives involving the kernel of Kohlrausch-Williams-Watts function are considered in details.

Definition 3.21 *Let* $\alpha > 0,\ 1+\kappa > \alpha > \kappa,\ \kappa = [\alpha]+1,\ -\infty < a < b < \infty$ *and* $\lambda \in \mathbb{C}$*. The left-sided general fractional derivative involving the kernel of Kohlrausch-Williams-Watts function with power law in the sense of Riemann-Liouville type is defined as*

$${}^{RL}_{Kww}D^{\alpha,\kappa,\lambda}_{a+}f(t) = \frac{d^{\kappa}}{dt^{\kappa}}\int_a^t e^{-\lambda(t-\tau)^{\alpha}}f(\tau)\,d\tau, \tag{3.120}$$

and the right-sided general fractional derivative involving the kernel of Kohlrausch-Williams-Watts function with power law in the sense of Riemann-Liouville type as

$${}^{RL}_{Kww}D^{\alpha,\kappa,\lambda}_{b-}f(t) = (-1)^{\kappa}\frac{d^{\kappa}}{dt^{\kappa}}\int_t^b e^{-\lambda(\tau-t)^{\alpha}}f(\tau)\,d\tau. \tag{3.121}$$

Definition 3.22 *The left-sided general fractional derivative involving the kernel of Kohlrausch-Williams-Watts function with power law in the sense of Liouville-Sonine-Caputo type is defined as*

$${}^{LSC}_{Kww}D^{\alpha,\kappa,\lambda}_{a+}f(t) = \int_a^t e^{-\lambda(t-\tau)^{\alpha}}f^{(\kappa)}(\tau)\,d\tau, \tag{3.122}$$

and the right-sided general fractional derivative involving the kernel of Kohlrausch-Williams-Watts function with power law in the sense of Liouville-Sonine-Caputo type as

$$ {}^{LSC}_{Kww}D^{\alpha,\kappa,\lambda}_{b-} f(t) = (-1)^{\kappa} \int_{t}^{b} e^{-\lambda(\tau-t)^{\alpha}} f^{(\kappa)}(\tau)\, d\tau. \tag{3.123} $$

Definition 3.23 *For* $\alpha > 0,\ 1+\kappa > \alpha > \kappa,\ \kappa = [\alpha]+1,\ \lambda \in \mathbb{C},\ a = -\infty$ *and* $b = \infty$*, the left-sided general fractional derivative involving the kernel of Kohlrausch-Williams-Watts function with power law in the sense of Riemann-Liouville type is defined as*

$$ {}^{RL}_{Kww}D^{\alpha,\kappa,\lambda}_{+} f(t) = \frac{d^{\kappa}}{dt^{\kappa}} \int_{-\infty}^{t} e^{-\lambda(t-\tau)^{\alpha}} f(\tau)\, d\tau, \tag{3.124} $$

and the right-sided general fractional derivative involving the kernel of Kohlrausch-Williams-Watts function with power law in the sense of Riemann-Liouville type as

$$ {}^{RL}_{Kww}D^{\alpha,\kappa,\lambda}_{-} f(t) = (-1)^{\kappa} \frac{d^{\kappa}}{dt^{\kappa}} \int_{t}^{\infty} e^{-\lambda(\tau-t)^{\alpha}} f(\tau)\, d\tau. \tag{3.125} $$

Definition 3.24 *For* $\alpha > 0,\ 1+\kappa > \alpha > \kappa,\ \kappa = [\alpha]+1,\ \lambda \in \mathbb{C},\ a = -\infty$ *and* $b = \infty$*, the left-sided general fractional derivative involving the kernel of Kohlrausch-Williams-Watts function with power law in the sense of Liouville-Sonine-Caputo type is defined as*

$$ {}^{LSC}_{Kww}D^{\alpha,\kappa,\lambda}_{+} f(t) = \int_{-\infty}^{t} e^{-\lambda(t-\tau)^{\alpha}} f^{(\kappa)}(\tau)\, d\tau, \tag{3.126} $$

and the right-sided general fractional derivative involving the kernel of Kohlrausch-Williams-Watts function with power law in the sense of Liouville-Sonine-Caputo type as

$$ {}^{LSC}_{Kww}D^{\alpha,\kappa,\lambda}_{-} f(t) = (-1)^{\kappa} \int_{t}^{\infty} e^{-\lambda(\tau-t)^{\alpha}} f^{(\kappa)}(\tau)\, d\tau. \tag{3.127} $$

When $\kappa = 1$, the relationships

$$ {}^{LS}_{Kww}D^{\alpha,\lambda}_{a+} f(t) = {}^{RL}_{Kww}D^{\alpha,\lambda}_{a+} f(t) + e^{-\lambda(t-a)^{\alpha}} f(a) \tag{3.128} $$

and

$$ {}^{LS}_{Kww}D^{\alpha,\lambda}_{b-} f(t) = {}^{RL}_{Kww}D^{\alpha,\lambda}_{b-} f(t) + e^{-\lambda(b-t)^{\alpha}} f(b) \tag{3.129} $$

are valid.

In fact, the left-sided general fractional derivative involving the kernel of Kohlrausch-Williams-Watts function with power law in the sense of Liouville-Sonine type, introduced by Sun, Hao, Zhang and Baleanu in 2017, can be represented in the form (see [113]; also see [299]).

$$ {}_{Ks}^{LS}D_{a+}^{\alpha,\lambda} f(t) = J(\alpha) {}_{Kww}^{LS} D_{a+}^{\alpha,1,\lambda} f(t) = J(\alpha) \int_a^t e^{-\lambda(t-\tau)^{\alpha}} f^{(1)}(\tau)\, d\tau, \tag{3.130} $$

where $J(\beta) = \frac{\Gamma(1+\alpha)}{(1-\alpha)^{1/\alpha}}$ and $\lambda = \frac{\alpha}{1-\alpha}$.

3.4.1 Special cases: the kernel of Gauss function

As the special case, when $\kappa \geq 0$ and $\lambda \in \mathbb{C}$, the left-sided generalized derivative involving the kernel of Gauss function in the sense of Liouville-Sonine-Caputo type is defined as

$$ {}_{Gt}^{LSC} D_{a+}^{\kappa,\lambda} f(t) = \int_a^t e^{-\lambda(t-\tau)^2} f^{(\kappa)}(\tau)\, d\tau, \tag{3.131} $$

and the right-sided generalized derivative involving the kernel of Gauss function in the sense of Liouville-Sonine-Caputo type as

$$ {}_{Gt}^{LSC} D_{b-}^{\kappa,\lambda} f(t) = (-1)^{\kappa} \int_t^b e^{-\lambda(\tau-t)^2} f^{(\kappa)}(\tau)\, d\tau. \tag{3.132} $$

Definition 3.25 *The left-sided generalized derivative involving the kernel of Gauss function in the sense of Riemann-Liouville type is defined as*

$$ {}_{Gt}^{RL} D_{a+}^{\kappa,\lambda} f(t) = \frac{d^{\kappa}}{dt^{\kappa}} \int_a^t e^{-\lambda(t-\tau)^2} f(\tau)\, d\tau, \tag{3.133} $$

and the right-sided generalized derivative involving the kernel of Gauss function in the sense of Riemann-Liouville type as

$$ {}_{Gt}^{RL} D_{b-}^{\kappa,\lambda} f(t) = (-1)^{\kappa} \frac{d^{\kappa}}{dt^{\kappa}} \int_t^b e^{-\lambda(\tau-t)^2} f(\tau)\, d\tau. \tag{3.134} $$

When $\kappa = 1$, there are the relationships as follows:

$$ {}_{Gt}^{RL} D_{a+}^{\kappa,\lambda} f(t) = {}_{Gt}^{LSC} D_{a+}^{\kappa,\lambda} f(t) - e^{-\lambda(t-a)^2} f(a), \tag{3.135} $$

and

$$ {}_{Gt}^{RL} D_{b-}^{1,\lambda} f(t) = {}_{Gt}^{LSC} D_{b-}^{1,\lambda} f(t) - e^{-\lambda(b-t)^2} f(b). \tag{3.136} $$

Definition 3.26 *Let $\kappa \geq 1$, $-\infty < a < b < \infty$ and $\lambda \in \mathbb{C}$. The left-sided generalized integral, as the inverse operator of the generalized derivative involving the kernel of Gauss function with power law, is defined as*

$$ {}_{Gt}^{RL}I_{a+}^{\kappa,\lambda} f(t) = \int_a^t \frac{1}{(t-\tau)^{3-\kappa}} E_{2,\kappa-2}^{-1}\left(-\lambda (t-\tau)^2\right) f(\tau)\, d\tau, \tag{3.137} $$

and the right-sided generalized integral as

$$ {}_{Gt}^{RL}I_{b-}^{\kappa,\lambda} f(t) = (-1)^{\kappa} \int_t^b \frac{1}{(\tau-t)^{3-\kappa}} E_{2,\kappa-2}^{-1}\left(-\lambda (\tau-t)^2\right) f(\tau)\, d\tau. \tag{3.138} $$

Property 3.9 *Let $\kappa \geq 0$, $-\infty < a < b < \infty$ and $\lambda \in \mathbb{C}$. Then we have*

$$ L\left\{ {}_{Gt}^{RL}D_{0+}^{\kappa,\lambda} f(t) \right\} = s^{\kappa-1}\left(1+\lambda s^{-2}\right)^{-1} f(s), \tag{3.139} $$

$$ L\left\{ {}_{Gt}^{LSC}D_{0+}^{\kappa,\lambda} f(t) \right\} = s^{-1}\left(1+\lambda s^{-2}\right)^{-1} \left(s^{\kappa} f(s) - \sum_{j=1}^{\kappa} s^{\kappa-j} f^{(j-1)}(0) \right) \tag{3.140} $$

and

$$ L\left\{ {}_{Gt}^{RL}I_{0+}^{\kappa,\lambda} f(t) \right\} = s^{1-\kappa}\left(1+\lambda s^{-2}\right) f(s). \tag{3.141} $$

Note that, when $D^{\kappa}\left({}_{Gt}^{RL}I_{0+}^{\kappa,\lambda} f(t) \right)|_{t=0} = 0$, there is

$$ L\left\{ {}_{Ml}^{RL}D_{0+}^{\alpha,\kappa,\lambda} f(t) \right\} = s^{\kappa-1}\left(1+\lambda s^{-2}\right)^{-1} f(s) \tag{3.142} $$

and when $D^{\kappa}\left({}_{Gt}^{RL}I_{0+}^{\kappa,\lambda} f(t) \right)|_{t=0} \neq 0$, there is

$$ L\left\{ {}_{Gt}^{RL}D_{0+}^{\kappa,\lambda} f(t) \right\} = s^{\kappa-1}\left(1+\lambda s^{-2}\right)^{-1} f(s) - \sum_{\kappa=0}^{\kappa-1} s^{\kappa-\mu-1} \left(D^{\mu}\left[{}_{Gt}^{RL}I_{0+}^{\kappa,\lambda} f(+0) \right] \right). \tag{3.143} $$

In particular, when $\kappa = 1$, we have

$$ L\left\{ {}_{Gt}^{RL}D_{0+}^{1,\lambda} f(t) \right\} = \left(1+\lambda s^{-2}\right)^{-1} f(s) - {}_{Gt}^{RL}I_{0+}^{1,\lambda} f(+0). \tag{3.144} $$

Property 3.10 *Let $\kappa \geq 0$, $-\infty < a < b < \infty$, $\lambda \in \mathbb{C}$ and $f(t) \in L_l(a,b)$ $(1 \leq l < \infty)$. Then we have*

$$ \left({}_{Gt}^{RL}D_{a+}^{\kappa,\lambda} {}_{Gt}^{RL}I_{a+}^{\kappa,\lambda} f \right)(t) = f(t), \tag{3.145} $$

and

$$ \left({}_{Gt}^{RL}D_{+}^{\kappa,\lambda} {}_{Gt}^{RL}I_{+}^{\kappa,\lambda} f \right)(t) = f(t). \tag{3.146} $$

Property 3.11 *Let $\kappa \geq 0$, $-\infty < a < b < \infty$, $\lambda \in \mathbb{C}$ and $f(t) \in AC^{\kappa}(a,b)$. Then we have*

$$\left({}_{Gt}^{LCS}D_{a+}^{\kappa,\lambda}\, {}_{Gt}^{RL}I_{a+}^{\kappa,\lambda} f\right)(t) = f(t), \tag{3.147}$$

and

$$\left({}_{Gt}^{LCS}D_{b-}^{\alpha,\kappa,\lambda}\, {}_{Gt}^{RL}I_{b-}^{\alpha,\kappa,\lambda} f\right)(t) = f(t). \tag{3.148}$$

In fact, when $-\infty < a < b < \infty$, $\lambda \in \mathbb{C}$, $f(t) \in AC(a,b)$, $J(\alpha) = \frac{1+\alpha^2}{\sqrt{\pi^{\alpha}(1-\alpha)}}$ and $\lambda = \frac{\alpha}{1-\alpha}$, the left-sided generalized derivative involving the kernel of Gauss function in the sense of Liouville-Sonine-Caputo type can be expressed by (see [305])

$${}_{Gtl}^{LSC}D_{a+}^{1,\lambda} f(t) = J(\alpha)\int_a^t e^{-\lambda(t-\tau)^2} f^{(1)}(\tau)\, d\tau = \frac{1+\alpha^2}{\sqrt{\pi^{\alpha}(1-\alpha)}}\int_a^t e^{-\frac{\alpha}{1-\alpha}(t-\tau)^2} f^{(1)}(\tau)\, d\tau. \tag{3.149}$$

3.4.2 Special cases: the kernel of Gaussian-like function

On one hand, when $\kappa \geq 0$ and $\lambda \in \mathbb{C}$, the left-sided general fractional derivative involving the kernel of Gaussian-like function in the sense of Liouville-Sonine-Caputo type is defined as

$${}_{Gtl}^{LSC}D_{a+}^{\alpha,\kappa,\lambda} f(t) = \int_a^t e^{-\lambda(t-\tau)^{2\alpha}} f^{(\kappa)}(\tau)\, d\tau, \tag{3.150}$$

and the right-sided general fractional derivative involving the kernel of Gaussian-like function in the sense of Liouville-Sonine-Caputo type as

$${}_{Gtl}^{LSC}D_{b-}^{\alpha,\kappa,\lambda} f(t) = (-1)^{\kappa}\int_t^b e^{-\lambda(\tau-t)^{2\alpha}} f^{(\kappa)}(\tau)\, d\tau. \tag{3.151}$$

On the other hand, the left-sided general fractional derivative involving the kernel of Gaussian-like function in the sense of Riemann-Liouville type can be expressed as

$${}_{Gtl}^{RL}D_{a+}^{\alpha,\kappa,\lambda} f(t) = \frac{d^{\kappa}}{dt^{\kappa}}\int_a^t e^{-\lambda(t-\tau)^{2\alpha}} f(\tau)\, d\tau, \tag{3.152}$$

and the right-sided general fractional derivative involving the kernel of Gaussian-like function in the sense of Riemann-Liouville type as

$${}_{Gtl}^{RL}D_{b-}^{\alpha,\kappa,\lambda} f(t) = (-1)^{\kappa}\frac{d^{\kappa}}{dt^{\kappa}}\int_t^b e^{-\lambda(\tau-t)^{2\alpha}} f(\tau)\, d\tau. \tag{3.153}$$

When $\kappa = 1$, the formulations

$$ {}^{RL}_{Gtl}D^{\alpha,1,\lambda}_{a+} f(t) = {}^{LSC}_{Gtl}D^{\alpha,1,\lambda}_{a+} f(t) - e^{-\lambda(t-a)^{2\alpha}} f(a) \tag{3.154}$$

and

$$ {}^{RL}_{Gtl}D^{\alpha,1,\lambda}_{b-} f(t) = {}^{LSC}_{Gtl}D^{\alpha,1,\lambda}_{b-} f(t) - e^{-\lambda(b-t)^{2\alpha}} f(b) \tag{3.155}$$

holds.

As a special case, the left-sided general fractional derivative involving the kernel of Gaussian-like function in the sense of Liouville-Sonine type, introduced by Yang in 2017, is can be represented in the form ([299])

$$ {}^{LSC}_{Gtl}D^{\alpha,\lambda}_{a+} f(t) = J(\alpha)\, {}^{LSC}_{Gtl}D^{\alpha,1,\lambda}_{a+} f(t) = J(\alpha) \int_a^t e^{-\frac{\alpha}{1-\alpha}(t-\tau)^{2\alpha}} f^{(1)}(\tau)\, d\tau, \tag{3.156}$$

where $J(\alpha) = \frac{1+\alpha^2}{\sqrt{\pi^\alpha (1-\alpha)}}$ and $\lambda = \frac{\alpha}{1-\alpha}$.

3.5 General fractional derivatives in the Miller-Ross kernel

The definitions and properties of the general fractional derivatives in the Miller-Ross kernel and the general fractional integrals are presented as follows.

Definition 3.27 *Let $\alpha \in \mathbb{C}$, $1 > Re(\alpha) > 0$, $-\infty < a < b < \infty$ and $\lambda \in \mathbb{C}$. The left-sided general fractional derivative within the Miller-Ross kernel in the sense of Riemann-Liouville type is defined as*

$$ \left({}^{RL}_{MR}D^{\alpha,\lambda}_{a+} f \right)(t) = \frac{d}{dt}\left({}_{MR}I^{\alpha,\lambda}_{a+} f(t) \right) = \frac{d}{dt} \int_a^t \mathrm{M}_\alpha \left(-\lambda (t-\tau)^\alpha\right) f(\tau)\, d\tau, \tag{3.157}$$

and the right-sided general fractional derivative within the Miller-Ross kernel in the sense of Riemann-Liouville type as

$$ \left({}^{RL}_{MR}D^{\alpha,\lambda}_{b-} f \right)(t) = -\frac{d}{dt}\left({}_{MR}I^{\alpha,\lambda}_{b-} f(t) \right) = -\frac{d}{dt} \int_t^b \mathrm{M}_\alpha \left(-\lambda (\tau-t)^\alpha\right) f(\tau)\, d\tau, \tag{3.158}$$

where the left-sided and right-sided general fractional integral operators within the Miller-Ross kernel are defined as

$$ {}_{MR}I^{\alpha,\lambda}_{a+} f(t) = \int_a^t \mathrm{M}_\alpha \left(-\lambda (t-\tau)^\alpha\right) f(\tau)\, d\tau \tag{3.159}$$

and

$$ {}_{MR}I_{b-}^{\alpha,\lambda} f(t) = \int_t^b \mathrm{M}_\alpha\left(-\lambda(\tau - t)^\alpha\right) f(\tau)\, d\tau, \tag{3.160}$$

respectively.

Furthermore, when $\alpha \in \mathbb{C}$, $Re(\alpha) > 0$, $\kappa = [Re(\alpha)]+1$, $-\infty < a < b < \infty$ and $\lambda \in \mathbb{C}$, the left-sided general fractional derivative within the Miller-Ross kernel in the sense of Riemann-Liouville type is defined as

$$\left({}_{MR}^{RL}D_{a+}^{\alpha,\lambda} f\right)(t) = \frac{d^\kappa}{dt^\kappa}\left({}_{MR}I_{a+}^{\alpha,\lambda} f(t)\right) = \frac{d^\kappa}{dt^\kappa}\int_a^t \mathrm{M}_\alpha\left(-\lambda(t-\tau)^\alpha\right) f(\tau)\, d\tau, \tag{3.161}$$

and the right-sided general fractional derivative within the Miller-Ross kernel in the sense of Riemann-Liouville type as

$$\left({}_{MR}^{RL}D_{b-}^{\alpha,\lambda} f\right)(t) = \left(-\frac{d}{dt}\right)^\kappa \left({}_{MR}I_{b-}^{\alpha,\lambda} f(t)\right) = (-1)^\kappa \frac{d^\kappa}{dt^\kappa}\int_t^b \mathrm{M}_\alpha\left(-\lambda(\tau - t)^\alpha\right) f(\tau)\, d\tau. \tag{3.162}$$

On the other hand, when $\alpha \in \mathbb{C}$, $1 > Re(\alpha) > 0$, $-\infty < a < b < \infty$ and $\lambda \in \mathbb{C}$, the left-sided general fractional derivative within the Miller-Ross kernel in the sense of Liouville-Sonine type is defined as

$$\left({}_{MR}^{LS}D_{a+}^{\alpha,\lambda} f\right)(t) = {}_{MR}I_{a+}^{\alpha,\lambda}\left(f^{(1)}(t)\right) = \int_a^t \mathrm{M}_\alpha\left(-\lambda(t-\tau)^\alpha\right) f^{(1)}(\tau)\, d\tau, \tag{3.163}$$

and the right-sided general fractional derivative within the Miller-Ross kernel in the sense of Liouville-Sonine type as

$$\left({}_{MR}^{LS}D_{b-}^{\alpha,\lambda} f\right)(t) = {}_{MR}I_{b-}^{\alpha,\lambda}\left(-f^{(1)}(t)\right) = -\int_t^b \mathrm{M}_\alpha\left(-\lambda(\tau - t)^\alpha\right) f^{(1)}(\tau)\, d\tau. \tag{3.164}$$

Furthermore, when $\alpha \in \mathbb{C}$, $\kappa + 1 > Re(\alpha) > \kappa$, $-\infty < a < b < \infty$ and $\lambda \in \mathbb{C}$, the left-sided general fractional derivative within the Miller-Ross kernel in the sense of Liouville-Sonine-Caputo type is defined as

$$\left({}_{MR}^{LSC}D_{a+}^{\alpha,\lambda} f\right)(t) = {}_{MR}I_{a+}^{\alpha,\lambda}\left(f^{(\kappa)}(t)\right) = \int_a^t \mathrm{M}_\alpha\left(-\lambda(t-\tau)^\alpha\right) f^{(\kappa)}(\tau)\, d\tau, \tag{3.165}$$

and the right-sided general fractional derivative within the Miller-Ross kernel in the sense of Liouville-Sonine-Caputo type as

$$\left({}_{MR}^{LSC}D_{b-}^{\alpha,\lambda} f\right)(t) = {}_{MR}I_{b-}^{\alpha,\lambda}\left((-1)^\kappa f^{(\kappa)}(t)\right) = (-1)^\kappa \int_t^b \mathrm{M}_\alpha\left(-\lambda(\tau - t)^\alpha\right) f^{(\kappa)}(\tau)\, d\tau. \tag{3.166}$$

We now consider the inverse operator of the above general fractional derivatives within the Miller-Ross kernel as follows.

Definition 3.28 *Let $\alpha \in \mathbb{C}$, $\kappa + 1 > Re(\alpha) > \kappa$, $-\infty < a < b < \infty$ and $\lambda \in \mathbb{C}$. The left-sided general fractional integral is defined as*

$$\left({}_{MR}^{R}I_{a+}^{\alpha,\lambda} f\right)(t) = \int_a^t \frac{1}{(t-\tau)^{\alpha+2-\kappa}} E_{1,\kappa-\alpha-1}\left(-\lambda(t-\tau)\right) f(\tau)\, d\tau, \tag{3.167}$$

and the right-sided general fractional integral as

$$\left({}_{MR}^{R}I_{b-}^{\alpha,\lambda} f\right) = \int_t^b \frac{1}{(t-\tau)^{\alpha+2-\kappa}} E_{1,\kappa-\alpha-1}\left(-\lambda(\tau-t)\right) f(\tau)\, d\tau. \tag{3.168}$$

In particular, when $\alpha \in \mathbb{C}$, $1 > Re(\alpha) > 0$, $-\infty < a < b < \infty$ and $\lambda \in \mathbb{C}$, the left-sided general fractional integral is defined as

$$\left({}_{MR}^{R}I_{a+}^{\alpha,\lambda} f\right)(t) = \int_a^t \frac{1}{(t-\tau)^{\alpha+1}} E_{1,-\alpha}\left(-\lambda(t-\tau)\right) f(\tau)\, d\tau, \tag{3.169}$$

and the right-sided general fractional integral as

$$\left({}_{MR}^{R}I_{b-}^{\alpha,\lambda} f\right) = \int_t^b \frac{1}{(\tau-t)^{\alpha+1}} E_{1,-\alpha}\left(-\lambda(\tau-t)\right) f(\tau)\, d\tau. \tag{3.170}$$

As the direct results, the properties of the general fractional derivatives and the general fractional integrals are given as follows.

Property 3.12 *If $\kappa + 1 > \alpha > \kappa$, $-\infty < a < b < \infty$, $\lambda \in \mathbb{C}$ and $f(t) \in L_l(a,b)$ $(1 \le l < \infty)$, then we have*

$$L\left\{\left({}_{MR}^{RL}D_{0+}^{\alpha,\lambda} f\right)(t)\right\} = s^{\kappa-\alpha-1}\left(1+\lambda s^{-1}\right)^{-1} f(s), \tag{3.171}$$

$$L\left\{\left({}_{MR}^{RL}D_{0+}^{\alpha,\lambda} f\right)(t)\right\} = s^{-\alpha}\left(1+\lambda s^{-1}\right)^{-1} f(s), \tag{3.172}$$

$$L\left\{\left({}_{MR}^{LS}D_{0+}^{\alpha,\lambda} f\right)(t)\right\} = s^{-\alpha-1}\left(1+\lambda s^{-1}\right)^{-1}\left(sf(s) - f(0)\right), \tag{3.173}$$

$$L\left\{\left({}_{MR}^{LSC}D_{0+}^{\alpha,\lambda} f\right)(t)\right\} = s^{-\alpha-1}\left(1+\lambda s^{-1}\right)^{-1}\left(s^{\kappa} f(s) - \sum_{j=1}^{\kappa} s^{\kappa-j} f^{(j-1)}(0)\right), \tag{3.174}$$

$$L\left\{\left({}_{MR}^{R}I_{0+}^{\alpha,\lambda} f\right)(t)\right\} = s^{\alpha+1-\kappa}\left(1+\lambda s^{-1}\right) f(s) \tag{3.175}$$

and

$$L\left\{\left({}_{MR}I_{0+}^{\alpha,\lambda} f\right)(t)\right\} = s^{\alpha+1}\left(1+\lambda s^{-1}\right) f(s). \tag{3.176}$$

Note that, when $D^{\kappa}\left({}^{R}_{MR}I^{\alpha,\lambda}_{0+}f(t)\right)|_{t=0}=0$, there is

$$L\left\{{}^{RL}_{MR}D^{\alpha,\lambda}_{0+}f(t)\right\}=s^{\kappa-\gamma}\left(1+\lambda s^{-\alpha}\right)^{-1}f(s), \tag{3.177}$$

and when $D^{\kappa}\left({}^{R}_{MR}I^{\alpha,\lambda}_{0+}f(t)\right)|_{t=0}\neq 0$, there is

$$L\left\{{}^{RL}_{MR}D^{\alpha,\lambda}_{0+}f(t)\right\}=s^{\kappa-\gamma}\left(1+\lambda s^{-\alpha}\right)^{-1}f(s)-\sum_{\mu=0}^{\kappa-1}s^{\kappa-\mu-1}\left(D^{\mu}\left[{}^{R}_{MR}I^{\alpha,\lambda}_{0+}f(+0)\right]\right). \tag{3.178}$$

In particular, when $\kappa=1$, we have

$$L\left\{{}^{RL}_{MR}D^{\alpha,\lambda}_{0+}f(t)\right\}=s^{1-\gamma}\left(1+\lambda s^{-\alpha}\right)^{-1}f(s)-{}^{R}_{MR}I^{\alpha,\lambda}_{0+}f(+0), \tag{3.179}$$

and when $\kappa=1$ and $D\left({}^{R}_{MR}I^{\alpha,\lambda}_{0+}f(t)\right)|_{t=0}=0$, we have

$$L\left\{{}^{RL}_{MR}D^{\alpha,\lambda}_{0+}f(t)\right\}=s^{1-\gamma}\left(1+\lambda s^{-\alpha}\right)^{-1}f(s). \tag{3.180}$$

Property 3.13 *Let $\kappa+1>\alpha>\kappa$, $-\infty<a<b<\infty$, $\lambda\in\mathbb{C}$ and $f(t)\in L_l(a,b)$ $(1\leq l<\infty)$. Then we have*

$$\left({}^{RL}_{MR}D^{\alpha,\lambda}_{a+}{}^{R}_{MR}I^{\alpha,\lambda}_{a+}f\right)(t)=f(t), \tag{3.181}$$

$$\left({}^{RL}_{MR}D^{\alpha,\lambda}_{b-}{}^{R}_{MR}I^{\alpha,\lambda}_{b-}f\right)(t)=f(t), \tag{3.182}$$

$$\left({}^{RL}_{MR}D^{\alpha,\lambda}_{a+}{}^{R}_{MR}I^{\alpha,\lambda}_{a+}f\right)(t)=f(t), \tag{3.183}$$

and

$$\left({}^{RL}_{MR}D^{\alpha,\lambda}_{b-}{}^{R}_{MR}I^{\alpha,\lambda}_{b-}f\right)(t)=f(t). \tag{3.184}$$

Property 3.14 *If $-\infty<a<b<\infty$, $\lambda\in\mathbb{C}$ and $f(t)\in AC_{\kappa}(a,b)$ $(1\leq\kappa<\infty)$, then we have*

$$\left({}^{LS}_{MR}D^{\alpha,\lambda}_{a+}{}^{R}_{MR}I^{\alpha,\lambda}_{a+}f\right)(t)=f(t), \tag{3.185}$$

$$\left({}^{LS}_{MR}D^{\alpha,\lambda}_{b-}{}^{R}_{MR}I^{\alpha,\lambda}_{b-}f\right)(t)=f(t), \tag{3.186}$$

$$\left({}^{LSC}_{MR}D^{\alpha,\lambda}_{a+}{}^{R}_{MR}I^{\alpha,\lambda}_{a+}f\right)(t)=f(t), \tag{3.187}$$

and

$$\left({}^{LSC}_{MR}D^{\alpha,\lambda}_{b-}{}^{R}_{MR}I^{\alpha,\lambda}_{b-}f\right)(t)=f(t). \tag{3.188}$$

Definition 3.29 *Let $\alpha \in \mathbb{C}$, $1 > Re(\alpha) > 0$, $a = -\infty$, $b = \infty$ and $\lambda \in \mathbb{C}$. The left-sided general fractional derivative within the Miller-Ross kernel in the sense of Riemann-Liouville type is defined as*

$$\left({}^{RL}_{MR}D^{\alpha,\lambda}_{+}f\right)(t) = \frac{d}{dt}\left({}_{MR}I^{\alpha,\lambda}_{+}f(t)\right) = \frac{d}{dt}\int_{-\infty}^{t} \mathrm{M}_{\alpha}\left(-\lambda(t-\tau)^{\alpha}\right) f(\tau)\, d\tau, \tag{3.189}$$

and the right-sided general fractional derivative within the Miller-Ross kernel in the sense of Riemann-Liouville type as

$$\left({}^{RL}_{MR}D^{\alpha,\lambda}_{-}f\right)(t) = -\frac{d}{dt}\left({}_{MR}I^{\alpha,\lambda}_{-}f(t)\right) = -\frac{d}{dt}\int_{t}^{\infty} \mathrm{M}_{\alpha}\left(-\lambda(\tau-t)^{\alpha}\right) f(\tau)\, d\tau, \tag{3.190}$$

where the left-sided and right-sided general fractional integral operators within the Miller-Ross kernel are defined as

$${}_{MR}I^{\alpha,\lambda}_{+}f(t) = \int_{-\infty}^{t} \mathrm{M}_{\alpha}\left(-\lambda(t-\tau)^{\alpha}\right) f(\tau)\, d\tau \tag{3.191}$$

and

$${}_{MR}I^{\alpha,\lambda}_{-}f(t) = \int_{t}^{\infty} \mathrm{M}_{\alpha}\left(-\lambda(\tau-t)^{\alpha}\right) f(\tau)\, d\tau, \tag{3.192}$$

respectively.

Furthermore, when $\alpha \in \mathbb{C}$, $Re(\alpha) > 0$, $\kappa = [Re(\alpha)] + 1$, $a = -\infty$, $b = \infty$ and $\lambda \in \mathbb{C}$, the left-sided general fractional derivative within the Miller-Ross kernel in the sense of Riemann-Liouville type is defined as

$$\left({}^{RL}_{MR}D^{\alpha,\lambda}_{+}f\right)(t) = \frac{d^{\kappa}}{dt^{\kappa}}\left({}_{MR}I^{\alpha,\lambda}_{+}f(t)\right) = \frac{d^{\kappa}}{dt^{\kappa}}\int_{-\infty}^{t} \mathrm{M}_{\alpha}\left(-\lambda(t-\tau)^{\alpha}\right) f(\tau)\, d\tau, \tag{3.193}$$

and the right-sided general fractional derivative within the Miller-Ross kernel in the sense of Riemann-Liouville type as

$$\left({}^{RL}_{MR}D^{\alpha,\lambda}_{-}f\right)(t) = \left(-\frac{d}{dt}\right)^{\kappa}\left({}_{MR}I^{\alpha,\lambda}_{-}f(t)\right) = (-1)^{\kappa}\frac{d^{\kappa}}{dt^{\kappa}}\int_{t}^{\infty} \mathrm{M}_{\alpha}\left(-\lambda(\tau-t)^{\alpha}\right) f(\tau)\, d\tau. \tag{3.194}$$

Furthermore, when $\alpha \in \mathbb{C}$, $\kappa+1 > Re(\alpha) > \kappa$, $a = -\infty$, $b = \infty$ and $\lambda \in \mathbb{C}$, the left-sided general fractional derivative within the Miller-Ross kernel in the sense of Liouville-Sonine-Caputo type is defined as

$$\left({}^{LSC}_{MR}D^{\alpha,\lambda}_{+}f\right)(t) = {}_{MR}I^{\alpha,\lambda}_{+}\left(f^{(\kappa)}(t)\right) = \int_{-\infty}^{t} \mathrm{M}_{\alpha}\left(-\lambda(t-\tau)^{\alpha}\right) f^{(\kappa)}(\tau)\, d\tau \tag{3.195}$$

and the right-sided general fractional derivative within the Miller-Ross kernel in the sense of Liouville-Sonine-Caputo type as

$$\left({}^{LSC}_{MR}D^{\alpha,\lambda}_{-}f\right)(t)={}_{MR}I^{\alpha,\lambda}_{-}\left((-1)^{\kappa}f^{(\kappa)}(t)\right)=(-1)^{\kappa}\int\limits_{t}^{\infty}\mathrm{M}_{\alpha}(-\lambda(\tau-t)^{\alpha})\,f^{(\kappa)}(\tau)\,d\tau. \tag{3.196}$$

The inverse operators of the above fractional derivatives within the Miller-Ross kernel are considered as follows.

Definition 3.30 *Let $\alpha\in\mathbb{C}$, $\kappa+1>Re(\alpha)>\kappa$, $a=-\infty$, $b=\infty$ and $\lambda\in\mathbb{C}$. The left-sided general fractional integral is defined as*

$$\left({}^{R}_{MR}I^{\alpha,\lambda}_{+}f\right)(t)=\int\limits_{-\infty}^{t}\frac{1}{(t-\tau)^{\alpha+2-\kappa}}E_{1,\kappa-\alpha-1}\left(-\lambda\left(t-\tau\right)\right)f\left(\tau\right)d\tau \tag{3.197}$$

and the right-sided general fractional integral as

$$\left({}^{R}_{MR}I^{\alpha,\lambda}_{-}f\right)=\int\limits_{t}^{\infty}\frac{1}{(t-\tau)^{\alpha+2-\kappa}}E_{1,\kappa-\alpha-1}\left(-\lambda\left(\tau-t\right)\right)f\left(\tau\right)d\tau. \tag{3.198}$$

Definition 3.31 *Let $\kappa+1>\alpha>\kappa$, $a=-\infty$, $b=\infty$ and $\lambda\in\mathbb{C}$. The left-sided general fractional integral is defined as*

$$\left({}^{R}_{MR}I^{\alpha,\lambda}_{+}f\right)(t)=\int\limits_{-\infty}^{t}\frac{1}{(t-\tau)^{\alpha+2-\kappa}}E_{1,\kappa-\alpha-1}\left(-\lambda\left(t-\tau\right)\right)f\left(\tau\right)d\tau \tag{3.199}$$

and the right-sided general fractional integral as

$$\left({}^{R}_{A}I^{\alpha,\gamma,\lambda}_{-}f\right)=\int\limits_{t}^{-\infty}\frac{1}{(t-\tau)^{\alpha+2-\kappa}}E_{1,\kappa-\alpha-1}\left(-\lambda\left(\tau-t\right)\right)f\left(\tau\right)d\tau. \tag{3.200}$$

We now give the properties of the general fractional derivatives and the general fractional integrals as follows.

Property 3.15 *Let $\kappa+1>\alpha>\kappa$, $a=-\infty$, $b=\infty$, $\lambda\in\mathbb{C}$ and $f(t)\in L_l(-\infty,\infty)$ $(1\le l<\infty)$. Then we have*

$$\left({}^{RL}_{MR}D^{\alpha,\lambda}_{+}\,{}^{R}_{MR}I^{\alpha,\lambda}_{+}f\right)(t)=f\left(t\right), \tag{3.201}$$

$$\left({}^{RL}_{MR}D^{\alpha,\lambda}_{-}\,{}^{R}_{MR}I^{\alpha,\lambda}_{-}f\right)(t)=f\left(t\right), \tag{3.202}$$

$$\left({}^{RL}_{MR}D^{\alpha,\lambda}_{+}\,{}^{R}_{MR}I^{\alpha,\lambda}_{+}f\right)(t)=f\left(t\right), \tag{3.203}$$

and

$$\left({}^{RL}_{MR}D^{\alpha,\lambda}_{-}\,{}^{R}_{MR}I^{\alpha,\lambda}_{-}f\right)(t)=f\left(t\right). \tag{3.204}$$

Property 3.16 *If* $a = -\infty$, $b = \infty$, $\lambda \in \mathbb{C}$ *and* $f(t) \in AC_\kappa(-\infty,\infty)$ $(1 \leq \kappa < \infty)$, *then we have*

$$\left({}_{MR}^{LS}D_{+}^{\alpha,\lambda R}{}_{MR}I_{+}^{\alpha,\lambda}f\right)(t) = f(t), \tag{3.205}$$

$$\left({}_{MR}^{LS}D_{-}^{\alpha,\lambda R}{}_{MR}I_{-}^{\alpha,\lambda}f\right)(t) = f(t), \tag{3.206}$$

$$\left({}_{MR}^{LSC}D_{+}^{\alpha,\lambda R}{}_{MR}I_{+}^{\alpha,\lambda}f\right)(t) = f(t), \tag{3.207}$$

and

$$\left({}_{MR}^{LSC}D_{-}^{\alpha,\lambda R}{}_{MR}I_{-}^{\alpha,\lambda}f\right)(t) = f(t). \tag{3.208}$$

3.6 Hilfer type general fractional derivatives in the Miller-Ross kernel

The definitions of the Hilfer type general fractional derivatives containing the kernel of the Miller-Ross function are given as follows.

Definition 3.32 *Let* $1 > \alpha > 0$, $0 \leq \beta \leq 1$, $-\infty < a < b < \infty$, $\gamma \in \mathbb{C}$ *and* $\lambda \in \mathbb{C}$. *The left-sided Hilfer type general fractional derivative containing the kernel of the Miller-Ross function is defined as*

$$\begin{aligned}\left({}_{MR}^{Hi}D_{a+}^{\alpha,\beta,\lambda}f\right)(t) &= \int_a^t \mathrm{M}_\alpha\left(-\lambda(t-\tau)^\alpha\right)\left(D_{a+}^{\alpha,\beta}f(\tau)\right)d\tau \\ &= \int_a^t \mathrm{M}_\alpha\left(-\lambda(t-\tau)^\alpha\right)\left(I_{a+}^{\beta(1-\alpha)}\frac{d}{d\tau}\left(I_{a+}^{(1-\beta)(1-\alpha)}f\right)\right)(\tau)\,d\tau,\end{aligned} \tag{3.209}$$

and the right-sided Hilfer type general fractional derivative containing the kernel of the Miller-Ross function as

$$\begin{aligned}\left({}_{MR}^{Hi}D_{b-}^{\alpha,\beta,\lambda}f\right)(t) &= -\int_t^b \mathrm{M}_\alpha\left(-\lambda(\tau-t)^\alpha\right)\left(D_{b-}^{\alpha,\beta}f(\tau)\right)d\tau \\ &= -\int_t^b \mathrm{M}_\alpha\left(-\lambda(\tau-t)^\alpha\right)\left(I_{b-}^{\beta(1-\alpha)}\frac{d}{d\tau}\left(I_{b-}^{(1-\beta)(1-\alpha)}f\right)\right)(\tau)\,d\tau.\end{aligned} \tag{3.210}$$

Definition 3.33 *Let* $\kappa + 1 > \alpha > \kappa$, $0 \leq \beta \leq 1$, $-\infty < a < b < \infty$, $\gamma \in \mathbb{C}$ *and* $\lambda \in \mathbb{C}$. *The left-sided Hilfer type general fractional derivative containing the kernel of the Miller-Ross function is defined as*

$$\begin{aligned}\left({}_{MR}^{Hir}D_{a+}^{\alpha,\beta,\lambda}f\right)(t) &= \int_a^t \mathrm{M}_\alpha\left(-\lambda(t-\tau)^\alpha\right)\left(D_{a+}^{\alpha,\beta,\kappa}f(\tau)\right)d\tau \\ &= \int_a^t \mathrm{M}_\alpha\left(-\lambda(t-\tau)^\alpha\right)\left(I_{a+}^{\beta(\kappa-\alpha)}\frac{d^\kappa}{d\tau^\kappa}\left(I_{a+}^{(1-\beta)(\kappa-\alpha)}f\right)\right)(\tau)\,d\tau,\end{aligned} \tag{3.211}$$

and the right-sided Hilfer type general fractional derivative containing the kernel of the Miller-Ross function as

$$\begin{aligned}&\left({}_{MR}^{Hir}D_{b-}^{\alpha,\beta,\lambda}f\right)(t)=(-1)^{\kappa}\int_{t}^{b}\mathrm{M}_{\alpha}\left(-\lambda\left(\tau-t\right)^{\alpha}\right)\left(D_{b-}^{\alpha,\beta,\kappa}f\left(\tau\right)\right)d\tau\\&=(-1)^{\kappa}\int_{t}^{b}\mathrm{M}_{\alpha}\left(-\lambda\left(\tau-t\right)^{\alpha}\right)\left(I_{b-}^{\beta(\kappa-\alpha)}\frac{d^{\kappa}}{d\tau^{\kappa}}\left(I_{b-}^{(1-\beta)(\kappa-\alpha)}f\right)\right)(\tau)\,d\tau.\end{aligned}\tag{3.212}$$

Definition 3.34 *Let $\kappa+1>\alpha>\kappa$, $\beta=1$, $-\infty<a<b<\infty$, $\gamma\in\mathbb{C}$ and $\lambda\in\mathbb{C}$. The left-sided Liouville-Sonine-Caputo type general fractional derivative containing the kernel of the Miller-Ross function is defined as*

$$\begin{aligned}&\left({}_{MR}^{Hir}D_{a+}^{\alpha,1,\lambda}f\right)(t)=\int_{a}^{t}\mathrm{M}_{\alpha}\left(-\lambda\left(t-\tau\right)^{\alpha}\right)\left({}_{LSC}D_{a+}^{\alpha,\kappa}f\left(\tau\right)\right)d\tau\\&=\int_{a}^{t}\mathrm{M}_{\alpha}\left(-\lambda\left(t-\tau\right)^{\alpha}\right)\left(I_{a+}^{\kappa-\alpha}f^{(\kappa)}\left(\tau\right)\right)d\tau,\end{aligned}\tag{3.213}$$

and the right-sided Liouville-Sonine-Caputo type general fractional derivative containing the kernel of the Miller-Ross function as

$$\begin{aligned}&\left({}_{MR}^{Hir}D_{b-}^{\alpha,1,\lambda}f\right)(t)=(-1)^{\kappa}\int_{t}^{b}\mathrm{M}_{\alpha}\left(-\lambda\left(\tau-t\right)^{\alpha}\right)\left({}_{LSC}D_{b-}^{\alpha,\kappa}f\left(\tau\right)\right)d\tau\\&=(-1)^{\kappa}\int_{t}^{b}\mathrm{M}_{\alpha}\left(-\lambda\left(\tau-t\right)^{\alpha}\right)\left(I_{b-}^{\kappa-\alpha}f^{(\kappa)}\left(\tau\right)\right)(\tau)\,d\tau.\end{aligned}\tag{3.214}$$

Definition 3.35 *Let $\kappa+1>\alpha>\kappa$, $\beta=0$, $-\infty<a<b<\infty$, $\gamma\in\mathbb{C}$ and $\lambda\in\mathbb{C}$. The left-sided Riemann-Liouville type general fractional derivative containing the kernel of the Miller-Ross function is defined as*

$$\begin{aligned}&\left({}_{MR}^{Hir}D_{a+}^{\alpha,0,\lambda}f\right)(t)=\int_{a}^{t}\mathrm{M}_{\alpha}\left(-\lambda\left(t-\tau\right)^{\alpha}\right)\left(D_{a+}^{\alpha,\kappa}f\left(\tau\right)\right)d\tau\\&=\int_{a}^{t}\mathrm{M}_{\alpha}\left(-\lambda\left(t-\tau\right)^{\alpha}\right)\left(\frac{d^{\kappa}}{d\tau^{\kappa}}\left(I_{a+}^{\kappa-\alpha}f\right)(\tau)\right)d\tau\end{aligned}\tag{3.215}$$

and the right-sided Riemann-Liouville type general fractional derivative containing the kernel of the Miller-Ross function as

$$\begin{aligned}&\left({}_{MR}^{Hir}D_{b-}^{\alpha,0,\lambda}f\right)(t)=(-1)^{\kappa}\int_{t}^{b}\mathrm{M}_{\alpha}\left(-\lambda\left(\tau-t\right)^{\alpha}\right)\left(D_{b-}^{\alpha,\kappa}f\left(\tau\right)\right)d\tau\\&=(-1)^{\kappa}\int_{t}^{b}\mathrm{M}_{\alpha}\left(-\lambda\left(\tau-t\right)^{\alpha}\right)\left(\frac{d^{\kappa}}{d\tau^{\kappa}}\left(I_{b-}^{\kappa-\alpha}f\right)(\tau)\right)d\tau.\end{aligned}\tag{3.216}$$

3.7 General fractional derivatives in the one-parametric Lorenzo-Hartley kernel

The definitions and properties of the general fractional derivatives in the one-parametric Lorenzo-Hartley kernel and the general fractional integrals are given as follows.

Definition 3.36 *Let $\alpha \in \mathbb{C}$, $1 > Re(\alpha) > 0$, $-\infty < a < b < \infty$ and $\gamma \in \mathbb{C}$. The left-sided Riemann-Liouville type general fractional derivative involving the one-parametric Lorenzo-Hartley kernel is defined as*

$$\left({}_{A}^{RL}D_{a+}^{\alpha,\gamma} f\right)(t) = \tfrac{d}{dt}\left({}_{A}I_{a+}^{\alpha,\gamma} f(t)\right) = \tfrac{d}{dt}\int_a^t F_\alpha\left(-\gamma (t-\tau)^\alpha\right) f(\tau)\, d\tau, \qquad (3.217)$$

and the right-sided Riemann-Liouville type general fractional derivative involving the one-parametric Lorenzo-Hartley kernel as

$$\left({}_{A}^{RL}D_{b-}^{\alpha,\gamma} f\right)(t) = -\tfrac{d}{dt}\left({}_{A}I_{b-}^{\alpha,\gamma} f(t)\right) = -\tfrac{d}{dt}\int_t^b F_\alpha\left(-\gamma (\tau-t)^\alpha\right) f(\tau)\, d\tau, \qquad (3.218)$$

where the left-sided and right-sided general fractional integral operators involving the one-parametric Lorenzo-Hartley kernel are defined as

$${}_{A}I_{a+}^{\alpha,\gamma} f(t) = \int_a^t F_\alpha\left(-\gamma (t-\tau)^\alpha\right) f(\tau)\, d\tau \qquad (3.219)$$

and

$${}_{A}I_{b-}^{\alpha,\gamma} f(t) = \int_t^b F_\alpha\left(-\gamma (\tau-t)^\alpha\right) f(\tau)\, d\tau, \qquad (3.220)$$

respectively.

Definition 3.37 *Let $\alpha \in \mathbb{C}$, $Re(\alpha) > 0$, $\kappa = [Re(\alpha)] + 1$, $-\infty < a < b < \infty$ and $\gamma \in \mathbb{C}$. The left-sided Riemann-Liouville type general fractional derivative involving the one-parametric Lorenzo-Hartley kernel is defined as*

$$\left({}_{A}^{RL}D_{a+}^{\alpha,\gamma} f\right)(t) = \tfrac{d^\kappa}{dt^\kappa}\int_a^t F_\alpha\left(-\gamma (t-\tau)^\alpha\right) f(\tau)\, d\tau, \qquad (3.221)$$

and the right-sided Riemann-Liouville type general fractional derivative involving the one-parametric Lorenzo-Hartley kernel as

$$\left({}_{A}^{RL}D_{b-}^{\alpha,\gamma} f\right)(t) = (-1)^\kappa \tfrac{d^\kappa}{dt^\kappa}\int_t^b F_\alpha\left(-\gamma (\tau-t)^\alpha\right) f(\tau)\, d\tau. \qquad (3.222)$$

Definition 3.38 *Let $\alpha \in \mathbb{C}$, $1 > Re(\alpha) > 0$, $-\infty < a < b < \infty$ and $\gamma \in \mathbb{C}$. The left-sided Liouville-Sonine type general fractional derivative involving the one-parametric Lorenzo-Hartley kernel is defined as*

$$\left({}_{A}^{LS}D_{a+}^{\alpha,\gamma} f\right)(t) = {}_{A}I_{a+}^{\alpha,\gamma}\left(f^{(1)}(t)\right) = \int_{a}^{t} F_{\alpha}\left(-\gamma(t-\tau)^{\alpha}\right) f^{(1)}(\tau)\, d\tau, \quad (3.223)$$

and the right-sided Liouville-Sonine type general fractional derivative involving the one-parametric Lorenzo-Hartley kernel as

$$\left({}_{A}^{LS}D_{b-}^{\alpha,\lambda} f\right)(t) = {}_{A}I_{b-}^{\alpha,\gamma}\left(-f^{(1)}(t)\right) = -\int_{t}^{b} F_{\alpha}\left(-(\tau-t)^{\alpha}\right) f^{(1)}(\tau)\, d\tau. \quad (3.224)$$

Definition 3.39 *Let $\alpha \in \mathbb{C}$, $\kappa+1 > Re(\alpha) > \kappa$, $-\infty < a < b < \infty$ and $\gamma \in \mathbb{C}$. The left-sided Liouville-Sonine-Caputo type general fractional derivative involving the one-parametric Lorenzo-Hartley kernel is defined as*

$$\left({}_{A}^{LSC}D_{a+}^{\alpha,\gamma} f\right)(t) = \int_{a}^{t} F_{\alpha}\left(-\gamma(t-\tau)^{\alpha}\right) f^{(\kappa)}(\tau)\, d\tau, \quad (3.225)$$

and the right-sided Liouville-Sonine-Caputo type general fractional derivative involving the one-parametric Lorenzo-Hartley kernel as

$$\left({}_{A}^{LSC}D_{b-}^{\alpha,\lambda} f\right)(t) = (-1)^{\kappa}\int_{t}^{b} F_{\alpha}\left(-\gamma(\tau-t)^{\alpha}\right) f^{(\kappa)}(\tau)\, d\tau. \quad (3.226)$$

In this case, we give the following inverse operators:

Definition 3.40 *Let $\alpha \in \mathbb{C}$, $\kappa+1 > Re(\alpha) > \kappa$, $-\infty < a < b < \infty$ and $\gamma \in \mathbb{C}$. The left-sided general fractional integral in the complex parameters is defined as*

$$\left({}_{A}^{R}I_{a+}^{\alpha,\gamma} f\right)(t) = \int_{a}^{t} (t-\tau)^{\kappa-\alpha-1} E_{\alpha,\kappa-\alpha}^{-1}\left(-\gamma(t-\tau)^{\alpha}\right) f(\tau)\, d\tau \ \ (t > a) \quad (3.227)$$

and the right-sided general fractional integral in the complex parameters as

$$\left({}_{A}^{R}I_{b-}^{\alpha,\gamma} f\right) = (-1)^{\kappa}\int_{t}^{b} (\tau-t)^{\kappa-\alpha-1} E_{\alpha,\kappa-\alpha}^{-1}\left(-\gamma(\tau-t)^{\alpha}\right) f(\tau)\, d\tau \ \ (t < b). \quad (3.228)$$

Definition 3.41 *Let $\alpha \in \mathbb{C}$, $1 > Re(\alpha) > 0$, $-\infty < a < b < \infty$ and $\gamma \in \mathbb{C}$. The left-sided general fractional integral in the complex parameters is defined as*

$$\left({}_{A}^{R}I_{a+}^{\alpha,\gamma} f\right)(t) = \int_{a}^{t} \frac{E_{\alpha,1-\alpha}^{-1}\left(-\gamma(t-\tau)^{\alpha}\right)}{(t-\tau)^{\alpha}} f(\tau)\, d\tau \ \ (t > a) \quad (3.229)$$

and the right-sided general fractional integral in the complex parameters as

$$\left({}_{A}^{R}I_{b-}^{\alpha,\gamma} f\right) = -\int_{t}^{b} \frac{E_{\alpha,1-\alpha}^{-1}\left(-\gamma\left(\tau - t\right)^{\alpha}\right)}{\left(\tau - t\right)^{\alpha}} f\left(\tau\right) d\tau \ \ (t < b). \tag{3.230}$$

The properties of the general fractional derivatives involving the one-parametric Lorenzo-Hartley kernel and the general fractional integrals are presented as follows.

Property 3.17 *If* $\kappa + 1 > \alpha > \kappa$, $-\infty < a < b < \infty$, $\gamma \in \mathbb{C}$ *and* $f(t) \in L_l(a, b)$ $(1 \le l < \infty)$, *then we have*

$$L\left\{\left({}_{A}^{RL}D_{0+}^{\alpha,\gamma} f\right)(t)\right\} = s^{\kappa-\alpha}\left(1 + \gamma s^{-\alpha}\right)^{-1} f(s), \tag{3.231}$$

$$L\left\{\left({}_{A}^{RL}D_{0+}^{\alpha,\gamma} f\right)(t)\right\} = s^{1-\alpha}\left(1 + \gamma s^{-\alpha}\right)^{-1} f(s), \tag{3.232}$$

$$L\left\{\left({}_{A}^{LS}D_{0+}^{\alpha,\gamma} f\right)(t)\right\} = s^{-\alpha}\left(1 + \gamma s^{-\alpha}\right)^{-1}\left(sf(s) - f(0)\right), \tag{3.233}$$

$$L\left\{\left({}_{A}^{LSC}D_{0+}^{\alpha,\gamma} f\right)(t)\right\} = s^{-\alpha}\left(1 + \gamma s^{-\alpha}\right)^{-1}\left(s^{\kappa} f(s) - \sum_{j=1}^{\kappa} s^{\kappa-j} f^{(j-1)}(0)\right), \tag{3.234}$$

$$L\left\{\left({}_{A}^{R}I_{0+}^{\alpha,\gamma} f\right)(t)\right\} = s^{\alpha-\kappa}\left(1 + \gamma s^{-\alpha}\right) f(s) \tag{3.235}$$

and

$$L\left\{\left({}_{A}I_{0+}^{\alpha,\gamma} f\right)(t)\right\} = s^{\alpha-1}\left(1 + \gamma s^{-\alpha}\right) f(s), \tag{3.236}$$

where

$$L\left\{A_{\alpha,\gamma}(-t)\right\} = s^{-\alpha}\left(1 + \gamma s^{-\alpha}\right)^{-1}. \tag{3.237}$$

Note that, when $D^{\kappa}\left({}_{A}^{R}I_{0+}^{\alpha,\gamma} f(t)\right)|_{t=0} = 0$, there is

$$L\left\{{}_{A}^{RL}D_{0+}^{\alpha,\gamma} f(t)\right\} = s^{\kappa-\alpha}\left(1 + \gamma s^{-\alpha}\right)^{-1} f(s), \tag{3.238}$$

and when $D^{\kappa}\left({}_{A}^{R}I_{0+}^{\alpha,\gamma} f(t)\right)|_{t=0} \neq 0$, there is

$$L\left\{{}_{A}^{RL}D_{0+}^{\alpha,\gamma} f(t)\right\} = s^{\kappa-\alpha}\left(1 + \gamma s^{-\alpha}\right)^{-1} f(s) - \sum_{\mu=0}^{\kappa-1} s^{\kappa-\mu-1}\left(D^{\mu}\left[{}_{A}^{R}I_{0+}^{\alpha,\gamma} f(+0)\right]\right). \tag{3.239}$$

In particular, when $\kappa = 1$, we have

$$L\left\{{}_{A}^{RL}D_{0+}^{\alpha,\gamma} f(t)\right\} = s^{1-\alpha}\left(1 + \gamma s^{-\alpha}\right)^{-1} f(s) - {}_{A}^{R}I_{0+}^{\alpha,\gamma} f(+0), \tag{3.240}$$

and when $\kappa = 1$ and $D\left({}_{A}^{R}I_{0+}^{\alpha,\gamma} f(t)\right)|_{t=0} = 0$, we have

$$L\left\{{}_{A}^{RL}D_{0+}^{\alpha,\gamma} f(t)\right\} = s^{1-\alpha}\left(1 + \gamma s^{-\alpha}\right)^{-1} f(s). \tag{3.241}$$

Property 3.18 *Let* $\kappa + 1 > \alpha > \kappa$, $-\infty < a < b < \infty$, $\gamma \in \mathbb{C}$ *and* $f(t) \in L_l(a,b)$ $(1 \leq l < \infty)$. *Then we have*

$$\left({}^{RL}_{A}D^{\alpha,\gamma}_{a+} {}^{R}_{A}I^{\alpha,\gamma}_{a+} f \right)(t) = f(t), \tag{3.242}$$

$$\left({}^{RL}_{A}D^{\alpha,\gamma}_{b-} {}^{R}_{A}I^{\alpha,\gamma}_{b-} f \right)(t) = f(t), \tag{3.243}$$

$$\left({}^{RL}_{A}D^{\alpha,\gamma}_{a+} {}^{R}_{A}I^{\alpha,\gamma}_{a+} f \right)(t) = f(t), \tag{3.244}$$

and

$$\left({}^{RL}_{A}D^{\alpha,\gamma}_{b-} {}^{R}_{A}I^{\alpha,\gamma}_{b-} f \right)(t) = f(t). \tag{3.245}$$

Property 3.19 *If* $-\infty < a < b < \infty$, $\gamma \in \mathbb{C}$ *and* $f(t) \in AC_\kappa(a,b)$ $(1 \leq \kappa < \infty)$, *then we have*

$$\left({}^{LS}_{A}D^{\alpha,\gamma}_{a+} {}^{R}_{A}I^{\alpha,\gamma}_{a+} f \right)(t) = f(t), \tag{3.246}$$

$$\left({}^{LS}_{A}D^{\alpha,\gamma}_{b-} {}^{R}_{A}I^{\alpha,\gamma}_{b-} f \right)(t) = f(t), \tag{3.247}$$

$$\left({}^{LSC}_{A}D^{\alpha,\gamma}_{a+} {}^{R}_{A}I^{\alpha,\gamma}_{a+} f \right)(t) = f(t), \tag{3.248}$$

and

$$\left({}^{LSC}_{A}D^{\alpha,\gamma}_{b-} {}^{R}_{A}I^{\alpha,\gamma}_{b-} f \right)(t) = f(t). \tag{3.249}$$

3.8 Hilfer type general fractional derivatives involving the one-parametric Lorenzo-Hartley kernel

We now start with the definitions of the Hilfer type general fractional derivatives involving the one-parametric Lorenzo-Hartley kernel.

Definition 3.42 *Let* $1 > \alpha > 0$, $0 \leq \beta \leq 1$, $-\infty < a < b < \infty$ *and* $\gamma \in \mathbb{C}$. *The left-sided Hilfer type general fractional derivative involving the one-parametric Lorenzo-Hartley kernel is defined as*

$$\begin{aligned} \left({}^{Hi}_{A}D^{\alpha,\beta,\gamma}_{a+} f \right)(t) &= \int_a^t F_\alpha\left(-\gamma (t-\tau)^\alpha\right) \left(D^{\alpha,\beta}_{a+} f(\tau) \right) d\tau \\ &= \int_a^t F_\alpha\left(-\gamma (t-\tau)^\alpha\right) \left(I^{\beta(1-\alpha)}_{a+} \frac{d}{d\tau} \left(I^{(1-\beta)(1-\alpha)}_{a+} f \right) \right)(\tau)\, d\tau, \end{aligned} \tag{3.250}$$

and the right-sided Hilfer type general fractional derivative involving the one-parametric Lorenzo-Hartley kernel as

$$\begin{aligned} \left({}^{Hi}_{A}D^{\alpha,\beta,\gamma}_{b-} f \right)(t) &= -\int_t^b A_{\alpha,\gamma}\left(-\gamma (\tau-t)^\alpha\right) \left(D^{\alpha,\beta}_{b-} f(\tau) \right) d\tau \\ &= -\int_t^b A_{\alpha,\gamma}\left(-\gamma (\tau-t)^\alpha\right) \left(I^{\beta(1-\alpha)}_{b-} \frac{d}{d\tau} \left(I^{(1-\beta)(1-\alpha)}_{b-} f \right) \right)(\tau)\, d\tau. \end{aligned} \tag{3.251}$$

Definition 3.43 *Let $\kappa+1 > \alpha > \kappa$, $\kappa = [\alpha]+1$, $0 \leq \beta \leq 1$, $-\infty < a < b < \infty$ and $\gamma \in \mathbb{C}$. The left-sided Hilfer type general fractional derivative involving the one-parametric Lorenzo-Hartley kernel is defined as*

$$\begin{aligned}\left({}_{A}^{Hir}D_{a+}^{\alpha,\beta,\gamma} f\right)(t) &= \int_{a}^{t} F_{\alpha}\left(-\gamma (t-\tau)^{\alpha}\right) \left(D_{a+}^{\alpha,\beta,\kappa} f(\tau)\right) d\tau \\ &= \int_{a}^{t} F_{\alpha}\left(-\gamma (t-\tau)^{\alpha}\right) \left(I_{a+}^{\beta(\kappa-\alpha)} \frac{d^{\kappa}}{d\tau^{\kappa}} \left(I_{a+}^{(1-\beta)(\kappa-\alpha)} f\right)\right)(\tau)\, d\tau\end{aligned} \tag{3.252}$$

and the right-sided Hilfer type general fractional derivative involving the one-parametric Lorenzo-Hartley kernel as

$$\begin{aligned}\left({}_{A}^{Hir}D_{b-}^{\alpha,\beta,\gamma} f\right)(t) &= (-1)^{\kappa} \int_{t}^{b} F_{\alpha}\left(-\gamma (\tau-t)^{\alpha}\right) \left(D_{a+}^{\alpha,\beta,\kappa} f(\tau)\right) d\tau \\ &= (-1)^{\kappa} \int_{t}^{b} F_{\alpha}\left(-\gamma (\tau-t)^{\alpha}\right) \left(\left(I_{b-}^{\beta(\kappa-\alpha)} \frac{d^{\kappa}}{d\tau^{\kappa}} \left(I_{b-}^{(1-\beta)(\kappa-\alpha)} f\right)\right)(\tau)\right) d\tau.\end{aligned} \tag{3.253}$$

3.9 General fractional derivatives in the subcosine kernel via Mittag-Leffler function

The definitions and properties of the general fractional derivatives within the subcosine kernel via Mittag-Leffler function and the general fractional integrals are given as follows.

Definition 3.44 *Let $\alpha \in \mathbb{C}$, $1 > Re(\alpha) > 0$, $-\infty < a < b < \infty$ and $\lambda \in \mathbb{C}$. The left-sided Riemann-Liouville type general fractional derivative within the subcosine kernel via Mittag-Leffler function is defined as*

$$\left({}_{Scs}^{RL}D_{a+}^{\alpha,\lambda} f\right)(t) = \frac{d}{dt}\left({}_{Scs}I_{a+}^{\alpha,\lambda} f(t)\right) = \frac{d}{dt}\int_{a}^{t} LCos_{\alpha}\left(-\lambda (t-\tau)^{\alpha}\right) f(\tau)\, d\tau, \tag{3.254}$$

and the right-sided Riemann-Liouville type general fractional derivative within the subcosine kernel via Mittag-Leffler function as

$$\left({}_{Scs}^{RL}D_{b-}^{\alpha,\lambda} f\right)(t) = -\frac{d}{dt}\left({}_{Scs}I_{b-}^{\alpha,\lambda} f(t)\right) = -\frac{d}{dt}\int_{t}^{b} LCos_{\alpha}\left(-\lambda (\tau-t)^{\alpha}\right) f(\tau)\, d\tau, \tag{3.255}$$

where the left-sided and right-sided general fractional integral operators within the subcosine kernel are defined as

$${}_{Scs}I_{a+}^{\alpha,\lambda} f(t) = \int_{a}^{t} LCos_{\alpha}\left(-\lambda (t-\tau)^{\alpha}\right) f(\tau)\, d\tau \tag{3.256}$$

and

$$_{Scs}I_{b-}^{\alpha,\lambda}f(t)=\int_{t}^{b}LCos_{\alpha}\left(-\lambda\left(\tau-t\right)^{\alpha}\right)f(\tau)\,d\tau, \tag{3.257}$$

respectively.

Definition 3.45 *Let $\alpha\in\mathbb{C}$, $Re(\alpha)>0$, $\kappa=[Re(\alpha)]+1$, $-\infty<a<b<\infty$ and $\lambda\in\mathbb{C}$. The left-sided Riemann-Liouville type general fractional derivative within the subcosine kernel via Mittag-Leffler function is defined as*

$$\left({}_{Scs}^{RL}D_{a+}^{\alpha,\lambda}f\right)(t)=\frac{d^{\kappa}}{dt^{\kappa}}\left({}_{Scs}I_{a+}^{\alpha,\lambda}f(t)\right)=\frac{d^{\kappa}}{dt^{\kappa}}\int_{a}^{t}LCos_{\alpha}\left(-\lambda\left(t-\tau\right)^{\alpha}\right)f(\tau)\,d\tau, \tag{3.258}$$

and the right-sided Riemann-Liouville type general fractional derivative within the subcosine kernel via Mittag-Leffler function as

$$\begin{aligned}\left({}_{Scs}^{RL}D_{b-}^{\alpha,\lambda}f\right)(t)&=\left(-\frac{d}{dt}\right)^{\kappa}\left({}_{Scs}I_{b-}^{\alpha,\lambda}f(t)\right)\\&=(-1)^{\kappa}\frac{d^{\kappa}}{dt^{\kappa}}\int_{t}^{b}LCos_{\alpha}\left(-\lambda\left(\tau-t\right)^{\alpha}\right)f(\tau)\,d\tau.\end{aligned} \tag{3.259}$$

Definition 3.46 *Let $\alpha\in\mathbb{C}$, $1>Re(\alpha)>0$, $-\infty<a<b<\infty$ and $\lambda\in\mathbb{C}$. The left-sided Liouville-Sonine type general fractional derivative within the subcosine kernel via Mittag-Leffler function is defined as*

$$\left({}_{Scs}^{LS}D_{a+}^{\alpha,\lambda}f\right)(t)={}_{Scs}I_{a+}^{\alpha,\lambda}\left(f^{(1)}(t)\right)=\int_{a}^{t}LCos_{\alpha}\left(-\lambda\left(t-\tau\right)^{\alpha}\right)f^{(1)}(\tau)\,d\tau, \tag{3.260}$$

and the right-sided Liouville-Sonine type general fractional derivative within the subcosine kernel via Mittag-Leffler function as

$$\left({}_{Scs}^{LS}D_{b-}^{\alpha,\lambda}f\right)(t)={}_{Scs}I_{b-}^{\alpha,\lambda}\left(-f^{(1)}(t)\right)=-\int_{t}^{b}LCos_{\alpha}\left(-\lambda\left(\tau-t\right)^{\alpha}\right)f^{(1)}(\tau)\,d\tau. \tag{3.261}$$

Definition 3.47 *Let $\alpha\in\mathbb{C}$, $\kappa+1>Re(\alpha)>\kappa$, $-\infty<a<b<\infty$ and $\lambda\in\mathbb{C}$. The left-sided Liouville-Sonine-Caputo type general fractional derivative within the subcosine kernel via Mittag-Leffler function is defined as*

$$\left({}_{Scs}^{LSC}D_{a+}^{\alpha,\lambda}f\right)(t)={}_{Scs}I_{a+}^{\alpha,\lambda}\left(f^{(\kappa)}(t)\right)=\int_{a}^{t}LCos_{\alpha}\left(-\lambda\left(t-\tau\right)^{\alpha}\right)f^{(\kappa)}(\tau)\,d\tau, \tag{3.262}$$

and the right-sided Liouville-Sonine-Caputo type general fractional derivative within the subcosine kernel via Mittag-Leffler function as

$$\left({}_{Scs}^{LSC}D_{b-}^{\alpha,\lambda}f\right)(t)={}_{Scs}I_{b-}^{\alpha,\lambda}\left((-1)^{\kappa}f^{(\kappa)}(t)\right)=(-1)^{\kappa}\int_{t}^{b}LCos_{\alpha}\left(-\lambda\left(\tau-t\right)^{\alpha}\right)f^{(\kappa)}(\tau)\,d\tau. \tag{3.263}$$

The inverse operators of the above fractional derivatives within the sub-cosine kernel via Mittag-Leffler function are presented as follows.

Definition 3.48 *Let $\alpha \in \mathbb{C}$, $\kappa + 1 > Re(\alpha) > \kappa$, $-\infty < a < b < \infty$ and $\lambda \in \mathbb{C}$. As the inverse operator, the left-sided general fractional integral is defined as*

$$\left({}_{Scs}^{R}I_{a+}^{\alpha,\lambda} f\right)(t) = \int_{a}^{t} (t-\tau)^{\kappa-2} E_{2\alpha,\kappa-1}^{-1}\left(-\lambda (t-\tau)^{2\alpha}\right) f(\tau)\, d\tau \tag{3.264}$$

and the right-sided general fractional integral as

$$\left({}_{Scs}^{R}I_{b-}^{\alpha,\lambda} f\right) = \frac{1}{\Gamma(\kappa)} \int_{t}^{b} (\tau-t)^{\kappa-2} E_{2\alpha,\kappa-1}^{-1}\left(-\lambda (\tau-t)^{2\alpha}\right) f(\tau)\, d\tau. \tag{3.265}$$

In particular, when $\alpha \in \mathbb{C}$, $1 > Re(\alpha) > 0$, $-\infty < a < b < \infty$ and $\lambda \in \mathbb{C}$, the left-sided general fractional integral is defined as

$$\left({}_{Scs}^{R}I_{a+}^{\alpha,\lambda} f\right)(t) = \int_{a}^{t} \frac{1}{t-\tau} E_{2\alpha,0}^{-1}\left(-\lambda (t-\tau)^{2\alpha}\right) f(\tau)\, d\tau \tag{3.266}$$

and the right-sided general fractional integral as

$$\left({}_{Scs}^{R}I_{b-}^{\alpha,\lambda} f\right) = \frac{1}{\Gamma(\kappa)} \int_{t}^{b} \frac{1}{\tau-t} E_{2\alpha,0}^{-1}\left(-\lambda (\tau-t)^{2\alpha}\right) f(\tau)\, d\tau. \tag{3.267}$$

Here, the properties of the general fractional derivatives and the general fractional integrals are presented as follows.

Property 3.20 *If $\kappa + 1 > \alpha > \kappa$, $-\infty < a < b < \infty$, $\lambda \in \mathbb{C}$ and $f(t) \in L_l(a,b)$ $(1 \le l < \infty)$, then we have*

$$L\left\{\left({}_{Scs}^{RL}D_{0+}^{\alpha,\lambda} f\right)(t)\right\} = s^{\kappa-1}\left(1+\lambda^2 s^{-2\alpha}\right)^{-1} f(s), \tag{3.268}$$

$$L\left\{\left({}_{Scs}^{RL}D_{0+}^{\alpha,\lambda} f\right)(t)\right\} = \left(1+\lambda^2 s^{-2\alpha}\right)^{-1} f(s), \tag{3.269}$$

$$L\left\{\left({}_{Scs}^{LS}D_{0+}^{\alpha,\lambda} f\right)(t)\right\} = s^{-1}\left(1+\lambda^2 s^{-2\alpha}\right)^{-1}\left(sf(s) - f(0)\right), \tag{3.270}$$

$$L\left\{\left({}_{Scs}^{LSC}D_{0+}^{\alpha,\lambda} f\right)(t)\right\} = s^{-1}\left(1+\lambda^2 s^{-2\alpha}\right)^{-1}\left(s^{\kappa} f(s) - \sum_{j=1}^{\kappa} s^{\kappa-j} f^{(j-1)}(0)\right), \tag{3.271}$$

$$L\left\{\left({}_{Scs}^{R}I_{0+}^{\alpha,\lambda}f\right)(t)\right\} = \left(1+\lambda^2 s^{-2\alpha}\right) f(s), \tag{3.272}$$

and

$$L\left\{\left({}_{Scs}^{R}I_{0+}^{\alpha,\lambda}f\right)(t)\right\} = s^{1-\kappa}\left(1+\lambda^2 s^{-2\alpha}\right) f(s). \tag{3.273}$$

Note that, when $D^{\kappa}\left({}_{Scs}^{R}I_{0+}^{\alpha,\lambda}f(t)\right)|_{t=0} = 0$, there is

$$L\left\{{}_{Scs}^{RL}D_{0+}^{\alpha,\lambda}f(t)\right\} = s^{\kappa-1}\left(1+\lambda^2 s^{-2\alpha}\right)^{-1} f(s), \tag{3.274}$$

and when $D^{\kappa}\left({}_{Scs}^{R}I_{0+}^{\alpha,\lambda}f(t)\right)|_{t=0} \neq 0$, there is

$$L\left\{{}_{Scs}^{RL}D_{0+}^{\alpha,\lambda}f(t)\right\} = s^{\kappa-1}\left(1+\lambda^2 s^{-2\alpha}\right)^{-1} f(s) - \sum_{\mu=0}^{\kappa-1} s^{\kappa-\mu-1}\left(D^{\mu}\left[{}_{Scs}^{R}I_{0+}^{\alpha,\lambda}f(+0)\right]\right). \tag{3.275}$$

In particular, when $\kappa = 1$, we have

$$L\left\{{}_{Scs}^{RL}D_{0+}^{\alpha,\lambda}f(t)\right\} = \left(1+\lambda^2 s^{-2\alpha}\right)^{-1} f(s) - {}_{Scs}^{R}I_{0+}^{\alpha,\lambda}f(+0), \tag{3.276}$$

and when $\kappa = 1$ and $D\left({}_{Scs}^{R}I_{0+}^{\alpha,\lambda}f(t)\right)|_{t=0} = 0$, we have

$$L\left\{{}_{Scs}^{RL}D_{0+}^{\alpha,\lambda}f(t)\right\} = \left(1+\lambda^2 s^{-2\alpha}\right)^{-1} f(s). \tag{3.277}$$

Property 3.21 *Let $\kappa+1 > \alpha > \kappa$, $-\infty < a < b < \infty$, $\lambda \in \mathbb{C}$ and $f(t) \in L_l(a,b)$ $(1 \le l < \infty)$. Then we have*

$$\left({}_{Scs}^{RL}D_{a+}^{\alpha,\lambda}\,{}_{Scs}^{R}I_{a+}^{\alpha,\lambda}f\right)(t) = f(t), \tag{3.278}$$

$$\left({}_{Scs}^{RL}D_{b-}^{\alpha,\lambda}\,{}_{Scs}^{R}I_{b-}^{\alpha,\lambda}f\right)(t) = f(t), \tag{3.279}$$

$$\left({}_{Scs}^{RL}D_{a+}^{\alpha,\lambda}\,{}_{Scs}^{R}I_{a+}^{\alpha,\lambda}f\right)(t) = f(t), \tag{3.280}$$

and

$$\left({}_{Scs}^{RL}D_{b-}^{\alpha,\lambda}\,{}_{Scs}^{R}I_{b-}^{\alpha,\lambda}f\right)(t) = f(t). \tag{3.281}$$

Property 3.22 *If $-\infty < a < b < \infty$, $\lambda \in \mathbb{C}$ and $f(t) \in AC_{\kappa}(a,b)$ $(1 \le \kappa < \infty)$, then we have*

$$\left({}_{Scs}^{LS}D_{a+}^{\alpha,\lambda}\,{}_{Scs}^{R}I_{a+}^{\alpha,\lambda}f\right)(t) = f(t), \tag{3.282}$$

$$\left({}_{Scs}^{LS}D_{b-}^{\alpha,\lambda}\,{}_{Scs}^{R}I_{b-}^{\alpha,\lambda}f\right)(t) = f(t), \tag{3.283}$$

$$\left({}_{Scs}^{LSC}D_{a+}^{\alpha,\lambda}\,{}_{Scs}^{R}I_{a+}^{\alpha,\lambda}f\right)(t) = f(t), \tag{3.284}$$

and

$$\left({}_{Scs}^{LSC}D_{b-}^{\alpha,\lambda}\,{}_{Scs}^{R}I_{b-}^{\alpha,\lambda}f\right)(t) = f(t). \tag{3.285}$$

Definition 3.49 *Let $\alpha \in \mathbb{C}$, $1 > Re(\alpha) > 0$, $a = -\infty$, $b = \infty$ and $\lambda \in \mathbb{C}$. The left-sided Riemann-Liouville type general fractional derivative within the subcosine kernel via Mittag-Leffler function is defined as*

$$\left({}_{Scs}^{RL}D_{+}^{\alpha,\lambda}f\right)(t) = \frac{d}{dt}\left({}_{Scs}I_{+}^{\alpha,\lambda}f(t)\right) = \frac{d}{dt}\int_{-\infty}^{t} LCos_{\alpha}\left(-\lambda(t-\tau)^{\alpha}\right)f(\tau)\,d\tau, \tag{3.286}$$

and the right-sided Riemann-Liouville type general fractional derivative within the subcosine kernel via Mittag-Leffler function as

$$\left({}_{Scs}^{RL}D_{-}^{\alpha,\lambda}f\right)(t) = -\frac{d}{dt}\left({}_{Scs}I_{-}^{\alpha,\lambda}f(t)\right) = -\frac{d}{dt}\int_{t}^{\infty} LCos_{\alpha}\left(-\lambda(\tau-t)^{\alpha}\right)f(\tau)\,d\tau, \tag{3.287}$$

where the left-sided and right-sided general fractional integral operators within the subcosine kernel via Mittag-Leffler function are defined as

$${}_{Scs}I_{+}^{\alpha,\lambda}f(t) = \int_{-\infty}^{t} LCos_{\alpha}\left(-\lambda(t-\tau)^{\alpha}\right)f(\tau)\,d\tau \tag{3.288}$$

and

$${}_{Scs}I_{-}^{\alpha,\lambda}f(t) = \int_{t}^{\infty} LCos_{\alpha}\left(-\lambda(\tau-t)^{\alpha}\right)f(\tau)\,d\tau, \tag{3.289}$$

respectively.

More generally, when $\alpha \in \mathbb{C}$, $Re(\alpha) > 0$, $\kappa = [Re(\alpha)] + 1$, $a = -\infty$, $b = \infty$ and $\lambda \in \mathbb{C}$, the left-sided Riemann-Liouville type general fractional derivative within the subcosine kernel via Mittag-Leffler function is defined as

$$\left({}_{Scs}^{RL}D_{+}^{\alpha,\lambda}f\right)(t) = \frac{d^{\kappa}}{dt^{\kappa}}\int_{-\infty}^{t} LCos_{\alpha}\left(-\lambda(t-\tau)^{\alpha}\right)f(\tau)\,d\tau, \tag{3.290}$$

and the right-sided Riemann-Liouville type general fractional derivative within the subcosine kernel via Mittag-Leffler function as

$$\left({}_{Scs}^{RL}D_{-}^{\alpha,\lambda}f\right)(t) = (-1)^{\kappa}\frac{d^{\kappa}}{dt^{\kappa}}\int_{t}^{\infty} LCos_{\alpha}\left(-\lambda(\tau-t)^{\alpha}\right)f(\tau)\,d\tau. \tag{3.291}$$

On the other hand, when $\alpha \in \mathbb{C}$, $1 > Re(\alpha) > 0$, $a = -\infty$, $b = \infty$ and $\lambda \in \mathbb{C}$, the left-sided Liouville-Sonine type general fractional derivative within the subcosine kernel via Mittag-Leffler function is defined as

$$\left({}_{Scs}^{LS}D_{+}^{\alpha,\lambda}f\right)(t) = {}_{Scs}I_{+}^{\alpha,\lambda}\left(f^{(1)}(t)\right) = \int_{-\infty}^{t} LCos_{\alpha}\left(-\lambda(t-\tau)^{\alpha}\right)f^{(1)}(\tau)\,d\tau, \tag{3.292}$$

and the right-sided Liouville-Sonine type general fractional derivative within the subcosine kernel via Mittag-Leffler function as

$$\left({}_{Scs}^{LS}D_{-}^{\alpha,\lambda} f\right)(t) = {}_{Scs} I_{-}^{\alpha,\lambda} \left(-f^{(1)}(t)\right) = -\int_{t}^{-\infty} LCos_{\alpha}\left(-\lambda(\tau-t)^{\alpha}\right) f^{(1)}(\tau)\, d\tau. \tag{3.293}$$

More generally, when $\alpha \in \mathbb{C}$, $\kappa+1 > Re(\alpha) > \kappa$, $a=-\infty$, $b=\infty$ and $\lambda \in$ C, the left-sided Liouville-Sonine-Caputo type general fractional derivative within the subcosine kernel via Mittag-Leffler function is defined as

$$\left({}_{Scs}^{LSC}D_{+}^{\alpha,\lambda} f\right)(t) = {}_{Scs} I_{+}^{\alpha,\lambda} \left(f^{(\kappa)}(t)\right) = \int_{-\infty}^{t} LCos_{\alpha}\left(-\lambda(t-\tau)^{\alpha}\right) f^{(\kappa)}(\tau)\, d\tau, \tag{3.294}$$

and the right-sided Liouville-Sonine-Caputo type general fractional derivative within the subcosine kernel via Mittag-Leffler function as

$$\begin{aligned}\left({}_{Scs}^{LSC}D_{-}^{\alpha,\lambda} f\right)(t) &= {}_{Scs} I_{-}^{\alpha,\lambda} \left((-1)^{\kappa} f^{(\kappa)}(t)\right) \\ &= (-1)^{\kappa} \int_{t}^{\infty} LCos_{\alpha}\left(-\lambda(\tau-t)^{\alpha}\right) f^{(\kappa)}(\tau)\, d\tau. \end{aligned} \tag{3.295}$$

The inverse operators of the above general fractional derivatives within the subcosine kernel via Mittag-Leffler function can be presented as follows.

Definition 3.50 *Let $\alpha \in \mathbb{C}$, $\kappa+1 > Re(\alpha) > \kappa$, $a=-\infty$, $b=\infty$ and $\lambda \in \mathbb{C}$. The left-sided general fractional integral is defined as*

$$\left({}_{Scs}^{R}I_{+}^{\alpha,\lambda} f\right)(t) = \int_{-\infty}^{t} (t-\tau)^{\kappa-2} E_{2\alpha,\kappa-1}^{-1}\left(-\lambda(t-\tau)^{2\alpha}\right) f(\tau)\, d\tau, \tag{3.296}$$

and the right-sided general fractional integral as

$$\left({}_{Scs}^{R}I_{-}^{\alpha,\lambda} f\right) = \frac{1}{\Gamma(\kappa)} \int_{t}^{\infty} (\tau-t)^{\kappa-2} E_{2\alpha,\kappa-1}^{-1}\left(-\lambda(\tau-t)^{2\alpha}\right) f(\tau)\, d\tau. \tag{3.297}$$

In particular, when $\alpha \in \mathbb{C}$, $1 > Re(\alpha) > 0$, $a=-\infty$, $b=\infty$ and $\lambda \in \mathbb{C}$, the left-sided general fractional integral is defined as

$$\left({}_{Scs}^{R}I_{+}^{\alpha,\lambda} f\right)(t) = \int_{-\infty}^{t} \frac{1}{t-\tau} E_{2\alpha,0}^{-1}\left(-\lambda(t-\tau)^{2\alpha}\right) f(\tau)\, d\tau, \tag{3.298}$$

and the right-sided general fractional integral as

$$\left({}_{Scs}^{R}I_{-}^{\alpha,\lambda} f\right) = \frac{1}{\Gamma(\kappa)} \int_{t}^{\infty} \frac{1}{\tau-t} E_{2\alpha,0}^{-1}\left(-\lambda(\tau-t)^{2\alpha}\right) f(\tau)\, d\tau. \tag{3.299}$$

The properties of the general fractional derivatives and the general fractional integrals are given as follows.

Property 3.23 *Let* $\kappa + 1 > \alpha > \kappa$, $a = -\infty$, $b = \infty$, $\lambda \in \mathbb{C}$ *and* $f(t) \in L_l(-\infty, \infty)$ $(1 \leq l < \infty)$. *Then we have*

$$\left({}^{RL}_{Scs}D^{\alpha,\lambda}_{+} {}^{R}_{Scs}I^{\alpha,\lambda}_{+} f \right)(t) = f(t), \tag{3.300}$$

$$\left({}^{RL}_{Scs}D^{\alpha,\lambda}_{-} {}^{R}_{Scs}I^{\alpha,\lambda}_{-} f \right)(t) = f(t), \tag{3.301}$$

$$\left({}^{RL}_{Scs}D^{\alpha,\lambda}_{+} {}^{R}_{Scs}I^{\alpha,\lambda}_{+} f \right)(t) = f(t), \tag{3.302}$$

and

$$\left({}^{RL}_{Scs}D^{\alpha,\lambda}_{-} {}^{R}_{Scs}I^{\alpha,\lambda}_{-} f \right)(t) = f(t). \tag{3.303}$$

Property 3.24 *If* $a = -\infty$, $b = \infty$, $\lambda \in \mathbb{C}$ *and* $f(t) \in AC_\kappa(-\infty, \infty)$ $(1 \leq \kappa < \infty)$, *then we have*

$$\left({}^{LS}_{Scs}D^{\alpha,\lambda}_{+} {}^{R}_{Scs}I^{\alpha,\lambda}_{+} f \right)(t) = f(t), \tag{3.304}$$

$$\left({}^{LS}_{Scs}D^{\alpha,\lambda}_{-} {}^{R}_{Scs}I^{\alpha,\lambda}_{-} f \right)(t) = f(t), \tag{3.305}$$

$$\left({}^{LSC}_{Scs}D^{\alpha,\lambda}_{+} {}^{R}_{Scs}I^{\alpha,\lambda}_{+} f \right)(t) = f(t), \tag{3.306}$$

and

$$\left({}^{LSC}_{Scs}D^{\alpha,\lambda}_{-} {}^{R}_{Scs}I^{\alpha,\lambda}_{-} f \right)(t) = f(t). \tag{3.307}$$

3.10 Hilfer type general fractional derivatives within the subcosine kernel via Mittag-Leffler function

The definitions and properties of the Hilfer type general fractional derivatives within the subcosine kernel via Mittag-Leffler function are given as follows.

Definition 3.51 *Let* $1 > \alpha > 0$, $0 \leq \beta \leq 1$, $-\infty < a < b < \infty$, $\gamma \in \mathbb{C}$ *and* $\lambda \in \mathbb{C}$. *The left-sided Hilfer type general fractional derivative within the subcosine kernel via Mittag-Leffler function is defined as*

$$\begin{aligned} \left({}^{Hi}_{Scs}D^{\alpha,\beta,\lambda}_{a+} f \right)(t) &= \int_a^t LCos_\alpha\left(-\lambda (t-\tau)^\alpha\right) \left(D^{\alpha,\beta}_{a+} f(\tau) \right) d\tau \\ &= \int_a^t LCos_\alpha\left(-\lambda (t-\tau)^\alpha\right) \left(I^{\beta(1-\alpha)}_{a+} \frac{d}{d\tau} \left(I^{(1-\beta)(1-\alpha)}_{a+} f \right) \right)(\tau)\, d\tau, \end{aligned} \tag{3.308}$$

and the right-sided Hilfer type general fractional derivative within the subcosine kernel via Mittag-Leffler function as

$$\begin{aligned}\left({}^{Hi}_{Scs}D^{\alpha,\beta,\lambda}_{b-}f\right)(t) &= -\int_t^b LCos_\alpha\left(-\lambda(\tau-t)^\alpha\right)\left(D^{\alpha,\beta}_{b-}f(\tau)\right)d\tau \\ &= -\int_t^b LCos_\alpha\left(-\lambda(\tau-t)^\alpha\right)\left(I^{\beta(1-\alpha)}_{b-}\tfrac{d}{d\tau}\left(I^{(1-\beta)(1-\alpha)}_{b-}f\right)\right)(\tau)\,d\tau.\end{aligned} \tag{3.309}$$

Generally, when $\kappa+1>\alpha>\kappa$, $0\le\beta\le1$, $-\infty<a<b<\infty$, $\gamma\in\mathbb{C}$ and $\lambda\in\mathbb{C}$, the left-sided Hilfer type general fractional derivative within the subcosine kernel via Mittag-Leffler function is defined as

$$\begin{aligned}\left({}^{Hir}_{Scs}D^{\alpha,\beta,\lambda}_{a+}f\right)(t) &= \int_a^t LCos_\alpha\left(-\lambda(t-\tau)^\alpha\right)\left(D^{\alpha,\beta,\kappa}_{a+}f(\tau)\right)d\tau \\ &= \int_a^t LCos_\alpha\left(-\lambda(t-\tau)^\alpha\right)\left(I^{\beta(\kappa-\alpha)}_{a+}\tfrac{d^\kappa}{d\tau^\kappa}\left(I^{(1-\beta)(\kappa-\alpha)}_{a+}f\right)\right)(\tau)\,d\tau,\end{aligned} \tag{3.310}$$

and the right-sided Hilfer type general fractional derivative within the subcosine kernel via Mittag-Leffler function as

$$\begin{aligned}\left({}^{Hir}_{Scs}D^{\alpha,\beta,\lambda}_{b-}f\right)(t) &= (-1)^\kappa\int_t^b LCos_\alpha\left(-\lambda(\tau-t)^\alpha\right)\left(D^{\alpha,\beta,\kappa}_{b-}f(\tau)\right)d\tau \\ &= (-1)^\kappa\int_t^b LCos_\alpha\left(-\lambda(\tau-t)^\alpha\right)\left(I^{\beta(\kappa-\alpha)}_{b-}\tfrac{d^\kappa}{d\tau^\kappa}\left(I^{(1-\beta)(\kappa-\alpha)}_{b-}f\right)\right)(\tau)\,d\tau.\end{aligned} \tag{3.311}$$

Definition 3.52 *Let $\kappa+1>\alpha>\kappa$, $\beta=1$, $-\infty<a<b<\infty$, $\gamma\in\mathbb{C}$ and $\lambda\in\mathbb{C}$. The left-sided Liouville-Sonine-Caputo type general fractional derivative within the subcosine kernel via Mittag-Leffler function is defined as*

$$\left({}^{Hir}_{Scs}D^{\alpha,1,\lambda}_{a+}f\right)(t) = \int_a^t LCos_\alpha\left(-\lambda(t-\tau)^\alpha\right)\left(I^{\kappa-\alpha}_{a+}f^{(\kappa)}(\tau)\right)d\tau, \tag{3.312}$$

and the right-sided Liouville-Sonine-Caputo type general fractional derivative within the subcosine kernel via Mittag-Leffler function as

$$\left({}^{Hir}_{Scs}D^{\alpha,1,\lambda}_{b-}f\right)(t) = (-1)^\kappa\int_t^b LCos_\alpha\left(-\lambda(\tau-t)^\alpha\right)\left(I^{\kappa-\alpha}_{b-}f^{(\kappa)}(\tau)\right)(\tau)\,d\tau. \tag{3.313}$$

In particular, when $1>\alpha>0$, $\beta=0$, $-\infty<a<b<\infty$, $\gamma\in\mathbb{C}$ and $\lambda\in\mathbb{C}$, the left-sided Riemann-Liouville type general fractional derivative within the subcosine kernel via Mittag-Leffler function is defined as

$$\left({}^{Ili}_{Scs}D^{\alpha,0,\lambda}_{a+}f\right)(t) = \int_a^t LCos_\alpha\left(-\lambda(t-\tau)^\alpha\right)\left(\tfrac{d}{d\tau}\left(I^{1-\alpha}_{a+}f\right)\right)(\tau)\,d\tau, \tag{3.314}$$

and the right-sided Riemann-Liouville type general fractional derivative within the subcosine kernel via Mittag-Leffler function as

$$\left({}_{Scs}^{Hi}D_{b-}^{\alpha,0,\lambda}f\right)(t) = -\int_t^b LCos_\alpha\left(-\lambda\left(\tau-t\right)^\alpha\right)\left(\tfrac{d}{d\tau}\left(I_{b-}^{1-\alpha}f\right)\right)(\tau)\,d\tau. \quad (3.315)$$

Furthermore, when $\kappa+1>\alpha>\kappa$, $\beta=0$, $-\infty<a<b<\infty$, $\gamma\in\mathbb{C}$ and $\lambda\in\mathbb{C}$, the left-sided Riemann-Liouville type general fractional derivative within the subcosine kernel via Mittag-Leffler function is defined as

$$\left({}_{Scs}^{Hir}D_{a+}^{\alpha,0,\lambda}f\right)(t) = \int_a^t LCos_\alpha\left(-\lambda\left(t-\tau\right)^\alpha\right)\left(\tfrac{d^\kappa}{d\tau^\kappa}\left(I_{a+}^{\kappa-\alpha}f\right)(\tau)\right)d\tau, \quad (3.316)$$

and the right-sided Riemann-Liouville type general fractional derivative within the subcosine kernel via Mittag-Leffler function as

$$\left({}_{Scs}^{Hir}D_{b-}^{\alpha,0,\lambda}f\right)(t) = (-1)^\kappa\int_t^b LCos_\alpha\left(-\lambda\left(\tau-t\right)^\alpha\right)\left(\tfrac{d^\kappa}{d\tau^\kappa}\left(I_{b-}^{\kappa-\alpha}f\right)(\tau)\right)d\tau. \quad (3.317)$$

3.11 General fractional derivatives in the subsine kernel via Mittag-Leffler function

The definitions and properties of the general fractional derivatives within the subsine kernel via Mittag-Leffler function and the general fractional integrals are given as follows.

Definition 3.53 *Let $\alpha\in\mathbb{C}$, $1>Re(\alpha)>0$, $-\infty<a<b<\infty$ and $\lambda\in\mathbb{C}$. The left-sided Riemann-Liouville type general fractional derivative within the subsine kernel via Mittag-Leffler function is defined as*

$$\left({}_{Sls}^{RL}D_{a+}^{\alpha,\lambda}f\right)(t) = \frac{d}{dt}\left({}_{Sls}I_{a+}^{\alpha,\lambda}f(t)\right) = \frac{d}{dt}\int_a^t LSin_\alpha\left(-\lambda\left(t-\tau\right)^\alpha\right)f(\tau)\,d\tau, \quad (3.318)$$

and the right-sided Riemann-Liouville type general fractional derivative within the subsine kernel via Mittag-Leffler function as

$$\left({}_{Sls}^{RL}D_{b-}^{\alpha,\lambda}f\right)(t) = -\frac{d}{dt}\left({}_{Sls}I_{b-}^{\alpha,\lambda}f(t)\right) = -\frac{d}{dt}\int_t^b LSin_\alpha\left(-\lambda\left(\tau-t\right)^\alpha\right)f(\tau)\,d\tau, \quad (3.319)$$

where the left-sided and right-sided general fractional integral operators within the subsine kernel are defined as

$$_{Sls}I_{a+}^{\alpha,\lambda}f(t)=\int_a^t LSin_\alpha\left(-\lambda(t-\tau)^\alpha\right)f(\tau)\,d\tau \tag{3.320}$$

and

$$_{Sls}I_{b-}^{\alpha,\lambda}f(t)=\int_t^b LSin_\alpha\left(-\lambda(\tau-t)^\alpha\right)f(\tau)\,d\tau, \tag{3.321}$$

respectively.

Furthermore, when $\alpha\in\mathbb{C}$, $Re(\alpha)>0$, $\kappa=[Re(\alpha)]+1$, $-\infty<a<b<\infty$ and $\lambda\in\mathbb{C}$, the left-sided Riemann-Liouville type general fractional derivative within the subsine kernel via Mittag-Leffler function is defined as

$$\left({}_{Sls}^{RL}D_{a+}^{\alpha,\lambda}f\right)(t)=\frac{d^\kappa}{dt^\kappa}\left({}_{Sls}I_{a+}^{\alpha,\lambda}f(t)\right)=\frac{d^\kappa}{dt^\kappa}\int_a^t LSin_\alpha\left(-\lambda(t-\tau)^\alpha\right)f(\tau)\,d\tau, \tag{3.322}$$

and the right-sided Riemann-Liouville type general fractional derivative within the subsine kernel via Mittag-Leffler function as

$$\begin{aligned}\left({}_{Sls}^{RL}D_{b-}^{\alpha,\lambda}f\right)(t)&=\left(-\frac{d}{dt}\right)^\kappa\left({}_{Sls}I_{b-}^{\alpha,\lambda}f(t)\right)\\&=(-1)^\kappa\frac{d^\kappa}{dt^\kappa}\int_t^b LSin_\alpha\left(-\lambda(\tau-t)^\alpha\right)f(\tau)\,d\tau.\end{aligned} \tag{3.323}$$

Definition 3.54 *Let $\alpha\in\mathbb{C}$, $1>Re(\alpha)>0$, $-\infty<a<b<\infty$ and $\lambda\in\mathbb{C}$. The left-sided Liouville-Sonine type general fractional derivative within the subsine kernel via Mittag-Leffler function is defined as*

$$\left({}_{Sls}^{LS}D_{a+}^{\alpha,\lambda}f\right)(t)={}_{Sls}I_{a+}^{\alpha,\lambda}\left(f^{(1)}(t)\right)=\int_a^t LSin_\alpha\left(-\lambda(t-\tau)^\alpha\right)f^{(1)}(\tau)\,d\tau, \tag{3.324}$$

and the right-sided Liouville-Sonine type general fractional derivative within the subsine kernel via Mittag-Leffler function as

$$\left({}_{Sls}^{LS}D_{b-}^{\alpha,\lambda}f\right)(t)={}_{Sls}I_{b-}^{\alpha,\lambda}\left(-f^{(1)}(t)\right)=-\int_t^b LSin_\alpha\left(-\lambda(\tau-t)^\alpha\right)f^{(1)}(\tau)\,d\tau. \tag{3.325}$$

More generally, when $\alpha\in\mathbb{C}$, $\kappa+1>Re(\alpha)>\kappa$, $-\infty<a<b<\infty$ and $\lambda\in\mathbb{C}$, the left-sided Liouville-Sonine-Caputo type general fractional derivative within the subsine kernel via Mittag-Leffler function is defined as

$$\left({}_{Sls}^{LSC}D_{a+}^{\alpha,\lambda}f\right)(t)={}_{Sls}I_{a+}^{\alpha,\lambda}\left(f^{(\kappa)}(t)\right)=\int_a^t LSin_\alpha\left(-\lambda(t-\tau)^\alpha\right)f^{(\kappa)}(\tau)\,d\tau, \tag{3.326}$$

and the right-sided Liouville-Sonine-Caputo type general fractional derivative within the subsine kernel via Mittag-Leffler function as

$$\left({}_{Sls}^{LSC}D_{b-}^{\alpha,\lambda}f\right)(t) = {}_{Sls}I_{b-}^{\alpha,\lambda}\left((-1)^{\kappa}f^{(\kappa)}(t)\right)$$
$$= (-1)^{\kappa}\int_{t}^{b} LSin_{\alpha}\left(-\lambda(\tau-t)^{\alpha}\right)f^{(\kappa)}(\tau)\,d\tau. \qquad (3.327)$$

The inverse operators of the above fractional derivatives within the subsine kernel via Mittag-Leffler function are considered as follows.

Definition 3.55 *Let $\alpha \in \mathbb{C}$, $\kappa+1 > Re(\alpha) > \kappa$, $-\infty < a < b < \infty$ and $\lambda \in \mathbb{C}$. The left-sided general fractional integral is defined as*

$$\left({}_{Sls}^{R}I_{a+}^{\alpha,\lambda}f\right)(t) = \frac{1}{\lambda}\int_{a}^{t}(t-\tau)^{\kappa-\alpha-2}E_{2\alpha,\kappa-\alpha-1}^{-1}\left(-\lambda^{2}(t-\tau)^{2\alpha}\right)f(\tau)\,d\tau \qquad (3.328)$$

and the right-sided general fractional integral as

$$\left({}_{Sls}^{R}I_{b-}^{\alpha,\lambda}f\right) = \frac{1}{\lambda}\int_{t}^{b}(\tau-t)^{\kappa-\alpha-2}E_{2\alpha,\kappa-\alpha-1}^{-1}\left(-\lambda^{2}(\tau-t)^{2\alpha}\right)f(\tau)\,d\tau. \qquad (3.329)$$

In particular, when $\alpha \in \mathbb{C}$, $1 > Re(\alpha) > 0$, $-\infty < a < b < \infty$ and $\lambda \in \mathbb{C}$, the left-sided general fractional integral is defined as

$$\left({}_{Sls}^{R}I_{a+}^{\alpha,\lambda}f\right)(t) = \frac{1}{\lambda}\int_{a}^{t}\frac{1}{(t-\tau)^{\alpha+1}}E_{2\alpha,-\alpha}^{-1}\left(-\lambda^{2}(t-\tau)^{2\alpha}\right)f(\tau)\,d\tau \qquad (3.330)$$

and the right-sided general fractional integral as

$$\left({}_{Sls}^{R}I_{b-}^{\alpha,\lambda}f\right) = \frac{1}{\lambda}\int_{t}^{b}\frac{1}{(\tau-t)^{\alpha+1}}E_{2\alpha,-\alpha}^{-1}\left(-\lambda^{2}(\tau-t)^{2\alpha}\right)f(\tau)\,d\tau. \qquad (3.331)$$

The properties of the general fractional derivatives and the general fractional integrals are given as follows.

Property 3.25 *If $\kappa+1 > \alpha > \kappa$, $-\infty < a < b < \infty$, $\lambda \in \mathbb{C}$ and $f(t) \in L_{l}(a,b)$ $(1 \le l < \infty)$, then we have*

$$L\left\{\left({}_{Sls}^{RL}D_{0+}^{\alpha,\lambda}f\right)(t)\right\} = \lambda s^{\kappa-(1+\alpha)}\left(1+\lambda^{2}s^{-2\alpha}\right)^{-1}f(s), \qquad (3.332)$$

$$L\left\{\left({}_{Sls}^{RL}D_{0+}^{\alpha,\lambda}f\right)(t)\right\} = \lambda s^{1-(1+\alpha)}\left(1+\lambda^{2}s^{-2\alpha}\right)^{-1}f(s), \qquad (3.333)$$

$$L\left\{\left({}_{Sls}^{LS}D_{0+}^{\alpha,\lambda}f\right)(t)\right\} = \lambda s^{-(1+\alpha)}\left(1+\lambda^2 s^{-2\alpha}\right)^{-1}\left(sf(s)-f(0)\right), \tag{3.334}$$

and

$$L\left\{\left({}_{Sls}^{LSC}D_{0+}^{\alpha,\lambda}f\right)(t)\right\} = \lambda s^{-(1+\alpha)}\left(1+\lambda^2 s^{-2\alpha}\right)^{-1}\left(s^\kappa f(s) - \sum_{j=1}^{\kappa} s^{\kappa-j} f^{(j-1)}(0)\right), \tag{3.335}$$

$$L\left\{\left({}_{Sls}^{R}I_{0+}^{\alpha,\lambda}f\right)(t)\right\} = \lambda^{-1}s^{\alpha}\left(1+\lambda^2 s^{-2\alpha}\right)f(s) \quad (0<\alpha<1) \tag{3.336}$$

and

$$L\left\{\left({}_{Sls}^{R}I_{0+}^{\alpha,\lambda}f\right)(t)\right\} = \lambda^{-1}s^{\alpha+1-\kappa}\left(1+\lambda^2 s^{-2\alpha}\right)f(s) \quad (\kappa<\alpha<\kappa+1). \tag{3.337}$$

Note that, when $D^\kappa\left({}_{Sls}^{R}I_{0+}^{\alpha,\lambda}f(t)\right)|_{t=0}=0$, there is

$$L\left\{{}_{Sls}^{RL}D_{0+}^{\alpha,\lambda}f(t)\right\} = \lambda s^{\kappa-(1+\alpha)}\left(1+\lambda^2 s^{-2\alpha}\right)^{-1}f(s), \tag{3.338}$$

and when $D^\kappa\left({}_{Sls}^{R}I_{0+}^{\alpha,\lambda}f(t)\right)|_{t=0}\neq 0$, there is

$$L\left\{{}_{Sls}^{RL}D_{0+}^{\alpha,\lambda}f(t)\right\} = \lambda s^{\kappa-(1+\alpha)}\left(1+\lambda^2 s^{-2\alpha}\right)^{-1}f(s) - \sum_{\mu=0}^{\kappa-1} s^{\kappa-\mu-1}\left(D^\mu\left[{}_{Sls}^{R}I_{0+}^{\alpha,\lambda}f(+0)\right]\right). \tag{3.339}$$

In particular, when $\kappa=1$, we have

$$L\left\{{}_{Sls}^{RL}D_{0+}^{\alpha,\lambda}f(t)\right\} = \lambda s^{-\alpha}\left(1+\lambda^2 s^{-2\alpha}\right)^{-1}f(s) - {}_{Sls}^{R}I_{0+}^{\alpha,\lambda}f(+0), \tag{3.340}$$

and when $\kappa=1$ and $D\left({}_{Sls}^{R}I_{0+}^{\alpha,\lambda}f(t)\right)|_{t=0}=0$, we have

$$L\left\{{}_{Sls}^{RL}D_{0+}^{\alpha,\lambda}f(t)\right\} = \lambda s^{-\alpha}\left(1+\lambda^2 s^{-2\alpha}\right)^{-1}f(s). \tag{3.341}$$

Property 3.26 *Let $\kappa+1>\alpha>\kappa$, $-\infty<a<b<\infty$, $\lambda\in C$ and $f(t)\in L_l(a,b)$ $(1\le l<\infty)$. Then we have*

$$\left({}_{Sls}^{RL}D_{a+}^{\alpha,\lambda}\,{}_{Sls}^{R}I_{a+}^{\alpha,\lambda}f\right)(t) = f(t), \tag{3.342}$$

$$\left({}_{Sls}^{RL}D_{b-}^{\alpha,\lambda}\,{}_{Sls}^{R}I_{b-}^{\alpha,\lambda}f\right)(t) = f(t), \tag{3.343}$$

$$\left({}_{Sls}^{RL}D_{a+}^{\alpha,\lambda}\,{}_{Sls}^{R}I_{a+}^{\alpha,\lambda}f\right)(t) = f(t), \tag{3.344}$$

and

$$\left({}_{Sls}^{RL}D_{b-}^{\alpha,\lambda}\,{}_{Sls}^{R}I_{b-}^{\alpha,\lambda}f\right)(t) = f(t). \tag{3.345}$$

Property 3.27 *If* $-\infty < a < b < \infty$, $\lambda \in \mathbb{C}$ *and* $f(t) \in AC_\kappa(a,b)$ $(1 \leq \kappa < \infty)$, *then we have*

$$\left({}^{LS}_{Sls}D^{\alpha,\lambda R}_{a+}\, {}_{Sls}I^{\alpha,\lambda}_{a+} f \right)(t) = f(t), \tag{3.346}$$

$$\left({}^{LS}_{Sls}D^{\alpha,\lambda R}_{b-}\, {}_{Sls}I^{\alpha,\lambda}_{b-} f \right)(t) = f(t), \tag{3.347}$$

$$\left({}^{LSC}_{Sls}D^{\alpha,\lambda R}_{a+}\, {}_{Sls}I^{\alpha,\lambda}_{a+} f \right)(t) = f(t), \tag{3.348}$$

and

$$\left({}^{LSC}_{Sls}D^{\alpha,\lambda R}_{b-}\, {}_{Sls}I^{\alpha,\lambda}_{b-} f \right)(t) = f(t). \tag{3.349}$$

Definition 3.56 *Let* $\alpha \in \mathbb{C}$, $1 > Re(\alpha) > 0$, $a = -\infty$, $b = \infty$ *and* $\lambda \in \mathbb{C}$. *The left-sided Riemann-Liouville type general fractional derivative within the subsine kernel via Mittag-Leffler function is defined as*

$$\left({}^{RL}_{Sls}D^{\alpha,\lambda}_{+} f \right)(t) = \frac{d}{dt}\left({}_{Sls}I^{\alpha,\lambda}_{+} f(t) \right) = \frac{d}{dt} \int_{-\infty}^{t} LSin_\alpha\left(-\lambda (t-\tau)^\alpha\right) f(\tau)\, d\tau, \tag{3.350}$$

and the right-sided Riemann-Liouville type general fractional derivative within the subsine kernel via Mittag-Leffler function as

$$\left({}^{RL}_{Sls}D^{\alpha,\lambda}_{-} f \right)(t) = -\frac{d}{dt}\left({}_{Sls}I^{\alpha,\lambda}_{-} f(t) \right) = -\frac{d}{dt} \int_{t}^{\infty} LSin_\alpha\left(-\lambda (\tau-t)^\alpha\right) f(\tau)\, d\tau, \tag{3.351}$$

where the left-sided and right-sided general fractional integral operators within the subsine kernel via Mittag-Leffler function are defined as

$${}_{Sls}I^{\alpha,\lambda}_{+} f(t) = \int_{-\infty}^{t} LSin_\alpha\left(-\lambda (t-\tau)^\alpha\right) f(\tau)\, d\tau \tag{3.352}$$

and

$${}_{Sls}I^{\alpha,\lambda}_{-} f(t) = \int_{t}^{\infty} LSin_\alpha\left(-\lambda (\tau-t)^\alpha\right) f(\tau)\, d\tau, \tag{3.353}$$

respectively.

Generally, when $\alpha \in \mathbb{C}$, $Re(\alpha) > 0$, $\kappa = [Re(\alpha)] + 1$, $a = -\infty$, $b = \infty$ and $\lambda \in \mathbb{C}$, the left-sided Riemann-Liouville type general fractional derivative within the subsine kernel via Mittag-Leffler function is defined as

$$\left({}^{RL}_{Sls}D^{\alpha,\lambda}_{+} f \right)(t) = \frac{d^\kappa}{dt^\kappa}\left({}_{Sls}I^{\alpha,\lambda}_{+} f(t) \right) = \frac{d^\kappa}{dt^\kappa} \int_{-\infty}^{t} LSin_\alpha\left(-\lambda (t-\tau)^\alpha\right) f(\tau)\, d\tau, \tag{3.354}$$

and the right-sided Riemann-Liouville type general fractional derivative within the subsine kernel via Mittag-Leffler function as

$$\left({}_{Sls}^{RL}D_{-}^{\alpha,\lambda}f\right)(t)=\left(-\frac{d}{dt}\right)^{\kappa}\left({}_{Sls}I_{-}^{\alpha,\lambda}f(t)\right)=(-1)^{\kappa}\frac{d^{\kappa}}{dt^{\kappa}}\int_{t}^{\infty}LSin_{\alpha}\left(-\lambda\left(\tau-t\right)^{\alpha}\right)f(\tau)\,d\tau. \tag{3.355}$$

Definition 3.57 *Let $\alpha\in\mathbb{C}$, $1>Re(\alpha)>0$, $a=-\infty$, $b=\infty$ and $\lambda\in\mathbb{C}$. The left-sided Liouville-Sonine type general fractional derivative within the subsine kernel via Mittag-Leffler function is defined as*

$$\left({}_{Sls}^{LS}D_{+}^{\alpha,\lambda}f\right)(t)={}_{Sls}I_{+}^{\alpha,\lambda}\left(f^{(1)}(t)\right)=\int_{-\infty}^{t}LSin_{\alpha}\left(-\lambda\left(t-\tau\right)^{\alpha}\right)f^{(1)}(\tau)\,d\tau \tag{3.356}$$

and the right-sided Liouville-Sonine type general fractional derivative within the subsine kernel via Mittag-Leffler function as

$$\left({}_{Sls}^{LS}D_{-}^{\alpha,\lambda}f\right)(t)={}_{Sls}I_{-}^{\alpha,\lambda}\left(-f^{(1)}(t)\right)=-\int_{t}^{-\infty}LSin_{\alpha}\left(-\lambda\left(\tau-t\right)^{\alpha}\right)f^{(1)}(\tau)\,d\tau. \tag{3.357}$$

Furthermore, when $\alpha\in\mathbb{C}$, $\kappa+1>Re(\alpha)>\kappa$, $a=-\infty$, $b=\infty$ and $\lambda\in$ C, the left-sided Liouville-Sonine-Caputo type general fractional derivative within the subsine kernel via Mittag-Leffler function is defined as

$$\left({}_{Sls}^{LSC}D_{+}^{\alpha,\lambda}f\right)(t)=\int_{-\infty}^{t}LSin_{\alpha}\left(-\lambda\left(t-\tau\right)^{\alpha}\right)f^{(\kappa)}(\tau)\,d\tau \tag{3.358}$$

and the right-sided Liouville-Sonine-Caputo type general fractional derivative within the subsine kernel via Mittag-Leffler function as

$$\left({}_{Sls}^{LSC}D_{-}^{\alpha,\lambda}f\right)(t)=(-1)^{\kappa}\int_{t}^{\infty}LSin_{\alpha}\left(-\lambda\left(\tau-t\right)^{\alpha}\right)f^{(\kappa)}(\tau)\,d\tau. \tag{3.359}$$

The inverse operators of the above general fractional derivatives within the subsine kernel via Mittag-Leffler function are presented as follows.

Definition 3.58 *Let $\alpha\in\mathbb{C}$, $\kappa+1>Re(\alpha)>\kappa$, $a=-\infty$, $b=\infty$ and $\lambda\in\mathbb{C}$. The left-sided general fractional integral is defined as*

$$\left({}_{Sls}^{R}I_{+}^{\alpha,\lambda}f\right)(t)=\frac{1}{\lambda}\int_{-\infty}^{t}(t-\tau)^{\kappa-\alpha-2}E_{2\alpha,\kappa-\alpha-1}^{-1}\left(-\lambda^{2}\left(t-\tau\right)^{2\alpha}\right)f(\tau)\,d\tau \tag{3.360}$$

and the right-sided general fractional integral as

$$\left({}_{Sls}^{R}I_{-}^{\alpha,\lambda}f\right)=\frac{1}{\lambda}\int_{t}^{\infty}(\tau-t)^{\kappa-\alpha-2}E_{2\alpha,\kappa-\alpha-1}^{-1}\left(-\lambda^{2}\left(\tau-t\right)^{2\alpha}\right)f(\tau)\,d\tau. \tag{3.361}$$

In particular, when $\alpha \in \mathbb{C}$, $1 > Re(\alpha) > 0$, $a = -\infty$, $b = \infty$ and $\lambda \in \mathbb{C}$, the left-sided general fractional integral is defined as

$$\left({}_{Sls}^{R}I_{+}^{\alpha,\lambda} f\right)(t) = \frac{1}{\lambda} \int_{-\infty}^{t} \frac{1}{(t-\tau)^{\alpha+1}} E_{2\alpha,-\alpha}^{-1}\left(-\lambda^2 (t-\tau)^{2\alpha}\right) f(\tau)\, d\tau \quad (3.362)$$

and the right-sided general fractional integral as

$$\left({}_{Sls}^{R}I_{-}^{\alpha,\lambda} f\right) = \frac{1}{\lambda} \int_{t}^{\infty} \frac{1}{(\tau-t)^{\alpha+1}} E_{2\alpha,-\alpha}^{-1}\left(-\lambda^2 (\tau-t)^{2\alpha}\right) f(\tau)\, d\tau. \quad (3.363)$$

In this case, the properties of the general fractional derivatives and fractional integrals are given as follows.

Property 3.28 *Let* $\kappa + 1 > \alpha > \kappa$, $a = -\infty$, $b = \infty$, $\lambda \in \mathbb{C}$ *and* $f(t) \in L_l(-\infty,\infty)$ $(1 \le l < \infty)$. *Then we have*

$$\left({}_{Sls}^{RL}D_{+}^{\alpha,\lambda}\, {}_{Sls}^{R}I_{+}^{\alpha,\lambda} f\right)(t) = f(t), \quad (3.364)$$

$$\left({}_{Sls}^{RL}D_{-}^{\alpha,\lambda}\, {}_{Sls}^{R}I_{-}^{\alpha,\lambda} f\right)(t) = f(t), \quad (3.365)$$

$$\left({}_{Sls}^{RL}D_{+}^{\alpha,\lambda}\, {}_{Sls}^{R}I_{+}^{\alpha,\lambda} f\right)(t) = f(t), \quad (3.366)$$

and

$$\left({}_{Sls}^{RL}D_{-}^{\alpha,\lambda}\, {}_{Sls}^{R}I_{-}^{\alpha,\lambda} f\right)(t) = f(t). \quad (3.367)$$

Property 3.29 *If* $a = -\infty$, $b = \infty$, $\lambda \in \mathbb{C}$ *and* $f(t) \in AC_\kappa(-\infty,\infty)$ $(1 \le \kappa < \infty)$, *then we have*

$$\left({}_{Sls}^{LS}D_{+}^{\alpha,\lambda}\, {}_{Sls}^{R}I_{+}^{\alpha,\lambda} f\right)(t) = f(t), \quad (3.368)$$

$$\left({}_{Sls}^{LS}D_{-}^{\alpha,\lambda}\, {}_{Sls}^{R}I_{-}^{\alpha,\lambda} f\right)(t) = f(t), \quad (3.369)$$

$$\left({}_{Sls}^{LSC}D_{+}^{\alpha,\lambda}\, {}_{Sls}^{R}I_{+}^{\alpha,\lambda} f\right)(t) = f(t), \quad (3.370)$$

and

$$\left({}_{Sls}^{LSC}D_{-}^{\alpha,\lambda}\, {}_{Sls}^{R}I_{-}^{\alpha,\lambda} f\right)(t) = f(t). \quad (3.371)$$

3.12 Hilfer type general fractional derivatives within the subsine kernel via Mittag-Leffler function

The definitions and properties of the Hilfer type general fractional derivatives within the subsine kernel via Mittag-Leffler function are given as follows.

Definition 3.59 *Let $1 > \alpha > 0$, $0 \leq \beta \leq 1$, $-\infty < a < b < \infty$, $\gamma \in \mathbb{C}$ and $\lambda \in \mathbb{C}$. The left-sided Hilfer type general fractional derivative within the subsine kernel via Mittag-Leffler function is defined as*

$$\begin{aligned}\left({}_{Sls}^{Hi}D_{a+}^{\alpha,\beta,\lambda}f\right)(t) &= \int_a^t LSin_\alpha\left(-\lambda(t-\tau)^\alpha\right)\left(D_{a+}^{\alpha,\beta}f(\tau)\right)d\tau \\ &= \int_a^t LSin_\alpha\left(-\lambda(t-\tau)^\alpha\right)\left(I_{a+}^{\beta(1-\alpha)}\frac{d}{d\tau}\left(I_{a+}^{(1-\beta)(1-\alpha)}f\right)\right)(\tau)\,d\tau,\end{aligned} \tag{3.372}$$

and the right-sided Hilfer type general fractional derivative within the subsine kernel via Mittag-Leffler function as

$$\begin{aligned}\left({}_{Sls}^{Hi}D_{b-}^{\alpha,\beta,\lambda}f\right)(t) &= -\int_t^b LSin_\alpha\left(-\lambda(\tau-t)^\alpha\right)\left(D_{b-}^{\alpha,\beta}f(\tau)\right)d\tau \\ &= -\int_t^b LSin_\alpha\left(-\lambda(\tau-t)^\alpha\right)\left(I_{b-}^{\beta(1-\alpha)}\frac{d}{d\tau}\left(I_{b-}^{(1-\beta)(1-\alpha)}f\right)\right)(\tau)\,d\tau.\end{aligned} \tag{3.373}$$

Generally, when $\kappa+1 > \alpha > \kappa$, $0 \leq \beta \leq 1$, $-\infty < a < b < \infty$, $\gamma \in \mathbb{C}$ and $\lambda \in \mathbb{C}$, the left-sided Hilfer type general fractional derivative within the subsine kernel via Mittag-Leffler function is defined as

$$\begin{aligned}\left({}_{Sls}^{Hir}D_{a+}^{\alpha,\beta,\lambda}f\right)(t) &= \int_a^t LSin_\alpha\left(-\lambda(t-\tau)^\alpha\right)\left(D_{a+}^{\alpha,\beta,\kappa}f(\tau)\right)d\tau \\ &= \int_a^t LSin_\alpha\left(-\lambda(t-\tau)^\alpha\right)\left(I_{a+}^{\beta(\kappa-\alpha)}\frac{d^\kappa}{d\tau^\kappa}\left(I_{a+}^{(1-\beta)(\kappa-\alpha)}f\right)\right)(\tau)\,d\tau,\end{aligned} \tag{3.374}$$

and the right-sided Hilfer type general fractional derivative within the subsine kernel via Mittag-Leffler function as

$$\begin{aligned}\left({}_{Sls}^{Hir}D_{b-}^{\alpha,\beta,\lambda}f\right)(t) &= (-1)^\kappa\int_t^b LSin_\alpha\left(-\lambda(\tau-t)^\alpha\right)\left(D_{b-}^{\alpha,\beta,\kappa}f(\tau)\right)d\tau \\ &= (-1)^\kappa\int_t^b LSin_\alpha\left(-\lambda(\tau-t)^\alpha\right)\left(I_{b-}^{\beta(\kappa-\alpha)}\frac{d^\kappa}{d\tau^\kappa}\left(I_{b-}^{(1-\beta)(\kappa-\alpha)}f\right)\right)(\tau)\,d\tau.\end{aligned} \tag{3.375}$$

On one hand, when $\kappa+1 > \alpha > \kappa$, $\beta = 1$, $-\infty < a < b < \infty$, $\gamma \in \mathbb{C}$ and $\lambda \in \mathbb{C}$, the left-sided Liouville-Sonine-Caputo type general fractional derivative within the subsine kernel via Mittag-Leffler function is defined as

$$\begin{aligned}\left({}_{Sls}^{Hir}D_{a+}^{\alpha,1,\lambda}f\right)(t) &= \int_a^t LSin_\alpha\left(-\lambda(t-\tau)^\alpha\right)\left({}_{LSC}D_{a+}^{\alpha,\kappa}f(\tau)\right)d\tau \\ &= \int_a^t LSin_\alpha\left(-\lambda(t-\tau)^\alpha\right)\left(I_{a+}^{\kappa-\alpha}f^{(\kappa)}(\tau)\right)d\tau,\end{aligned} \tag{3.376}$$

and the right-sided Liouville-Sonine-Caputo type general fractional derivative within the subsine kernel via Mittag-Leffler function as

$$\begin{aligned}\left({}_{Sls}^{Hir}D_{b-}^{\alpha,1,\lambda}f\right)(t) &= (-1)^\kappa\int_t^b LSin_\alpha\left(-\lambda(\tau-t)^\alpha\right)\left({}_{LSC}D_{b-}^{\alpha,\kappa}f(\tau)\right)d\tau \\ &= (-1)^\kappa\int_t^b LSin_\alpha\left(-\lambda(\tau-t)^\alpha\right)\left(I_{b-}^{\kappa-\alpha}f^{(\kappa)}(\tau)\right)(\tau)\,d\tau.\end{aligned} \tag{3.377}$$

On the other hand, when $\kappa+1>\alpha>\kappa$, $\beta=0$, $-\infty<a<b<\infty$, $\gamma\in\mathbb{C}$ and $\lambda\in\mathbb{C}$, the left-sided Riemann-Liouville type general fractional derivative within the subsine kernel via Mittag-Leffler function is defined as

$$\begin{aligned}&\left({}_{Sls}^{Hir}D_{a+}^{\alpha,0,\lambda}f\right)(t)=\int_a^t LSin_\alpha\left(-\lambda\left(t-\tau\right)^\alpha\right)\left(D_{a+}^{\alpha,\kappa}f\left(\tau\right)\right)d\tau\\&=\int_a^t LSin_\alpha\left(-\lambda\left(t-\tau\right)^\alpha\right)\left(\tfrac{d^\kappa}{d\tau^\kappa}\left(I_{a+}^{\kappa-\alpha}f\right)(\tau)\right)d\tau,\end{aligned}\tag{3.378}$$

and the right-sided Riemann-Liouville type general fractional derivative within the subsine kernel via Mittag-Leffler function as

$$\begin{aligned}&\left({}_{Sls}^{Hir}D_{b-}^{\alpha,0,\lambda}f\right)(t)=(-1)^\kappa\int_t^b LSin_\alpha\left(-\lambda\left(\tau-t\right)^\alpha\right)\left(D_{b-}^{\alpha,\kappa}f\left(\tau\right)\right)d\tau\\&=(-1)^\kappa\int_t^b LSin_\alpha\left(-\lambda\left(\tau-t\right)^\alpha\right)\left(\tfrac{d^\kappa}{d\tau^\kappa}\left(I_{b-}^{\kappa-\alpha}f\right)(\tau)\right)d\tau.\end{aligned}\tag{3.379}$$

3.13 General fractional derivatives in the two-parametric Lorenzo-Hartley kernel

The definitions and properties of the general fractional derivatives in the two-parametric Lorenzo-Hartley kernel and the general fractional integrals are given as follows.

Definition 3.60 *Let $\alpha\in\mathbb{C}$, $1>Re(\alpha)>0$, $-\infty<a<b<\infty$ and $\lambda\in\mathbb{C}$. The left-sided Riemann-Liouville type general fractional derivative within the two-parametric Lorenzo-Hartley kernel is defined as*

$$\left({}_{HL}^{RL}D_{a+}^{\alpha,\upsilon,\lambda}f\right)(t)=\frac{d}{dt}\left({}_{HL}I_{a+}^{\alpha,\upsilon,\lambda}f\left(t\right)\right)=\frac{d}{dt}\int_a^t R_{\alpha,\upsilon}\left(-\lambda\left(t-\tau\right)^\alpha\right)f\left(\tau\right)d\tau,\tag{3.380}$$

and the right-sided Riemann-Liouville type general fractional derivative within the two-parametric Lorenzo-Hartley kernel as

$$\left({}_{HL}^{RL}D_{b-}^{\alpha,\upsilon,\lambda}f\right)(t)=-\frac{d}{dt}\left({}_{HL}I_{b-}^{\alpha,\upsilon,\lambda}f\left(t\right)\right)=-\frac{d}{dt}\int_t^b R_{\alpha,\upsilon}\left(-\lambda\left(\tau-t\right)^\alpha\right)f\left(\tau\right)d\tau,\tag{3.381}$$

where the left-sided and right-sided general fractional integral operators within the two-parametric Lorenzo-Hartley kernel are defined as

$${}_{HL}I_{a+}^{\alpha,\upsilon,\lambda}f\left(t\right)=\int_a^t R_{\alpha,\upsilon}\left(-\lambda\left(t-\tau\right)^\alpha\right)f\left(\tau\right)d\tau\tag{3.382}$$

and

$$ {}_{HL}I_{b-}^{\alpha,\upsilon,\lambda} f(t) = \int_t^b R_{\alpha,\upsilon}\left(-\lambda(\tau - t)^{\alpha}\right) f(\tau)\, d\tau, \tag{3.383}$$

respectively.

More generally, when $\alpha \in \mathbb{C}$, $Re(\alpha) > 0$, $\kappa = [Re(\alpha)] + 1$, $-\infty < a < b < \infty$ and $\lambda \in \mathbb{C}$, the left-sided Riemann-Liouville type general fractional derivative within the two-parametric Lorenzo-Hartley kernel is defined as

$$\left({}_{HL}^{RL}D_{a+}^{\alpha,\upsilon,\lambda} f\right)(t) = \frac{d^{\kappa}}{dt^{\kappa}}\left({}_{HL}I_{a+}^{\alpha,\upsilon,\lambda} f(t)\right) = \frac{d^{\kappa}}{dt^{\kappa}} \int_a^t R_{\alpha,\upsilon}\left(-\lambda(t - \tau)^{\alpha}\right) f(\tau)\, d\tau, \tag{3.384}$$

and the right-sided Riemann-Liouville type general fractional derivative within the two-parametric Lorenzo-Hartley kernel as

$$\left({}_{HL}^{RL}D_{b-}^{\alpha,\upsilon,\lambda} f\right)(t) = \left(-\frac{d}{dt}\right)^{\kappa}\left({}_{HL}I_{b-}^{\alpha,\upsilon,\lambda} f(t)\right) = (-1)^{\kappa} \frac{d^{\kappa}}{dt^{\kappa}} \int_t^b R_{\alpha,\upsilon}\left(-\lambda(\tau - t)^{\alpha}\right) f(\tau)\, d\tau. \tag{3.385}$$

Definition 3.61 *Let $\alpha \in \mathbb{C}$, $1 > Re(\alpha) > 0$, $-\infty < a < b < \infty$ and $\lambda \in \mathbb{C}$. The left-sided Liouville-Sonine type general fractional derivative within the two-parametric Lorenzo-Hartley kernel is defined as*

$$\left({}_{HL}^{LS}D_{a+}^{\alpha,\upsilon,\lambda} f\right)(t) = {}_{HL}I_{a+}^{\alpha,\upsilon,\lambda}\left(f^{(1)}(t)\right) = \int_a^t R_{\alpha,\upsilon}\left(-\lambda(t - \tau)^{\alpha}\right) f^{(1)}(\tau)\, d\tau, \tag{3.386}$$

and the right-sided Liouville-Sonine type general fractional derivative within the two-parametric Lorenzo-Hartley kernel as

$$\left({}_{HL}^{LS}D_{b-}^{\alpha,\upsilon,\lambda} f\right)(t) = {}_{HL}I_{b-}^{\alpha,\upsilon,\lambda}\left(-f^{(1)}(t)\right) = -\int_t^b R_{\alpha,\upsilon}\left(-\lambda(\tau - t)^{\alpha}\right) f^{(1)}(\tau)\, d\tau. \tag{3.387}$$

More generally, for $\alpha \in \mathbb{C}$, $\kappa + 1 > Re(\alpha) > \kappa$, $-\infty < a < b < \infty$ and $\lambda \in \mathbb{C}$, the left-sided Liouville-Sonine-Caputo type general fractional derivative within the two-parametric Lorenzo-Hartley kernel is defined as

$$\left({}_{HL}^{LSC}D_{a+}^{\alpha,\upsilon,\lambda} f\right)(t) = {}_{HL}I_{a+}^{\alpha,\upsilon,\lambda}\left(f^{(\kappa)}(t)\right) = \int_a^t R_{\alpha,\upsilon}\left(-\lambda(t - \tau)^{\alpha}\right) f^{(\kappa)}(\tau)\, d\tau, \tag{3.388}$$

and the right-sided Liouville-Sonine-Caputo type general fractional derivative within the two-parametric Lorenzo-Hartley kernel as

$$\begin{aligned}&\left({}_{HL}^{LSC}D_{b-}^{\alpha,\upsilon,\lambda} f\right)(t)\\ &= {}_{HL}I_{b-}^{\alpha,\upsilon,\lambda}\left((-1)^{\kappa} f^{(\kappa)}(t)\right) = (-1)^{\kappa}\int_t^b R_{\alpha,\upsilon}\left(-\lambda(\tau - t)^{\alpha}\right) f^{(\kappa)}(\tau)\, d\tau.\end{aligned} \tag{3.389}$$

The inverse operators of the above general fractional derivatives within the two-parametric Lorenzo-Hartley kernel are shown as follows.

Definition 3.62 *Let $\alpha \in \mathbb{C}$, $\kappa + 1 > Re(\alpha) > \kappa$, $-\infty < a < b < \infty$ and $\lambda \in \mathbb{C}$. The left-sided general fractional integral is defined as*

$$\left({}_{HL}^{R}I_{a+}^{\alpha,\upsilon,\lambda} f\right)(t) = \int_{a}^{t} (t-\tau)^{\kappa-(\alpha-\upsilon)-1} E_{\alpha,\kappa-(\alpha-\upsilon)}^{-1} \left(-\gamma (t-\tau)^{\alpha}\right) f(\tau)\, d\tau \tag{3.390}$$

and the right-sided general fractional integral as

$$\left({}_{MR}^{R}I_{b-}^{\alpha,\upsilon,\lambda} f\right) = \int_{t}^{b} (\tau - t)^{\kappa-(\alpha-\upsilon)-1} E_{\alpha,\kappa-(\alpha-\upsilon)}^{-1} \left(-\gamma (\tau - t)^{\alpha}\right) f(\tau)\, d\tau. \tag{3.391}$$

In particular, when $\alpha \in \mathbb{C}$, $1 > Re(\alpha) > 0$, $-\infty < a < b < \infty$ and $\lambda \in \mathbb{C}$, the left-sided general fractional integral is defined as

$$\left({}_{HL}^{R}I_{a+}^{\alpha,\upsilon,\lambda} f\right)(t) = \int_{a}^{t} \frac{1}{(t-\tau)^{\alpha-\upsilon}} E_{\alpha,1-(\alpha-\upsilon)}^{-1} \left(-\gamma (t-\tau)^{\alpha}\right) f(\tau)\, d\tau \tag{3.392}$$

and the right-sided general fractional integral as

$$\left({}_{MR}^{R}I_{b-}^{\alpha,\upsilon,\lambda} f\right) = \int_{t}^{b} \frac{1}{(\tau - t)^{\alpha-\upsilon}} E_{\alpha,\kappa-(\alpha-\upsilon)}^{-1} \left(-\gamma (\tau - t)^{\alpha}\right) f(\tau)\, d\tau. \tag{3.393}$$

The properties of the general fractional derivatives and the general fractional integrals are given as follows.

Property 3.30 *If $\kappa + 1 > \alpha > \kappa$, $-\infty < a < b < \infty$, $\lambda \in \mathbb{C}$ and $f(t) \in L_l(a,b)$ $(1 \le l < \infty)$, then we have*

$$L\left\{\left({}_{HL}^{RL}D_{0+}^{\alpha,\upsilon,\lambda} f\right)(t)\right\} = s^{\kappa-(\alpha-\upsilon)} \left(1 + \lambda s^{-\alpha}\right)^{-1} f(s), \tag{3.394}$$

$$L\left\{\left({}_{HL}^{RL}D_{0+}^{\alpha,\upsilon,\lambda} f\right)(t)\right\} = s^{\kappa-(\alpha-\upsilon)} \left(1 + \lambda s^{-\alpha}\right)^{-1} f(s), \tag{3.395}$$

$$L\left\{\left({}_{HL}^{LS}D_{0+}^{\alpha,\upsilon,\lambda} f\right)(t)\right\} = s^{-(\alpha-\upsilon)} \left(1 - \lambda s^{-\alpha}\right)^{-1} \left(sf(s) - f(0)\right), \tag{3.396}$$

$$L\left\{\left({}_{HL}^{LSC}D_{0+}^{\alpha,\upsilon,\lambda} f\right)(t)\right\} = s^{-(\alpha-\upsilon)} \left(1 - \lambda s^{-\alpha}\right)^{-1} \left(s^{\kappa} f(s) - \sum_{j=1}^{\kappa} s^{\kappa-j} f^{(j-1)}(0)\right), \tag{3.397}$$

$$L\left\{\left({}_{HL}^{R}I_{0+}^{\alpha,\upsilon,\lambda} f\right)(t)\right\} = s^{(\alpha-\upsilon)-\kappa} \left(1 - \lambda s^{-\alpha}\right) f(s) \tag{3.398}$$

and

$$L\left\{\left({}_{HL}^{R}I_{0+}^{\alpha,\upsilon,\lambda} f\right)(t)\right\} = s^{(\alpha-\upsilon)-1} \left(1 - \lambda s^{-\alpha}\right) f(s). \tag{3.399}$$

Note that, when $D^{\kappa}\left({}_{HL}^{R}I_{0+}^{\alpha,\upsilon,\lambda}f(t)\right)|_{t=0}=0$, there is

$$L\left\{{}_{HL}^{RL}D_{0+}^{\alpha,\upsilon,\lambda}f(t)\right\}=s^{\kappa-(\alpha-\upsilon)}\left(1+\lambda s^{-\alpha}\right)^{-1}f(s), \tag{3.400}$$

and when $D^{\kappa}\left({}_{HL}^{R}I_{0+}^{\alpha,\upsilon,\lambda}f(t)\right)|_{t=0}\neq 0$, there is

$$L\left\{{}_{HL}^{RL}D_{0+}^{\alpha,\upsilon,\lambda}f(t)\right\}=s^{\kappa-(\alpha-\upsilon)}\left(1+\lambda s^{-\alpha}\right)^{-1}f(s)-\sum_{\mu=0}^{\kappa-1}s^{\kappa-\mu-1}\left(D^{\mu}\left[{}_{HL}^{R}I_{0+}^{\alpha,\upsilon,\lambda}f(+0)\right]\right). \tag{3.401}$$

In particular, when $\kappa=1$, we have

$$L\left\{{}_{HL}^{RL}D_{0+}^{\alpha,\upsilon,\lambda}f(t)\right\}=s^{1-(\alpha-\upsilon)}\left(1+\lambda s^{-\alpha}\right)^{-1}f(s)-{}_{MR}^{R}I_{0+}^{\alpha,\lambda}f(+0), \tag{3.402}$$

and when $\kappa=1$ and $D\left({}_{HL}^{R}I_{0+}^{\alpha,\upsilon,\lambda}f(t)\right)|_{t=0}=0$, we have

$$L\left\{{}_{HL}^{RL}D_{0+}^{\alpha,\upsilon,\lambda}f(t)\right\}=s^{1-(\alpha-\upsilon)}\left(1+\lambda s^{-\alpha}\right)^{-1}f(s). \tag{3.403}$$

Property 3.31 *Let $\kappa+1>\alpha>\kappa$, $-\infty<a<b<\infty$, $\lambda\in\mathbb{C}$ and $f(t)\in L_l(a,b)$ $(1\leq l<\infty)$. Then we have*

$$\left({}_{HL}^{RL}D_{a+}^{\alpha,\lambda}{}_{HL}^{R}I_{a+}^{\alpha,\lambda}f\right)(t)=f(t), \tag{3.404}$$

$$\left({}_{HL}^{RL}D_{b-}^{\alpha,\lambda}{}_{HL}^{R}I_{b-}^{\alpha,\lambda}f\right)(t)=f(t), \tag{3.405}$$

$$\left({}_{HL}^{RL}D_{a+}^{\alpha,\lambda}{}_{HL}^{R}I_{a+}^{\alpha,\lambda}f\right)(t)=f(t), \tag{3.406}$$

and

$$\left({}_{HL}^{RL}D_{b-}^{\alpha,\lambda}{}_{HL}^{R}I_{b-}^{\alpha,\lambda}f\right)(t)=f(t). \tag{3.407}$$

Property 3.32 *If $-\infty<a<b<\infty$, $\lambda\in\mathbb{C}$ and $f(t)\in AC_{\kappa}(a,b)$ $(1\leq\kappa<\infty)$, then we have*

$$\left({}_{HL}^{LS}D_{a+}^{\alpha,\lambda}{}_{HL}^{R}I_{a+}^{\alpha,\lambda}f\right)(t)=f(t), \tag{3.408}$$

$$\left({}_{HL}^{LS}D_{b-}^{\alpha,\lambda}{}_{HL}^{R}I_{b-}^{\alpha,\lambda}f\right)(t)=f(t), \tag{3.409}$$

$$\left({}_{HL}^{LSC}D_{a+}^{\alpha,\lambda}{}_{HL}^{R}I_{a+}^{\alpha,\lambda}f\right)(t)=f(t), \tag{3.410}$$

and

$$\left({}_{HL}^{LSC}D_{b-}^{\alpha,\lambda}{}_{HL}^{R}I_{b-}^{\alpha,\lambda}f\right)(t)=f(t). \tag{3.411}$$

Definition 3.63 *Let $\alpha \in \mathbb{C}$, $1 > Re(\alpha) > 0$, $a = -\infty$, $b = \infty$ and $\lambda \in \mathbb{C}$. The left-sided Riemann-Liouville type general fractional derivative within the two-parametric Lorenzo-Hartley kernel is defined as*

$$\left({}_{HL}^{RL}D_{+}^{\alpha,\upsilon,\lambda}f\right)(t) = \frac{d}{dt}\left({}_{HL}I_{+}^{\alpha,\upsilon,\lambda}f(t)\right) = \frac{d}{dt}\int_{-\infty}^{t} R_{\alpha,\upsilon}\left(-\lambda(t-\tau)^{\alpha}\right)f(\tau)\,d\tau, \tag{3.412}$$

and the right-sided Riemann-Liouville type general fractional derivative within the two-parametric Lorenzo-Hartley kernel as

$$\left({}_{HL}^{RL}D_{-}^{\alpha,\upsilon,\lambda}f\right)(t) = -\frac{d}{dt}\left({}_{HL}I_{-}^{\alpha,\upsilon,\lambda}f(t)\right) = -\frac{d}{dt}\int_{t}^{\infty} R_{\alpha,\upsilon}\left(-\lambda(\tau-t)^{\alpha}\right)f(\tau)\,d\tau, \tag{3.413}$$

where the left-sided and right-sided general fractional integral operators within the two-parametric Lorenzo-Hartley kernel are defined as

$${}_{HL}I_{+}^{\alpha,\upsilon,\lambda}f(t) = \int_{-\infty}^{t} R_{\alpha,\upsilon}\left(-\lambda(t-\tau)^{\alpha}\right)f(\tau)\,d\tau \tag{3.414}$$

and

$${}_{HL}I_{-}^{\alpha,\upsilon,\lambda}f(t) = \int_{t}^{\infty} R_{\alpha,\upsilon}\left(-\lambda(\tau-t)^{\alpha}\right)f(\tau)\,d\tau, \tag{3.415}$$

respectively.

For $\alpha \in \mathbb{C}$, $Re(\alpha) > 0$, $\kappa = [Re(\alpha)] + 1$, $a = -\infty$, $b = \infty$ and $\lambda \in \mathbb{C}$, the left-sided Riemann-Liouville type general fractional derivative within the two-parametric Lorenzo-Hartley kernel is defined as

$$\left({}_{HL}^{RL}D_{+}^{\alpha,\upsilon,\lambda}f\right)(t) = \frac{d^{\kappa}}{dt^{\kappa}}\int_{-\infty}^{t} R_{\alpha,\upsilon}\left(-\lambda(t-\tau)^{\alpha}\right)f(\tau)\,d\tau, \tag{3.416}$$

and the right-sided Riemann-Liouville type general fractional derivative within the two-parametric Lorenzo-Hartley kernel as

$$\left({}_{HL}^{RL}D_{-}^{\alpha,\upsilon,\lambda}f\right)(t) = (-1)^{\kappa}\frac{d^{\kappa}}{dt^{\kappa}}\int_{t}^{\infty} R_{\alpha,\upsilon}\left(-\lambda(\tau-t)^{\alpha}\right)f(\tau)\,d\tau. \tag{3.417}$$

Definition 3.64 *Let $\alpha \in \mathbb{C}$, $1 > Re(\alpha) > 0$, $a = -\infty$, $b = \infty$ and $\lambda \in \mathbb{C}$. The left-sided Liouville-Sonine type general fractional derivative within the two-parametric Lorenzo-Hartley kernel is defined as*

$$\left({}_{HL}^{LS}D_{+}^{\alpha,\upsilon,\lambda}f\right)(t) = {}_{HL}I_{+}^{\alpha,\upsilon,\lambda}\left(f^{(1)}(t)\right) = \int_{-\infty}^{t} R_{\alpha,\upsilon}\left(-\lambda(t-\tau)^{\alpha}\right)f^{(1)}(\tau)\,d\tau, \tag{3.418}$$

and the right-sided Liouville-Sonine type general fractional derivative within the two-parametric Lorenzo-Hartley kernel as

$$\left({}_{HL}^{LS}D_{-}^{\alpha,\upsilon,\lambda}f\right)(t) = {}_{HL}I_{-}^{\alpha,\upsilon,\lambda}\left(-f^{(1)}(t)\right) = -\int_{t}^{-\infty} R_{\alpha,\upsilon}\left(-\lambda(\tau-t)^{\alpha}\right)f^{(1)}(\tau)\,d\tau. \tag{3.419}$$

Definition 3.65 *For $\alpha \in \mathbb{C}$, $\kappa+1 > Re(\alpha) > \kappa$, $a = -\infty$, $b = \infty$ and $\lambda \in \mathbb{C}$, the left-sided Liouville-Sonine-Caputo type general fractional derivative within the two-parametric Lorenzo-Hartley kernel is defined as*

$$\left({}_{HL}^{LSC}D_{+}^{\alpha,\upsilon,\lambda}f\right)(t) = {}_{HL}I_{+}^{\alpha,\upsilon,\lambda}\left(f^{(\kappa)}(t)\right) = \int_{-\infty}^{t} R_{\alpha,\upsilon}\left(-\lambda(t-\tau)^{\alpha}\right)f^{(\kappa)}(\tau)\,d\tau, \tag{3.420}$$

and the right-sided Liouville-Sonine-Caputo type general fractional derivative within the two-parametric Lorenzo-Hartley kernel as

$$\begin{aligned}\left({}_{HL}^{LSC}D_{-}^{\alpha,\upsilon,\lambda}f\right)(t) &= {}_{HL}I_{-}^{\alpha,\upsilon,\lambda}\left((-1)^{\kappa}f^{(\kappa)}(t)\right)\\ &= (-1)^{\kappa}\int_{t}^{\infty} R_{\alpha,\upsilon}\left(-\lambda(\tau-t)^{\alpha}\right)f^{(\kappa)}(\tau)\,d\tau.\end{aligned} \tag{3.421}$$

The inverse operators of the above general fractional derivatives within the two-parametric Lorenzo-Hartley kernel can be considered as follows.

Definition 3.66 *Let $\alpha \in \mathbb{C}$, $\kappa+1 > Re(\alpha) > \kappa$, $a = -\infty$, $b = \infty$ and $\lambda \in \mathbb{C}$. The left-sided general fractional integral is defined as*

$$\left({}_{HL}^{R}I_{+}^{\alpha,\upsilon,\lambda}f\right)(t) = \int_{-\infty}^{t} (t-\tau)^{\kappa-(\alpha-\upsilon)-1} E_{\alpha,\kappa-(\alpha-\upsilon)}^{-1}\left(-\gamma(t-\tau)^{\alpha}\right)f(\tau)\,d\tau, \tag{3.422}$$

and the right-sided general fractional integral as

$$\left({}_{MR}^{R}I_{-}^{\alpha,\upsilon,\lambda}f\right) = \int_{t}^{\infty} (\tau-t)^{\kappa-(\alpha-\upsilon)-1} E_{\alpha,\kappa-(\alpha-\upsilon)}^{-1}\left(-\gamma(\tau-t)^{\alpha}\right)f(\tau)\,d\tau. \tag{3.423}$$

In particular, when $\alpha \in \mathbb{C}$, $1 > Re(\alpha) > 0$, $a = -\infty$, $b = \infty$ and $\lambda \in \mathbb{C}$, the left-sided general fractional integral is defined as

$$\left({}_{HL}^{R}I_{+}^{\alpha,\upsilon,\lambda}f\right)(t) = \int_{-\infty}^{t} \frac{1}{(t-\tau)^{\alpha-\upsilon}} E_{\alpha,1-(\alpha-\upsilon)}^{-1}\left(-\gamma(t-\tau)^{\alpha}\right)f(\tau)\,d\tau, \tag{3.424}$$

and the right-sided general fractional integral as

$$\left({}_{MR}^{R}I_{-}^{\alpha,\upsilon,\lambda}f\right) = \int_{t}^{\infty} \frac{1}{(\tau-t)^{\alpha-\upsilon}} E_{\alpha,\kappa-(\alpha-\upsilon)}^{-1}\left(-\gamma(\tau-t)^{\alpha}\right)f(\tau)\,d\tau. \tag{3.425}$$

The properties of the general fractional derivatives and the general fractional integrals are given as follows.

Property 3.33 *Let* $\kappa+1>\alpha>\kappa$, $a=-\infty$, $b=\infty$, $\lambda\in\mathbb{C}$ *and* $f(t)\in L_l(-\infty,\infty)$ $(1\leq l<\infty)$. *Then we have*

$$\left({}_{HL}^{RL}D_{+}^{\alpha,\upsilon,\lambda}\,{}_{HL}^{R}I_{+}^{\alpha,\upsilon,\lambda}f\right)(t)=f(t), \tag{3.426}$$

$$\left({}_{HL}^{RL}D_{-}^{\alpha,\upsilon,\lambda}\,{}_{HL}^{R}I_{-}^{\alpha,\upsilon,\lambda}f\right)(t)=f(t), \tag{3.427}$$

$$\left({}_{HL}^{RL}D_{+}^{\alpha,\upsilon,\lambda}\,{}_{HL}^{R}I_{+}^{\alpha,\upsilon,\lambda}f\right)(t)=f(t), \tag{3.428}$$

and

$$\left({}_{HL}^{RL}D_{-}^{\alpha,\upsilon,\lambda}\,{}_{HL}^{R}I_{-}^{\alpha,\upsilon,\lambda}f\right)(t)=f(t). \tag{3.429}$$

Property 3.34 *If* $a=-\infty$, $b=\infty$, $\lambda\in\mathbb{C}$ *and* $f(t)\in AC_{\kappa}(-\infty,\infty)$ $(1\leq\kappa<\infty)$, *then we have*

$$\left({}_{HL}^{LS}D_{+}^{\alpha,\upsilon,\lambda}\,{}_{HL}^{R}I_{+}^{\alpha,\upsilon,\lambda}f\right)(t)=f(t), \tag{3.430}$$

$$\left({}_{HL}^{LS}D_{-}^{\alpha,\upsilon,\lambda}\,{}_{HL}^{R}I_{-}^{\alpha,\upsilon,\lambda}f\right)(t)=f(t), \tag{3.431}$$

$$\left({}_{HL}^{LSC}D_{+}^{\alpha,\upsilon,\lambda}\,{}_{HL}^{R}I_{+}^{\alpha,\upsilon,\lambda}f\right)(t)=f(t), \tag{3.432}$$

and

$$\left({}_{HL}^{LSC}D_{-}^{\alpha,\upsilon,\lambda}\,{}_{HL}^{R}I_{-}^{\alpha,\upsilon,\lambda}f\right)(t)=f(t). \tag{3.433}$$

3.14 General fractional derivatives in the Gorenflo-Mainardi kernel via Wiman Function

The definitions and properties of the general fractional derivatives in the Gorenflo-Mainardi kernel via Wiman function and general fractional integrals are given as follows.

Definition 3.67 *Let* $\alpha\in\mathbb{C}$, $1>Re(\alpha)>0$, $-\infty<a<b<\infty$, $\gamma\in\mathbb{C}$ *and* $\lambda\in\mathbb{C}$. *The left-sided general fractional derivative containing the kernel of the Gorenflo-Mainardi function via Wiman function in the sense of Riemann-Liouville type is defined as*

$$\left({}_{GM}^{RL}D_{a+}^{\alpha,\gamma,\lambda}f\right)(t)=\frac{d}{dt}\left({}_{GM}I_{a+}^{\alpha,\gamma,\lambda}f(t)\right)=\frac{d}{dt}\int_{a}^{t}G_{\alpha,\gamma}\left(-\lambda(t-\tau)^{\alpha}\right)f(\tau)\,d\tau, \tag{3.434}$$

and the right-sided general fractional derivative containing the kernel of the Gorenflo-Mainardi function via Wiman function in the sense of Riemann-Liouville type as

$$\left({}_{GM}^{RL}D_{b-}^{\alpha,\gamma,\lambda}f\right)(t) = -\tfrac{d}{dt}\left({}_{GM}I_{b-}^{\alpha,\gamma,\lambda}f(t)\right) = -\tfrac{d}{dt}\int_t^b G_{\alpha,\gamma}\left(-\lambda(\tau-t)^{\alpha}\right)f(\tau)\,d\tau, \tag{3.435}$$

where the left-sided and right-sided general fractional integral operators containing the kernel of the Gorenflo-Mainardi function via Wiman function are defined as

$${}_{GM}I_{a+}^{\alpha,\gamma,\lambda}f(t) = \int_a^t G_{\alpha,\gamma}\left(-\lambda(t-\tau)^{\alpha}\right)f(\tau)\,d\tau \tag{3.436}$$

and

$${}_{GM}I_{b-}^{\alpha,\gamma,\lambda}f(t) = \int_t^b G_{\alpha,\gamma}\left(-\lambda(\tau-t)^{\alpha}\right)f(\tau)\,d\tau, \tag{3.437}$$

respectively.

For $\alpha \in \mathbb{C}$, $Re(\alpha) > 0$, $\kappa = [Re(\alpha)] + 1$, $-\infty < a < b < \infty$, $\gamma \in \mathbb{C}$ and $\lambda \in \mathbb{C}$, the left-sided general fractional derivative containing the kernel of the Gorenflo-Mainardi function via Wiman function in the sense of Riemann-Liouville type is defined as

$$\left({}_{GM}^{RL}D_{a+}^{\alpha,\gamma,\lambda}f\right)(t) = \tfrac{d^{\kappa}}{dt^{\kappa}}\int_a^t G_{\alpha,\gamma}\left(-\lambda(t-\tau)^{\alpha}\right)f(\tau)\,d\tau, \tag{3.438}$$

and the right-sided general fractional derivative containing the kernel of the Gorenflo-Mainardi function via Wiman function in the sense of Riemann-Liouville type as

$$\left({}_{GM}^{RL}D_{b-}^{\alpha,\gamma,\lambda}f\right)(t) = (-1)^{\kappa}\tfrac{d^{\kappa}}{dt^{\kappa}}\int_t^b G_{\alpha,\gamma}\left(-\lambda(\tau-t)^{\alpha}\right)f(\tau)\,d\tau. \tag{3.439}$$

Definition 3.68 *Let $\alpha \in \mathbb{C}$, $1 > Re(\alpha) > 0$, $-\infty < a < b < \infty$, $\gamma \in \mathbb{C}$ and $\lambda \in \mathbb{C}$. The left-sided general fractional derivative containing the kernel of the Gorenflo-Mainardi function via Wiman function in the sense of Liouville-Sonine type is defined as*

$$\left({}_{GM}^{LS}D_{a+}^{\alpha,\gamma,\lambda}f\right)(t) = {}_{GM}I_{a+}^{\alpha,\gamma,\lambda}\left(f^{(1)}(t)\right) = \int_a^t G_{\alpha,\gamma}\left(-\lambda(t-\tau)^{\alpha}\right)f^{(1)}(\tau)\,d\tau, \tag{3.440}$$

and the right-sided general fractional derivative containing the kernel of the Gorenflo-Mainardi function via Wiman function in the sense of Liouville-Sonine type as

$$\left({}_{GM}^{LS}D_{b-}^{\alpha,\gamma,\lambda}f\right)(t) = {}_{GM}I_{b-}^{\alpha,\gamma,\lambda}\left(-f^{(1)}(t)\right) = -\int_t^b G_{\alpha,\gamma}\left(-\lambda(\tau-t)^{\alpha}\right)f^{(1)}(\tau)\,d\tau. \tag{3.441}$$

More generally, for $\alpha \in \mathbb{C}$, $\kappa + 1 > Re(\alpha) > \kappa$, $-\infty < a < b < \infty$, $\gamma \in \mathbb{C}$ and $\lambda \in \mathbb{C}$, the left-sided general fractional derivative containing the kernel of the Gorenflo-Mainardi function via Wiman function in the sense of Liouville-Sonine-Caputo type is defined as

$$\left({}_{GM}^{LSC}D_{a+}^{\alpha,\gamma,\lambda} f\right)(t) = \int_a^t G_{\alpha,\gamma}\left(-\lambda (t-\tau)^{\alpha}\right) f^{(\kappa)}(\tau)\, d\tau, \tag{3.442}$$

and the right-sided general fractional derivative containing the kernel of the Gorenflo-Mainardi function via Wiman function in the sense of Liouville-Sonine-Caputo type as

$$\left({}_{GM}^{LSC}D_{b-}^{\alpha,\gamma,\lambda} f\right)(t) = (-1)^{\kappa} \int_t^b G_{\alpha,\gamma}\left(-\lambda (\tau-t)^{\alpha}\right) f^{(\kappa)}(\tau)\, d\tau. \tag{3.443}$$

Definition 3.69 *For $\alpha \in \mathbb{C}$, $\kappa + 1 > Re(\alpha) > \kappa$, $-\infty < a < b < \infty$, $\gamma \in \mathbb{C}$ and $\lambda \in \mathbb{C}$, the left-sided general fractional integral is defined as*

$$\left({}_{GM}^{R}I_{a+}^{\alpha,\gamma,\lambda} f\right)(t) = \int_a^t (t-\tau)^{\kappa-\gamma-1} E_{\alpha,\kappa-\gamma}^{-1}\left(-\lambda (t-\tau)^{\alpha}\right) f(\tau)\, d\tau, \tag{3.444}$$

and the right-sided general fractional integral as

$$\left({}_{GM}^{R}I_{b-}^{\alpha,\gamma,\lambda} f\right) = (-1)^{\kappa} \int_t^b (\tau-t)^{\kappa-\gamma-1} E_{\alpha,\kappa-\gamma}^{-1}\left(-\lambda (\tau-t)^{\alpha}\right) f(\tau)\, d\tau. \tag{3.445}$$

In particular, for $\alpha \in \mathbb{C}$, $1 > Re(\alpha) > 0$, $-\infty < a < b < \infty$, $\gamma \in \mathbb{C}$ and $\lambda \in \mathbb{C}$, the left-sided general fractional integral is defined as

$$\left({}_{GM}^{R}I_{a+}^{\alpha,\gamma,\lambda} f\right)(t) = \int_a^t (t-\tau)^{-\gamma} E_{\alpha,1-\gamma}^{-1}\left(-\lambda (t-\tau)^{\alpha}\right) f(\tau)\, d\tau, \tag{3.446}$$

and the right-sided general fractional integral as

$$\left({}_{GM}^{R}I_{b-}^{\alpha,\gamma,\lambda} f\right) = -\int_t^b (\tau-t)^{-\gamma} E_{\alpha,1-\gamma}^{-1}\left(-\lambda (\tau-t)^{\alpha}\right) f(\tau)\, d\tau. \tag{3.447}$$

The properties of the general fractional derivatives and the general fractional integrals are given as follows.

Property 3.35 *If $\kappa + 1 > \alpha > \kappa$, $-\infty < a < b < \infty$, $\gamma \in \mathbb{C}$, $\lambda \in \mathbb{C}$ and $f(t) \in L_l(a,b)$ $(1 \le l < \infty)$, then we have*

$$L\left\{\left({}_{GM}^{RL}D_{0+}^{\alpha,\gamma,\lambda} f\right)(t)\right\} = s^{\kappa-\gamma}\left(1+\lambda s^{-\alpha}\right)^{-1} f(s), \tag{3.448}$$

$$L\left\{\left({}^{RL}_{GM}D^{\alpha,\gamma,\lambda}_{0+}f\right)(t)\right\} = s^{1-\gamma}\left(1+\lambda s^{-\alpha}\right)^{-1} f(s), \tag{3.449}$$

$$L\left\{\left({}^{LS}_{GM}D^{\alpha,\gamma,\lambda}_{0+}f\right)(t)\right\} = s^{-\gamma}\left(1+\lambda s^{-\alpha}\right)^{-1}\left(sf(s)-f(0)\right), \tag{3.450}$$

$$L\left\{\left({}^{LSC}_{GM}D^{\alpha,\gamma,\lambda}_{0+}f\right)(t)\right\} = s^{-\gamma}\left(1+\lambda s^{-\alpha}\right)^{-1}\left(s^{\kappa}f(s)-\sum_{j=1}^{\kappa}s^{\kappa-j}f^{(j-1)}(0)\right), \tag{3.451}$$

$$L\left\{\left({}^{R}_{GM}I^{\alpha,\gamma,\lambda}_{0+}f\right)(t)\right\} = s^{\gamma-\kappa}\left(1+\lambda s^{-\alpha}\right)f(s) \tag{3.452}$$

and

$$L\left\{\left({}_{GM}I^{\alpha,\gamma,\lambda}_{0+}f\right)(t)\right\} = s^{-\gamma}\left(1+\lambda s^{-\alpha}\right)f(s). \tag{3.453}$$

Note that, when $D^{\kappa}\left({}^{R}_{GM}I^{\alpha,\gamma,\lambda}_{0+}f(t)\right)|_{t=0} = 0$, there is

$$L\left\{{}^{RL}_{GM}D^{\alpha,\gamma,\lambda}_{0+}f(t)\right\} = s^{\kappa-\gamma}\left(1+\lambda s^{-\alpha}\right)^{-1} f(s). \tag{3.454}$$

and when $D^{\kappa}\left({}^{R}_{GM}I^{\alpha,\gamma,\lambda}_{0+}f(t)\right)|_{t=0} \neq 0$, there is

$$L\left\{{}^{RL}_{GM}D^{\alpha,\gamma,\lambda}_{0+}f(t)\right\} = s^{\kappa-\gamma}\left(1+\lambda s^{-\alpha}\right)^{-1} f(s) - \sum_{\mu=0}^{\kappa-1}s^{\kappa-\mu-1}\left(D^{\mu}\left[{}^{R}_{GM}I^{\alpha,\gamma,\lambda}_{0+}f(+0)\right]\right). \tag{3.455}$$

In particular, for $\kappa = 1$ we have

$$L\left\{{}^{RL}_{GM}D^{\alpha,\gamma,\lambda}_{0+}f(t)\right\} = s^{1-\gamma}\left(1+\lambda s^{-\alpha}\right)^{-1} f(s) - {}^{R}_{GM}I^{\alpha,\gamma,\lambda}_{0+}f(+0), \tag{3.456}$$

and, for $\kappa = 1$ and $D\left({}^{R}_{GM}I^{\alpha,\gamma,\lambda}_{0+}f(t)\right)|_{t=0} = 0$ we have

$$L\left\{{}^{RL}_{GM}D^{\alpha,\gamma,\lambda}_{0+}f(t)\right\} = s^{1-\gamma}\left(1+\lambda s^{-\alpha}\right)^{-1} f(s). \tag{3.457}$$

For the details of the general fractional derivatives and the general fractional integrals, see [301, 302].

Property 3.36 *Let $\kappa+1 > \alpha > \kappa$, $-\infty < a < b < \infty$, $\gamma \in \mathbb{C}$, $\lambda \in \mathbb{C}$ and $f(t) \in L_l(a,b)$ $(1 \le l < \infty)$. Then we have*

$$\left({}^{RL}_{GM}D^{\alpha,\gamma,\lambda}_{a+}\,{}^{R}_{GM}I^{\alpha,\gamma,\lambda}_{a+}f\right)(t) = f(t), \tag{3.458}$$

$$\left({}^{RL}_{GM}D^{\alpha,\gamma,\lambda}_{b-}\,{}^{R}_{GM}I^{\alpha,\gamma,\lambda}_{b-}f\right)(t) = f(t), \tag{3.459}$$

$$\left({}^{RL}_{GM}D^{\alpha,\gamma,\lambda}_{a+}\,{}^{R}_{GM}I^{\alpha,\gamma,\lambda}_{a+}f\right)(t) = f(t), \tag{3.460}$$

and

$$\left({}^{RL}_{GM}D^{\alpha,\gamma,\lambda}_{b-}\,{}^{R}_{GM}I^{\alpha,\gamma,\lambda}_{b-}f\right)(t) = f(t). \tag{3.461}$$

Property 3.37 *If $-\infty < a < b < \infty$, $\gamma \in \mathbb{C}$, $\lambda \in \mathbb{C}$ and $f(t) \in AC_\kappa(a,b)$ $(1 \le \kappa < \infty)$, then we have*

$$\left({}_{GM}^{LS}D_{a+}^{\alpha,\gamma,\lambda R}{}_{GM}I_{a+}^{\alpha,\gamma,\lambda} f\right)(t) = f(t), \tag{3.462}$$

$$\left({}_{GM}^{LS}D_{b-}^{\alpha,\gamma,\lambda R}{}_{GM}I_{b-}^{\alpha,\gamma,\lambda} f\right)(t) = f(t), \tag{3.463}$$

$$\left({}_{GM}^{LSC}D_{a+}^{\alpha,\gamma,\lambda R}{}_{GM}I_{a+}^{\alpha,\gamma,\lambda} f\right)(t) = f(t), \tag{3.464}$$

and

$$\left({}_{GM}^{LSC}D_{b-}^{\alpha,\gamma,\lambda R}{}_{GM}I_{b-}^{\alpha,\gamma,\lambda} f\right)(t) = f(t). \tag{3.465}$$

Definition 3.70 *Let $\alpha \in \mathbb{C}$, $1 > Re(\alpha) > 0$, $a = -\infty$, $b = \infty$, $\gamma \in \mathbb{C}$ and $\lambda \in \mathbb{C}$. The left-sided general fractional derivative containing the kernel of the Gorenflo-Mainardi function via Wiman function in the sense of Riemann-Liouville type is defined as*

$$\left({}_{GM}^{RL}D_{+}^{\alpha,\gamma,\lambda} f\right)(t) = \frac{d}{dt}\left({}_{GM}I_{+}^{\alpha,\gamma,\lambda} f(t)\right) = \frac{d}{dt}\int_{-\infty}^{t} G_{\alpha,\gamma}\left(-\lambda(t-\tau)^{\alpha}\right) f(\tau)\, d\tau, \tag{3.466}$$

and the right-sided general fractional derivative containing the kernel of the Gorenflo-Mainardi function via Wiman function in the sense of Riemann-Liouville type as

$$\left({}_{GM}^{RL}D_{-}^{\alpha,\gamma,\lambda} f\right)(t) = -\frac{d}{dt}\left({}_{GM}I_{-}^{\alpha,\gamma,\lambda} f(t)\right) = -\frac{d}{dt}\int_{t}^{\infty} G_{\alpha,\gamma}\left(-\lambda(\tau-t)^{\alpha}\right) f(\tau)\, d\tau, \tag{3.467}$$

where the left-sided and right-sided general fractional integral operators containing the kernel of the Gorenflo-Mainardi function via Wiman function are defined as

$${}_{GM}I_{+}^{\alpha,\gamma,\lambda} f(t) = \int_{-\infty}^{t} G_{\alpha,\gamma}\left(-\lambda(t-\tau)^{\alpha}\right) f(\tau)\, d\tau \tag{3.468}$$

and

$${}_{GM}I_{-}^{\alpha,\gamma,\lambda} f(t) = \int_{t}^{\infty} G_{\alpha,\gamma}\left(-\lambda(\tau-t)^{\alpha}\right) f(\tau)\, d\tau, \tag{3.469}$$

respectively.

On the other hand, when $\alpha \in \mathbb{C}$, $Re(\alpha) > 0$, $\kappa = [Re(\alpha)] + 1$,$a = -\infty$, $b = \infty$, $\gamma \in \mathbb{C}$ and $\lambda \in \mathbb{C}$, the left-sided general fractional derivative containing the kernel of the Gorenflo-Mainardi function via Wiman function in the sense of Riemann-Liouville type is defined as

$$\left({}_{GM}^{RL}D_{+}^{\alpha,\gamma,\lambda} f\right)(t) = \frac{d^\kappa}{dt^\kappa}\int_{-\infty}^{t} G_{\alpha,\gamma}\left(-\lambda(t-\tau)^{\alpha}\right) f(\tau)\, d\tau, \tag{3.470}$$

and the right-sided general fractional derivative containing the kernel of the Gorenflo-Mainardi function via Wiman function in the sense of Riemann-Liouville type as

$$\left({}_{GM}^{RL}D_{-}^{\alpha,\gamma,\lambda}f\right)(t) = (-1)^{\kappa}\frac{d^{\kappa}}{dt^{\kappa}}\int_{t}^{\infty} G_{\alpha,\gamma}\left(-\lambda\left(\tau - t\right)^{\alpha}\right) f\left(\tau\right) d\tau. \tag{3.471}$$

Definition 3.71 *For $\alpha \in \mathbb{C}$, $1 > Re(\alpha) > 0$, $a = -\infty$, $b = \infty$, $\gamma \in \mathbb{C}$ and $\lambda \in \mathbb{C}$, the left-sided general fractional derivative containing the kernel of the Gorenflo-Mainardi function via Wiman function in the sense of Liouville-Sonine type is defined as*

$$\left({}_{GM}^{LS}D_{+}^{\alpha,\gamma,\lambda}f\right)(t) = \int_{-\infty}^{t} G_{\alpha,\gamma}\left(-\lambda\left(t-\tau\right)^{\alpha}\right) f^{(1)}\left(\tau\right) d\tau, \tag{3.472}$$

and the right-sided general fractional derivative containing the kernel of the Gorenflo-Mainardi function via Wiman function in the sense of Liouville-Sonine type as

$$\left({}_{GM}^{LS}D_{-}^{\alpha,\gamma,\lambda}f\right)(t) = -\int_{t}^{\infty} G_{\alpha,\gamma}\left(-\lambda\left(\tau - t\right)^{\alpha}\right) f^{(1)}\left(\tau\right) d\tau. \tag{3.473}$$

For $\alpha \in \mathbb{C}$, $\kappa + 1 > Re(\alpha) > \kappa$, $a = -\infty$, $b = \infty$, $\gamma \in \mathbb{C}$ and $\lambda \in \mathbb{C}$, the left-sided general fractional derivative containing the kernel of the Gorenflo-Mainardi function via Wiman function in the sense of Liouville-Sonine-Caputo type is defined as

$$\left({}_{GM}^{LSC}D_{+}^{\alpha,\gamma,\lambda}f\right)(t) = \int_{-\infty}^{t} G_{\alpha,\gamma}\left(-\lambda\left(t-\tau\right)^{\alpha}\right) f^{(\kappa)}\left(\tau\right) d\tau, \tag{3.474}$$

and the right-sided general fractional derivative containing the kernel of the Gorenflo-Mainardi function via Wiman function in the sense of Liouville-Sonine-Caputo type as

$$\left({}_{GM}^{LSC}D_{-}^{\alpha,\gamma,\lambda}f\right)(t) = (-1)^{\kappa}\int_{t}^{\infty} G_{\alpha,\gamma}\left(-\lambda\left(\tau - t\right)^{\alpha}\right) f^{(\kappa)}\left(\tau\right) d\tau. \tag{3.475}$$

Their inverse operators are presented as follows.

Definition 3.72 *Let $\alpha \in \mathbb{C}$, $\kappa + 1 > Re(\alpha) > \kappa$, $a = -\infty$, $b = \infty$, $\gamma \in \mathbb{C}$ and $\lambda \in \mathbb{C}$. The left-sided general fractional integral is defined as*

$$\left({}_{GM}^{R}I_{+}^{\alpha,\gamma,\lambda}f\right)(t) = \int_{-\infty}^{t} \left(t-\tau\right)^{\kappa-\gamma-1} E_{\alpha,\kappa-\gamma}^{-1}\left(-\lambda\left(t-\tau\right)^{\alpha}\right) f\left(\tau\right) d\tau, \tag{3.476}$$

and the right-sided general fractional integral as

$$\left({}_{GM}^{R}I_{-}^{\alpha,\gamma,\lambda}f\right) = (-1)^{\kappa}\int_{t}^{-\infty} \left(\tau - t\right)^{\kappa-\gamma-1} E_{\alpha,\kappa-\gamma}^{-1}\left(-\lambda\left(\tau - t\right)^{\alpha}\right) f\left(\tau\right) d\tau. \tag{3.477}$$

In particular, for $\alpha \in \mathbb{C}$, $1 > Re(\alpha) > 0$, $a = -\infty$, $b = \infty$, $\gamma \in \mathbb{C}$ and $\lambda \in \mathbb{C}$, the left-sided general fractional integral is defined as

$$\left({}_{GM}^{R}I_{+}^{\alpha,\gamma,\lambda} f\right)(t) = \int_{-\infty}^{t} (t-\tau)^{-\gamma} E_{\alpha,1-\gamma}^{-1}\left(-\lambda (t-\tau)^{\alpha}\right) f(\tau)\, d\tau \tag{3.478}$$

and the right-sided general fractional integral as

$$\left({}_{GM}^{R}I_{-}^{\alpha,\gamma,\lambda} f\right) = -\int_{t}^{\infty} (\tau-t)^{-\gamma} E_{\alpha,1-\gamma}^{-1}\left(-\lambda (\tau-t)^{\alpha}\right) f(\tau)\, d\tau. \tag{3.479}$$

The properties of the general fractional derivatives and the general fractional integrals can be given as follows.

Property 3.38 *Let $\kappa + 1 > \alpha > \kappa$, $a = -\infty$, $b = \infty$, $\gamma \in \mathbb{C}$, $\lambda \in \mathbb{C}$ and $f(t) \in L_l(-\infty, \infty)$ $(1 \le l < \infty)$. Then we have*

$$\left({}_{GM}^{RL}D_{+}^{\alpha,\gamma,\lambda}\,{}_{GM}^{R}I_{+}^{\alpha,\gamma,\lambda} f\right)(t) = f(t), \tag{3.480}$$

$$\left({}_{GM}^{RL}D_{-}^{\alpha,\gamma,\lambda}\,{}_{GM}^{R}I_{-}^{\alpha,\gamma,\lambda} f\right)(t) = f(t), \tag{3.481}$$

$$\left({}_{GM}^{RL}D_{+}^{\alpha,\gamma,\lambda}\,{}_{GM}^{R}I_{+}^{\alpha,\gamma,\lambda} f\right)(t) = f(t), \tag{3.482}$$

and

$$\left({}_{GM}^{RL}D_{-}^{\alpha,\gamma,\lambda}\,{}_{GM}^{R}I_{-}^{\alpha,\gamma,\lambda} f\right)(t) = f(t). \tag{3.483}$$

Property 3.39 *If $a = -\infty$, $b = \infty$, $\gamma \in \mathbb{C}$, $\lambda \in \mathbb{C}$ and $f(t) \in AC_\kappa(-\infty, \infty)$ $(1 \le \kappa < \infty)$, then we have*

$$\left({}_{GM}^{LS}D_{+}^{\alpha,\gamma,\lambda}\,{}_{GM}^{R}I_{+}^{\alpha,\gamma,\lambda} f\right)(t) = f(t), \tag{3.484}$$

$$\left({}_{GM}^{LS}D_{-}^{\alpha,\gamma,\lambda}\,{}_{GM}^{R}I_{-}^{\alpha,\gamma,\lambda} f\right)(t) = f(t), \tag{3.485}$$

$$\left({}_{GM}^{LSC}D_{+}^{\alpha,\gamma,\lambda}\,{}_{GM}^{R}I_{+}^{\alpha,\gamma,\lambda} f\right)(t) = f(t), \tag{3.486}$$

and

$$\left({}_{GM}^{LSC}D_{-}^{\alpha,\gamma,\lambda}\,{}_{GM}^{R}I_{-}^{\alpha,\gamma,\lambda} f\right)(t) = f(t). \tag{3.487}$$

3.15 Hilfer type general fractional derivatives involving the kernel of the Gorenflo-Mainardi function

The definitions and properties of the Hilfer type general fractional derivatives involving the kernel of the Gorenflo-Mainardi function via Wiman function can be given as follows.

Definition 3.73 *Let $1 > \alpha > 0$, $0 \le \beta \le 1$, $-\infty < a < b < \infty$, $\gamma \in \mathbb{C}$ and $\lambda \in \mathbb{C}$. The left-sided Hilfer type general fractional derivative containing the kernel of the Gorenflo-Mainardi function via Wiman function is defined as*

$$\begin{aligned}\left({}_{GM}^{Hi}D_{a+}^{\alpha,\beta,\gamma,\lambda} f\right)(t) &= \int_a^t G_{\alpha,\gamma}\left(-\lambda (t-\tau)^{\alpha}\right)\left(D_{a+}^{\alpha,\beta} f(\tau)\right) d\tau \\ &= \int_a^t G_{\alpha,\gamma}\left(-\lambda (t-\tau)^{\alpha}\right)\left(I_{a+}^{\beta(1-\alpha)} \frac{d}{d\tau}\left(I_{a+}^{(1-\beta)(1-\alpha)} f\right)\right)(\tau)\, d\tau,\end{aligned} \tag{3.488}$$

and the right-sided Hilfer type general fractional derivative containing the kernel of the Gorenflo-Mainardi function via Wiman function as

$$\begin{aligned}\left({}_{GM}^{Hi}D_{b-}^{\alpha,\beta,\gamma,\lambda} f\right)(t) &= -\int_t^b G_{\alpha,\gamma}\left(-\lambda (\tau-t)^{\alpha}\right)\left(D_{b-}^{\alpha,\beta} f(\tau)\right) d\tau \\ &= -\int_t^b G_{\alpha,\gamma}\left(-\lambda (\tau-t)^{\alpha}\right)\left(I_{b-}^{\beta(1-\alpha)} \frac{d}{d\tau}\left(I_{b-}^{(1-\beta)(1-\alpha)} f\right)\right)(\tau)\, d\tau.\end{aligned} \tag{3.489}$$

On the other hand, for $\kappa+1 > \alpha > \kappa$, $0 \le \beta \le 1$, $-\infty < a < b < \infty$, $\gamma \in \mathbb{C}$ and $\lambda \in \mathbb{C}$, the left-sided Hilfer type general fractional derivative containing the kernel of the Gorenflo-Mainardi function via Wiman function is defined as

$$\left({}_{GM}^{Hir}D_{a+}^{\alpha,\beta,\gamma,\lambda} f\right)(t) = \int_a^t G_{\alpha,\gamma}\left(-\lambda (t-\tau)^{\alpha}\right)\left(I_{a+}^{\beta(\kappa-\alpha)} \frac{d^{\kappa}}{d\tau^{\kappa}}\left(I_{a+}^{(1-\beta)(\kappa-\alpha)} f\right)\right)(\tau)\, d\tau, \tag{3.490}$$

and the right-sided Hilfer type general fractional derivative containing the kernel of the Gorenflo-Mainardi function via Wiman function as

$$\begin{aligned}&\left({}_{GM}^{Hir}D_{b-}^{\alpha,\beta,\gamma,\lambda} f\right)(t) \\ &= (-1)^{\kappa} \int_t^b G_{\alpha,\gamma}\left(-\lambda (\tau-t)^{\alpha}\right)\left(I_{b-}^{\beta(\kappa-\alpha)} \frac{d^{\kappa}}{d\tau^{\kappa}}\left(I_{b-}^{(1-\beta)(\kappa-\alpha)} f\right)\right)(\tau)\, d\tau.\end{aligned} \tag{3.491}$$

For $1 > \alpha > 0$, $\beta = 0$, $-\infty < a < b < \infty$, $\gamma \in \mathbb{C}$ and $\lambda \in \mathbb{C}$, the left-sided Riemann-Liouville type general fractional derivative containing the kernel of the Gorenflo-Mainardi function via Wiman function is defined as

$$\left({}_{GM}^{Hi}D_{a+}^{\alpha,0,\gamma,\lambda} f\right)(t) = \int_a^t G_{\alpha,\gamma}\left(-\lambda (t-\tau)^{\alpha}\right)\left(\frac{d}{d\tau}\left(I_{a+}^{1-\alpha} f\right)\right)(\tau)\, d\tau, \tag{3.492}$$

and the right-sided Riemann-Liouville type general fractional derivative containing the kernel of the Gorenflo-Mainardi function via Wiman function as

$$\left({}_{GM}^{Hi}D_{b-}^{\alpha,0,\gamma,\lambda}f\right)(t)=-\int\limits_{t}^{b}G_{\alpha,\gamma}\left(-\lambda\left(\tau-t\right)^{\alpha}\right)\left(\tfrac{d}{d\tau}\left(I_{b-}^{1-\alpha}f\right)\right)(\tau)\,d\tau. \tag{3.493}$$

More generally, for $\kappa+1>\alpha>\kappa$, $\beta=0$, $-\infty<a<b<\infty$, $\gamma\in\mathbb{C}$ and $\lambda\in\mathbb{C}$, the left-sided Riemann-Liouville type general fractional derivative containing the kernel of the Gorenflo-Mainardi function via Wiman function is defined as

$$\left({}_{GM}^{Hir}D_{a+}^{\alpha,0,\gamma,\lambda}f\right)(t)=\int\limits_{a}^{t}G_{\alpha,\gamma}\left(-\lambda\left(t-\tau\right)^{\alpha}\right)\left(\tfrac{d^{\kappa}}{d\tau^{\kappa}}\left(I_{a+}^{\kappa-\alpha}f\right)(\tau)\right)d\tau, \tag{3.494}$$

and the right-sided Riemann-Liouville type general fractional derivative containing the kernel of the Gorenflo-Mainardi function via Wiman function as

$$\left({}_{GM}^{Hir}D_{b-}^{\alpha,0,\gamma,\lambda}f\right)(t)=(-1)^{\kappa}\int\limits_{t}^{b}G_{\alpha,\gamma}\left(-\lambda\left(\tau-t\right)^{\alpha}\right)\left(\tfrac{d^{\kappa}}{d\tau^{\kappa}}\left(I_{b-}^{\kappa-\alpha}f\right)(\tau)\right)d\tau. \tag{3.495}$$

3.16 General fractional derivatives in the generalized Prabhakar kernel

The definitions and properties of the general fractional derivatives within the generalized Prabhakar kernel via Prabhakar function and the general fractional integrals, which are proposed by Kilbas, Saigo and Saxena (see [97]; also see [101]), are presented as follows.

Definition 3.74 *Let $\alpha\in\mathbb{C}$, $1>Re(\alpha)>0$, $-\infty<a<b<\infty$, $\varphi\in\mathbb{C}$, $\upsilon\in\mathbb{C}$ and $\lambda\in\mathbb{C}$. The left-sided Riemann-Liouville type general fractional derivative within the generalized Prabhakar kernel via Prabhakar function is defined by*

$$\left({}_{GP}^{RL}D_{a+}^{\alpha,\upsilon,\varphi,\lambda}f\right)(t)=\tfrac{d}{dt}\left({}_{GP}I_{a+}^{\alpha,\upsilon,\varphi,\lambda}f(t)\right), \tag{3.496}$$

and the right-sided Riemann-Liouville type general fractional derivative within the generalized Prabhakar kernel via Prabhakar function by

$$\left({}_{GP}^{RL}D_{b-}^{\alpha,\upsilon,\varphi,\lambda}f\right)(t)=-\tfrac{d}{dt}\left({}_{GP}I_{b-}^{\alpha,\upsilon,\varphi,\lambda}f(t)\right), \tag{3.497}$$

where the left-sided and right-sided general fractional integral operators within the Prabhakar kernel via Prabhakar function are defined as

$${}_{GP}I_{a+}^{\alpha,\upsilon,\varphi,\lambda}f(t)=\int\limits_{a}^{t}(t-\tau)^{\upsilon-1}E_{\alpha,\upsilon}^{\varphi}\left(-\lambda\left(t-\tau\right)^{\alpha}\right)f(\tau)\,d\tau \tag{3.498}$$

and

$$ {}_{GP}I_{b-}^{\alpha,\upsilon,\varphi,\lambda} f(t) = \int_t^b (\tau - t)^{\upsilon-1} E_{\alpha,\upsilon}^{\varphi} (-\lambda (\tau - t)^{\alpha}) f(\tau)\, d\tau, \qquad (3.499) $$

respectively.

More generally, for $\alpha \in \mathbb{C}$, $Re(\alpha) > 0$, $\kappa = [Re(\alpha)] + 1$, $-\infty < a < b < \infty$, $\varphi \in \mathbb{C}$, $\upsilon \in \mathbb{C}$ and $\lambda \in \mathbb{C}$, the left-sided Riemann-Liouville type general fractional derivative within the generalized Prabhakar kernel via Prabhakar function is defined as (see [97]; also see [101])

$$ \left({}_{GP}^{RL}D_{a+}^{\alpha,\upsilon,\varphi,\lambda} f\right)(t) = \frac{d^{\kappa}}{dt^{\kappa}} \int_a^t (t - \tau)^{\upsilon-1} E_{\alpha,\upsilon}^{\varphi} (-\lambda (t - \tau)^{\alpha}) f(\tau)\, d\tau, \qquad (3.500) $$

and the right-sided Riemann-Liouville type general fractional derivative within the generalized Prabhakar kernel via Prabhakar function as

$$ \left({}_{GP}^{RL}D_{b-}^{\alpha,\upsilon,\varphi,\lambda} f\right)(t) = (-1)^{\kappa} \frac{d^{\kappa}}{dt^{\kappa}} \int_t^b (\tau - t)^{\upsilon-1} E_{\alpha,\upsilon}^{\varphi} (-\lambda (\tau - t)^{\alpha}) f(\tau)\, d\tau, \qquad (3.501) $$

Definition 3.75 *Let $\alpha \in \mathbb{C}$, $1 > Re(\alpha) > 0$, $-\infty < a < b < \infty$, $\varphi \in \mathbb{C}$, $\upsilon \in \mathbb{C}$ and $\lambda \in \mathbb{C}$. The left-sided Liouville-Sonine type general fractional derivative within the generalized Prabhakar kernel via Prabhakar function is defined as*

$$ \left({}_{GP}^{LS}D_{a+}^{\alpha,\upsilon,\varphi,\lambda} f\right)(t) = \int_a^t (t - \tau)^{\upsilon-1} E_{\alpha,\upsilon}^{\varphi} (-\lambda (t - \tau)^{\alpha}) f^{(1)}(\tau)\, d\tau, \qquad (3.502) $$

and the right-sided Liouville-Sonine type general fractional derivative within the generalized Prabhakar kernel via Prabhakar function as

$$ \left({}_{GP}^{LS}D_{b-}^{\alpha,\upsilon,\varphi,\lambda} f\right)(t) = -\int_t^b (\tau - t)^{\upsilon-1} E_{\alpha,\upsilon}^{\varphi} (-\lambda (\tau - t)^{\alpha}) f^{(1)}(\tau)\, d\tau. \qquad (3.503) $$

Generally, for $\alpha \in \mathbb{C}$, $\kappa + 1 > Re(\alpha) > \kappa$, $-\infty < a < b < \infty$, $\varphi \in \mathbb{C}$, $\upsilon \in \mathbb{C}$ and $\lambda \in \mathbb{C}$, the left-sided Liouville-Sonine-Caputo type general fractional derivative within the generalized Prabhakar kernel via Prabhakar function is defined as (see [101])

$$ \left({}_{GP}^{LSC}D_{a+}^{\alpha,\varphi,\upsilon,\lambda} f\right)(t) = \int_a^t (t - \tau)^{\upsilon-1} E_{\alpha,\upsilon}^{\varphi} (-\lambda (t - \tau)^{\alpha}) f^{(\kappa)}(\tau)\, d\tau, \qquad (3.504) $$

and the right-sided Liouville-Sonine-Caputo type general fractional derivative within the generalized Prabhakar kernel via Prabhakar function as

$$ \left({}_{GP}^{LSC}D_{b-}^{\alpha,\varphi,\upsilon,\lambda} f\right)(t) = (-1)^{\kappa} \int_t^b (\tau - t)^{\upsilon-1} E_{\alpha,\upsilon}^{\varphi} (-\lambda (\tau - t)^{\alpha}) f^{(\kappa)}(\tau)\, d\tau. \qquad (3.505) $$

The inverse operators of the above fractional derivatives within the generalized Prabhakar kernel via Prabhakar function are presented as follows.

Definition 3.76 *Let* $\alpha \in \mathbb{C}$, $\kappa + 1 > Re(\alpha) > \kappa$, $-\infty < a < b < \infty$, $\varphi \in \mathbb{C}$, $\upsilon \in \mathrm{C}$ *and* $\lambda \in \mathbb{C}$.

As the inverse operator, the left-sided general fractional integral via Prabhakar function is defined by (see [97]; also see [101])

$$\left({}_{GP}^{R}I_{a+}^{\alpha,\varphi,\upsilon,\lambda} f\right)(t) = \int_a^t (t-\tau)^{\kappa-\upsilon-1} E_{\alpha,\kappa-\upsilon}^{-\varphi}\left(-\lambda (t-\tau)^{\alpha}\right) f(\tau)\, d\tau \quad (3.506)$$

and the right-sided general fractional integral via Prabhakar function by

$$\left({}_{GP}^{R}I_{b-}^{\alpha,\varphi,\upsilon,\lambda} f\right) = \int_t^b (\tau-t)^{\kappa-\upsilon-1} E_{\alpha,\kappa-\upsilon}^{-\varphi}\left(-\lambda (\tau-t)^{\alpha}\right) f(\tau)\, d\tau. \quad (3.507)$$

In particular, for $\alpha \in \mathbb{C}$, $1 > Re(\alpha) > 0$, $-\infty < a < b < \infty$, $\varphi \in \mathbb{C}$, $\upsilon \in \mathbb{C}$ and $\lambda \in \mathbb{C}$, the left-sided general fractional integral via Prabhakar function is defined as

$$\left({}_{GP}^{R}I_{a+}^{\alpha,\varphi,\upsilon,\lambda} f\right)(t) = \int_a^t \frac{1}{(t-\tau)^{\upsilon}} E_{\alpha,\kappa-\upsilon}^{-\varphi}\left(-\lambda (t-\tau)^{\alpha}\right) f(\tau)\, d\tau \quad (3.508)$$

and the right-sided general fractional integral via Prabhakar function as

$$\left({}_{GP}^{R}I_{b-}^{\alpha,\varphi,\upsilon,\lambda} f\right) = \int_t^b \frac{1}{(\tau-t)^{\upsilon}} E_{\alpha,\kappa-\upsilon}^{-\varphi}\left(-\lambda (\tau-t)^{\alpha}\right) f(\tau)\, d\tau. \quad (3.509)$$

The properties of the general fractional derivatives and the general fractional integrals based on the Prabhakar function are given as follows.

Property 3.40 *If* $\kappa + 1 > \alpha > \kappa$, $-\infty < a < b < \infty$, $\varphi \in \mathbb{C}$, $\upsilon \in \mathbb{C}$, $\lambda \in \mathbb{C}$ *and* $f(t) \in L_l(a,b)$ $(1 \le l < \infty)$, *then we have (see [101])*

$$L\left\{\left({}_{GP}^{RL}D_{0+}^{\alpha,\upsilon,\varphi,\lambda} f\right)(t)\right\} = s^{\kappa-\upsilon}\left(1+\lambda s^{-\alpha}\right)^{-\varphi} f(s), \quad (3.510)$$

$$L\left\{\left({}_{GP}^{RL}D_{0+}^{\alpha,\upsilon,\varphi,\lambda} f\right)(t)\right\} = s^{1-\upsilon}\left(1+\lambda s^{-\alpha}\right)^{-\varphi} f(s), \quad (3.511)$$

$$L\left\{\left({}_{GP}^{LS}D_{0+}^{\alpha,\upsilon,\varphi,\lambda} f\right)(t)\right\} = s^{-\upsilon}\left(1+\lambda s^{-\alpha}\right)^{-\varphi}\left(sf(s) - f(0)\right), \quad (3.512)$$

$$L\left\{\left({}_{GP}^{LSC}D_{0+}^{\alpha,\upsilon,\varphi,\lambda} f\right)(t)\right\} = s^{-\upsilon}\left(1+\lambda s^{-\alpha}\right)^{-\varphi}\left(s^{\kappa} f(s) - \sum_{j=1}^{\kappa} s^{\kappa-j} f^{(j-1)}(0)\right), \quad (3.513)$$

$$L\left\{\left({}^{R}_{GP}I_{0+}^{\alpha,\upsilon,\varphi,\lambda}f\right)(t)\right\} = s^{\upsilon-\kappa}\left(1+\lambda s^{-\alpha}\right)^{\varphi} f(s), \tag{3.514}$$

and

$$L\left\{\left({}^{R}_{GP}I_{0+}^{\alpha,\upsilon,\varphi,\lambda}f\right)(t)\right\} = s^{\upsilon-1}\left(1+\lambda s^{-\alpha}\right)^{\varphi} f(s). \tag{3.515}$$

Note that, when $D^{\kappa}\left({}^{R}_{GP}I_{0+}^{\alpha,\upsilon,\varphi,\lambda}f(t)\right)|_{t=0} = 0$, there is

$$L\left\{{}^{RL}_{GP}D_{0+}^{\alpha,\upsilon,\varphi,\lambda}f(t)\right\} = s^{\kappa-1}\left(1+\lambda s^{-\alpha}\right)^{-1} f(s). \tag{3.516}$$

and when $D^{\kappa}\left({}^{R}_{GP}I_{0+}^{\alpha,\upsilon,\varphi,\lambda}f(t)\right)|_{t=0} \neq 0$, there is

$$L\left\{{}^{RL}_{GP}D_{0+}^{\alpha,\upsilon,\varphi,\lambda}f(t)\right\} = s^{\kappa-1}\left(1+\lambda s^{-\alpha}\right)^{-1} f(s) - \sum_{\mu=0}^{\kappa-1} s^{\kappa-\mu-1}\left(D^{\mu}\left[{}^{RL}_{Ml}I_{0+}^{\alpha,\kappa,\lambda}f(+0)\right]\right). \tag{3.517}$$

In particular, for $\kappa = 1$ we have

$$L\left\{{}^{RL}_{GP}D_{0+}^{\alpha,\upsilon,\varphi,\lambda}f(t)\right\} = \left(1+\lambda s^{-\alpha}\right)^{-1} f(s) - {}^{R}_{GP}I_{0+}^{\alpha,\upsilon,\varphi,\lambda}f(+0). \tag{3.518}$$

and for $\kappa = 1$ and $D\left({}^{R}_{GP}I_{0+}^{\alpha,\upsilon,\varphi,\lambda}f(t)\right)|_{t=0} = 0$ we have

$$L\left\{{}^{RL}_{GP}D_{0+}^{\alpha,\upsilon,\varphi,\lambda}f(t)\right\} = \left(1+\lambda s^{-\alpha}\right)^{-1} f(s). \tag{3.519}$$

Property 3.41 *Let $\kappa+1 > \alpha > \kappa$, $-\infty < a < b < \infty$, $\varphi \in$ C, $\upsilon \in \mathbb{C}$, $\lambda \in \mathbb{C}$ and $f(t) \in L_l(a,b)$ $(1 \le l < \infty)$. Then we have (see [97]; also see [101])*

$$\left({}^{RL}_{GP}D_{a+}^{\alpha,\upsilon,\varphi,\lambda}\,{}^{R}_{GP}I_{a+}^{\alpha,\upsilon,\varphi,\lambda}f\right)(t) = f(t), \tag{3.520}$$

$$\left({}^{RL}_{GP}D_{b-}^{\alpha,\upsilon,\varphi,\lambda}\,{}^{R}_{GP}I_{b-}^{\alpha,\upsilon,\varphi,\lambda}f\right)(t) = f(t), \tag{3.521}$$

$$\left({}^{RL}_{GP}D_{a+}^{\alpha,\upsilon,\varphi,\lambda}\,{}^{R}_{GP}I_{a+}^{\alpha,\upsilon,\varphi,\lambda}f\right)(t) = f(t), \tag{3.522}$$

and

$$\left({}^{RL}_{GP}D_{b-}^{\alpha,\upsilon,\varphi,\lambda}\,{}^{R}_{GP}I_{b-}^{\alpha,\upsilon,\varphi,\lambda}f\right)(t) = f(t). \tag{3.523}$$

Property 3.42 *If $-\infty < a < b < \infty$, $\varphi \in \mathbb{C}$, $\upsilon \in \mathbb{C}$, $\lambda \in \mathbb{C}$ and $f(t) \in AC_{\kappa}(a,b)$ $(1 \le \kappa < \infty)$, then we have*

$$\left({}^{LS}_{GP}D_{a+}^{\alpha,\upsilon,\varphi,\lambda}\,{}^{R}_{GP}I_{a+}^{\alpha,\upsilon,\varphi,\lambda}f\right)(t) = f(t), \tag{3.524}$$

$$\left({}^{LS}_{GP}D_{b-}^{\alpha,\upsilon,\varphi,\lambda}\,{}^{R}_{GP}I_{b-}^{\alpha,\upsilon,\varphi,\lambda}f\right)(t) = f(t), \tag{3.525}$$

$$\left({}^{LSC}_{GP}D_{a+}^{\alpha,\upsilon,\varphi,\lambda}\,{}^{R}_{GP}I_{a+}^{\alpha,\upsilon,\varphi,\lambda}f\right)(t) = f(t), \tag{3.526}$$

and

$$\left({}^{LSC}_{GP}D_{b-}^{\alpha,\upsilon,\varphi,\lambda}\,{}^{R}_{GP}I_{b-}^{\alpha,\upsilon,\varphi,\lambda}f\right)(t) = f(t). \tag{3.527}$$

Definition 3.77 *Let $\alpha \in \mathbb{C}$, $1 > Re(\alpha) > 0$, $a = -\infty$, $b = \infty$, $\varphi \in \mathbb{C}$, $\upsilon \in \mathbb{C}$ and $\lambda \in \mathbb{C}$. The left-sided Riemann-Liouville type general fractional derivative within the generalized Prabhakar kernel via Prabhakar function is defined as*

$$\begin{aligned}\left({}_{GP}^{RL}D_{+}^{\alpha,\upsilon,\varphi,\lambda}f\right)(t) &= \tfrac{d}{dt}\left({}_{GP}I_{+}^{\alpha,\upsilon,\varphi,\lambda}f(t)\right)\\ &= \tfrac{d}{dt}\int\limits_{-\infty}^{t}(t-\tau)^{\upsilon-1}E_{\alpha,\upsilon}^{\varphi}\left(-\lambda(t-\tau)^{\alpha}\right)f(\tau)\,d\tau,\end{aligned} \tag{3.528}$$

and the right-sided Riemann-Liouville type general fractional derivative within the generalized Prabhakar kernel via Prabhakar function as

$$\begin{aligned}\left({}_{GP}^{RL}D_{-}^{\alpha,\upsilon,\varphi,\lambda}f\right)(t) &= -\tfrac{d}{dt}\left({}_{GP}I_{-}^{\alpha,\upsilon,\varphi,\lambda}f(t)\right)\\ &= -\tfrac{d}{dt}\int\limits_{t}^{\infty}(\tau-t)^{\upsilon-1}E_{\alpha,\upsilon}^{\varphi}\left(-\lambda(\tau-t)^{\alpha}\right)f(\tau)\,d\tau,\end{aligned} \tag{3.529}$$

where the left-sided and right-sided general fractional integral operators within the Prabhakar kernel via Prabhakar function are defined as

$${}_{GP}I_{+}^{\alpha,\upsilon,\varphi,\lambda}f(t) = \int\limits_{-\infty}^{t}(t-\tau)^{\upsilon-1}E_{\alpha,\upsilon}^{\varphi}\left(-\lambda(t-\tau)^{\alpha}\right)f(\tau)\,d\tau \tag{3.530}$$

and

$${}_{GP}I_{-}^{\alpha,\upsilon,\varphi,\lambda}f(t) = \int\limits_{t}^{\infty}(\tau-t)^{\upsilon-1}E_{\alpha,\upsilon}^{\varphi}\left(-\lambda(\tau-t)^{\alpha}\right)f(\tau)\,d\tau, \tag{3.531}$$

respectively.

More generally, for $\alpha \in \mathbb{C}$, $Re(\alpha) > 0$, $\kappa = [Re(\alpha)] + 1$, $a = -\infty$, $b = \infty$, $\varphi \in \mathbb{C}$, $\upsilon \in \mathbb{C}$ and $\lambda \in \mathbb{C}$, the left-sided Riemann-Liouville type general fractional derivative within the generalized Prabhakar kernel via Prabhakar function is defined as

$$\left({}_{GP}^{RL}D_{+}^{\alpha,\upsilon,\varphi,\lambda}f\right)(t) = \tfrac{d^{\kappa}}{dt^{\kappa}}\int\limits_{-\infty}^{t}(t-\tau)^{\upsilon-1}E_{\alpha,\upsilon}^{\varphi}\left(-\lambda(t-\tau)^{\alpha}\right)f(\tau)\,d\tau, \tag{3.532}$$

and the right-sided Riemann-Liouville type general fractional derivative within the generalized Prabhakar kernel via Prabhakar function as

$$\left({}_{GP}^{RL}D_{-}^{\alpha,\upsilon,\varphi,\lambda}f\right)(t) = (-1)^{\kappa}\tfrac{d^{\kappa}}{dt^{\kappa}}\int\limits_{t}^{\infty}(\tau-t)^{\upsilon-1}E_{\alpha,\upsilon}^{\varphi}\left(-\lambda(\tau-t)^{\alpha}\right)f(\tau)\,d\tau. \tag{3.533}$$

Definition 3.78 *For $\alpha \in \mathbb{C}$, $1 > Re(\alpha) > 0$, $a = -\infty$, $b = \infty$, $\varphi \in \mathbb{C}$, $\upsilon \in \mathbb{C}$ and $\lambda \in \mathbb{C}$, the left-sided Liouville-Sonine type general fractional derivative within the generalized Prabhakar kernel via Prabhakar function is defined as*

$$\begin{aligned}\left({}_{GP}^{LS}D_{+}^{\alpha,\upsilon,\varphi,\lambda}f\right)(t) &= {}_{GP}I_{+}^{\alpha,\upsilon,\varphi,\lambda}\left(f^{(1)}(t)\right)\\ &= \int_{-\infty}^{t}(t-\tau)^{\upsilon-1}E_{\alpha,\upsilon}^{\varphi}\left(-\lambda(t-\tau)^{\alpha}\right)f^{(1)}(\tau)\,d\tau,\end{aligned} \tag{3.534}$$

and the right-sided Liouville-Sonine type general fractional derivative within the generalized Prabhakar kernel via Prabhakar function as

$$\begin{aligned}\left({}_{GP}^{LS}D_{-}^{\alpha,\upsilon,\varphi,\lambda}f\right)(t) &= {}_{GP}I_{-}^{\alpha,\upsilon,\varphi,\lambda}\left(-f^{(1)}(t)\right)\\ &= -\int_{t}^{-\infty}(\tau-t)^{\upsilon-1}E_{\alpha,\upsilon}^{\varphi}\left(-\lambda(\tau-t)^{\alpha}\right)f^{(1)}(\tau)\,d\tau.\end{aligned} \tag{3.535}$$

More generally, when $\alpha \in \mathbb{C}$, $\kappa+1 > Re(\alpha) > \kappa$, $a = -\infty$, $b = \infty$, $\varphi \in \mathbb{C}$, $\upsilon \in \mathbb{C}$ and $\lambda \in \mathbb{C}$, the left-sided Liouville-Sonine-Caputo type general fractional derivative within the generalized Prabhakar kernel via Prabhakar function is defined as

$$\begin{aligned}\left({}_{GP}^{LSC}D_{+}^{\alpha,\upsilon,\varphi,\lambda}f\right)(t) &= {}_{GP}I_{+}^{\alpha,\upsilon,\varphi,\lambda}\left(f^{(\kappa)}(t)\right)\\ &= \int_{-\infty}^{t}(t-\tau)^{\upsilon-1}E_{\alpha,\upsilon}^{\varphi}\left(-\lambda(t-\tau)^{\alpha}\right)f^{(\kappa)}(\tau)\,d\tau,\end{aligned} \tag{3.536}$$

and the right-sided Liouville-Sonine-Caputo type general fractional derivative within the generalized Prabhakar kernel via Prabhakar function as

$$\begin{aligned}\left({}_{GP}^{LSC}D_{-}^{\alpha,\upsilon,\varphi,\lambda}f\right)(t) &= {}_{GP}I_{-}^{\alpha,\upsilon,\varphi,\lambda}\left((-1)^{\kappa}f^{(\kappa)}(t)\right)\\ &= (-1)^{\kappa}\int_{t}^{-\infty}(\tau-t)^{\upsilon-1}E_{\alpha,\upsilon}^{\varphi}\left(-\lambda(\tau-t)^{\alpha}\right)f^{(\kappa)}(\tau)\,d\tau.\end{aligned} \tag{3.537}$$

The inverse operators of the above general fractional derivatives within the generalized Prabhakar kernel via Prabhakar function can be considered as follows.

Definition 3.79 *Let $\alpha \in \mathbb{C}$, $\kappa+1 > Re(\alpha) > \kappa$, $a = -\infty$, $b = \infty$, $\varphi \in \mathbb{C}$, $\upsilon \in \mathbb{C}$ and $\lambda \in \mathbb{C}$. The left-sided general fractional integral via Prabhakar function is defined as*

$$\left({}_{GP}^{R}I_{+}^{\alpha,\varphi,\upsilon,\lambda}f\right)(t) = \int_{-\infty}^{t}(t-\tau)^{\kappa-\upsilon-1}E_{\alpha,\kappa-\upsilon}^{-\varphi}\left(-\lambda(t-\tau)^{\alpha}\right)f(\tau)\,d\tau, \tag{3.538}$$

and the right-sided general fractional integral via Prabhakar function as

$$\left({}_{GP}^{R}I_{-}^{\alpha,\varphi,\upsilon,\lambda}f\right) = \int_{t}^{\infty}(\tau-t)^{\kappa-\upsilon-1}E_{\alpha,\kappa-\upsilon}^{-\varphi}\left(-\lambda(\tau-t)^{\alpha}\right)f(\tau)\,d\tau. \tag{3.539}$$

In particular, when $\alpha \in \mathbb{C}$, $1 > Re(\alpha) > 0$, $a = -\infty$, $b = \infty$, $\varphi \in \mathbb{C}$, $\upsilon \in \mathbb{C}$ and $\lambda \in \mathbb{C}$, the left-sided general fractional integral via Prabhakar function is defined as

$$\left({}_{GP}^{R}I_{+}^{\alpha,\varphi,\upsilon,\lambda} f\right)(t) = \int_{-\infty}^{t} \frac{1}{(t-\tau)^{\upsilon}} E_{\alpha,\kappa-\upsilon}^{-\varphi}\left(-\lambda (t-\tau)^{\alpha}\right) f(\tau)\, d\tau, \tag{3.540}$$

and the right-sided general fractional integral via Prabhakar function as

$$\left({}_{GP}^{R}I_{-}^{\alpha,\varphi,\upsilon,\lambda} f\right) = \int_{t}^{\infty} \frac{1}{(\tau-t)^{\upsilon}} E_{\alpha,\kappa-\upsilon}^{-\varphi}\left(-\lambda (\tau-t)^{\alpha}\right) f(\tau)\, d\tau. \tag{3.541}$$

The properties of the general fractional derivatives and the general fractional integrals via Prabhakar function are presented as follows.

Property 3.43 *Let $\kappa + 1 > \alpha > \kappa$, $a = -\infty$, $b = \infty$, $\varphi \in \mathbb{C}$, $\upsilon \in \mathbb{C}$ and $\lambda \in \mathbb{C}$ and $f(t) \in L_l(-\infty, \infty)$ $(1 \le l < \infty)$. Then we have*

$$\left({}_{GP}^{RL}D_{+}^{\alpha,\upsilon,\varphi,\lambda}\,{}_{GP}^{R}I_{+}^{\alpha,\upsilon,\varphi,\lambda} f\right)(t) = f(t), \tag{3.542}$$

$$\left({}_{GP}^{RL}D_{-}^{\alpha,\upsilon,\varphi,\lambda}\,{}_{GP}^{R}I_{-}^{\alpha,\upsilon,\varphi,\lambda} f\right)(t) = f(t), \tag{3.543}$$

$$\left({}_{GP}^{RL}D_{+}^{\alpha,\upsilon,\varphi,\lambda}\,{}_{GP}^{R}I_{+}^{\alpha,\upsilon,\varphi,\lambda} f\right)(t) = f(t), \tag{3.544}$$

and

$$\left({}_{GP}^{RL}D_{-}^{\alpha,\upsilon,\varphi,\lambda}\,{}_{GP}^{R}I_{-}^{\alpha,\upsilon,\varphi,\lambda} f\right)(t) = f(t). \tag{3.545}$$

Property 3.44 *If $a = -\infty$,$b = \infty$, $\varphi \in \mathbb{C}$, $\upsilon \in \mathbb{C}$, $\lambda \in \mathbb{C}$ and $f(t) \in AC_{\kappa}(-\infty, \infty)$ $(1 \le \kappa < \infty)$, then we have*

$$\left({}_{GP}^{LS}D_{+}^{\alpha,\upsilon,\varphi,\lambda}\,{}_{GP}^{R}I_{+}^{\alpha,\upsilon,\varphi,\lambda} f\right)(t) = f(t), \tag{3.546}$$

$$\left({}_{GP}^{LS}D_{-}^{\alpha,\upsilon,\varphi,\lambda}\,{}_{GP}^{R}I_{-}^{\alpha,\upsilon,\varphi,\lambda} f\right)(t) = f(t), \tag{3.547}$$

$$\left({}_{GP}^{LSC}D_{+}^{\alpha,\upsilon,\varphi,\lambda}\,{}_{GP}^{R}I_{+}^{\alpha,\upsilon,\varphi,\lambda} f\right)(t) = f(t), \tag{3.548}$$

and

$$\left({}_{GP}^{LSC}D_{-}^{\alpha,\upsilon,\varphi,\lambda}\,{}_{GP}^{R}I_{-}^{\alpha,\upsilon,\varphi,\lambda} f\right)(t) = f(t). \tag{3.549}$$

3.17 General fractional derivatives involving the generalized Prabhakar kernel with respect to another function

The definitions and properties of the general fractional derivatives and general fractional integrals involving the generalized Prabhakar kernel via Prabhakar function with respect to another function are presented as follows.

Definition 3.80 *Let $1 > \alpha > 0$, $-\infty < a < b < \infty$, $\varphi \in \mathbb{C}$, $\upsilon \in \mathbb{C}$, $h^{(1)}(t) > 0$, $\lambda \in \mathbb{C}$ and $h^{(1)}(t) > 0$. The left-sided Riemann-Liouville type general fractional derivative involving the generalized Prabhakar kernel via Prabhakar function with respect to another function is defined as*

$$\begin{aligned}&\left({}_{GP}^{RL}D_{a+,h}^{\alpha,\upsilon,\varphi,\lambda}f\right)(t)=\left(\tfrac{1}{h^{(1)}(t)}\tfrac{d}{dt}\right)\left({}_{GP}I_{a+,h}^{\alpha,\upsilon,\varphi,\lambda}f(t)\right)\\&=\left(\tfrac{1}{h^{(1)}(t)}\tfrac{d}{dt}\right)\int_a^t\left(h(t)-h(\tau)\right)^{\upsilon-1}E_{\alpha,\upsilon}^{\varphi}\left(-\lambda\left(h(t)-h(\tau)\right)^{\alpha}\right)f(\tau)\,h^{(1)}(\tau)\,d\tau,\end{aligned} \tag{3.550}$$

and the right-sided Riemann-Liouville type general fractional derivative involving the generalized Prabhakar kernel via Prabhakar function with respect to another function as

$$\begin{aligned}&\left({}_{GP}^{RL}D_{b-,h}^{\alpha,\upsilon,\varphi,\lambda}f\right)(t)=\left(-\tfrac{1}{h^{(1)}(t)}\tfrac{d}{dt}\right)\left({}_{GP}I_{b-,h}^{\alpha,\upsilon,\varphi,\lambda}f(t)\right)\\&=\left(-\tfrac{1}{h^{(1)}(t)}\tfrac{d}{dt}\right)\int_t^b\left(h(\tau)-h(t)\right)^{\upsilon-1}E_{\alpha,\upsilon}^{\varphi}\left(-\lambda\left(h(\tau)-h(t)\right)^{\alpha}\right)h^{(1)}(\tau)\,f(\tau)\,d\tau,\end{aligned} \tag{3.551}$$

where the left-sided and right-sided general fractional integral operators involving the generalized Prabhakar kernel via Prabhakar function with respect to another function are represented as

$${}_{GP}I_{a+,h}^{\alpha,\upsilon,\varphi,\lambda}f(t)=\int_a^t\left(h(t)-h(\tau)\right)^{\upsilon-1}E_{\alpha,\upsilon}^{\varphi}\left(-\lambda\left(h(t)-h(\tau)\right)^{\alpha}\right)h^{(1)}(\tau)\,f(\tau)\,d\tau, \tag{3.552}$$

and

$${}_{GP}I_{b-,h}^{\alpha,\upsilon,\varphi,\lambda}f(t)=\int_t^b\left(h(\tau)-h(t)\right)^{\upsilon-1}E_{\alpha,\upsilon}^{\varphi}\left(-\lambda\left(h(\tau)-h(t)\right)^{\alpha}\right)h^{(1)}(\tau)\,f(\tau)\,d\tau, \tag{3.553}$$

respectively.

Definition 3.81 *For $\kappa+1 > \alpha > \kappa$, $-\infty < a < b < \infty$, $\varphi \in \mathbb{C}$, $h^{(1)}(t) > 0$, $\upsilon \in \mathbb{C}$ and $\lambda \in \mathbb{C}$, the left-sided Liouville-Sonine-Caputo type general fractional derivative involving the generalized Prabhakar kernel via Prabhakar* *function with respect to another function is defined as*

$$\left({}_{GP}^{LSC}D_{a+,h}^{\alpha,\varphi,\upsilon,\lambda}f\right)(t) = {}_{GP}I_{a+,h}^{\alpha,\varphi,\upsilon,\lambda}\left(\left(\frac{1}{h^{(1)}(t)}\frac{d}{dt}\right)^{\kappa}f_h^{(\kappa)}(t)\right)$$
$$= \int_a^t (h(t)-h(\tau))^{\upsilon-1}E_{\alpha,\upsilon}^{\varphi}(-\lambda(h(t)-h(\tau))^{\alpha})\left(\frac{1}{h^{(1)}(\tau)}\frac{d}{d\tau}\right)^{\kappa}f_h^{(\kappa)}(\tau)\,d\tau, \tag{3.554}$$

and the right-sided Liouville-Sonine-Caputo type general fractional derivative involving the generalized Prabhakar kernel via Prabhakar function with respect to another function as

$$\left({}_{GP}^{LSC}D_{b-,h}^{\alpha,\varphi,\upsilon,\lambda}f\right)(t) = {}_{GP}I_{b-,h}^{\alpha,\varphi,\upsilon,\lambda}\left(\left(-\frac{1}{h^{(1)}(t)}\frac{d}{dt}\right)^{\kappa}f_h^{(\kappa)}(t)\right)$$
$$= \int_t^b (h(\tau)-h(t))^{\upsilon-1}E_{\alpha,\upsilon}^{\varphi}(-\lambda(h(\tau)-h(t))^{\alpha})\left(-\frac{1}{h^{(1)}(\tau)}\frac{d}{d\tau}\right)^{\kappa}f_h^{(\kappa)}(\tau)\,d\tau. \tag{3.555}$$

Definition 3.82 *Let $\kappa+1>\alpha>\kappa$, $-\infty<a<b<\infty$, $\varphi\in\mathbb{C}$, $h^{(1)}(t)>0$, $\upsilon\in\mathbb{C}$ and $\lambda\in\mathbb{C}$. The left-sided general fractional integral via Prabhakar function with respect to another function is defined as*

$$\left({}_{GP}^{R}I_{a+,h}^{\alpha,\varphi,\upsilon,\lambda}f\right)(t)$$
$$= \int_a^t (h(t)-h(\tau))^{\kappa-\upsilon-1}E_{\alpha,\kappa-\upsilon}^{-\varphi}(-\lambda(h(t)-h(\tau))^{\alpha})h^{(1)}(\tau)f(\tau)\,d\tau, \tag{3.556}$$

and the right-sided general fractional integral via Prabhakar function with respect to another function as

$$\left({}_{GP}^{R}I_{b-,h}^{\alpha,\varphi,\upsilon,\lambda}f\right)$$
$$= \int_t^b (h(\tau)-h(t))^{\kappa-\upsilon-1}E_{\alpha,\kappa-\upsilon}^{-\varphi}(-\lambda(h(\tau)-h(t))^{\alpha})h^{(1)}(\tau)f(\tau)\,d\tau. \tag{3.557}$$

In particular, when $1>\alpha>0$, $-\infty<a<b<\infty$, $\varphi\in\mathbb{C}$, $h^{(1)}(t)>0$, $\upsilon\in\mathbb{C}$ and $\lambda\in\mathbb{C}$, the left-sided general fractional integral via Prabhakar function with respect to another function is defined as

$$\left({}_{GP}^{R}I_{a+,h}^{\alpha,\varphi,\upsilon,\lambda}f\right)(t)$$
$$= \int_a^t \frac{1}{(h(t)-h(\tau))^{\upsilon}}E_{\alpha,1-\upsilon}^{-\varphi}(-\lambda(h(t)-h(\tau))^{\alpha})h^{(1)}(\tau)f(\tau)\,d\tau, \tag{3.558}$$

and the right-sided general fractional integral via Prabhakar function with respect to another function as

$$\left({}_{GP}^{R}I_{b-,h}^{\alpha,\varphi,\upsilon,\lambda}f\right)$$
$$= \int_t^b \frac{1}{(h(\tau)-h(t))^{\upsilon}}E_{\alpha,1-\upsilon}^{-\varphi}(-\lambda(h(\tau)-h(t))^{\alpha})h^{(1)}(\tau)f(\tau)\,d\tau. \tag{3.559}$$

The properties of the general fractional derivatives and general fractional integrals with respect to another function are presented as follows.

Property 3.45 *Let* $\kappa+1>\alpha>\kappa$, $-\infty<a<b<\infty$, $\varphi\in\mathbb{C}$, $\upsilon\in\mathbb{C}$,$h^{(1)}(t)>0$, $\lambda\in\mathbb{C}$ *and* $f(t)\in L_l(a,b)$ $(1\leq l<\infty)$. *Then we have*

$$\left({}_{GP}^{RL}D_{a+,h}^{\alpha,\upsilon,\varphi,\lambda}\,{}_{GP}^{R}I_{a+,h}^{\alpha,\upsilon,\varphi,\lambda}f\right)(t)=f(t), \tag{3.560}$$

$$\left({}_{GP}^{RL}D_{b-,h}^{\alpha,\upsilon,\varphi,\lambda}\,{}_{GP}^{R}I_{b-,h}^{\alpha,\upsilon,\varphi,\lambda}f\right)(t)=f(t), \tag{3.561}$$

$$\left({}_{GP}^{RL}D_{a+,h}^{\alpha,\upsilon,\varphi,\lambda}\,{}_{GP}^{R}I_{a+,h}^{\alpha,\upsilon,\varphi,\lambda}f\right)(t)=f(t), \tag{3.562}$$

and

$$\left({}_{GP}^{RL}D_{b-,h}^{\alpha,\upsilon,\varphi,\lambda}\,{}_{GP}^{R}I_{b-,h}^{\alpha,\upsilon,\varphi,\lambda}f\right)(t)=f(t). \tag{3.563}$$

Property 3.46 *If* $-\infty<a<b<\infty$, $\varphi\in\mathbb{C}$, $\upsilon\in\mathbb{C}$, $h^{(1)}(t)>0$, $\lambda\in\mathbb{C}$ *and* $f(t)\in AC_\kappa(a,b)$ $(1\leq\kappa<\infty)$, *then we have*

$$\left({}_{GP}^{LS}D_{a+,h}^{\alpha,\upsilon,\varphi,\lambda}\,{}_{GP}^{R}I_{a+,h}^{\alpha,\upsilon,\varphi,\lambda}f\right)(t)=f(t), \tag{3.564}$$

$$\left({}_{GP}^{LS}D_{b-,h}^{\alpha,\upsilon,\varphi,\lambda}\,{}_{GP}^{R}I_{b-,h}^{\alpha,\upsilon,\varphi,\lambda}f\right)(t)=f(t), \tag{3.565}$$

$$\left({}_{GP}^{LSC}D_{a+,h}^{\alpha,\upsilon,\varphi,\lambda}\,{}_{GP}^{R}I_{a+,h}^{\alpha,\upsilon,\varphi,\lambda}f\right)(t)=f(t), \tag{3.566}$$

and

$$\left({}_{GP}^{LSC}D_{b-,h}^{\alpha,\upsilon,\varphi,\lambda}\,{}_{GP}^{R}I_{b-,h}^{\alpha,\upsilon,\varphi,\lambda}f\right)(t)=f(t). \tag{3.567}$$

3.18 Hilfer type general fractional derivatives within the Prabhakar kernel

The definitions and properties of the Hilfer type general fractional derivatives within the generalized Prabhakar kernel via Prabhakar function are given as follows.

Definition 3.83 *Let* $1>\alpha>0$, $0\leq\beta\leq1$, $-\infty<a<b<\infty$, $\gamma\in\mathbb{C}$, $\varphi\in\mathbb{C}$, $\upsilon\in\mathbb{C}$ *and* $\lambda\in\mathbb{C}$. *The left-sided Hilfer type general fractional derivative within the generalized Prabhakar kernel via Prabhakar function is defined as (see [101])*

$$\begin{aligned}\left({}_{GP}^{Hi}D_{a+}^{\alpha,\beta,\upsilon,\varphi,\lambda}f\right)(t)&=\int_a^t(t-\tau)^{\upsilon-1}E_{\alpha,\upsilon}^{\varphi}\left(-\lambda(t-\tau)^{\alpha}\right)\left(D_{a+}^{\alpha,\beta}f(\tau)\right)d\tau\\&=\int_a^t(t-\tau)^{\upsilon-1}E_{\alpha,\upsilon}^{\varphi}\left(-\lambda(t-\tau)^{\alpha}\right)\left(I_{a+}^{\beta(1-\alpha)}\frac{d}{d\tau}\left(I_{a+}^{(1-\beta)(1-\alpha)}f\right)\right)(\tau)\,d\tau,\end{aligned} \tag{3.568}$$

and the right-sided Hilfer type general fractional derivative within the generalized Prabhakar kernel via Prabhakar function as

$$\left({}_{GP}^{Hi}D_{b-}^{\alpha,\beta,\upsilon,\varphi,\lambda}f\right)(t)=-\int_t^b(\tau-t)^{\upsilon-1}E_{\alpha,\upsilon}^{\varphi}\left(-\lambda(\tau-t)^{\alpha}\right)\left(D_{b-}^{\alpha,\beta}f(\tau)\right)d\tau$$
$$=-\int_t^b(\tau-t)^{\upsilon-1}E_{\alpha,\upsilon}^{\varphi}\left(-\lambda(\tau-t)^{\alpha}\right)\left(I_{b-}^{\beta(1-\alpha)}\frac{d}{d\tau}\left(I_{b-}^{(1-\beta)(1-\alpha)}f\right)\right)(\tau)\,d\tau. \tag{3.569}$$

Generally, when $\kappa+1>\alpha>\kappa$, $0\leq\beta\leq1$, $-\infty<a<b<\infty$, $\gamma\in\mathbb{C}$, $\varphi\in\mathbb{C}$, $\upsilon\in\mathbb{C}$ and $\lambda\in\mathbb{C}$, the left-sided Hilfer type general fractional derivative within the generalized Prabhakar kernel via Prabhakar function is defined as

$$\left({}_{GP}^{Hir}D_{a+}^{\alpha,\beta,\upsilon,\varphi,\lambda}f\right)(t)$$
$$=\int_a^t(t-\tau)^{\upsilon-1}E_{\alpha,\upsilon}^{\varphi}\left(-\lambda(t-\tau)^{\alpha}\right)\left(I_{a+}^{\beta(\kappa-\alpha)}\frac{d^{\kappa}}{d\tau^{\kappa}}\left(I_{a+}^{(1-\beta)(\kappa-\alpha)}f\right)\right)(\tau)\,d\tau, \tag{3.570}$$

and the right-sided Hilfer type general fractional derivative within the generalized Prabhakar kernel via Prabhakar function as

$$\left({}_{GP}^{Hir}D_{b-}^{\alpha,\beta,\upsilon,\varphi,\lambda}f\right)(t)$$
$$=(-1)^{\kappa}\int_t^b(\tau-t)^{\upsilon-1}E_{\alpha,\upsilon}^{\varphi}\left(-\lambda(\tau-t)^{\alpha}\right)\left(I_{b-}^{\beta(\kappa-\alpha)}\frac{d^{\kappa}}{d\tau^{\kappa}}\left(I_{b-}^{(1-\beta)(\kappa-\alpha)}f\right)\right)(\tau)\,d\tau. \tag{3.571}$$

In particular, for $1>\alpha>0$, $\beta=1$, $-\infty<a<b<\infty$, $\gamma\in\mathbb{C}$, $\varphi\in\mathbb{C}$, $\upsilon\in\mathbb{C}$ and $\lambda\in\mathbb{C}$, the left-sided Hilfer type general fractional derivative within the generalized Prabhakar kernel via Prabhakar function is defined as (see [101])

$$\left({}_{GP}^{Hi}D_{a+}^{\alpha,1,\upsilon,\varphi,\lambda}f\right)(t)=\int_a^t(t-\tau)^{\upsilon-1}E_{\alpha,\upsilon}^{\varphi}\left(-\lambda(t-\tau)^{\alpha}\right)\left(I_{a+}^{1-\alpha}f^{(1)}(\tau)\right)d\tau, \tag{3.572}$$

and the right-sided Hilfer type general fractional derivative within the generalized Prabhakar kernel via Prabhakar function as

$$\left({}_{GP}^{Hi}D_{b-}^{\alpha,1,\upsilon,\varphi,\lambda}f\right)(t)=-\int_t^b(\tau-t)^{\upsilon-1}E_{\alpha,\upsilon}^{\varphi}\left(-\lambda(\tau-t)^{\alpha}\right)\left(I_{b-}^{1-\alpha}f^{(1)}(\tau)\right)(\tau)\,d\tau. \tag{3.573}$$

Furthermore, for $\kappa+1>\alpha>\kappa$, $\beta=1$, $-\infty<a<b<\infty$, $\gamma\in\mathbb{C}$, $\varphi\in\mathbb{C}$, $\upsilon\in\mathbb{C}$ and $\lambda\in\mathbb{C}$, the left-sided Hilfer type general fractional derivative within the generalized Prabhakar kernel via Prabhakar function is defined as

$$\left({}_{GP}^{Hir}D_{a+}^{\alpha,1,\upsilon,\varphi,\lambda}f\right)(t)=\int_a^t(t-\tau)^{\upsilon-1}E_{\alpha,\upsilon}^{\varphi}\left(-\lambda(t-\tau)^{\alpha}\right)\left(I_{a+}^{\kappa-\alpha}\left(f^{(\kappa)}(\tau)\right)\right)d\tau, \tag{3.574}$$

and the right-sided Hilfer type general fractional derivative within the generalized Prabhakar kernel via Prabhakar function as

$$\left({}_{GP}^{Hir}D_{b-}^{\alpha,1,\upsilon,\varphi,\lambda}f\right)(t)=(-1)^{\kappa}\int_{t}^{b}(\tau-t)^{\upsilon-1}E_{\alpha,\upsilon}^{\varphi}(-\lambda(\tau-t)^{\alpha})\left(I_{b-}^{\beta(\kappa-\alpha)}f^{(\kappa)}(\tau)\right)d\tau. \tag{3.575}$$

In particular, for $1>\alpha>0$, $\beta=0$, $-\infty<a<b<\infty$, $\gamma\in\mathbb{C}$, $\varphi\in\mathbb{C}$, $\upsilon\in\mathbb{C}$ and $\lambda\in\mathbb{C}$, the left-sided Hilfer type general fractional derivative within the generalized Prabhakar kernel via Prabhakar function is defined as

$$\left({}_{GP}^{Hi}D_{a+}^{\alpha,0,\upsilon,\varphi,\lambda}f\right)(t)=\int_{a}^{t}(t-\tau)^{\upsilon-1}E_{\alpha,\upsilon}^{\varphi}\left(-\lambda(t-\tau)^{\alpha}\right)\left(\tfrac{d}{d\tau}\left(I_{a+}^{1-\alpha}f\right)(\tau)\right)d\tau, \tag{3.576}$$

and the right-sided Hilfer type general fractional derivative within the generalized Prabhakar kernel via Prabhakar function as

$$\left({}_{GP}^{Hi}D_{b-}^{\alpha,0,\upsilon,\varphi,\lambda}f\right)(t)=-\int_{t}^{b}(\tau-t)^{\upsilon-1}E_{\alpha,\upsilon}^{\varphi}\left(-\lambda(\tau-t)^{\alpha}\right)\left(\tfrac{d}{d\tau}\left(I_{b-}^{1-\alpha}f(\tau)\right)\right)d\tau. \tag{3.577}$$

Furthermore, when $\kappa+1>\alpha>\kappa$, $\beta=0$, $-\infty<a<b<\infty$, $\gamma\in\mathbb{C}$, $\varphi\in\mathbb{C}$, $\upsilon\in\mathbb{C}$ and $\lambda\in\mathbb{C}$, the left-sided Hilfer type general fractional derivative within the generalized Prabhakar kernel via Prabhakar function is defined as

$$\begin{aligned}\left({}_{GP}^{Hir}D_{a+}^{\alpha,0,\upsilon,\varphi,\lambda}f\right)(t)&=\int_{a}^{t}(t-\tau)^{\upsilon-1}E_{\alpha,\upsilon}^{\varphi}\left(-\lambda(t-\tau)^{\alpha}\right)\left(D_{a+}^{\alpha,0,\kappa}f(\tau)\right)d\tau\\&=\int_{a}^{t}(t-\tau)^{\upsilon-1}E_{\alpha,\upsilon}^{\varphi}\left(-\lambda(t-\tau)^{\alpha}\right)\left(\tfrac{d^{\kappa}}{d\tau^{\kappa}}\left(I_{a+}^{\kappa-\alpha}f(\tau)\right)\right)d\tau,\end{aligned} \tag{3.578}$$

and the right-sided Hilfer type general fractional derivative within the generalized Prabhakar kernel via Prabhakar function as

$$\begin{aligned}\left({}_{GP}^{Hir}D_{b-}^{\alpha,0,\upsilon,\varphi,\lambda}f\right)(t)&=(-1)^{\kappa}\int_{t}^{b}(\tau-t)^{\upsilon-1}E_{\alpha,\upsilon}^{\varphi}\left(-\lambda(\tau-t)^{\alpha}\right)\left(D_{b-}^{\alpha,0,\kappa}f(\tau)\right)d\tau\\&=(-1)^{\kappa}\int_{t}^{b}(\tau-t)^{\upsilon-1}E_{\alpha,\upsilon}^{\varphi}\left(-\lambda(\tau-t)^{\alpha}\right)\left(\tfrac{d^{\kappa}}{d\tau^{\kappa}}\left(I_{b-}^{\kappa-\alpha}f(\tau)\right)\right)d\tau.\end{aligned} \tag{3.579}$$

3.19 Hilfer type general fractional derivatives within the Prabhakar kernel with respect to another function

The definitions and properties of the Hilfer type general fractional derivatives involving the generalized Prabhakar kernel via Prabhakar function with respect to another function are given as follows.

Definition 3.84 *Let* $1 > \alpha > 0$, $0 \leq \beta \leq 1$, $-\infty < a < b < \infty$, $\gamma \in \mathbb{C}$, $\varphi \in \mathbb{C}$, $\upsilon \in \mathbb{C}$ *and* $\lambda \in \mathbb{C}$. *The left-sided Hilfer type general fractional derivative involving the generalized Prabhakar kernel via Prabhakar function with respect to another function is defined as*

$$\left({}_{GP}^{Hi}D_{a+,h}^{\alpha,\beta,\upsilon,\varphi,\lambda} f\right)(t) = \int_a^t (h(t) - h(\tau))^{\upsilon-1} E_{\alpha,\upsilon}^{\varphi}\left(-\lambda (h(t) - h(\tau))^{\alpha}\right) \left(D_{a+,h}^{\alpha,\beta} f(\tau)\right) h^{(1)}(\tau)\, d\tau, \tag{3.580}$$

and the right-sided Hilfer type general fractional derivative involving the generalized Prabhakar kernel via Prabhakar function with respect to another function as

$$\left({}_{GP}^{Hi}D_{b-,h}^{\alpha,\beta,\upsilon,\varphi,\lambda} f\right)(t) = -\int_t^b (h(\tau) - h(t))^{\upsilon-1} E_{\alpha,\upsilon}^{\varphi}\left(-\lambda (h(\tau) - h(t))^{\alpha}\right) \left(D_{b-,h}^{\alpha,\beta} f(\tau)\right) h^{(1)}(\tau)\, d\tau. \tag{3.581}$$

More generally, when $\kappa + 1 > \alpha > \kappa$, $0 \leq \beta \leq 1$, $-\infty < a < b < \infty$, $\gamma \in \mathbb{C}$, $\varphi \in \mathbb{C}$, $\upsilon \in \mathbb{C}$ and $\lambda \in \mathbb{C}$, the left-sided Hilfer type general fractional derivative involving the generalized Prabhakar kernel via Prabhakar function with respect to another function is defined as

$$\left({}_{GP}^{Hir}D_{a+,h}^{\alpha,\beta,\upsilon,\varphi,\lambda} f\right)(t) = \int_a^t (h(t) - h(\tau))^{\upsilon-1} E_{\alpha,\upsilon}^{\varphi}\left(-\lambda (h(t) - h(\tau))^{\alpha}\right) \left(D_{a+,h}^{\alpha,\beta,\kappa} f(\tau)\right) h^{(1)}(\tau)\, d\tau, \tag{3.582}$$

and the right-sided Hilfer type general fractional derivative involving the generalized Prabhakar kernel via Prabhakar function with respect to another function as

$$\left({}_{GP}^{Hir}D_{b-,h}^{\alpha,\beta,\upsilon,\varphi,\lambda} f\right)(t) = (-1)^{\kappa} \int_t^b (h(\tau) - h(t))^{\upsilon-1} E_{\alpha,\upsilon}^{\varphi}\left(-\lambda (h(\tau) - h(t))^{\alpha}\right) \left(D_{b-,h}^{\alpha,\beta,\kappa} f(\tau)\right) h^{(1)}(\tau)\, d\tau. \tag{3.583}$$

3.20 Applications

3.20.1 Relaxation models within general fractional derivatives

In this section, we present the general fractional derivatives to model the relaxation models in complex phenomenon. To begin with, we illustrate the open problems on the relaxation models within the general fractional derivatives involving the special functions.

Example 3.1 *The mathematical model of the relaxation involving the Liouville-Sonine-Caputo type general fractional derivative within the kernel of Mittag-Leffler function can be suggested as follows*

$$ {}_{Mlt}^{RL}D_{0+}^{\alpha,1,\lambda} f(t) + \gamma f(t) = 0 $$

subjected to the initial condition

$$ f(0+) = h, \tag{3.584} $$

where λ, γ and h are the constants.

Example 3.2 *The mathematical model of the relaxation involving the Liouville-Sonine-Caputo type general fractional derivative within the kernel of Mittag-Leffler function can be illustrated as*

$$ {}_{Mle}^{LSC}D_{0+}^{\alpha,1,\lambda} f(t) + \gamma f(t) = 0 $$

subjected to the initial condition

$$ f(0+) = h, \tag{3.585} $$

where λ, γ and h are the constants.

Example 3.3 *The mathematical model of the relaxation involving the Liouville-Sonine-Caputo type general fractional derivative within the kernel of Mittag-Leffler function can be shown as follows:*

$$ {}_{Ml}^{LSC}D_{0+}^{\alpha,1,\lambda} f(t) + \gamma f(t) = 0 $$

subjected to the initial condition

$$ f(0+) = h, \tag{3.586} $$

where λ, γ and h are the constants.

Example 3.4 *The mathematical model of the relaxation involving the Liouville-Sonine-Caputo type general fractional derivative within the kernel of Mittag-Leffler function can be given as follows:*

$$ {}_{Ml}^{LSC}D_{0+}^{1,\lambda;\nu} f(t) + \gamma f(t) = 0, \tag{3.587} $$

subjected to the initial condition

$$ f(0+) = h, \tag{3.588} $$

where λ, γ and h are the constants.

Example 3.5 *The mathematical model of the relaxation within the general fractional derivative in the sense of Richard type can be shown as follows:*

$$ {}_{Ml}^{Rzh}D_{[a,b]}^{\alpha,1,\lambda} f(t) + \gamma f(t) = 0 \tag{3.589} $$

where λ and γ are the constants.

Example 3.6 *The mathematical model of the relaxation within the general fractional derivative in the sense of Feller type can be written as*

$$ {}_{Mlz}^{Rzj}D_{[a,b]}^{\alpha,1,\lambda} f(t) + \gamma f(t) = 0 \tag{3.590} $$

where λ and γ are the constants.

Example 3.7 *The mathematical model of the relaxation within the general fractional derivative in the sense of Richard type can be presented as*

$$ {}_{Mlj}^{Rzl}D_{[a,b]}^{\alpha,\lambda} f(t) + \gamma f(t) = 0 \tag{3.591} $$

where λ and γ are the constants.

Example 3.8 *The mathematical model of the relaxation based on the general fractional derivative involving the kernel of Mittag-Leffler function in the sense of Riemann-Liouville type with respect to another function can be represented as*

$$ {}_{Ml}^{RL}D_{0+,h}^{\alpha,\lambda} f(t) + \gamma f(t) = 0 \tag{3.592} $$

where λ and γ are the constants, $h^{(1)}(t) > 0$.

Example 3.9 *The mathematical model of the relaxation based on the general fractional derivative involving the kernel of Mittag-Leffler function with power law in the sense of Liouville-Sonine-Caputo type with respect to another function can be written as*

$$ {}_{Ml}^{LSC}D_{0+,h}^{\alpha,\lambda} f(t) + \gamma f(t) = 0 \tag{3.593} $$

where λ and γ are the constants and $h^{(1)}(t) > 0$.

Example 3.10 *The mathematical model of the relaxation based on the general fractional derivative containing the kernel of the Gorenflo-Mainardi function via Wiman function in the sense of Riemann-Liouville type can be represented as*

$$ \left({}_{GM}^{RL}D_{0+}^{\alpha,\gamma,\lambda} f \right)(t) + \mu f(t) = 0, \tag{3.594} $$

where λ , μ and γ are the constants.

Example 3.11 *The mathematical model of the relaxation based on the general fractional derivative containing the kernel of the Gorenflo-Mainardi function via Wiman function in the sense of Liouville-Sonine type can be represented as*

$$\left({}_{GM}^{LS}D_{0+}^{\alpha,\gamma,\lambda}f\right)(t)+\mu f(t)=0, \tag{3.595}$$

where λ , μ *and* γ *are the constants.*

Example 3.12 *The mathematical model of the relaxation based on the Hilfer type general fractional derivative containing the kernel of the Gorenflo-Mainardi function via Wiman function can be given as*

$$\left({}_{GM}^{Hi}D_{0+}^{\alpha,\beta,\gamma,\lambda}f\right)(t)+\mu f(t)=0, \tag{3.596}$$

where λ , μ *and* γ *are the constants and* $0\leq\beta\leq1$.

Example 3.13 *The mathematical model of the relaxation based on the general fractional derivative within the Miller-Ross kernel in the sense of Riemann-Liouville type can be represented as*

$$\left({}_{MR}^{RL}D_{0+}^{\alpha,\lambda}f\right)(t)+\mu f(t)=0, \tag{3.597}$$

where λ *and* μ *are the constants.*

Example 3.14 *The mathematical model of the relaxation based on the general fractional derivative within the Miller-Ross kernel in the sense of Liouville-Sonine type can be written as*

$$\left({}_{MR}^{LS}D_{0+}^{\alpha,\lambda}f\right)(t)+\mu f(t)=0, \tag{3.598}$$

where λ *and* μ *are the constants.*

Example 3.15 *The mathematical model of the relaxation based on the Hilfer type general fractional derivative containing the kernel of the Miller-Ross function can be given as*

$$\left({}_{MR}^{Hi}D_{0+}^{\alpha,\beta,\lambda}f\right)(t)+\mu f(t)=0, \tag{3.599}$$

where λ *and* μ *are the constants and* $0\leq\beta\leq1$.

Example 3.16 *The mathematical model of the relaxation based on the Riemann-Liouville type general fractional derivative involving the two-parametric Lorenzo-Hartley kernel can be represented as*

$$\left({}_{HL}^{RL}D_{0+}^{\alpha,\upsilon,\lambda}f\right)(t)+\mu f(t)=0, \tag{3.600}$$

where λ *and* μ *are the constants and* $\upsilon>0$.

Example 3.17 *The mathematical model of the relaxation based on the Liouville-Sonine type general fractional derivative involving the two-parametric Lorenzo-Hartley kernel can be written as*

$$\left({}_{HL}^{LS}D_{0+}^{\alpha,\upsilon,\lambda} f \right)(t) + \mu f(t) = 0, \tag{3.601}$$

where γ *and* μ *are the constants and* $\upsilon > 0$.

Example 3.18 *The mathematical model of the relaxation based on the Riemann-Liouville type general fractional derivative involving the one-parametric Lorenzo-Hartley kernel can be represented as*

$$\left({}_{A}^{RL}D_{0+}^{\alpha,\gamma} f \right)(t) + \mu f(t) = 0, \tag{3.602}$$

where γ *and* μ *are the constants.*

Example 3.19 *The mathematical model of the relaxation based on the Liouville-Sonine type general fractional derivative involving the one-parametric Lorenzo-Hartley kernel can be written as*

$$\left({}_{A}^{LS}D_{0+}^{\alpha,\gamma} f \right)(t) + \mu f(t) = 0, \tag{3.603}$$

where γ *and* μ *are the constants.*

Example 3.20 *The mathematical model of the relaxation based on the Hilfer type general fractional derivative involving the one-parametric Lorenzo-Hartley kernel can be given as*

$$\left({}_{A}^{Hi}D_{0+}^{\alpha,\beta,\gamma} f \right)(t) + \mu f(t) = 0, \tag{3.604}$$

where γ , $0 \leq \beta \leq 1$ *and* μ *are the constants.*

Note that there are many mathematical models for the relaxation involving the special functions, such as the Davidson-Cole model (see [306]), Havriliak-Negami model (see [307]), Jurlewicz-Trzmiel-Weron model (see [308]), Oliveira-Mainardi-Vaz model (see [309]) and so on.

3.20.2 Rheological models within general fractional derivatives

In this section, we present the mathematical models of the rheological materials based on the general fractional derivatives involving the special functions. The open problems for the rheological models in general fractional derivatives are discussed in detail. To begin with the ideas, we present the mathematical models for the rheological materials as follows.

Example 3.21 *The mathematical model for the law of deformation based on the Liouville-Sonine-Caputo type general fractional derivative within the kernel of Mittag-Leffler function is represented in the form*

$$\sigma(t) = \xi {}_{Mlt}^{RL}D_{0+}^{\alpha,1,\lambda}\varepsilon(t) \quad (0 < \alpha < 1), \tag{3.605}$$

which can be written as

$$\sigma(t) = \xi \frac{d}{dt}\int_0^t E_{2\alpha}\left(-\lambda(t-\tau)^{2\alpha}\right)\varepsilon(\tau)\,d\tau \quad (0 < \alpha < 1), \tag{3.606}$$

where α, λ and ξ are the material constants, $\sigma(t)$ is the stress, and $\varepsilon(t)$ is the strain.

Example 3.22 *The mathematical model for the law of deformation based on the Liouville-Sonine-Caputo type general fractional derivative within the kernel of Mittag-Leffler function can be written as*

$$\sigma(t) = \xi {}_{Ml}^{LSC}D_{0+}^{\alpha,1,\lambda}\varepsilon(t) \quad (0 < \alpha < 1), \tag{3.607}$$

which can be represented in the form

$$\sigma(t) = \xi \int_0^t E_{\alpha}\left(-\lambda(t-\tau)^{\alpha}\right)\varepsilon^{(1)}(\tau)\,d\tau \quad (0 < \alpha < 1), \tag{3.608}$$

where α $(0 < \alpha < 1)$, λ and ξ are the material constants, $\sigma(t)$ is the stress, and $\varepsilon(t)$ is the strain.

In particular, when $\lambda = 1$, the mathematical model for the law of deformation based on the Liouville-Sonine-Caputo type general fractional derivative within the kernel of Mittag-Leffler function can be expressed as (see [301, 302])

$$\sigma(t) = \xi {}_{Ml}^{LSC}D_{0+}^{\alpha,1,1}\varepsilon(t) \quad (0 < \alpha < 1), \tag{3.609}$$

which is expressed as

$$\sigma(t) = \xi \int_0^t E_{\alpha}\left(-(t-\tau)^{\alpha}\right)\varepsilon^{(1)}(\tau)\,d\tau \quad (0 < \alpha < 1), \tag{3.610}$$

where α $(0 < \alpha < 1)$, λ and ξ are the material constants, $\sigma(t)$ is the stress, and $\varepsilon(t)$ is the strain.

Example 3.23 *The mathematical model for the law of deformation based on the Riemann-Liouville type general fractional derivative within the kernel of Mittag-Leffler function can be given as*

$$\sigma(t) = \xi {}_{Ml}^{RL}D_{0+}^{\alpha,\lambda}\varepsilon(t) \quad (0 < \alpha < 1), \tag{3.611}$$

which yields that

$$\sigma(t) = \xi \frac{d}{dt} \int_0^t E_\alpha\left(-\lambda (t-\tau)^\alpha\right) \varepsilon(\tau)\, d\tau, \tag{3.612}$$

where α $(0 < \alpha < 1)$, λ *and* ξ *are the material constants,* $\sigma(t)$ *is the stress, and* $\varepsilon(t)$ *is the strain.*

In particular, when $\lambda = 1$, the mathematical model for the law of deformation based on the Riemann-Liouville type general fractional derivative within the kernel of Mittag-Leffler function can be given as (see [302])

$$\sigma(t) = \xi {}_{Ml}^{RL} D_{0+}^{\alpha,1} \varepsilon(t) \quad (0 < \alpha < 1), \tag{3.613}$$

which becomes

$$\sigma(t) = \xi \frac{d}{dt} \int_0^t E_\alpha\left(-(t-\tau)^\alpha\right) \varepsilon(\tau)\, d\tau, \tag{3.614}$$

where α $(0 < \alpha < 1)$, λ and ξ are the material constants, $\sigma(t)$ is the stress, and $\varepsilon(t)$ is the strain.

Example 3.24 *The mathematical model for the law of deformation based on the general fractional derivative involving the kernel of Mittag-Leffler function with the negative power law in the sense of Riemann-Liouville type can be written as*

$$\sigma(t) = \xi {}_{Mlr}^{RL} D_{0+}^{\alpha,1,\lambda} \varepsilon(t) \quad (0 < \alpha < 1), \tag{3.615}$$

which can be expressed as

$$\sigma(t) = \xi \frac{d}{dt} \int_0^t E_\alpha\left(-\lambda |t-\tau|^\alpha\right) \varepsilon(\tau)\, d\tau, \tag{3.616}$$

where α $(0 < \alpha < 1)$, λ *and* ξ *are the material constants,* $\sigma(t)$ *is the stress, and* $\varepsilon(t)$ *is the strain.*

Example 3.25 *The mathematical model for the law of deformation based on the general fractional derivative involving the kernel of Mittag-Leffler function with the negative power law in the sense of Liouville-Sonine-Caputo type can be written as*

$$\sigma(t) = \xi {}_{Mlr}^{LSC} D_{a+}^{\alpha,1,\lambda} \varepsilon(t) \quad (0 < \alpha < 1), \tag{3.617}$$

which implies that

$$\sigma(t) = \xi \int_0^t E_\alpha\left(-\lambda |t-\tau|^\alpha\right) \varepsilon^{(1)}(\tau)\, d\tau, \tag{3.618}$$

where α $(0 < \alpha < 1)$, λ *and* ξ *are the material constants,* $\sigma(t)$ *is the stress, and* $\varepsilon(t)$ *is the strain.*

Example 3.26 *The mathematical model for the law of deformation based on the Liouville-Sonine-Caputo type general fractional derivative within the kernel of Mittag-Leffler function with respect to another function can be expressed as*

$$\sigma(t) = \xi_{Ml}^{LSC} D_{0+,h}^{\alpha,1,\lambda} \varepsilon(t) \quad (0 < \alpha < 1), \tag{3.619}$$

which becomes

$$\sigma(t) = \xi \int_0^t E_\alpha\left(-\lambda\left(h(t) - h(\tau)\right)^\alpha\right) \left(\frac{1}{h^{(1)}(\tau)} \varepsilon_h^{(1)}(\tau)\right)(\tau)\, h^{(1)}(\tau)\, d\tau \quad (0<\alpha<1), \tag{3.620}$$

where α $(0 < \alpha < 1)$, λ *and* ξ *are the material constants,* $h^{(1)}(t) > 0$, $\sigma(t)$ *is the stress, and* $\varepsilon(t)$ *is the strain.*

Example 3.27 *The mathematical model for the law of deformation based on the Riemann-Liouville type general fractional derivative within the kernel of Mittag-Leffler function with respect to another function can be written as*

$$\sigma(t) = \xi_{Ml}^{RL} D_{0+,h}^{\alpha,1,\lambda} \varepsilon(t) \quad (0 < \alpha < 1), \tag{3.621}$$

which can be repeated as

$$\sigma(t) = \xi \left(\frac{1}{h^{(1)}(t)} \frac{d}{dt}\right) \int_0^t E_\alpha\left(-\lambda\left(h(t) - h(\tau)\right)^\alpha\right) f(\tau)\, h^{(1)}(\tau)\, d\tau, \tag{3.622}$$

where α $(0 < \alpha < 1)$, λ *and* ξ *are the material constants,* $h^{(1)}(t) > 0$, $\sigma(t)$ *is the stress, and* $\varepsilon(t)$ *is the strain.*

Example 3.28 *The mathematical model for the law of deformation based on the general fractional derivative containing the kernel of the Gorenflo-Mainardi function via Wiman function in the sense of Riemann-Liouville type can be illustrated as*

$$\sigma(t) = \xi \left({}_{GM}^{RL} D_{0+}^{\alpha,\gamma,\lambda} \varepsilon\right)(t) \quad (0 < \alpha < 1), \tag{3.623}$$

which is rewritten as

$$\sigma(t) = \xi \frac{d}{dt} \int_0^t G_{\alpha,\gamma}\left(-\lambda (t - \tau)^\alpha\right) \varepsilon(\tau)\, d\tau, \tag{3.624}$$

where α $(0 < \alpha < 1)$, γ, λ *and* ξ *are the material constants,* $\sigma(t)$ *is the stress, and* $\varepsilon(t)$ *is the strain.*

In particular, when $\lambda = 1$, the mathematical model for the law of deformation based on the general fractional derivative containing the kernel of the Gorenflo-Mainardi function via Wiman function in the sense of Riemann-Liouville type can be illustrated as

$$\sigma(t) = \xi \left({}^{RL}_{GM}D^{\alpha,\gamma,\lambda}_{0+} \varepsilon \right)(t) \quad (0 < \alpha < 1), \tag{3.625}$$

which is rewritten as

$$\sigma(t) = \xi \frac{d}{dt} \int_0^t G_{\alpha,\gamma}\left(-(t-\tau)^{\alpha}\right) \varepsilon(\tau)\, d\tau, \tag{3.626}$$

where α $(0 < \alpha < 1)$, γ and ξ are the material constants, $\sigma(t)$ is the stress, and $\varepsilon(t)$ is the strain.

Example 3.29 *The mathematical model for the law of deformation based on the general fractional derivative containing the kernel of the Gorenflo-Mainardi function via Wiman function in the sense of Liouville-Sonine type can be given as*

$$\sigma(t) = \xi \int_0^t G_{\alpha,\gamma}\left(-\lambda (t-\tau)^{\alpha}\right) \varepsilon^{(1)}(\tau)\, d\tau, \tag{3.627}$$

where α $(0 < \alpha < 1)$, γ, λ *and* ξ *are the material constants,* $\sigma(t)$ *is the stress, and* $\varepsilon(t)$ *is the strain.*

In particular, when $\lambda = 1$, the mathematical model for the law of deformation based on the general fractional derivative containing the kernel of the Gorenflo-Mainardi function via Wiman function in the sense of Liouville-Sonine type can be given as (see [302])

$$\sigma(t) = \xi \int_0^t G_{\alpha,\gamma}\left(-(t-\tau)^{\alpha}\right) \varepsilon^{(1)}(\tau)\, d\tau, \tag{3.628}$$

where α $(0 < \alpha < 1)$, γ and ξ are the material constants, $\sigma(t)$ is the stress, and $\varepsilon(t)$ is the strain.

Example 3.30 *The mathematical model for the law of deformation based on the Hilfer type general fractional derivative containing the kernel of the Gorenflo-Mainardi function via Wiman function can be illustrated as*

$$\sigma(t) = \xi \int_0^t G_{\alpha,\gamma}\left(-\lambda (t-\tau)^{\alpha}\right) \left(I^{\beta(1-\alpha)}_{a+} \frac{d}{d\tau} \left(I^{(1-\beta)(1-\alpha)}_{a+} \varepsilon \right) \right)(\tau)\, d\tau, \tag{3.629}$$

where α $(0 < \alpha < 1)$, β $(0 \leq \alpha \leq 1)$, γ, λ *and* ξ *are the material constants,* $\sigma(t)$ *is the stress, and* $\varepsilon(t)$ *is the strain.*

Example 3.31 *The mathematical model for the law of deformation based on the general fractional derivative within the Miller-Ross kernel in the sense of Riemann-Liouville type can be illustrated as*

$$\sigma(t) = \xi \frac{d}{dt} \int_0^t \mathrm{M}_\alpha\left(-\lambda(t-\tau)^\alpha\right) \varepsilon(\tau)\, d\tau, \tag{3.630}$$

where α $(0 < \alpha < 1)$, λ *and* ξ *are the material constants,* $\sigma(t)$ *is the stress, and* $\varepsilon(t)$ *is the strain.*

Example 3.32 *The mathematical model for the law of deformation based on the general fractional derivative within the Miller-Ross kernel in the sense of Liouville-Sonine type can be considered as*

$$\sigma(t) = \xi \int_0^t \mathrm{M}_\alpha\left(-\lambda(t-\tau)^\alpha\right) \varepsilon^{(1)}(\tau)\, d\tau, \tag{3.631}$$

where α $(0 < \alpha < 1)$, λ *and* ξ *are the material constants,* $\sigma(t)$ *is the stress, and* $\varepsilon(t)$ *is the strain.*

Example 3.33 *The mathematical model for the law of deformation based on the Hilfer type general fractional derivative containing the kernel of the Miller-Ross function can be illustrated as*

$$\sigma(t) = \xi \int_0^t \mathrm{M}_\alpha\left(-\lambda(t-\tau)^\alpha\right) \left(I_{a+}^{\beta(1-\alpha)} \frac{d}{d\tau}\left(I_{a+}^{(1-\beta)(1-\alpha)}\varepsilon\right)\right)(\tau)\, d\tau, \tag{3.632}$$

where α $(0 < \alpha < 1)$, β $(0 \le \alpha \le 1)$, γ, λ *and* ξ *are the material constants,* $\sigma(t)$ *is the stress, and* $\varepsilon(t)$ *is the strain.*

Example 3.34 *The mathematical model for the law of deformation based on the Riemann-Liouville type general fractional derivative involving the one-parametric Lorenzo-Hartley kernel can be illustrated as*

$$\sigma(t) = \xi \frac{d}{dt} \int_0^t F_\alpha\left(-\lambda(t-\tau)^\alpha\right) \varepsilon(\tau)\, d\tau, \tag{3.633}$$

where α $(0 < \alpha < 1)$, λ *and* ξ *are the material constants,* $\sigma(t)$ *is the stress, and* $\varepsilon(t)$ *is the strain.*

Example 3.35 *The mathematical model for the law of deformation based on the Liouville-Sonine type general fractional derivative involving the one-parametric Lorenzo-Hartley kernel can be considered as*

$$\sigma(t) = \xi \int_0^t F_\alpha\left(-\lambda(t-\tau)^\alpha\right) \varepsilon^{(1)}(\tau)\, d\tau, \tag{3.634}$$

where α $(0 < \alpha < 1)$, λ *and* ξ *are the material constants,* $\sigma(t)$ *is the stress, and* $\varepsilon(t)$ *is the strain.*

Example 3.36 *The mathematical model for the law of deformation based on the Hilfer type general fractional derivative involving the one-parametric Lorenzo-Hartley kernel can be illustrated as*

$$\sigma(t) = \xi \int_0^t F_\alpha\left(-\lambda(t-\tau)^\alpha\right)\left(I_{a+}^{\beta(1-\alpha)}\frac{d}{d\tau}\left(I_{a+}^{(1-\beta)(1-\alpha)}\varepsilon\right)\right)(\tau)\,d\tau, \quad (3.635)$$

where α $(0 < \alpha < 1)$, β $(0 \le \alpha \le 1)$, γ, λ *and* ξ *are the material constants,* $\sigma(t)$ *is the stress, and* $\varepsilon(t)$ *is the strain.*

Example 3.37 *The mathematical model for the law of deformation based on the Riemann-Liouville type general fractional derivative within the generalized Prabhakar kernel via Prabhakar function can be illustrated as*

$$\sigma(t) = \xi \frac{d}{dt}\int_0^t (t-\tau)^{\upsilon-1} E_{\alpha,\upsilon}^{\varphi}\left(-\lambda(t-\tau)^\alpha\right)\varepsilon(\tau)\,d\tau, \quad (3.636)$$

where α $(0 < \alpha < 1)$, λ *and* ξ *are the material constants,* $\sigma(t)$ *is the stress, and* $\varepsilon(t)$ *is the strain.*

Example 3.38 *The mathematical model for the law of deformation based on the Liouville-Sonine type general fractional derivative within the generalized Prabhakar kernel via Prabhakar function can be considered as*

$$\sigma(t) = \xi \int_0^t (t-\tau)^{\upsilon-1} E_{\alpha,\upsilon}^{\varphi}\left(-\lambda(t-\tau)^\alpha\right)\varepsilon^{(1)}(\tau)\,d\tau, \quad (3.637)$$

where α $(0 < \alpha < 1)$, λ *and* ξ *are the material constants,* $\sigma(t)$ *is the stress, and* $\varepsilon(t)$ *is the strain.*

Example 3.39 *The mathematical model for the law of deformation based on the Hilfer type general fractional derivative within the generalized Prabhakar kernel via Prabhakar function can be illustrated as*

$$\sigma(t) = \xi \int_0^t (t-\tau)^{\upsilon-1} E_{\alpha,\upsilon}^{\varphi}\left(-\lambda(t-\tau)^\alpha\right)\left(I_{a+}^{\beta(1-\alpha)}\frac{d}{d\tau}\left(I_{a+}^{(1-\beta)(1-\alpha)}\varepsilon\right)\right)(\tau)\,d\tau, \quad (3.638)$$

where α $(0 < \alpha < 1)$, β $(0 \le \alpha \le 1)$, γ, λ *and* ξ *are the material constants,* $\sigma(t)$ *is the stress, and* $\varepsilon(t)$ *is the strain.*

Example 3.40 *The mathematical model for the law of deformation based on the Riemann-Liouville type general fractional derivative involving the generalized Prabhakar kernel via Prabhakar function with respect to another function can be illustrated as*

$$\sigma(t) = \xi\left(\frac{1}{h^{(1)}(t)}\frac{d}{dt}\right)\int_0^t (h(t)-h(\tau))^{\upsilon-1} E^{\varphi}_{\alpha,\upsilon}\left(-\lambda(h(t)-h(\tau))^{\alpha}\right)\varepsilon(\tau)\, h^{(1)}(\tau)\, d\tau, \tag{3.639}$$

where α $(0<\alpha<1)$, λ *and* ξ *are the material constants,* $\sigma(t)$ *is the stress, and* $\varepsilon(t)$ *is the strain.*

Example 3.41 *The mathematical model for the law of deformation based on the Liouville-Sonine type general fractional derivative involving the generalized Prabhakar kernel via Prabhakar function with respect to another function can be illustrated as*

$$\sigma(t) = \xi\int_0^t (h(t)-h(\tau))^{\upsilon-1} E^{\varphi}_{\alpha,\upsilon}\left(-\lambda(h(t)-h(\tau))^{\alpha}\right)\left(\frac{1}{h^{(1)}(\tau)}\frac{d}{d\tau}\right)\varepsilon_h^{(1)}(\tau)\, d\tau, \tag{3.640}$$

where α $(0<\alpha<1)$, λ *and* ξ *are the material constants,* $\sigma(t)$ *is the stress, and* $\varepsilon(t)$ *is the strain.*

Example 3.42 *The mathematical model for the law of deformation based on the Hilfer type general fractional derivative involving the generalized Prabhakar kernel via Prabhakar function with respect to another function can be illustrated as*

$$\sigma(t) = \xi\int_a^t (h(t)-h(\tau))^{\upsilon-1} E^{\varphi}_{\alpha,\upsilon}\left(-\lambda(h(t)-h(\tau))^{\alpha}\right)\left(D^{\alpha,\beta}_{a+,h}\varepsilon(\tau)\right) d\tau, \tag{3.641}$$

where α $(0<\alpha<1)$, β $(0\le\alpha\le1)$, γ, λ *and* ξ *are the material constants,* $\sigma(t)$ *is the stress, and* $\varepsilon(t)$ *is the strain.*

Chapter 4

Fractional Derivatives of Variable Order and Applications

In this chapter, we investigate the variable-order fractional calculus and the variable-order general fractional calculus involving the kernel of the special functions, such as the Mittag-Leffler function, weak power-law function, Lorenzo-Hartley function, Wiman function, Miller-Ross function and Prabhakar function. The Hilfer type fractional derivatives of variable order, the Riesz type fractional derivatives of variable order, the Feller type fractional derivatives of variable order and the Richard type fractional derivatives of variable order are also considered. The general fractional derivatives and general fractional integrals of the variable order involving the special functions are also discussed. The mathematical models for the relaxation and rheological models are addressed in detail.

4.1 Fractional derivatives of variable order involving the kernel of the singular power function

In this section, we introduce the fractional integrals and fractional derivatives of variable order involving the kernel of the singular power function.

Definition 4.1 *Let $\alpha(t) > 0$ and $-\infty < a < b < \infty$. The left-sided Riemann-Liouville type fractional integral of variable order is defined as (see [95])*

$$\left({}_{RL}I_{a+}^{\alpha(\cdot)} f\right)(t) = \int_a^t \frac{1}{\Gamma(\alpha(\tau))} \frac{f(\tau)}{(t-\tau)^{1-\alpha(\tau)}} d\tau, \tag{4.1}$$

and the right-sided Riemann-Liouville type fractional integral of variable order as

$$\left({}_{RL}I_{b-}^{\alpha(\cdot)} f\right)(t) = \int_t^b \frac{1}{\Gamma(\alpha(\tau))} \frac{f(\tau)}{(\tau-t)^{1-\alpha(\tau)}} d\tau. \tag{4.2}$$

Definition 4.2 *Let $\kappa < \alpha(t) < 1 + \kappa$ and $-\infty < a < b < \infty$. The left-sided Riemann-Liouville type fractional derivative of variable order is defined as*

$$\left({}_{RL}D_{a+}^{\alpha(\cdot)} f\right)(t) = \left(\tfrac{d}{dt}\right)^{\kappa} \left({}_{RL}I_{a+}^{\kappa-\alpha(\cdot)} f\right)(t) = \tfrac{d^{\kappa}}{dt^{\kappa}} \int_a^t \tfrac{1}{\Gamma(\kappa-\alpha(\tau))} \tfrac{f(\tau)}{(t-\tau)^{\alpha(\tau)-\kappa+1}} d\tau, \tag{4.3}$$

and the right-sided Riemann-Liouville type fractional derivative of variable order as

$$\left({}_{RL}D_{b-}^{\alpha(\cdot)} f\right)(t) = \left(-\tfrac{d}{dt}\right)^{\kappa} \left({}_{RL}I_{b-}^{\alpha(\cdot)} f\right)(t) = \left(-\tfrac{d}{dt}\right)^{\kappa} \int_t^b \tfrac{1}{\Gamma(\kappa-\alpha(\tau))} \tfrac{f(\tau)}{(\tau-t)^{\alpha(\tau)-\kappa+1}} d\tau. \tag{4.4}$$

Definition 4.3 *Let $\kappa < \alpha(t) < 1 + \kappa$ and $-\infty < a < b < \infty$. The left-sided Riemann-Liouville type fractional derivative of variable order is defined as (see [310])*

$$\left({}_{RL}^{CGS}D_{a+}^{\alpha(\cdot)} f\right)(t) = \frac{1}{\Gamma(\kappa-\alpha(\tau))} \frac{d^{\kappa}}{dt^{\kappa}} \int_a^t \frac{f(\tau)}{(t-\tau)^{\alpha(\tau)-\kappa+1}} d\tau, \tag{4.5}$$

and the right-sided Riemann-Liouville type fractional derivative of variable order as

$$\left({}_{RL}^{CGS}D_{b-}^{\alpha(\cdot)} f\right)(t) = \frac{1}{\Gamma(\kappa-\alpha(\tau))} \left(-\frac{d}{dt}\right)^{\kappa} \int_t^b \frac{f(\tau)}{(\tau-t)^{\alpha(\tau)-\kappa+1}} d\tau. \tag{4.6}$$

In particular, when $0 < \alpha(t) < 1$ and $-\infty < a < b < \infty$, the left-sided Riemann-Liouville type fractional derivative of variable order is defined as (see [310])

$$\left({}_{RL}D_{a+}^{\alpha(\cdot)} f\right)(t) = \tfrac{d}{dt} \left({}_{RL}I_{a+}^{1-\alpha(\cdot)} f\right)(t) = \tfrac{d}{dt} \int_a^t \tfrac{1}{\Gamma(1-\alpha(\tau))} \tfrac{f(\tau)}{(t-\tau)^{\alpha(\tau)}} d\tau, \tag{4.7}$$

and the right-sided Riemann-Liouville type fractional derivative of variable order as

$$\left({}_{RL}D_{b-}^{\alpha(\cdot)} f\right)(t) = -\tfrac{d}{dt} \left(I_{b-}^{\alpha(\cdot)} f\right)(t) = -\tfrac{d}{dt} \int_t^b \tfrac{1}{\Gamma(1-\alpha(\tau))} \tfrac{f(\tau)}{(\tau-t)^{\alpha(\tau)}} d\tau. \tag{4.8}$$

Definition 4.4 *Let $\alpha(t) \geq 0$ and $\kappa = [\alpha(t)] + 1$. The left-sided Liouville-Sonine-Caputo type fractional derivative of variable order is defined as (see [95])*

$${}_{LSC}D_{+a}^{\alpha(\cdot)} f(t) = I_{+a}^{\alpha(\cdot)} \left[f^{(\kappa)}(t)\right] = \int_a^t \tfrac{1}{\Gamma(\kappa-\alpha(\tau))} \tfrac{f^{(\kappa)}(\tau)}{(t-\tau)^{\alpha(\tau)-\kappa+1}} d\tau, \tag{4.9}$$

and the right-sided Liouville-Sonine-Caputo type fractional derivative of variable order as

$$_{LSC}D_{b-}^{\alpha(\cdot)}f(x)=I_{b-}^{\alpha(\cdot)}\left[f^{(\kappa)}(x)\right]=(-1)^{\kappa}\int_{t}^{b}\frac{1}{\Gamma(\kappa-\alpha(\tau))}\frac{f^{(\kappa)}(\tau)}{(\tau-t)^{\alpha(\tau)-\kappa+1}}d\tau. \tag{4.10}$$

In particular, when $0<\alpha(t)<1$, the left-sided Liouville-Sonine fractional derivative of variable order as (see [95])

$$_{LS}D_{+a}^{\alpha(\cdot)}f(t)=I_{+a}^{\alpha(\cdot)}\left[f^{(1)}(t)\right]=\int_{a}^{t}\frac{1}{\Gamma(1-\alpha(\tau))}\frac{f^{(1)}(\tau)}{(t-\tau)^{\alpha(\tau)}}d\tau, \tag{4.11}$$

the right-sided Liouville-Sonine fractional derivative of variable order as

$$_{LS}D_{b-}^{\alpha(\cdot)}f(t)=I_{b-}^{\alpha(\cdot)}\left[f^{(1)}(t)\right]=-\int_{t}^{b}\frac{1}{\Gamma(1-\alpha(\tau))}\frac{f^{(1)}(\tau)}{(\tau-t)^{\alpha(\tau)}}d\tau. \tag{4.12}$$

More generally, when $0<\alpha(\tau,t)<1$, the left-sided Riemann-Liouville type fractional integral of variable order is defined as (see [311]; also see [312])

$$\left({}^{HL}_{RLq}I_{a+}^{\alpha(\cdot)}f\right)(t)=\int_{a}^{t}\frac{1}{\Gamma(\alpha(\tau,t))}\frac{f(\tau)}{(t-\tau)^{1-\alpha(\tau,t)}}d\tau, \tag{4.13}$$

the right-sided Riemann-Liouville type fractional integral of variable order as (see [312])

$$\left({}^{HL}_{RLq}I_{b-}^{\alpha(\cdot)}f\right)(t)=\int_{t}^{b}\frac{1}{\Gamma(\alpha(\tau,t))}\frac{f(\tau)}{(\tau-t)^{1-\alpha(\tau,t)}}d\tau, \tag{4.14}$$

the left-sided Riemann-Liouville type fractional derivative of variable order is defined as (see [312])

$$\left({}_{RLq}D_{a+}^{\alpha(\cdot)}f\right)(t)=\frac{d}{dt}\left({}_{RLq}I_{a+}^{1-\alpha(\cdot)}f\right)(t)=\frac{d}{dt}\int_{a}^{t}\frac{1}{\Gamma(1-\alpha(\tau,t))}\frac{f(\tau)}{(t-\tau)^{\alpha(\tau,t)}}d\tau, \tag{4.15}$$

the right-sided Riemann-Liouville type fractional derivative of variable order as

$$\left({}_{RLq}D_{b-}^{\alpha(\cdot)}f\right)(t)=-\frac{d}{dt}\left({}_{RLq}I_{b-}^{\alpha(\cdot)}f\right)(t)=-\frac{d}{dt}\int_{t}^{b}\frac{1}{\Gamma(1-\alpha(\tau,t))}\frac{f(\tau)}{(\tau-t)^{\alpha(\tau,t)}}d\tau, \tag{4.16}$$

the left-sided Riemann-Liouville type fractional derivative of variable order is defined as (see [313])

$$\left({}_{RLq}D_{a+}^{\alpha(\cdot)}f\right)(t)=\frac{1}{\Gamma(1-\alpha(\tau,t))}\frac{d}{dt}\int_{a}^{t}\frac{f(\tau)}{(t-\tau)^{\alpha(\tau,t)}}d\tau, \tag{4.17}$$

the right-sided Riemann-Liouville type fractional derivative of variable order as (see [313])

$$\left({}_{RLq}D_{b-}^{\alpha(\cdot)}f\right)(t)=\frac{1}{\Gamma\left(1-\alpha\left(\tau,t\right)\right)}\left(-\frac{d}{dt}\right)\int\limits_{t}^{b}\frac{f\left(\tau\right)}{\left(\tau-t\right)^{\alpha(\tau,t)}}d\tau, \tag{4.18}$$

the left-sided Liouville-Sonine fractional derivative of variable order as (see [312])

$${}_{LSq}D_{+a}^{\alpha(\cdot)}f\left(t\right)={}_{RLq}\,I_{+a}^{\alpha(\cdot)}\left[f^{(1)}\left(t\right)\right]=\int\limits_{a}^{t}\frac{1}{\Gamma(1-\alpha(\tau,t))}\frac{f^{(1)}(\tau)}{(t-\tau)^{\alpha(\tau,t)}}d\tau, \tag{4.19}$$

and the right-sided Liouville-Sonine fractional derivative of variable order as (see [312])

$${}_{LSq}D_{b-}^{\alpha(\cdot)}f\left(t\right)={}_{RLq}\,I_{b-}^{\alpha(\cdot)}\left[f^{(1)}\left(t\right)\right]=-\int\limits_{t}^{b}\frac{1}{\Gamma(1-\alpha(\tau,t))}\frac{f^{(1)}(\tau)}{(\tau-t)^{\alpha(\tau,t)}}d\tau. \tag{4.20}$$

In particular, when $\alpha\left(\tau,t\right)=\alpha\left(t\right)$, the left-sided Riemann-Liouville type fractional integral of variable order is defined as (see [93])

$$\left({}_{RLqs}I_{a+}^{\alpha(\cdot)}f\right)(t)=\int\limits_{a}^{t}\frac{1}{\Gamma(\alpha(t))}\frac{f(\tau)}{(t-\tau)^{1-\alpha(t)}}d\tau=\frac{1}{\Gamma(\alpha(t))}\int\limits_{a}^{t}\frac{f(\tau)}{(t-\tau)^{1-\alpha(t)}}d\tau, \tag{4.21}$$

the right-sided Riemann-Liouville type fractional integral of variable order as

$$\left({}_{RLqs}I_{b-}^{\alpha(\cdot)}f\right)(t)=\int\limits_{t}^{b}\frac{1}{\Gamma(\alpha(t))}\frac{f(\tau)}{(\tau-t)^{1-\alpha(t)}}d\tau=\frac{1}{\Gamma(\alpha(t))}\int\limits_{t}^{b}\frac{f(\tau)}{(\tau-t)^{1-\alpha(t)}}d\tau, \tag{4.22}$$

the left-sided Riemann-Liouville type fractional derivative of variable order is defined as (see [93])

$$\left({}_{RLqs}^{SR}D_{a+}^{\alpha(\cdot)}f\right)(t)=\frac{1}{\Gamma\left(1-\alpha\left(t\right)\right)}\frac{d}{dt}\int\limits_{a}^{t}\frac{f\left(\tau\right)}{\left(t-\tau\right)^{\alpha(t)}}d\tau, \tag{4.23}$$

the right-sided Riemann-Liouville type fractional derivative of variable order as

$$\left({}_{RLqs}D_{b-}^{\alpha(\cdot)}f\right)(t)=-\frac{1}{\Gamma\left(1-\alpha\left(t\right)\right)}\frac{d}{dt}\int\limits_{t}^{b}\frac{f\left(\tau\right)}{\left(t-\tau\right)^{\alpha(t)}}d\tau, \tag{4.24}$$

the left-sided Riemann-Liouville type fractional derivative of variable order is defined as

$$\left({}_{RLqs}D_{a+}^{\alpha(\cdot)}f\right)(t)=\frac{d}{dt}\left({}_{RLqs}I_{a+}^{1-\alpha(\cdot)}f\right)(t)=\frac{d}{dt}\left(\frac{1}{\Gamma(1-\alpha(t))}\int\limits_{a}^{t}\frac{f(\tau)}{(t-\tau)^{\alpha(t)}}d\tau\right), \tag{4.25}$$

the right-sided Riemann-Liouville type fractional derivative of variable order as

$$\left({}_{RLqs}D_{b-}^{\alpha(\cdot)}f\right)(t) = -\frac{d}{dt}\left({}_{RLqs}I_{b-}^{\alpha(\cdot)}f\right)(t) = -\frac{d}{dt}\left(\frac{1}{\Gamma(1-\alpha(t))}\int_t^b \frac{f(\tau)}{(\tau-t)^{\alpha(t)}}d\tau\right), \tag{4.26}$$

the left-sided Liouville-Sonine fractional derivative of variable order as (see [314]; with the complement function, see [95])

$${}_{LSqs}D_{+a}^{\alpha(\cdot)}f(t) = {}_{RLqs}I_{+a}^{\alpha(\cdot)}\left[f^{(1)}(t)\right] = \frac{1}{\Gamma(1-\alpha(t))}\int_a^t \frac{f^{(1)}(\tau)}{(t-\tau)^{\alpha(t)}}d\tau, \tag{4.27}$$

and the right-sided Liouville-Sonine fractional derivative of variable order as

$${}_{LSqs}D_{b-}^{\alpha(\cdot)}f(t) = {}_{RLqs}I_{b-}^{\alpha(\cdot)}\left[f^{(1)}(t)\right] = -\frac{1}{\Gamma(1-\alpha(t))}\int_t^b \frac{f^{(1)}(\tau)}{(\tau-t)^{\alpha(t)}}d\tau. \tag{4.28}$$

For more information of the theory and applications of the variable-order fractional calculus within the kernel of the singular power function, see [315, 316, 317, 318, 319, 320, 321, 322, 323].

4.2 Riesz, Feller and Richard types fractional derivatives of variable order

In this section, we introduce the Riesz type fractional derivatives of variable order (see [313]), proposed by Zayernouri and Karniadakis in 2015, and give the Feller and Richard types fractional derivatives of variable order based on the Riemann-Liouville fractional derivatives. Meanwhile, we present the definitions of the Riesz, Feller and Richard types fractional derivatives of variable order based on the Liouville-Sonine fractional derivatives.

Definition 4.5 *The Riesz type fractional derivatives of variable order based on the Riemann-Liouville fractional derivative of variable order is defined as (see [313])*

$$\begin{aligned}\left({}_{ZK}^{Ri}D_{[a,b]}^{\alpha(\cdot)}f\right)(t) &= -\frac{1}{2\cos(\pi\alpha(t)/2)}\times\left(\left({}_{RLqs}D_{a+}^{\alpha(\cdot)}f\right)(t)+\left({}_{RLqs}D_{b-}^{\alpha(\cdot)}f\right)(t)\right)\\ &= -\frac{1}{2\cos(\pi\alpha(t)/2)}\times\frac{1}{\Gamma(1-\alpha(t))}\frac{d}{dt}\int_a^t \frac{f(\tau)}{(t-\tau)^{\alpha(t)}}d\tau\\ &+\frac{1}{2\cos(\pi\alpha(t)/2)}\times\frac{1}{\Gamma(1-\alpha(t))}\frac{d}{dt}\int_t^b \frac{f(\tau)}{(\tau-t)^{\alpha(t)}}d\tau,\end{aligned} \tag{4.29}$$

where

$$\left({}_{RLqs}D_{a+}^{\alpha(\cdot)}f\right)(t) = \frac{1}{\Gamma(1-\alpha(t))}\frac{d}{dt}\int_a^t \frac{f(\tau)}{(t-\tau)^{\alpha(t)}}d\tau \tag{4.30}$$

and

$$\left({}_{RLqs}D_{b-}^{\alpha(\cdot)}f\right)(t) = -\frac{1}{\Gamma(1-\alpha(t))}\frac{d}{dt}\int_t^b \frac{f(\tau)}{(\tau-t)^{\alpha(t)}}d\tau. \tag{4.31}$$

More generally, the Riesz type fractional derivatives of variable order based on the Riemann-Liouville fractional derivative of variable order is defined as

$$\begin{aligned}&\left({}_{GZK}^{Ri}D_{[a,b]}^{\alpha(\cdot)}f\right)(t) = -\tfrac{1}{2\cos(\pi\alpha(t)/2)} \times \left(\left({}_{RLq}D_{a+}^{\alpha(\cdot)}f\right)(t) + \left({}_{RLq}D_{b-}^{\alpha(\cdot)}f\right)(t)\right)\\&= -\tfrac{1}{2\cos(\pi\alpha(t)/2)} \times \tfrac{d}{dt}\int_a^t \tfrac{1}{\Gamma(1-\alpha(\tau,t))}\tfrac{f(\tau)}{(t-\tau)^{\alpha(\tau,t)}}d\tau\\&+\tfrac{1}{2\cos(\pi\alpha(t)/2)} \times \tfrac{d}{dt}\int_t^b \tfrac{1}{\Gamma(1-\alpha(\tau,t))}\tfrac{f(\tau)}{(\tau-t)^{\alpha(\tau,t)}}d\tau,\end{aligned} \tag{4.32}$$

where

$$\left({}_{RLq}D_{a+}^{\alpha(\cdot)}f\right)(t) = \frac{d}{dt}\int_a^t \frac{1}{\Gamma(1-\alpha(\tau,t))}\frac{f(\tau)}{(t-\tau)^{\alpha(\tau,t)}}d\tau, \tag{4.33}$$

and

$$\left({}_{RLq}D_{b-}^{\alpha(\cdot)}f\right)(t) = -\frac{d}{dt}\int_t^b \frac{1}{\Gamma(1-\alpha(\tau,t))}\frac{f(\tau)}{(\tau-t)^{\alpha(\tau,t)}}d\tau. \tag{4.34}$$

In particular, when $\alpha(\tau,t) = \alpha(t)$, the Riesz type fractional derivatives of variable order based on the Riemann-Liouville type fractional derivative of variable order is defined as

$$\begin{aligned}&\left({}_{SZK}^{Ri}D_{[a,b]}^{\alpha(\cdot)}f\right)(t) = -\tfrac{1}{2\cos(\pi\alpha(t)/2)} \times \left(\left({}_{RL}D_{a+}^{\alpha(\cdot)}f\right)(t) + \left({}_{RL}D_{b-}^{\alpha(\cdot)}f\right)(t)\right)\\&= -\tfrac{1}{2\cos(\pi\alpha(t)/2)} \times \tfrac{d}{dt}\int_a^t \tfrac{1}{\Gamma(1-\alpha(\tau))}\tfrac{f(\tau)}{(t-\tau)^{\alpha(\tau)}}d\tau\\&+\tfrac{1}{2\cos(\pi\alpha(t)/2)} \times \tfrac{d}{dt}\int_t^b \tfrac{1}{\Gamma(1-\alpha(\tau))}\tfrac{f(\tau)}{(\tau-t)^{\alpha(\tau)}}d\tau,\end{aligned} \tag{4.35}$$

where

$$\left({}_{RL}D_{a+}^{\alpha(\cdot)}f\right)(t) = \frac{d}{dt}\int_a^t \frac{1}{\Gamma(1-\alpha(\tau))}\frac{f(\tau)}{(t-\tau)^{\alpha(\tau)}}d\tau, \tag{4.36}$$

and

$$\left({}_{RL}D_{b-}^{\alpha(\cdot)}f\right)(t) = -\frac{d}{dt}\int_t^b \frac{1}{\Gamma(1-\alpha(\tau))}\frac{f(\tau)}{(\tau-t)^{\alpha(\tau)}}d\tau. \tag{4.37}$$

Definition 4.6 *The Riesz type fractional derivatives of variable order based on the Liouville-Sonine fractional derivative of variable order is defined as*

$$\begin{aligned}&\left({}^{Ri}_{LSqs}D^{\alpha(\cdot)}_{[a,b]}f\right)(t) = -\tfrac{1}{2\cos(\pi\alpha(t)/2)} \times \left(\left({}_{LSqs}D^{\alpha(\cdot)}_{a+}f\right)(t) + \left({}_{LSqs}D^{\alpha(\cdot)}_{b-}f\right)(t)\right)\\&= -\tfrac{1}{2\cos(\pi\alpha(t)/2)} \times \tfrac{1}{\Gamma(1-\alpha(t))}\int_a^t \tfrac{f^{(1)}(\tau)}{(t-\tau)^{\alpha(t)}}d\tau\\&+\tfrac{1}{2\cos(\pi\alpha(t)/2)} \times \tfrac{1}{\Gamma(1-\alpha(t))}\int_t^b \tfrac{f^{(1)}(\tau)}{(\tau-t)^{\alpha(t)}}d\tau,\end{aligned} \tag{4.38}$$

where

$$\left({}_{LSqs}D^{\alpha(\cdot)}_{a+}f\right)(t) = \frac{1}{\Gamma(1-\alpha(t))}\int_a^t \frac{f^{(1)}(\tau)}{(t-\tau)^{\alpha(t)}}d\tau \tag{4.39}$$

and

$$\left({}_{LSqs}D^{\alpha(\cdot)}_{b-}f\right)(t) = -\frac{1}{\Gamma(1-\alpha(t))}\int_t^b \frac{f^{(1)}(\tau)}{(\tau-t)^{\alpha(t)}}d\tau. \tag{4.40}$$

More generally, the Riesz type fractional derivatives of variable order based on the Liouville-Sonine fractional derivative of variable order is defined as

$$\begin{aligned}&\left({}^{Ri}_{GLSq}D^{\alpha(\cdot)}_{[a,b]}f\right)(t) = -\tfrac{1}{2\cos(\pi\alpha(t)/2)} \times \left(\left({}_{LSq}D^{\alpha(\cdot)}_{a+}f\right)(t) + \left({}_{LSq}D^{\alpha(\cdot)}_{b-}f\right)(t)\right)\\&= -\tfrac{1}{2\cos(\pi\alpha(t)/2)} \times \int_a^t \tfrac{1}{\Gamma(1-\alpha(\tau,t))}\tfrac{f^{(1)}(\tau)}{(t-\tau)^{\alpha(\tau,t)}}d\tau\\&+\tfrac{1}{2\cos(\pi\alpha(t)/2)} \times \int_t^b \tfrac{1}{\Gamma(1-\alpha(\tau,t))}\tfrac{f^{(1)}(\tau)}{(\tau-t)^{\alpha(\tau,t)}}d\tau,\end{aligned} \tag{4.41}$$

where

$$\left({}_{LSq}D^{\alpha(\cdot)}_{a+}f\right)(t) = \int_a^t \frac{1}{\Gamma(1-\alpha(\tau,t))}\frac{f^{(1)}(\tau)}{(t-\tau)^{\alpha(\tau,t)}}d\tau, \tag{4.42}$$

and

$$\left({}_{RLq}D^{\alpha(\cdot)}_{b-}f\right)(t) = -\int_t^b \frac{1}{\Gamma(1-\alpha(\tau,t))}\frac{f^{(1)}(\tau)}{(\tau-t)^{\alpha(\tau,t)}}d\tau. \tag{4.43}$$

In particular, when $\alpha(\tau,t) = \alpha(t)$, the Riesz type fractional derivatives of variable order based on the Liouville-Sonine fractional derivative of variable order is defined as

$$\begin{aligned}&\left({}^{Ri}_{GLC}D^{\alpha(\cdot)}_{[a,b]}f\right)(t) = -\tfrac{1}{2\cos(\pi\alpha(t)/2)} \times \left(\left({}_{GLC}D^{\alpha(\cdot)}_{a+}f\right)(t) + \left({}_{GLC}D^{\alpha(\cdot)}_{b-}f\right)(t)\right)\\&= -\tfrac{1}{2\cos(\pi\alpha(t)/2)} \times \int_a^t \tfrac{1}{\Gamma(1-\alpha(\tau))}\tfrac{f^{(1)}(\tau)}{(t-\tau)^{\alpha(\tau)}}d\tau\\&+\tfrac{1}{2\cos(\pi\alpha(t)/2)} \times \int_t^b \tfrac{1}{\Gamma(1-\alpha(\tau))}\tfrac{f^{(1)}(\tau)}{(\tau-t)^{\alpha(\tau)}}d\tau,\end{aligned} \tag{4.44}$$

where

$$\left({}_{GLC}D_{a+}^{\alpha(\cdot)}f\right)(t)=\int_a^t\frac{1}{\Gamma\left(1-\alpha\left(\tau\right)\right)}\frac{f^{(1)}\left(\tau\right)}{\left(t-\tau\right)^{\alpha(\tau)}}d\tau, \tag{4.45}$$

and

$$\left({}_{GLC}D_{b-}^{\alpha(\cdot)}f\right)(t)=-\int_t^b\frac{1}{\Gamma\left(1-\alpha\left(\tau\right)\right)}\frac{f^{(1)}\left(\tau\right)}{\left(\tau-t\right)^{\alpha(\tau)}}d\tau. \tag{4.46}$$

Definition 4.7 *The Feller type fractional derivatives of variable order based on the Riemann-Liouville fractional derivative of variable order is defined as*

$$\begin{aligned}
&\left({}_{ZK}^{Fe}D_{[a,b]}^{\alpha(\cdot)}f\right)(t)\\
&=-\left(H_+\left(\vartheta,\alpha\left(t\right)\right)\left({}_{RLqs}D_{a+}^{\alpha(\cdot)}f\right)(t)+H_-\left(\vartheta,\alpha\left(t\right)\right)\left({}_{RLqs}D_{b-}^{\alpha(\cdot)}f\right)(t)\right)\\
&=-\frac{\sin((\alpha(t)+\vartheta)\pi/2)}{\sin(\pi\vartheta)}\times\frac{1}{\Gamma(1-\alpha(t))}\frac{d}{dt}\int_a^t\frac{f(\tau)}{(t-\tau)^{\alpha(t)}}d\tau\\
&+\frac{\sin((\alpha(t)-\vartheta)\pi/2)}{\sin(\pi\vartheta)}\times\frac{1}{\Gamma(1-\alpha(t))}\frac{d}{dt}\int_t^b\frac{f(\tau)}{(\tau-t)^{\alpha(t)}}d\tau,
\end{aligned} \tag{4.47}$$

where

$$\left({}_{RLqs}D_{a+}^{\alpha(\cdot)}f\right)(t)=\frac{1}{\Gamma\left(1-\alpha\left(t\right)\right)}\frac{d}{dt}\int_a^t\frac{f\left(\tau\right)}{\left(t-\tau\right)^{\alpha(t)}}d\tau \tag{4.48}$$

and

$$\left({}_{RLqs}D_{b-}^{\alpha(\cdot)}f\right)(t)=-\frac{1}{\Gamma\left(1-\alpha\left(t\right)\right)}\frac{d}{dt}\int_t^b\frac{f\left(\tau\right)}{\left(\tau-t\right)^{\alpha(t)}}d\tau. \tag{4.49}$$

Definition 4.8 *The Feller type fractional derivatives of variable order based on the Liouville-Sonine fractional derivative of variable order is defined as*

$$\begin{aligned}
&\left({}_{ZKLC}^{Fe}D_{[a,b]}^{\alpha(\cdot)}f\right)(t)\\
&=-\left(H_+\left(\vartheta,\alpha\left(t\right)\right)\left({}_{RLqs}D_{a+}^{\alpha(\cdot)}f\right)(t)+H_-\left(\vartheta,\alpha\left(t\right)\right)\left({}_{RLqs}D_{b-}^{\alpha(\cdot)}f\right)(t)\right)\\
&=-\frac{\sin((\alpha(t)+\vartheta)\pi/2)}{\sin(\pi\vartheta)}\times\frac{1}{\Gamma(1-\alpha(t))}\int_a^t\frac{f^{(1)}(\tau)}{(t-\tau)^{\alpha(t)}}d\tau\\
&+\frac{\sin((\alpha(t)-\vartheta)\pi/2)}{\sin(\pi\vartheta)}\times\frac{1}{\Gamma(1-\alpha(t))}\int_t^b\frac{f^{(1)}(\tau)}{(\tau-t)^{\alpha(t)}}d\tau,
\end{aligned} \tag{4.50}$$

where

$$\left({}_{LSqs}D_{a+}^{\alpha(\cdot)}f\right)(t)=\frac{1}{\Gamma\left(1-\alpha\left(t\right)\right)}\int_a^t\frac{f^{(1)}\left(\tau\right)}{\left(t-\tau\right)^{\alpha(t)}}d\tau \tag{4.51}$$

and

$$\left({}_{LSqs}D_{b-}^{\alpha(\cdot)}f\right)(t) = -\frac{1}{\Gamma(1-\alpha(t))}\int_t^b \frac{f^{(1)}(\tau)}{(\tau-t)^{\alpha(t)}}d\tau. \tag{4.52}$$

Definition 4.9 *The Richard type fractional derivatives of variable order based on the Riemann-Liouville fractional derivative of variable order is defined as*

$$\begin{aligned}\left({}^{Rc}_{ZK}D_{[a,b]}^{\alpha(\cdot)}f\right)(t) &= -\tfrac{1}{2\cos(\pi\alpha(t)/2)} \times \left(\left({}_{RLqs}D_{a+}^{\alpha(\cdot)}f\right)(t) + \left({}_{RLqs}D_{b-}^{\alpha(\cdot)}f\right)(t)\right)\\ &= -\tfrac{1}{2\cos(\pi\alpha(t)/2)} \times \tfrac{1}{\Gamma(1-\alpha(t))}\tfrac{d}{dt}\int_a^t \tfrac{f(\tau)}{(t-\tau)^{\alpha(t)}}d\tau\\ &+\tfrac{1}{2\cos(\pi\alpha(t)/2)} \times \tfrac{1}{\Gamma(1-\alpha(t))}\tfrac{d}{dt}\int_t^b \tfrac{f(\tau)}{(\tau-t)^{\alpha(t)}}d\tau,\end{aligned} \tag{4.53}$$

where

$$\left({}_{RLqs}D_{a+}^{\alpha(\cdot)}f\right)(t) = \frac{1}{\Gamma(1-\alpha(t))}\frac{d}{dt}\int_a^t \frac{f(\tau)}{(t-\tau)^{\alpha(t)}}d\tau \tag{4.54}$$

and

$$\left({}_{RLqs}D_{b-}^{\alpha(\cdot)}f\right)(t) = -\frac{1}{\Gamma(1-\alpha(t))}\frac{d}{dt}\int_t^b \frac{f(\tau)}{(\tau-t)^{\alpha(t)}}d\tau. \tag{4.55}$$

Definition 4.10 *The Richard type fractional derivatives of variable order based on the Liouville-Sonine fractional derivative of variable order is defined as*

$$\begin{aligned}\left({}^{Rc}_{ZKLC}D_{[a,b]}^{\alpha(\cdot)}f\right)(t) &= -\tfrac{1}{2\cos(\pi\alpha(t)/2)} \times \left(\left({}_{RLqs}D_{a+}^{\alpha(\cdot)}f\right)(t) + \left({}_{RLqs}D_{b-}^{\alpha(\cdot)}f\right)(t)\right)\\ &= -\tfrac{1}{2\cos(\pi\alpha(t)/2)} \times \tfrac{1}{\Gamma(1-\alpha(t))}\int_a^t \tfrac{f^{(1)}(\tau)}{(t-\tau)^{\alpha(t)}}d\tau\\ &+\tfrac{1}{2\cos(\pi\alpha(t)/2)} \times \tfrac{1}{\Gamma(1-\alpha(t))}\int_t^b \tfrac{f^{(1)}(\tau)}{(\tau-t)^{\alpha(t)}}d\tau,\end{aligned} \tag{4.56}$$

where

$$\left({}_{LSqs}D_{a+}^{\alpha(\cdot)}f\right)(t) = \frac{1}{\Gamma(1-\alpha(t))}\int_a^t \frac{f^{(1)}(\tau)}{(t-\tau)^{\alpha(t)}}d\tau \tag{4.57}$$

and

$$\left({}_{LSqs}D_{b-}^{\alpha(\cdot)}f\right)(t) = -\frac{1}{\Gamma(1-\alpha(t))}\int_t^b \frac{f^{(1)}(\tau)}{(\tau-t)^{\alpha(t)}}d\tau. \tag{4.58}$$

4.3 Hilfer type fractional derivatives of variable order involving the kernel of the singular power function

In this section, we introduce the Hilfer type fractional derivatives of variable order involving the kernel of the singular power function.

Definition 4.11 *Let $\kappa < \alpha(\tau,t) < 1+\kappa$, $\kappa = [\alpha(\tau,t)]+1, 0 \le \beta \le 1$, $f \in L_\kappa(a,b)$ and $-\infty < a < b < \infty$. The left-sided variable-order Hilfer fractional derivative is defined as*

$$ {}^{Hi}_{Hq}D^{\alpha(\cdot),\beta,\kappa}_{a+} f(t) = \left({}^{HL}_{Hq}I^{\beta(\kappa-\alpha(\cdot))}_{a+} \frac{d^\kappa}{dt^\kappa} \left({}^{HL}_{Hq}I^{(1-\beta)(\kappa-\alpha(\cdot))}_{a+} f \right) \right)(t), \tag{4.59} $$

and the right-sided variable-order Hilfer fractional derivative as

$$ {}^{Hi}_{Hq}D^{\alpha(\cdot),\beta,\kappa}_{b-} f(t) = \left((-1)^\kappa\, {}^{HL}_{Hq}I^{\beta(\kappa-\alpha(\cdot))}_{b-} \frac{d^\kappa}{dt^\kappa} \left({}^{HL}_{Hq}I^{(1-\beta)(\kappa-\alpha(\cdot))}_{b-} f \right) \right)(t), \tag{4.60} $$

$$ \left({}^{HL}_{Hq}I^{\beta(\kappa-\alpha(\cdot))}_{a+} f \right)(t) = \int_a^t \frac{1}{\Gamma(\beta(\kappa-\alpha(\tau,t)))} \frac{f(\tau)}{(t-\tau)^{1-\beta(\kappa-\alpha(\tau,t))}} d\tau, \tag{4.61} $$

$$ \left({}^{HL}_{Hq}I^{\beta(\kappa-\alpha(\cdot))}_{b-} f \right)(t) = \int_t^b \frac{1}{\Gamma(\beta(\kappa-\alpha(\tau,t)))} \frac{f(\tau)}{(\tau-t)^{1-\beta(\kappa-\alpha(\tau,t))}} d\tau, \tag{4.62} $$

$$ \left({}^{HL}_{Hq}I^{(1-\beta)(\kappa-\alpha(\cdot))}_{a+} \right) f(t) = \int_a^t \frac{1}{\Gamma((1-\beta)(\kappa-\alpha(\tau,t)))} \frac{f(\tau)}{(t-\tau)^{1-(1-\beta)(\kappa-\alpha(\tau,t))}} d\tau \tag{4.63} $$

and

$$ \left({}^{HL}_{Hq}I^{(1-\beta)(\kappa-\alpha(\cdot))}_{b-} f \right)(t) = \int_t^b \frac{1}{\Gamma((1-\beta)(\kappa-\alpha(\tau,t)))} \frac{f(\tau)}{(\tau-t)^{1-(1-\beta)(\kappa-\alpha(\tau,t))}} d\tau. \tag{4.64} $$

Definition 4.12 *Let $\kappa < \alpha(t) < 1+\kappa$, $\kappa = [\alpha(t)]+1$, $0 \le \beta \le 1$, $f \in L_\kappa(a,b)$ and $-\infty < a < b < \infty$. The left-sided variable-order Hilfer fractional derivative is defined as*

$$ {}^{Hi}_{H}D^{\alpha(\cdot),\beta,\kappa}_{a+} f(t) = \left({}^{HL}_{H}I^{\beta(\kappa-\alpha(\cdot))}_{a+} \frac{d^\kappa}{dt^\kappa} \left({}^{HL}_{H}I^{(1-\beta)(\kappa-\alpha(\cdot))}_{a+} f \right) \right)(t), \tag{4.65} $$

and the right-sided variable-order Hilfer fractional derivative as

$$ {}^{Hi}_{H}D^{\alpha(\cdot),\beta,\kappa}_{b-} f(t) = \left((-1)^\kappa\, {}^{HL}_{H}I^{\beta(\kappa-\alpha(\cdot))}_{b-} \frac{d^\kappa}{dt^\kappa} \left({}^{HL}_{H}I^{(1-\beta)(\kappa-\alpha(\cdot))}_{b-} f \right) \right)(t), \tag{4.66} $$

$$\left({}_{H}^{HL}I_{a+}^{\beta(\kappa-\alpha(\cdot))}f\right)(t)=\int_{a}^{t}\frac{1}{\Gamma\left(\beta\left(\kappa-\alpha\left(\tau\right)\right)\right)}\frac{f\left(\tau\right)}{\left(t-\tau\right)^{1-\beta(\kappa-\alpha(\tau))}}d\tau, \tag{4.67}$$

$$\left({}_{H}^{HL}I_{b-}^{\beta(\kappa-\alpha(\cdot))}f\right)(t)=\int_{t}^{b}\frac{1}{\Gamma\left(\beta\left(\kappa-\alpha\left(\tau\right)\right)\right)}\frac{f\left(\tau\right)}{\left(\tau-t\right)^{1-\beta(\kappa-\alpha(\tau))}}d\tau, \tag{4.68}$$

$$\left({}_{H}^{HL}I_{a+}^{(1-\beta)(\kappa-\alpha(\cdot))}\right)f\left(t\right)=\int_{a}^{t}\frac{1}{\Gamma\left(\left(1-\beta\right)\left(\kappa-\alpha\left(\tau\right)\right)\right)}\frac{f\left(\tau\right)}{\left(t-\tau\right)^{1-(1-\beta)(\kappa-\alpha(\tau))}}d\tau \tag{4.69}$$

and

$$\left({}_{H}^{HL}I_{b-}^{(1-\beta)(\kappa-\alpha(\cdot))}f\right)(t)=\int_{t}^{b}\frac{1}{\Gamma\left(\left(1-\beta\right)\left(\kappa-\alpha\left(\tau\right)\right)\right)}\frac{f\left(\tau\right)}{\left(\tau-t\right)^{1-(1-\beta)(\kappa-\alpha(\tau))}}d\tau. \tag{4.70}$$

Definition 4.13 *Let* $0<\alpha\left(t\right)<1$, $0\leq\beta\leq1$, $f\in L_1\left(a,b\right)$ *and* $-\infty<a<b<\infty$. *The left-sided variable-order Hilfer type fractional derivative is defined as*

$${}_{Sqs}^{Hi}D_{a+}^{\alpha(\cdot),\beta}f\left(t\right)=\left({}_{RLqs}^{HL}I_{a+}^{\beta(1-\alpha(\cdot))}\frac{d}{dt}\left({}_{RLqs}^{HL}I_{a+}^{(1-\beta)(1-\alpha(\cdot))}f\right)\right)(t), \tag{4.71}$$

and the right-sided variable-order Hilfer type fractional derivative as

$${}_{Sqs}^{Hi}D_{b-}^{\alpha(\cdot),\beta}f\left(t\right)=\left(-{}_{RLqs}^{HL}I_{b-}^{\beta(1-\alpha(\cdot))}\frac{d}{dt}\left({}_{RLqs}^{HL}I_{b-}^{(1-\beta)(1-\alpha(\cdot))}f\right)\right)(t), \tag{4.72}$$

where

$$\left({}_{RLqs}^{HL}I_{a+}^{\beta(1-\alpha(\cdot))}f\right)(t)=\frac{1}{\Gamma(\beta(1-\alpha(t)))}\int_{a}^{t}\frac{f(\tau)}{(t-\tau)^{1-\beta(1-\alpha(t))}}d\tau, \tag{4.73}$$

$$\left({}_{RLqs}^{HL}I_{b-}^{\beta(1-\alpha(\cdot))}f\right)(t)=\frac{1}{\Gamma(\beta(1-\alpha(t)))}\int_{t}^{b}\frac{f(\tau)}{(\tau-t)^{1-\beta(1-\alpha(t))}}d\tau, \tag{4.74}$$

$$\left({}_{RLqs}^{HL}I_{a+}^{(1-\beta)(1-\alpha(\cdot))}\right)f\left(t\right)=\frac{1}{\Gamma((1-\beta)(1-\alpha(t)))}\int_{a}^{t}\frac{f(\tau)}{(t-\tau)^{1-(1-\beta)(1-\alpha(t))}}d\tau, \tag{4.75}$$

and

$$\left({}_{RLqs}^{HL}I_{b-}^{(1-\beta)(1-\alpha(\cdot))}f\right)(t)=\frac{1}{\Gamma((1-\beta)(1-\alpha(t)))}\int_{t}^{b}\frac{f(\tau)}{(\tau-t)^{1-(1-\beta)(1-\alpha(t))}}d\tau. \tag{4.76}$$

Definition 4.14 *Let $\kappa < \alpha(t) < 1+\kappa$, $\kappa = [\alpha(t)]+1$, $0 \leq \beta \leq 1$, $f \in L_\kappa(a,b)$ and $-\infty < a < b < \infty$. The left-sided variable-order Hilfer fractional derivative is defined as*

$$ {}^{Hi}_{Hqs}D^{\alpha(\cdot),\beta,\kappa}_{a+} f(t) = \left({}^{HL}_{Hqs}I^{\beta(\kappa-\alpha(\cdot))}_{a+} \frac{d^\kappa}{dt^\kappa} \left({}^{HL}_{Hqs}I^{(1-\beta)(\kappa-\alpha(\cdot))}_{a+} f \right) \right)(t), \tag{4.77} $$

and the right-sided variable-order Hilfer fractional derivative as

$$ {}^{Hi}_{Hqs}D^{\alpha(\cdot),\beta,\kappa}_{b-} f(t) = \left((-1)^\kappa \, {}^{HL}_{Hqs}I^{\beta(\kappa-\alpha(\cdot))}_{b-} \frac{d^\kappa}{dt^\kappa} \left({}^{HL}_{Hqs}I^{(1-\beta)(\kappa-\alpha(\cdot))}_{b-} f \right) \right)(t), \tag{4.78} $$

$$ \left({}^{HL}_{Hqs}I^{\beta(\kappa-\alpha(\cdot))}_{a+} f \right)(t) = \frac{1}{\Gamma(\beta(\kappa-\alpha(t)))} \int_a^t \frac{f(\tau)}{(t-\tau)^{1-\beta(\kappa-\alpha(t))}} d\tau, \tag{4.79} $$

$$ \left({}^{HL}_{Hqs}I^{\beta(\kappa-\alpha(\cdot))}_{b-} f \right)(t) = \frac{1}{\Gamma(\beta(\kappa-\alpha(t)))} \int_t^b \frac{f(\tau)}{(\tau-t)^{1-\beta(\kappa-\alpha(t))}} d\tau, \tag{4.80} $$

$$ \left({}^{HL}_{Hqs}I^{(1-\beta)(\kappa-\alpha(\cdot))}_{a+} \right) f(t) = \frac{1}{\Gamma((1-\beta)(\kappa-\alpha(t)))} \int_a^t \frac{f(\tau)}{(t-\tau)^{1-(1-\beta)(\kappa-\alpha(t))}} d\tau, \tag{4.81} $$

and

$$ \left({}^{HL}_{Hqs}I^{(1-\beta)(\kappa-\alpha(\cdot))}_{b-} f \right)(t) = \frac{1}{\Gamma((1-\beta)(\kappa-\alpha(t)))} \int_t^b \frac{f(\tau)}{(\tau-t)^{1-(1-\beta)(\kappa-\alpha(t))}} d\tau. \tag{4.82} $$

4.4 Tempered fractional derivatives of variable order involving the kernel of the weak power-law function

In this section, the variable-order general fractional derivatives and the variable-order tempered fractional integrals involving the kernel of the weak power-law function are given as follows.

Definition 4.15 *let $\kappa < \alpha(t) < \kappa+1$, $-\infty < a < b < \infty$ and $\lambda \in \mathbb{R}$. The left-sided variable-order Riemann-Liouville type tempered fractional derivative is defined as*

$$ \left({}^{RL}_{Cp}D^{\alpha(\cdot),\lambda}_{a+} f \right)(t) = \frac{d^\kappa}{dt^\kappa} \int_a^t \frac{1}{\Gamma(\kappa-\alpha(\tau))} \frac{f(\tau)}{(t-\tau)^{\alpha(\tau)-\kappa+1}} e^{-\lambda(t-\tau)} d\tau, \tag{4.83} $$

and the right-sided variable-order Riemann-Liouville type tempered fractional derivative as

$$ \left({}^{RL}_{Cp}D^{\alpha(\cdot),\lambda}_{b-} f \right)(t) = (-1)^\kappa \frac{d^\kappa}{dt^\kappa} \int_t^b \frac{1}{\Gamma(\kappa-\alpha(\tau))} \frac{f(\tau)}{(\tau-t)^{\alpha(\tau)-\kappa+1}} e^{-\lambda(\tau-t)} d\tau. \tag{4.84} $$

Definition 4.16 *Let $\alpha(t) > 0$, $\kappa = [\alpha(t)] + 1$, $-\infty < a < b < \infty$ and $\lambda \in \mathbb{R}$. The left-sided variable-order Liouville-Sonine-Caputo type tempered fractional derivative is defined as*

$$\left({}^{LSC}_{Cp}D^{\alpha(\cdot),\lambda}_{a+} f\right)(t) = \int_a^t \frac{1}{\Gamma(\kappa - \alpha(\tau))} \frac{e^{-\lambda(t-\tau)}}{(t-\tau)^{\alpha(\tau)-\kappa+1}} f^{(\kappa)}(\tau)\, d\tau, \tag{4.85}$$

and the right-sided variable-order Liouville-Sonine-Caputo type tempered fractional derivative as

$$\left({}^{LSC}_{Cp}D^{\alpha(\cdot),\lambda}_{b-} f\right)(t) = (-1)^{\kappa} \int_t^b \frac{1}{\Gamma(\kappa - \alpha(\tau))} \frac{e^{-\lambda(\tau-t)}}{(\tau-t)^{\alpha(\tau)-\kappa+1}} f^{(\kappa)}(\tau)\, d\tau, \tag{4.86}$$

respectively.

Definition 4.17 *Let $\kappa + 1 > \alpha(t) > \kappa$, $-\infty < a < b < \infty$ and $\lambda \in \mathbb{R}$. The left-sided variable-order tempered fractional integral is defined as*

$$\left({}^{RL}_{Cp}I^{\alpha(\cdot),\lambda}_{a+} f\right)(t) = \int_a^t \frac{1}{(t-\tau)^{1-\alpha(\tau)}} E^{\alpha(\tau)-\kappa}_{1,\alpha(\tau)}(-\lambda(t-\tau))\, f(\tau)\, d\tau, \tag{4.87}$$

and the right-sided variable-order tempered fractional integral as

$$\left({}^{RL}_{Cp}I^{\alpha(\cdot),\lambda}_{b-} f\right)(t) = \int_t^b \frac{1}{(\tau-t)^{1-\alpha(\tau)}} E^{\alpha(\tau)-\kappa}_{1,\alpha(\tau)}(-\lambda(\tau-t))\, f(\tau)\, d\tau. \tag{4.88}$$

4.5 General fractional derivatives of variable order involving the kernel of Mittag-Leffler function

In this section, we present the general fractional derivatives and general fractional integrals of variable order involving the kernel of Mittag-Leffler function.

Definition 4.18 *Let $\alpha(t) > 0$, $\kappa = [\alpha(t)] + 1$, $\kappa < \alpha(t) < \kappa + 1$ and $\lambda \in \mathbb{R}$. The left-sided Liouville-Sonine-Caputo type variable-order general fractional derivative involving the kernel of Mittag-Leffler function is defined by*

$$^{LSC}_{Mlt}D^{\alpha(\cdot),\kappa,\lambda}_{a+} f(t) = \int_a^t E_{2\alpha(\tau)}\left(-\lambda(t-\tau)^{2\alpha(\tau)}\right) f^{(\kappa)}(\tau)\, d\tau, \tag{4.89}$$

and the right-sided Liouville-Sonine-Caputo type variable-order general fractional derivative involving the kernel of Mittag-Leffler function by

$$ {}_{Mlt}^{LSC}D_{b-}^{\alpha(\cdot),\kappa,\lambda} f(t) = (-1)^{\kappa} \int_t^b E_{2\alpha(\tau)}\left(-\lambda(\tau-t)^{2\alpha(\tau)}\right) f^{(\kappa)}(\tau)\, d\tau. \quad (4.90) $$

Definition 4.19 *Let $\alpha(t) > 0$, $\kappa = [\alpha(t)] + 1$, $\kappa < \alpha(t) < \kappa + 1$ and $\lambda \in \mathbb{R}$. The left-sided Riemann-Liouville type variable-order general fractional derivative involving the kernel of Mittag-Leffler function is defined by*

$$ {}_{Mlt}^{RL}D_{a+}^{\alpha(\cdot),\kappa,\lambda} f(t) = \frac{d^{\kappa}}{dt^{\kappa}} \int_a^t E_{2\alpha(\tau)}\left(-\lambda(t-\tau)^{2\alpha(\tau)}\right) f(\tau)\, d\tau, \quad (4.91) $$

and the right-sided Riemann-Liouville type variable-order general fractional derivative involving the kernel of Mittag-Leffler function by

$$ {}_{Mlt}^{RL}D_{b-}^{\alpha(\cdot),\kappa,\lambda} f(t) = (-1)^{\kappa} \frac{d^{\kappa}}{dt^{\kappa}} \int_t^b E_{2\alpha(\tau)}\left(-\lambda(\tau-t)^{2\alpha(\tau)}\right) f(\tau)\, d\tau. \quad (4.92) $$

Definition 4.20 *Let $\alpha(t) > 0$, $1 + \kappa > \alpha(t) > \kappa$, $\kappa = [\alpha(t)] + 1$, $-\infty < a < b < \infty$ and $\lambda \in \mathbb{R}$. The left-sided variable-order fractional integral is defined as*

$$ {}_{Mlt}^{RL}I_{a+}^{\alpha(\cdot),\kappa,\lambda} f(t) = \int_a^t \frac{1}{(t-\tau)^{2\alpha(\tau)+1-\kappa}} E_{2\alpha(\tau),\kappa-2\alpha(\tau)}^{-1}\left(-\lambda(t-\tau)^{2\alpha(\tau)}\right) f(\tau)\, d\tau, \quad (4.93) $$

and the right-sided variable-order fractional integral as

$$ {}_{Mlt}^{RL}I_{b-}^{\alpha(\cdot),\kappa,\lambda} f(t) = (-1)^{\kappa} \int_t^b \frac{1}{(\tau-t)^{2\alpha(\tau)+1-\kappa}} E_{2\alpha(\tau),\kappa-2\alpha(\tau)}^{-1}\left(-\lambda(\tau-t)^{2\alpha(\tau)}\right) f(\tau)\, d\tau. \quad (4.94) $$

In fact, the general fractional derivatives and general fractional integrals of variable order involving the kernel of Mittag-Leffler function with the normalization function was discussed in [110].

We now consider the general fractional derivatives and general fractional integrals of variable order via Mittag-Leffler function as follows.

Definition 4.21 *Let $\alpha(t) > 0$, $\kappa = [\alpha(t)] + 1$, $\kappa < \alpha(t) < \kappa + 1$ and $\lambda \in \mathbb{R}$. The left-sided Liouville-Sonine-Caputo type variable-order general fractional derivative with the Mittag-Leffler function is defined by*

$$ {}_{Ml}^{LSC}D_{a+}^{\alpha(\cdot),\kappa,\lambda} f(t) = \int_a^t E_{\alpha(\tau)}\left(-\lambda(t-\tau)^{\alpha(\tau)}\right) f^{(\kappa)}(\tau)\, d\tau, \quad (4.95) $$

and the right-sided Liouville-Sonine-Caputo type variable-order general fractional derivative with the Mittag-Leffler function by

$$ {}_{Ml}^{LSC}D_{b-}^{\alpha(\cdot),\kappa,\lambda} f(t) = (-1)^{\kappa} \int_t^b E_{\alpha(\tau)}\left(-\lambda(\tau-t)^{\alpha(\tau)}\right) f^{(\kappa)}(\tau)\,d\tau. \tag{4.96} $$

Definition 4.22 *Let $\alpha(t) > 0$, $\kappa = [\alpha(t)] + 1$, $\kappa < \alpha(t) < \kappa + 1$ and $\lambda \in \mathbb{R}$. The left-sided Riemann-Liouville type variable-order general fractional derivative with the Mittag-Leffler function is defined as*

$$ {}_{Ml}^{RL}D_{a+}^{\alpha(\cdot),\kappa,\lambda} f(t) = \frac{d^{\kappa}}{dt^{\kappa}} \int_a^t E_{\alpha(\tau)}\left(-\lambda(t-\tau)^{\alpha(\tau)}\right) f(\tau)\,d\tau, \tag{4.97} $$

and the right-sided Riemann-Liouville type variable-order general fractional derivative with the Mittag-Leffler function as

$$ {}_{Ml}^{RL}D_{b-}^{\alpha(\cdot),\kappa,\lambda} f(t) = (-1)^{\kappa} \frac{d^{\kappa}}{dt^{\kappa}} \int_t^b E_{\alpha(\tau)}\left(-\lambda(\tau-t)^{\alpha(\tau)}\right) f(\tau)\,d\tau. \tag{4.98} $$

Definition 4.23 *Let $\alpha(t) > 0$, $1 + \kappa > \alpha(t) > \kappa$, $\kappa = [\alpha(t)] + 1$, $-\infty < a < b < \infty$ and $\lambda \in \mathbb{R}$. The left-sided variable-order fractional integral is defined as*

$$ {}_{Ml}^{RL}I_{a+}^{\alpha(\cdot),\kappa,\lambda} f(t) = \int_a^t \frac{1}{(t-\tau)^{\alpha(\tau)+1-\kappa}} E_{\alpha(\tau),\kappa-\alpha(\tau)}^{-1}\left(-\lambda(t-\tau)^{\alpha(\tau)}\right) f(\tau)\,d\tau, \tag{4.99} $$

and the right-sided variable-order fractional integral as

$$ {}_{Ml}^{RL}I_{b-}^{\alpha(\cdot),\kappa,\lambda} f(t) = (-1)^{\kappa} \int_t^b \frac{1}{(\tau-t)^{\alpha(\tau)+1-\kappa}} E_{\alpha(\tau),\kappa-\alpha(\tau)}^{-1}\left(-\lambda(\tau-t)^{\alpha(\tau)}\right) f(\tau)\,d\tau. \tag{4.100} $$

We have to claim that the Liouville-Sonine type general fractional derivative of variable order involving the kernel of Mittag-Leffler function with the normalization function was discussed in [110]. Meanwhile, the Riemann-Liouville type variable-order general fractional derivative with the Mittag-Leffler function is the special case of the Sousa and de Oliveira results (see [324]).

4.6 General fractional derivatives of variable order involving the kernel of Wiman function

In this section, we present the variable-order general fractional calculus containing the kernel of the Gorenflo-Mainardi function via Wiman function.

Definition 4.24 *Let $\alpha(t) > 0$, $\kappa = [\alpha(t)] + 1$, $-\infty < a < b < \infty$, $\gamma \in \mathbb{R}$ and $\lambda \in \mathbb{R}$. The left-sided Riemann-Liouville type variable-order general fractional derivative containing the kernel of the Gorenflo-Mainardi function via Wiman function is defined as*

$$\left({}_{GM}^{RL}D_{a+}^{\alpha(\cdot),\gamma,\lambda} f\right)(t) = \frac{d^\kappa}{dt^\kappa}\left({}_{GM}I_{a+}^{\alpha(\cdot),\gamma,\lambda} f(t)\right), \tag{4.101}$$

and the right-sided Riemann-Liouville type variable-order general fractional derivative containing the kernel of the Gorenflo-Mainardi function via Wiman function as

$$\left({}_{GM}^{RL}D_{b-}^{\alpha(\cdot),\gamma,\lambda} f\right)(t) = \left(-\frac{d}{dt}\right)^\kappa\left({}_{GM}I_{b-}^{\alpha(\cdot),\gamma,\lambda} f(t)\right), \tag{4.102}$$

where the left-sided and right-sided variable-order general fractional integral operators containing the kernel of the Gorenflo-Mainardi function via Wiman function can be written as

$${}_{GM}I_{a+}^{\alpha(\cdot),\gamma,\lambda} f(t) = \int_a^t G_{\alpha(\tau),\gamma}\left(-\lambda(t-\tau)^{\alpha(\tau)}\right) f(\tau)\, d\tau \tag{4.103}$$

and

$${}_{GM}I_{b-}^{\alpha(\cdot),\gamma,\lambda} f(t) = \int_t^b G_{\alpha(\tau),\gamma}\left(-\lambda(\tau-t)^{\alpha(\tau)}\right) f(\tau)\, d\tau, \tag{4.104}$$

respectively.

Definition 4.25 *Let $1 > \alpha(t) > 0$, $-\infty < a < b < \infty$, $\gamma \in \mathbb{R}$ and $\lambda \in \mathbb{R}$. The left-sided Liouville-Sonine type variable-order general fractional derivative containing the kernel of the Gorenflo-Mainardi function via Wiman function is defined by*

$$\left({}_{GM}^{LS}D_{a+}^{\alpha(\cdot),\gamma,\lambda} f\right)(t) = \int_a^t G_{\alpha(\tau),\gamma}\left(-\lambda(t-\tau)^{\alpha(\tau)}\right) f^{(1)}(\tau)\, d\tau, \tag{4.105}$$

and the right-sided Liouville-Sonine type variable-order general fractional derivative containing the kernel of the Gorenflo-Mainardi function via Wiman function by

$$\left({}_{GM}^{LS}D_{b-}^{\alpha(\cdot),\gamma,\lambda} f\right)(t) = -\int_t^b G_{\alpha(\tau),\gamma}\left(-\lambda(\tau-t)^{\alpha(\tau)}\right) f^{(1)}(\tau)\, d\tau. \tag{4.106}$$

For $\kappa + 1 > \alpha(t) > \kappa$, $-\infty < a < b < \infty$, $\gamma \in \mathbb{R}$ and $\lambda \in \mathbb{R}$, the left-sided Liouville-Sonine-Caputo type variable-order general fractional derivative containing the kernel of the Gorenflo-Mainardi function via Wiman function is defined by

$$\left({}_{GM}^{LSC}D_{a+}^{\alpha(\cdot),\gamma,\lambda}f\right)(t) = \int_a^t G_{\alpha(\tau),\gamma}\left(-\lambda(t-\tau)^{\alpha(\tau)}\right) f^{(\kappa)}(\tau)\,d\tau, \tag{4.107}$$

and the right-sided Liouville-Sonine-Caputo type variable-order general fractional derivative containing the kernel of the Gorenflo-Mainardi function via Wiman function by

$$\left({}_{GM}^{LSC}D_{b-}^{\alpha(\cdot),\gamma,\lambda}f\right)(t) = (-1)^{\kappa}\int_t^b G_{\alpha(\tau),\gamma}\left(-\lambda(\tau-t)^{\alpha(\tau)}\right) f^{(\kappa)}(\tau)\,d\tau. \tag{4.108}$$

Definition 4.26 *Let $\kappa + 1 > \alpha(t) > \kappa$, $-\infty < a < b < \infty$, $\gamma \in \mathbb{R}$ and $\lambda \in \mathbb{R}$. The left-sided variable-order general fractional integral is defined as*

$$\left({}_{GM}^{R}I_{a+}^{\alpha(\cdot),\gamma,\lambda}f\right)(t) = \int_a^t (t-\tau)^{\kappa-\gamma-1} E_{\alpha(\tau),\kappa-\gamma}^{-1}\left(-\lambda(t-\tau)^{\alpha(\tau)}\right) f(\tau)\,d\tau \tag{4.109}$$

and the right-sided variable-order general fractional integral as

$$\left({}_{GM}^{R}I_{b-}^{\alpha(\cdot),\gamma,\lambda}f\right) = (-1)^{\kappa}\int_t^b (\tau-t)^{\kappa-\gamma-1} E_{\alpha(\tau),\kappa-\gamma}^{-1}\left(-\lambda(\tau-t)^{\alpha(\tau)}\right) f(\tau)\,d\tau. \tag{4.110}$$

More generally, we have the following definitions of the variable-order general fractional derivative containing the kernel of the Gorenflo-Mainardi function via Wiman function and the variable-order general fractional integrals as follows.

Definition 4.27 *Let $1 > \alpha(t,\tau) > 0$, $-\infty < a < b < \infty$, $\gamma \in \mathbb{R}$ and $\lambda \in \mathbb{R}$. The left-sided Riemann-Liouville type variable-order general fractional derivative containing the kernel of the Gorenflo-Mainardi function via Wiman function is defined as*

$$\left({}_{GMq}^{RL}D_{a+}^{\alpha(\cdot),\gamma,\lambda}f\right)(t) = \tfrac{d}{dt}\left({}_{GMq}I_{a+}^{\alpha(\cdot),\gamma,\lambda}f(t)\right), \tag{4.111}$$

and the right-sided Riemann-Liouville type variable-order general fractional derivative containing the kernel of the Gorenflo-Mainardi function via Wiman function as

$$\left({}_{GMq}^{RL}D_{b-}^{\alpha(\cdot),\gamma,\lambda}f\right)(t) = \left(-\tfrac{d}{dt}\right)\left({}_{GMq}I_{b-}^{\alpha(\cdot),\gamma,\lambda}f(t)\right), \tag{4.112}$$

where the left-sided and right-sided variable-order general fractional integral operators containing the kernel of the Gorenflo-Mainardi function via Wiman function can be written as

$$ {}_{GMq}I_{a+}^{\alpha(\cdot),\gamma,\lambda} f(t) = \int_a^t G_{\alpha(t,\tau),\gamma}\left(-\lambda (t-\tau)^{\alpha(t,\tau)}\right) f(\tau)\, d\tau \tag{4.113} $$

and

$$ {}_{GMq}I_{b-}^{\alpha(\cdot),\gamma,\lambda} f(t) = \int_t^b G_{\alpha(t,\tau),\gamma}\left(-\lambda (\tau-t)^{\alpha(t,\tau)}\right) f(\tau)\, d\tau, \tag{4.114} $$

respectively.

Definition 4.28 *Let $1 > \alpha(t,\tau) > 0$, $-\infty < a < b < \infty$, $\gamma \in \mathbb{R}$ and $\lambda \in \mathbb{R}$. The left-sided Liouville-Sonine type variable-order general fractional derivative containing the kernel of the Gorenflo-Mainardi function via Wiman function is defined by*

$$ \left({}_{GMq}^{LS}D_{a+}^{\alpha(\cdot),\gamma,\lambda} f\right)(t) = {}_{GMq}I_{a+}^{\alpha(\cdot),\gamma,\lambda}\left(f^{(1)}(t)\right), \tag{4.115} $$

and the right-sided Liouville-Sonine type variable-order general fractional derivative containing the kernel of the Gorenflo-Mainardi function via Wiman function by

$$ \left({}_{GMq}^{LS}D_{b-}^{\alpha(\cdot),\gamma,\lambda} f\right)(t) = {}_{GMq}I_{b-}^{\alpha(\cdot),\gamma,\lambda}\left(-f^{(1)}(t)\right). \tag{4.116} $$

Let $1 > \alpha(t,\tau) > 0$, $-\infty < a < b < \infty$, $\gamma \in \mathbb{R}$ and $\lambda \in \mathbb{R}$. The left-sided variable-order general fractional integral is defined as

$$ \left({}_{GMq}^{R}I_{a+}^{\alpha(\cdot),\gamma,\lambda} f\right)(t) = \int_a^t (t-\tau)^{-\gamma} E_{\alpha(t,\tau),1-\gamma}^{-1}\left(-\lambda (t-\tau)^{\alpha(t,\tau)}\right) f(\tau)\, d\tau, \tag{4.117} $$

and the right-sided variable-order general fractional integral as

$$ \left({}_{GMq}^{R}I_{b-}^{\alpha(\cdot),\gamma,\lambda} f\right) = -\int_t^b (\tau-t)^{-\gamma} E_{\alpha(t,\tau),1-\gamma}^{-1}\left(-\lambda (\tau-t)^{\alpha(t,\tau)}\right) f(\tau)\, d\tau. \tag{4.118} $$

Definition 4.29 *Let $1 > \alpha(t) > 0$, $-\infty < a < b < \infty$, $\gamma \in \mathbb{R}$ and $\lambda \in \mathbb{R}$. The left-sided Liouville-Sonine type variable-order general fractional derivative containing the kernel of the Gorenflo-Mainardi function via Wiman function is defined by*

$$ \left({}_{GMqs}^{LS}D_{a+}^{\alpha(\cdot),\gamma,\lambda} f\right)(t) = \int_a^t G_{\alpha(t),\gamma}\left(-\lambda (t-\tau)^{\alpha(t)}\right) f^{(1)}(\tau)\, d\tau, \tag{4.119} $$

and the right-sided Liouville-Sonine type variable-order general fractional derivative containing the kernel of the Gorenflo-Mainardi function via Wiman function by

$$\left({}^{LS}_{GMqs}D^{\alpha(\cdot),\gamma,\lambda}_{b-}f\right)(t) = -\int_t^b G_{\alpha(t),\gamma}\left(-\lambda(\tau-t)^{\alpha(t)}\right)f^{(1)}(\tau)\,d\tau. \tag{4.120}$$

Definition 4.30 *Let* $1 > \alpha(t) > 0$, $-\infty < a < b < \infty$, $\gamma \in \mathbb{R}$ *and* $\lambda \in \mathbb{R}$. *The left-sided variable-order general fractional integral is defined as*

$$\left({}^{R}_{GMqs}I^{\alpha(\cdot),\gamma,\lambda}_{a+}f\right)(t) = \int_a^t (t-\tau)^{-\gamma} E^{-1}_{\alpha(t),1-\gamma}\left(-\lambda(t-\tau)^{\alpha(t)}\right)f(\tau)\,d\tau, \tag{4.121}$$

and the right-sided variable-order general fractional integral as

$$\left({}^{R}_{GMqs}I^{\alpha(\cdot),\gamma,\lambda}_{b-}f\right) = -\int_t^b (\tau-t)^{-\gamma} E^{-1}_{\alpha(t),1-\gamma}\left(-\lambda(\tau-t)^{\alpha(t)}\right)f(\tau)\,d\tau. \tag{4.122}$$

4.7 General fractional derivatives of variable order involving the one-parametric Lorenzo-Hartley function

In this section, we present the variable-order general fractional calculus involving the one-parametric Lorenzo-Hartley function.

Definition 4.31 *Let* $1 > \alpha(t) > 0$, $-\infty < a < b < \infty$ *and* $\gamma \in \mathbb{R}$. *The left-sided Riemann-Liouville type variable-order general fractional derivative involving the one-parametric Lorenzo-Hartley kernel is defined as*

$$\left({}^{RL}_{A}D^{\alpha(\cdot),\gamma}_{a+}f\right)(t) = \tfrac{d}{dt}\left({}_{A}I^{\alpha(\cdot),\gamma}_{a+}f(t)\right), \tag{4.123}$$

and the right-sided Riemann-Liouville type variable-order general fractional derivative involving the one-parametric Lorenzo-Hartley kernel as

$$\left({}^{RL}_{A}D^{\alpha(\cdot),\gamma}_{b-}f\right)(t) = \left(-\tfrac{d}{dt}\right)\left({}_{A}I^{\alpha(\cdot),\gamma}_{b-}f(t)\right), \tag{4.124}$$

where the left-sided and right-sided variable-order general fractional integral operators involving the one-parametric Lorenzo-Hartley kernel can be written as

$${}_{A}I^{\alpha(\cdot),\gamma}_{a+}f(t) = \int_a^t F_{\alpha(\tau)}\left(-\gamma(t-\tau)^{\alpha(\tau)}\right)f(\tau)\,d\tau \tag{4.125}$$

and

$$ {}_{A}I_{b-}^{\alpha(\cdot),\gamma} f(t) = \int_{t}^{b} F_{\alpha(\tau)}\left(-\gamma(\tau-t)^{\alpha(\tau)}\right) f(\tau)\, d\tau, \tag{4.126} $$

respectively.

More generally, when $\alpha(t) > 0$, $\kappa = [\alpha(t)] + 1$, $-\infty < a < b < \infty$ and $\gamma \in \mathbb{C}$, the left-sided Riemann-Liouville type variable-order general fractional derivative involving the one-parametric Lorenzo-Hartley kernel is defined as

$$ \left({}_{A}^{RL}D_{a+}^{\alpha(\cdot),\gamma} f\right)(t) = \frac{d^{\kappa}}{dt^{\kappa}} \int_{a}^{t} F_{\alpha(\tau)}\left(-\gamma(t-\tau)^{\alpha(\tau)}\right) f(\tau)\, d\tau, \tag{4.127} $$

and the right-sided Riemann-Liouville type variable-order general fractional derivative involving the one-parametric Lorenzo-Hartley kernel as

$$ \left({}_{A}^{RL}D_{b-}^{\alpha(\cdot),\gamma} f\right)(t) = \left(-\frac{d}{dt}\right)^{\kappa} \int_{t}^{b} F_{\alpha(\tau)}\left(-\gamma(\tau-t)^{\alpha(\tau)}\right) f(\tau)\, d\tau. \tag{4.128} $$

Definition 4.32 *Let $1 > \alpha(t) > 0$, $-\infty < a < b < \infty$ and $\gamma \in \mathbb{C}$. The left-sided Liouville-Sonine type variable-order general fractional derivative involving the one-parametric Lorenzo-Hartley kernel is defined as*

$$ \left({}_{A}^{LS}D_{a+}^{\alpha(\cdot),\gamma} f\right)(t) = \int_{a}^{t} F_{\alpha(\tau)}\left(-\gamma(t-\tau)^{\alpha(\tau)}\right) f^{(1)}(\tau)\, d\tau, \tag{4.129} $$

and the right-sided Liouville-Sonine type variable-order general fractional derivative involving the one-parametric Lorenzo-Hartley kernel as

$$ \left({}_{A}^{LS}D_{b-}^{\alpha(\cdot),\lambda} f\right)(t) = -\int_{t}^{b} F_{\alpha(\tau)}\left(-\gamma(\tau-t)^{\alpha(\tau)}\right) f^{(1)}(\tau)\, d\tau. \tag{4.130} $$

More generally, when $\kappa + 1 > \alpha(t) > \kappa$, $-\infty < a < b < \infty$ and $\gamma \in \mathbb{C}$, the left-sided Liouville-Sonine-Caputo type variable-order general fractional derivative involving the one-parametric Lorenzo-Hartley kernel is defined as

$$ \left({}_{A}^{LSC}D_{a+}^{\alpha(\cdot),\gamma} f\right)(t) = \int_{a}^{t} F_{\alpha(\tau)}\left(-\gamma(t-\tau)^{\alpha(\tau)}\right) f^{(\kappa)}(\tau)\, d\tau, \tag{4.131} $$

and the right-sided Liouville-Sonine-Caputo type variable-order general fractional derivative involving the one-parametric Lorenzo-Hartley kernel as

$$ \left({}_{A}^{LSC}D_{b-}^{\alpha(\cdot),\lambda} f\right)(t) = (-1)^{\kappa} \int_{t}^{b} F_{\alpha(\tau)}\left(-\gamma(\tau-t)^{\alpha(\tau)}\right) f^{(\kappa)}(\tau)\, d\tau. \tag{4.132} $$

Definition 4.33 *Let $\kappa + 1 > \alpha(t) > \kappa$, $-\infty < a < b < \infty$ and $\gamma \in \mathbb{C}$. The left-sided variable-order general fractional integral in the real parameters is defined as*

$$\left({}_{A}^{R}I_{a+}^{\alpha(\cdot),\gamma} f\right)(t) = \int_{a}^{t} (t-\tau)^{\kappa-\alpha(\tau)-1} E_{\alpha(\tau),\kappa-\alpha(\tau)}^{-1} \left(-\gamma (t-\tau)^{\alpha(\tau)}\right) f(\tau)\, d\tau, \tag{4.133}$$

and the right-sided variable-order general fractional integral in the real parameters as

$$\left({}_{A}^{R}I_{b-}^{\alpha(\cdot),\gamma} f\right) = (-1)^{\kappa} \int_{t}^{b} (\tau-t)^{\kappa-\alpha(\tau)-1} E_{\alpha(\tau),\kappa-\alpha(\tau)}^{-1} \left(-\gamma (\tau-t)^{\alpha(\tau)}\right) f(\tau)\, d\tau. \tag{4.134}$$

4.8 General fractional derivatives of variable order involving the Miller-Ross function

In this section, we present the variable-order general fractional calculus involving the Miller-Ross function.

Definition 4.34 *Let $1 > \alpha(t) > 0$, $-\infty < a < b < \infty$ and $\lambda \in \mathbb{R}$. The left-sided variable-order general fractional derivative within the Miller-Ross kernel in the sense of Riemann-Liouville type is defined as*

$$\left({}_{MR}^{RL}D_{a+}^{\alpha(\cdot),\lambda} f\right)(t) = \tfrac{d}{dt}\left({}_{MR}I_{a+}^{\alpha(\cdot),\lambda} f(t)\right), \tag{4.135}$$

and the right-sided general fractional derivative within the Miller-Ross kernel in the sense of Riemann-Liouville type as

$$\left({}_{MR}^{RL}D_{b-}^{\alpha(\cdot),\lambda} f\right)(t) = \left(-\tfrac{d}{dt}\right)\left({}_{MR}I_{b-}^{\alpha(\cdot),\lambda} f(t)\right), \tag{4.136}$$

where the left-sided and right-sided variable-order general fractional integral operators within the Miller-Ross kernel are defined as

$${}_{MR}I_{a+}^{\alpha(\cdot),\lambda} f(t) = \int_{a}^{t} \mathrm{M}_{\alpha(\tau)} \left(-\lambda (t-\tau)^{\alpha(\tau)}\right) f(\tau)\, d\tau \tag{4.137}$$

and

$${}_{MR}I_{b-}^{\alpha(\cdot),\lambda} f(t) = \int_{t}^{b} \mathrm{M}_{\alpha(\tau)} \left(-\lambda (\tau-t)^{\alpha(\tau)}\right) f(\tau)\, d\tau, \tag{4.138}$$

respectively.

For $\alpha(t) > 0$, $\kappa = [\alpha(t)] + 1$, $-\infty < a < b < \infty$ and $\lambda \in \mathbb{R}$, the left-sided variable-order general fractional derivative within the Miller-Ross kernel in the sense of Riemann-Liouville type is defined as

$$\left({}^{RL}_{MR}D^{\alpha(\cdot),\lambda}_{a+}f\right)(t) = \frac{d^{\kappa}}{dt^{\kappa}}\int\limits_{a}^{t} \mathrm{M}_{\alpha(\tau)}\left(-\lambda(t-\tau)^{\alpha(\tau)}\right)f(\tau)\,d\tau, \tag{4.139}$$

and the right-sided variable-order general fractional derivative within the Miller-Ross kernel in the sense of Riemann-Liouville type as

$$\left({}^{RL}_{MR}D^{\alpha(\cdot),\lambda}_{b-}f\right)(t) = \left(-\frac{d}{dt}\right)^{\kappa}\int\limits_{t}^{b} \mathrm{M}_{\alpha(\tau)}\left(-\lambda(\tau-t)^{\alpha(\tau)}\right)f(\tau)\,d\tau. \tag{4.140}$$

Definition 4.35 *Let $1 > \alpha(t) > 0$, $-\infty < a < b < \infty$ and $\lambda \in \mathbb{R}$. The left-sided variable-order general fractional derivative within the Miller-Ross kernel in the sense of Liouville-Sonine type is defined as*

$$\left({}^{LS}_{MR}D^{\alpha(\cdot),\lambda}_{a+}f\right)(t) = \int\limits_{a}^{t} \mathrm{M}_{\alpha(\tau)}\left(-\lambda(t-\tau)^{\alpha(\tau)}\right)f^{(1)}(\tau)\,d\tau, \tag{4.141}$$

and the right-sided variable-order general fractional derivative within the Miller-Ross kernel in the sense of Liouville-Sonine type as

$$\left({}^{LS}_{MR}D^{\alpha(\cdot),\lambda}_{b-}f\right)(t) = -\int\limits_{t}^{b} \mathrm{M}_{\alpha(\tau)}\left(-\lambda(\tau-t)^{\alpha(\tau)}\right)f^{(1)}(\tau)\,d\tau. \tag{4.142}$$

For $\kappa+1 > \alpha(t) > \kappa$, $-\infty < a < b < \infty$ and $\lambda \in \mathbb{R}$, the left-sided variable-order general fractional derivative within the Miller-Ross kernel in the sense of Liouville-Sonine-Caputo type is defined as

$$\left({}^{LSC}_{MR}D^{\alpha(\cdot),\lambda}_{a+}f\right)(t) = \int\limits_{a}^{t} \mathrm{M}_{\alpha(\tau)}\left(-\lambda(t-\tau)^{\alpha(\tau)}\right)f^{(\kappa)}(\tau)\,d\tau, \tag{4.143}$$

and the right-sided variable-order general fractional derivative within the Miller-Ross kernel in the sense of Liouville-Sonine-Caputo type as

$$\left({}^{LSC}_{MR}D^{\alpha(\cdot),\lambda}_{b-}f\right)(t) = (-1)^{\kappa}\int\limits_{t}^{b} \mathrm{M}_{\alpha(\tau)}\left(-\lambda(\tau-t)^{\alpha(\tau)}\right)f^{(\kappa)}(\tau)\,d\tau. \tag{4.144}$$

Definition 4.36 *Let $1 > \alpha(t) > 0$, $-\infty < a < b < \infty$ and $\lambda \in \mathbb{R}$. The left-sided variable-order general fractional integral is defined as*

$$\left({}^{R}_{MR}I^{\alpha(\cdot),\lambda}_{a+}f\right)(t) = \int\limits_{a}^{t} \frac{1}{(t-\tau)^{\alpha(\tau)+1}} E_{1,-\alpha(\tau)}\left(-\lambda(t-\tau)\right)f(\tau)\,d\tau, \tag{4.145}$$

and the right-sided variable-order general fractional integral as

$$\left({}^{R}_{MR}I^{\alpha(\cdot),\gamma,\lambda}_{b-}f\right) = \int\limits_{t}^{b} \frac{1}{(\tau-t)^{\alpha(\tau)+1}} E_{1,-\alpha(\tau)}\left(-\lambda(\tau-t)\right)f(\tau)\,d\tau. \tag{4.146}$$

More generally, when $\kappa+1>\alpha>\kappa$, $-\infty<a<b<\infty$ and $\lambda\in\mathbb{R}$, the left-sided variable-order general fractional integral is defined as

$$\left({}_{MR}^{R}I_{a+}^{\alpha(\cdot),\lambda}f\right)(t)=\int_a^t \frac{1}{(t-\tau)^{\alpha(\tau)+2-\kappa}}E_{1,\kappa-\alpha(\tau)-1}\left(-\lambda\left(t-\tau\right)\right)f\left(\tau\right)d\tau, \tag{4.147}$$

and the right-sided variable-order general fractional integral as

$$\left({}_{MR}^{R}I_{b-}^{\alpha(\cdot),\gamma,\lambda}f\right)=\int_t^b \frac{1}{(t-\tau)^{\alpha(\tau)+2-\kappa}}E_{1,\kappa-\alpha(\tau)-1}\left(-\lambda\left(\tau-t\right)\right)f\left(\tau\right)d\tau. \tag{4.148}$$

4.9 General fractional derivatives of variable order involving the Prabhakar function

In this section, we present the variable-order general fractional calculus involving the Prabhakar function.

Definition 4.37 *Let $1>\alpha(t)>0$, $-\infty<a<b<\infty$, $\varphi\in\mathbb{R}$, $\upsilon\in\mathbb{R}$ and $\lambda\in\mathbb{R}$. The left-sided Riemann-Liouville type variable-order general fractional derivative within the generalized Prabhakar kernel via Prabhakar function is defined as*

$$\left({}_{GP}^{RL}D_{a+}^{\alpha(\cdot),\upsilon,\varphi,\lambda}f\right)(t)=\tfrac{d}{dt}\left({}_{GP}I_{a+}^{\alpha(\cdot),\upsilon,\varphi,\lambda}f\left(t\right)\right), \tag{4.149}$$

and the right-sided Riemann-Liouville type variable-order general fractional derivative within the generalized Prabhakar kernel via Prabhakar function as

$$\left({}_{GP}^{RL}D_{b-}^{\alpha(\cdot),\upsilon,\varphi,\lambda}f\right)(t)=-\tfrac{d}{dt}\int_t^b\left(\tau-t\right)^{\upsilon-1}E_{\alpha(\tau),\upsilon}^{\varphi}\left(-\lambda\left(\tau-t\right)^{\alpha(\tau)}\right)f\left(\tau\right)d\tau, \tag{4.150}$$

where the left-sided and right-sided variable-order general fractional integral operators within the Prabhakar kernel via Prabhakar function are represented in the forms

$${}_{GP}I_{a+}^{\alpha(\cdot),\upsilon,\varphi,\lambda}f\left(t\right)=\int_a^t\left(t-\tau\right)^{\upsilon-1}E_{\alpha(\tau),\upsilon}^{\varphi}\left(-\lambda\left(t-\tau\right)^{\alpha(\tau)}\right)f\left(\tau\right)d\tau \tag{4.151}$$

and

$${}_{GP}I_{b-}^{\alpha(\cdot),\upsilon,\varphi,\lambda}f\left(t\right)=\int_t^b\left(\tau-t\right)^{\upsilon-1}E_{\alpha(\tau),\upsilon}^{\varphi}\left(-\lambda\left(\tau-t\right)^{\alpha(\tau)}\right)f\left(\tau\right)d\tau, \tag{4.152}$$

respectively.

Definition 4.38 *For $\alpha(t) > 0$, $\kappa = [\alpha(t)] + 1$, $-\infty < a < b < \infty$, $\varphi \in \mathbb{R}$, $\upsilon \in \mathbb{R}$ and $\lambda \in \mathbb{R}$, the left-sided Riemann-Liouville type variable-order general fractional derivative within the generalized Prabhakar kernel via Prabhakar function is defined as*

$$\left({}_{GP}^{RL}D_{a+}^{\alpha(\cdot),\upsilon,\varphi,\lambda} f\right)(t) = \frac{d^\kappa}{dt^\kappa} \int_a^t (t-\tau)^{\upsilon-1} E_{\alpha(\tau),\upsilon}^{\varphi}\left(-\lambda (t-\tau)^{\alpha(\tau)}\right) f(\tau)\, d\tau, \tag{4.153}$$

and the right-sided Riemann-Liouville type variable-order general fractional derivative within the generalized Prabhakar kernel via Prabhakar function as

$$\left({}_{GP}^{RL}D_{b-}^{\alpha(\cdot),\upsilon,\varphi,\lambda} f\right)(t) = \left(-\frac{d}{dt}\right)^\kappa \int_t^b (\tau-t)^{\upsilon-1} E_{\alpha(\tau),\upsilon}^{\varphi}\left(-\lambda (\tau-t)^{\alpha(\tau)}\right) f(\tau)\, d\tau. \tag{4.154}$$

Definition 4.39 *For$\kappa + 1 > \alpha(t) > \kappa$, $-\infty < a < b < \infty$, $\varphi \in \mathbb{R}$, $\upsilon \in \mathbb{R}$ and $\lambda \in \mathbb{R}$, the left-sided Liouville-Sonine-Caputo type variable-order general fractional derivative within the generalized Prabhakar kernel via Prabhakar function is defined as*

$$\left({}_{GP}^{LSC}D_{a+}^{\alpha(\cdot),\varphi,\upsilon,\lambda} f\right)(t) = \int_a^t (t-\tau)^{\upsilon-1} E_{\alpha(\tau),\upsilon}^{\varphi}\left(-\lambda (t-\tau)^{\alpha(\tau)}\right) f^{(\kappa)}(\tau)\, d\tau, \tag{4.155}$$

and the right-sided Liouville-Sonine-Caputo type variable-order general fractional derivative within the generalized Prabhakar kernel via Prabhakar function as

$$\left({}_{GP}^{LSC}D_{b-}^{\alpha(\cdot),\varphi,\upsilon,\lambda} f\right)(t) = (-1)^\kappa \int_t^b (\tau-t)^{\upsilon-1} E_{\alpha(\tau),\upsilon}^{\varphi}\left(-\lambda (\tau-t)^{\alpha(\tau)}\right) f^{(\kappa)}(\tau)\, d\tau. \tag{4.156}$$

Definition 4.40 *For $\kappa+1 > \alpha(t) > \kappa$, $-\infty < a < b < \infty$, $\varphi \in \mathbb{R}$, $\upsilon \in \mathbb{R}$ and $\lambda \in \mathbb{R}$, the left-sided variable-order general fractional integral via Prabhakar function is defined as*

$$\left({}_{GP}^{R}I_{a+}^{\alpha(\cdot),\varphi,\upsilon,\lambda} f\right)(t) = \int_a^t (t-\tau)^{\kappa-\upsilon-1} E_{\alpha(\tau),\kappa-\upsilon}^{-\varphi}\left(-\lambda (t-\tau)^{\alpha(\tau)}\right) f(\tau)\, d\tau \tag{4.157}$$

and the right-sided variable-order general fractional integral via Prabhakar function as

$$\left({}_{GP}^{R}I_{b-}^{\alpha(\cdot),\varphi,\upsilon,\lambda} f\right) = \int_t^b (\tau-t)^{\kappa-\upsilon-1} E_{\alpha(\tau),\kappa-\upsilon}^{-\varphi}\left(-\lambda (\tau-t)^{\alpha(\tau)}\right) f(\tau)\, d\tau. \tag{4.158}$$

4.10 Applications

4.10.1 Relaxation models within variable-order fractional derivatives

Now, we give the open problems on the relaxation models within the variable-order fractional derivatives.

Example 4.1 *The mathematical model of the relaxation involving the Riemann-Liouville type fractional derivative of variable order can be given as*

$$\left({}_{RL}D_{0+}^{\alpha(\cdot)}f\right)(t)+\gamma f(t)=0, \tag{4.159}$$

which deduces that

$$\frac{d}{dt}\int_0^t \frac{1}{\Gamma(1-\alpha(\tau))}\frac{f(\tau)}{(\tau-t)^{\alpha(\tau)}}d\tau+\gamma f(t)=0, \tag{4.160}$$

where γ is the relaxation constant.

Example 4.2 *The mathematical model of the relaxation involving the Liouville-Sonine type fractional derivative of variable order can be given as*

$${}_{LS}D_{0+}^{\alpha(\cdot)}f(t)+\gamma f(t)=0, \tag{4.161}$$

which becomes

$$\int_0^t \frac{1}{\Gamma(1-\alpha(\tau))}\frac{f^{(1)}(\tau)}{(t-\tau)^{\alpha(\tau)}}d\tau+\gamma f(t)=0, \tag{4.162}$$

where γ is the relaxation constant.

Example 4.3 *The mathematical model of the relaxation based on the Hilfer type fractional derivative of variable order involving the kernel of the singular power function is written as follows*

$$\left({}_{RLq}^{HL}I_{0+}^{\beta(1-\alpha(\cdot))}\frac{d}{dt}\left({}_{RLq}^{HL}I_{0+}^{(1-\beta)(1-\alpha(\cdot))}f\right)\right)(t)+\gamma f(t)=0, \tag{4.163}$$

where γ is the relaxation constant and β is the constant.

Example 4.4 *The mathematical model of the relaxation based on the Hilfer type fractional derivative of variable order involving the kernel of the singular power function is written as follows*

$$\left({}_{RLq}^{HL}I_{0+}^{\beta(1-\alpha(\cdot))}\frac{d}{dt}\left({}_{RLq}^{HL}I_{0+}^{(1-\beta)(1-\alpha(\cdot))}f\right)\right)(t)+\gamma f(t)=0, \tag{4.164}$$

where γ is the relaxation constant and β is the constant.

Example 4.5 *The mathematical model of the relaxation based on the variable-order Riemann-Liouville type tempered fractional derivative is presented as follows*

$$\frac{d}{dt}\int_0^t \frac{1}{\Gamma(1-\alpha(\tau))}\frac{f(\tau)}{(t-\tau)^{\alpha(\tau)}}e^{-\lambda(t-\tau)}d\tau+\gamma f(t)=0, \tag{4.165}$$

where γ is the relaxation constant and λ is the constant.

Example 4.6 *The mathematical model of the relaxation based on the variable-order Liouville-Sonine type tempered fractional derivative is represented in the form:*

$$\int_0^t \frac{1}{\Gamma(1-\alpha(\tau))}\frac{f^{(1)}(\tau)}{(t-\tau)^{\alpha(\tau)}}e^{-\lambda(t-\tau)}d\tau+\gamma f(t)=0, \tag{4.166}$$

where γ is the relaxation constant and λ is the constant.

Example 4.7 *The mathematical model of the relaxation based on the Liouville-Sonine type variable-order general fractional derivative with the Mittag-Leffler function is given as*

$$\int_0^t E_{\alpha(\tau)}\left(-\lambda(t-\tau)^{\alpha(\tau)}\right)f^{(1)}(\tau)\,d\tau+\gamma f(t)=0, \tag{4.167}$$

where γ is the relaxation constant and λ is the constant.

Example 4.8 *The mathematical model of the relaxation based on the Riemann-Liouville type variable-order general fractional derivative with the Mittag-Leffler function is given as*

$$\frac{d}{dt}\int_0^t E_{\alpha(\tau)}\left(-\lambda(t-\tau)^{\alpha(\tau)}\right)f(\tau)\,d\tau+\gamma f(t)=0, \tag{4.168}$$

where γ is the relaxation constant and λ is the constant.

Example 4.9 *The mathematical model of the relaxation based on the Liouville-Sonine type variable-order general fractional derivative containing the kernel of the Gorenflo-Mainardi function via Wiman function is given as*

$$\int_0^t G_{\alpha(\tau),\gamma}\left(-\lambda(t-\tau)^{\alpha(\tau)}\right) f^{(1)}(\tau)\,d\tau + \gamma f(t) = 0, \tag{4.169}$$

where γ is the relaxation constant and λ is the constant.

Example 4.10 *The mathematical model of the relaxation based on the Riemann-Liouville type variable-order general fractional derivative containing the kernel of the Gorenflo-Mainardi function via Wiman function is expressed as*

$$\frac{d}{dt}\int_0^t G_{\alpha(\tau),\gamma}\left(-\lambda(t-\tau)^{\alpha(\tau)}\right) f(\tau)\,d\tau + \gamma f(t) = 0, \tag{4.170}$$

where γ is the relaxation constant and λ is the constant.

Example 4.11 *The mathematical model of the relaxation based on the Riemann-Liouville type variable-order general fractional derivative involving the one-parametric Lorenzo-Hartley kernel is given as*

$$\frac{d}{dt}\int_0^t F_{\alpha(\tau)}\left(-\gamma(t-\tau)^{\alpha(\tau)}\right) f(\tau)\,d\tau + \gamma f(t) = 0, \tag{4.171}$$

where γ is the relaxation constant and λ is the constant.

Example 4.12 *The mathematical model of the relaxation based on the Liouville-Sonine type variable-order general fractional derivative involving the one-parametric Lorenzo-Hartley kernel is expressed as*

$$\int_0^t F_{\alpha(\tau)}\left(-\gamma(t-\tau)^{\alpha(\tau)}\right) f^{(1)}(\tau)\,d\tau + \gamma f(t) = 0, \tag{4.172}$$

where γ is the relaxation constant and λ is the constant.

Example 4.13 *The mathematical model of the relaxation based on the variable-order general fractional derivative within the Miller-Ross kernel in the sense of Riemann-Liouville type is given as*

$$\frac{d}{dt}\int_0^t \mathrm{M}_{\alpha(\tau)}\left(-\lambda(t-\tau)^{\alpha(\tau)}\right) f(\tau)\,d\tau + \gamma f(t) = 0, \tag{4.173}$$

where γ is the relaxation constant and λ is the constant.

Example 4.14 *The mathematical model of the relaxation based on the variable-order general fractional derivative within the Miller-Ross kernel in the sense of Liouville-Sonine type is expressed as*

$$\int_0^t \mathrm{M}_{\alpha(\tau)}\left(-\lambda(t-\tau)^{\alpha(\tau)}\right) f^{(1)}(\tau)\,d\tau + \gamma f(t) = 0, \tag{4.174}$$

where γ is the relaxation constant and λ is the constant.

Example 4.15 *The mathematical model of the relaxation based on the Riemann-Liouville type variable-order general fractional derivative within the generalized Prabhakar kernel via Prabhakar function is given as*

$$\frac{d}{dt}\int_0^t (t-\tau)^{\upsilon-1} E^{\varphi}_{\alpha(\tau),\upsilon}\left(-\lambda(t-\tau)^{\alpha(\tau)}\right) f(\tau)\,d\tau + \gamma f(t) = 0, \tag{4.175}$$

where γ is the relaxation constant and both φ and υ are the constants.

Example 4.16 *The mathematical model of the relaxation based on the variable-order general fractional derivative within the Miller-Ross kernel in the sense of Liouville-Sonine type is expressed as*

$$\int_0^t (t-\tau)^{\upsilon-1} E^{\varphi}_{\alpha(\tau),\upsilon}\left(-\lambda(t-\tau)^{\alpha(\tau)}\right) f^{(1)}(\tau)\,d\tau + \gamma f(t) = 0, \tag{4.176}$$

where γ is the relaxation constant and both φ and υ are the constants.

4.10.2 Rheological models within variable-order fractional derivatives

In this section, we present the mathematical models of the rheological materials based on the variable-order fractional derivatives. The open problems for the rheological models in variable-order fractional derivatives are presented in detail. To start with the ideas, we illustrate the mathematical models for the rheological materials as follows.

Example 4.17 *The mathematical model for the law of deformation based on the Riemann-Liouville type fractional derivative of variable order is represented in the form (see [325])*

$$\sigma(t) = \xi\left({}_{RL}D^{\alpha(\cdot)}_{0+}\varepsilon\right)(t), \tag{4.177}$$

which can be written as

$$\sigma(t) = \xi \frac{d}{dt} \int_0^t \frac{1}{\Gamma(1-\alpha(\tau))} \frac{\varepsilon(\tau)}{(\tau - t)^{\alpha(\tau)}} d\tau, \tag{4.178}$$

where ξ is the material constant, $\sigma(t)$ is the stress, and $\varepsilon(t)$ is the strain.

Example 4.18 *The mathematical model for the law of deformation based on the Liouville-Sonine type fractional derivative of variable order is represented in the form (see [326])*

$$\sigma(t) = \xi \left({}_{LS}D_{0+}^{\alpha(\cdot)} \varepsilon \right)(t), \tag{4.179}$$

which leads to

$$\sigma(t) = \xi \int_0^t \frac{1}{\Gamma(1-\alpha(\tau))} \frac{\varepsilon^{(1)}(\tau)}{(\tau - t)^{\alpha(\tau)}} d\tau, \tag{4.180}$$

where ξ is the material constant, $\sigma(t)$ is the stress, and $\varepsilon(t)$ is the strain.

Example 4.19 *The mathematical model for the law of deformation based on the Hilfer type fractional derivative of variable order involving the kernel of the singular power function is written as follows*

$$\sigma(t) = \xi \left({}_{RLq}^{HL} I_{0+}^{\beta(1-\alpha(\cdot))} \frac{d}{dt} \left({}_{RLq}^{HL} I_{0+}^{(1-\beta)(1-\alpha(\cdot))} \varepsilon \right) \right)(t), \tag{4.181}$$

where ξ is the material constant, $\sigma(t)$ is the stress, $\varepsilon(t)$ is the strain and β is the constant.

Example 4.20 *The mathematical model for the law of deformation based on the Hilfer type fractional derivative of variable order involving the kernel of the singular power function is written as follows*

$$\sigma(t) = \xi \left({}_{RLq}^{HL} I_{0+}^{\beta(1-\alpha(\cdot))} \frac{d}{dt} \left({}_{RLq}^{HL} I_{0+}^{(1-\beta)(1-\alpha(\cdot))} \varepsilon \right) \right)(t), \tag{4.182}$$

where ξ is the material constant, $\sigma(t)$ is the stress, $\varepsilon(t)$ is the strain and β is the constant.

Example 4.21 *The mathematical model for the law of deformation based on the variable-order Riemann-Liouville type tempered fractional derivative is presented as follows*

$$\sigma(t) = \xi \frac{d}{dt} \int_0^t \frac{1}{\Gamma(1-\alpha(\tau))} \frac{\varepsilon(\tau)}{(t - \tau)^{\alpha(\tau)}} e^{-\lambda(t-\tau)} d\tau, \tag{4.183}$$

where ξ is the material constant, $\sigma(t)$ is the stress, $\varepsilon(t)$ is the strain and λ is the constant.

Example 4.22 *The mathematical model for the law of deformation based on the variable-order Liouville-Sonine type tempered fractional derivative is represented in the form:*

$$\sigma(t) = \xi \int_0^t \frac{1}{\Gamma(1-\alpha(\tau))} \frac{\varepsilon^{(1)}(\tau)}{(t-\tau)^{\alpha(\tau)}} e^{-\lambda(t-\tau)} d\tau, \tag{4.184}$$

where ξ is the material constant, $\sigma(t)$ is the stress, $\varepsilon(t)$ is the strain and λ is the constant.

Example 4.23 *The mathematical model for the law of deformation based on the Liouville-Sonine type variable-order general fractional derivative with the Mittag-Leffler function is given as*

$$\sigma(t) = \xi \int_0^t E_{\alpha(\tau)}\left(-\lambda(t-\tau)^{\alpha(\tau)}\right) \varepsilon^{(1)}(\tau) d\tau, \tag{4.185}$$

where ξ is the material constant, $\sigma(t)$ is the stress, $\varepsilon(t)$ is the strain and λ is the constant.

Example 4.24 *The mathematical model for the law of deformation based on the Riemann-Liouville type variable-order general fractional derivative with the Mittag-Leffler function is given as*

$$\sigma(t) = \xi \frac{d}{dt} \int_0^t E_{\alpha(\tau)}\left(-\lambda(t-\tau)^{\alpha(\tau)}\right) \varepsilon(\tau) d\tau, \tag{4.186}$$

where ξ is the material constant, $\sigma(t)$ is the stress, $\varepsilon(t)$ is the strain and λ is the constant.

Example 4.25 *The mathematical model for the law of deformation based on the Liouville-Sonine type variable-order general fractional derivative containing the kernel of the Gorenflo-Mainardi function via Wiman function is given as*

$$\sigma(t) = \xi \int_0^t G_{\alpha(\tau),\gamma}\left(-\lambda(t-\tau)^{\alpha(\tau)}\right) \varepsilon^{(1)}(\tau) d\tau, \tag{4.187}$$

where ξ is the material constant, $\sigma(t)$ is the stress, $\varepsilon(t)$ is the strain and λ is the constant.

Example 4.26 *The mathematical model for the law of deformation based on the Riemann-Liouville type variable-order general fractional derivative containing the kernel of the Gorenflo-Mainardi function via Wiman function is expressed as*

$$\sigma(t) = \xi \frac{d}{dt} \int_0^t G_{\alpha(\tau),\gamma}\left(-\lambda(t-\tau)^{\alpha(\tau)}\right) \varepsilon(\tau)\, d\tau, \tag{4.188}$$

where ξ is the material constant, $\sigma(t)$ is the stress, $\varepsilon(t)$ is the strain and λ is the constant.

Example 4.27 *The mathematical model for the law of deformation based on the Riemann-Liouville type variable-order general fractional derivative involving the one-parametric Lorenzo-Hartley kernel is given as*

$$\sigma(t) = \xi \frac{d}{dt} \int_0^t F_{\alpha(\tau)}\left(-\gamma(t-\tau)^{\alpha(\tau)}\right) \varepsilon(\tau)\, d\tau, \tag{4.189}$$

where ξ is the material constant, $\sigma(t)$ is the stress, $\varepsilon(t)$ is the strain and λ is the constant.

Example 4.28 *The mathematical model for the law of deformation based on the Liouville-Sonine type variable-order general fractional derivative involving the one-parametric Lorenzo-Hartley kernel is expressed as*

$$\sigma(t) = \xi \int_0^t F_{\alpha(\tau)}\left(-\gamma(t-\tau)^{\alpha(\tau)}\right) \varepsilon^{(1)}(\tau)\, d\tau, \tag{4.190}$$

where ξ is the material constant, $\sigma(t)$ is the stress, $\varepsilon(t)$ is the strain and λ is the constant.

Example 4.29 *The mathematical model for the law of deformation based on the variable-order general fractional derivative within the Miller-Ross kernel in the sense of Riemann-Liouville type is given as*

$$\sigma(t) = \xi \frac{d}{dt} \int_0^t \mathrm{M}_{\alpha(\tau)}\left(-\lambda(t-\tau)^{\alpha(\tau)}\right) \varepsilon(\tau)\, d\tau, \tag{4.191}$$

where ξ is the material constant, $\sigma(t)$ is the stress, $\varepsilon(t)$ is the strain and λ is the constant.

Example 4.30 *The mathematical model for the law of deformation based on the variable-order general fractional derivative within the Miller-Ross kernel in the sense of Liouville-Sonine type is expressed as*

$$\sigma(t) = \xi \int_0^t \mathrm{M}_{\alpha(\tau)}\left(-\lambda(t-\tau)^{\alpha(\tau)}\right) \varepsilon^{(1)}(\tau)\, d\tau, \tag{4.192}$$

where ξ is the material constant, $\sigma(t)$ is the stress, $\varepsilon(t)$ is the strain and λ is the constant.

Example 4.31 *The mathematical model for the law of deformation based on the Riemann-Liouville type variable-order general fractional derivative within the generalized Prabhakar kernel via Prabhakar function is given as*

$$\sigma(t) = \xi \frac{d}{dt} \int_0^t (t-\tau)^{\upsilon-1} E_{\alpha(\tau),\upsilon}^{\varphi} \left(-\lambda (t-\tau)^{\alpha(\tau)}\right) \varepsilon(\tau)\, d\tau, \tag{4.193}$$

where ξ is the material constant, $\sigma(t)$ is the stress, $\varepsilon(t)$ is the strain, λ is the constant, and both φ and υ are the constants.

Example 4.32 *The mathematical model for the law of deformation based on the variable-order general fractional derivative within the Miller-Ross kernel in the sense of Liouville-Sonine type is expressed as*

$$\sigma(t) = \xi \int_0^t (t-\tau)^{\upsilon-1} E_{\alpha(\tau),\upsilon}^{\varphi} \left(-\lambda (t-\tau)^{\alpha(\tau)}\right) \varepsilon^{(1)}(\tau)\, d\tau, \tag{4.194}$$

where ξ is the material constant, $\sigma(t)$ is the stress, $\varepsilon(t)$ is the strain, λ is the constant, and both φ and υ are the constants.

Chapter 5

Fractional Derivatives of Variable Order with Respect to Another Function and Applications

In this chapter, we introduce the variable-order fractional calculus with respect to another function, first proposed by Yang and Machado in 2017 [114], and the variable-order general fractional calculus involving the Mittag-Leffler function with respect to another function, proposed by Sousa and Oliveira in 2018 for the first time [324]. The variable-order general fractional derivatives containing the kernel of the special functions, such as the Mittag-Leffler function and weak power-law function are presented. The definitions of the variable-order fractional derivatives and the variable-order general fractional integrals with respect to another function are introduced. The Hilfer type fractional derivatives of variable order with respect to another function and the Riesz type fractional derivatives of variable order with respect to another function are also proposed. The mathematical models for the relaxation and rheological models are discussed in detail.

5.1 Fractional derivatives of variable order with respect to another function

In this section, we recall the fractional integrals and fractional derivatives of variable order with respect to another function.

Definition 5.1 *Let $\alpha(t) > 0$, $-\infty < a < b < \infty$ and $h^{(1)}(t) > 0$. The left-sided variable order Riemann-Liouville type fractional integral with respect to another function is defined as*

$$\left({}_{RL}I_{a+,h}^{\alpha(\cdot)} f\right)(t) = \int_a^t \frac{h^{(1)}(\tau)}{\Gamma(\alpha(\tau))} \frac{f(\tau)}{(h(t) - h(\tau))^{1-\alpha(\tau)}} d\tau, \qquad (5.1)$$

and the right-sided variable-order Riemann-Liouville type fractional integral with respect to another function as

$$\left({}_{RL}I^{\alpha(\cdot)}_{b-,h}f\right)(t)=\int_t^b \frac{h^{(1)}(\tau)}{\Gamma(\alpha(\tau))}\frac{f(\tau)}{(h(\tau)-h(t))^{1-\alpha(\tau)}}d\tau. \tag{5.2}$$

Definition 5.2 *Let $\kappa<\alpha(t)<1+\kappa$, $-\infty<a<b<\infty$ and $h^{(1)}(t)>0$. The left-sided Riemann-Liouville type fractional derivative of variable order with respect to another function is defined as (see [114])*

$$\left({}^{CGS}_{RL}D^{\alpha(\cdot)}_{a+,h}f\right)(t)=\frac{1}{\Gamma(\kappa-\alpha(\tau))}\left(\frac{1}{h^{(1)}(t)}\frac{d}{dt}\right)^{\kappa}\int_a^t \frac{h^{(1)}(\tau)f(\tau)}{(h(t)-h(\tau))^{\alpha(\tau)-\kappa+1}}d\tau, \tag{5.3}$$

and the right-sided Riemann-Liouville type fractional derivative of variable order with respect to another function as

$$\left({}^{CGS}_{RL}D^{\alpha(\cdot)}_{b-,h}f\right)(t)=\frac{1}{\Gamma(\kappa-\alpha(\tau))}\left(-\frac{1}{h^{(1)}(t)}\frac{d}{dt}\right)^{\kappa}\int_t^b \frac{h^{(1)}(\tau)f(\tau)}{(h(\tau)-h(t))^{\alpha(\tau)-\kappa+1}}d\tau. \tag{5.4}$$

In particular, when $0<\alpha(t)<1$, $-\infty<a<b<\infty$ and $h^{(1)}(t)>0$, the left-sided Riemann-Liouville type fractional derivative of variable order with respect to another function is defined as (see [114]; also see [324])

$$\left({}_{RL}D^{\alpha(\cdot)}_{a+,h}f\right)(t)=\left(\frac{1}{h^{(1)}(t)}\frac{d}{dt}\right)\int_a^t \frac{h^{(1)}(\tau)}{\Gamma(1-\alpha(\tau))}\frac{f(\tau)}{(h(t)-h(\tau))^{\alpha(\tau)}}d\tau, \tag{5.5}$$

and the right-sided Riemann-Liouville type fractional derivative of variable order with respect to another function as (see [114])

$$\left({}_{RL}D^{\alpha(\cdot)}_{b-,h}f\right)(t)=\left(-\frac{1}{h^{(1)}(t)}\frac{d}{dt}\right)\int_t^b \frac{h^{(1)}(\tau)}{\Gamma(1-\alpha(\tau))}\frac{f(\tau)}{(h(\tau)-h(t))^{\alpha(\tau)}}d\tau. \tag{5.6}$$

Definition 5.3 *Let $\alpha(t)\geq 0$, $\kappa=[\alpha(t)]+1$ and $h^{(1)}(t)>0$. The left-sided Liouville-Sonine-Caputo type fractional derivative of variable order with respect to another function is defined as*

$${}_{LSC}D^{\alpha(\cdot)}_{a+,h}f(t)=\int_a^t \frac{1}{\Gamma(\kappa-\alpha(\tau))}\frac{h^{(1)}(\tau)}{(h(t)-h(\tau))^{\alpha(\tau)-\kappa+1}}\left(\left(\frac{1}{h^{(1)}(t)}\frac{d}{dt}\right)^{\kappa}f_h^{(\kappa)}(\tau)\right)d\tau, \tag{5.7}$$

and the right-sided Liouville-Sonine-Caputo type fractional derivative of variable order with respect to another function as

$${}_{LSC}D^{\alpha(\cdot)}_{b-,h}f(t)=(-1)^{\kappa}\int_t^b \frac{1}{\Gamma(\kappa-\alpha(\tau))}\frac{h^{(1)}(\tau)}{(h(\tau)-h(t))^{\alpha(\tau)-\kappa+1}}\left(\left(-\frac{1}{h^{(1)}(\tau)}\frac{d}{d\tau}\right)^{\kappa}f_h^{(\kappa)}(\tau)\right)d\tau. \tag{5.8}$$

In particular, when $0 < \alpha(t) < 1$ and $h^{(1)}(t) > 0$, the left-sided Liouville-Sonine fractional derivative of variable order with respect to another function as (see [114]; also see [324])

$$ {}_{LS}D^{\alpha(\cdot)}_{a+,h}f(t) = \int_a^t \frac{1}{\Gamma(1-\alpha(\tau))}\frac{h^{(1)}(\tau)}{(h(t)-h(\tau))^{\alpha(\tau)}}\left(\left(\frac{1}{h^{(1)}(\tau)}\frac{d}{d\tau}\right)f_h^{(1)}(\tau)\right)d\tau, \quad (5.9) $$

the right-sided Liouville-Sonine fractional derivative of variable order with respect to another function as

$$ {}_{LS}D^{\alpha(\cdot)}_{b-,h}f(t) = \int_t^b \frac{1}{\Gamma(1-\alpha(\tau))}\frac{h^{(1)}(\tau)}{(h(\tau)-h(t))^{\alpha(\tau)}}\left(\left(-\frac{1}{h^{(1)}(\tau)}\frac{d}{d\tau}\right)f_h^{(1)}(\tau)\right)d\tau. \quad (5.10) $$

More generally, when $0 < \alpha(\tau,t) < 1$ and $h^{(1)}(t) > 0$, the left-sided Riemann-Liouville type fractional integral of variable order with respect to another function is defined as (see [114])

$$ \left({}^{HL}_{RLq}I^{\alpha(\cdot)}_{a+,h}f\right)(t) = \int_a^t \frac{h^{(1)}(\tau)}{\Gamma(\alpha(\tau,t))}\frac{f(\tau)}{(h(t)-h(\tau))^{1-\alpha(\tau,t)}}d\tau, \quad (5.11) $$

the right-sided Riemann-Liouville type fractional integral of variable order with respect to another function as (see [114])

$$ \left({}^{HL}_{RLq}I^{\alpha(\cdot)}_{b-,h}f\right)(t) = \int_t^b \frac{h^{(1)}(\tau)}{\Gamma(\alpha(\tau,t))}\frac{f(\tau)}{(h(\tau)-h(t))^{1-\alpha(\tau,t)}}d\tau, \quad (5.12) $$

the left-sided Riemann-Liouville type fractional derivative of variable order with respect to another function is defined as (see [114])

$$ \left({}_{RLq}D^{\alpha(\cdot)}_{a+,h}f\right)(t) = \left(\frac{1}{h^{(1)}(t)}\frac{d}{dt}\right)\int_a^t \frac{h^{(1)}(\tau)}{\Gamma(1-\alpha(\tau,t))}\frac{f(\tau)}{(h(t)-h(\tau))^{\alpha(\tau,t)}}d\tau, \quad (5.13) $$

the right-sided Riemann-Liouville type fractional derivative of variable order with respect to another function as

$$ \left({}_{RLq}D^{\alpha(\cdot)}_{b-,h}f\right)(t) = \left(-\frac{1}{h^{(1)}(t)}\frac{d}{dt}\right)\int_t^b \frac{h^{(1)}(\tau)}{\Gamma(1-\alpha(\tau,t))}\frac{f(\tau)}{(h(\tau)-h(t))^{\alpha(\tau,t)}}d\tau, \quad (5.14) $$

the left-sided Riemann-Liouville type fractional derivative of variable order with respect to another function is defined as

$$ \left({}_{RLq}D^{\alpha(\cdot)}_{a+,h}f\right)(t) = \frac{1}{\Gamma(1-\alpha(\tau,t))}\left(\frac{1}{h^{(1)}(t)}\frac{d}{dt}\right)\int_a^t \frac{h^{(1)}(\tau)f(\tau)}{(h(t)-h(\tau))^{\alpha(\tau,t)}}d\tau, \quad (5.15) $$

the right-sided Riemann-Liouville type fractional derivative of variable order with respect to another function as

$$\left({}_{RLq}D_{b-,h}^{\alpha(\cdot)}f\right)(t)=\frac{1}{\Gamma(1-\alpha(\tau,t))}\left(-\frac{1}{h^{(1)}(t)}\frac{d}{dt}\right)\int_t^b\frac{h^{(1)}(\tau)f(\tau)}{(h(\tau)-h(t))^{\alpha(\tau,t)}}d\tau, \tag{5.16}$$

the left-sided Liouville-Sonine fractional derivative of variable order with respect to another function as

$${}_{LSq}D_{a+,h}^{\alpha(\cdot)}f(x)=\int_a^t\frac{h^{(1)}(\tau)}{\Gamma(1-\alpha(\tau,t))}\frac{f^{(1)}(t)}{(h(t)-h(\tau))^{\alpha(\tau,t)}}dt, \tag{5.17}$$

and the right-sided Liouville-Sonine fractional derivative of variable order with respect to another function as

$${}_{LSq}D_{b-,h}^{\alpha(\cdot)}f(t)=-\int_t^b\frac{h^{(1)}(\tau)}{\Gamma(1-\alpha(\tau,t))}\frac{f^{(1)}(t)}{(h(\tau)-h(t))^{\alpha(\tau,t)}}dt. \tag{5.18}$$

In particular, when $\alpha(\tau,t)=\alpha(t)$, the left-sided Riemann-Liouville type fractional integral of variable order with respect to another function is defined as

$$\left({}_{RLqs}I_{a+,h}^{\alpha(\cdot)}f\right)(t)=\frac{1}{\Gamma(\alpha(t))}\int_a^t\frac{h^{(1)}(\tau)f(\tau)}{(h(t)-h(\tau))^{1-\alpha(t)}}d\tau, \tag{5.19}$$

the right-sided Riemann-Liouville type fractional integral of variable order with respect to another function as

$$\left({}_{RLqs}I_{b-,h}^{\alpha(\cdot)}f\right)(t)=\frac{1}{\Gamma(\alpha(t))}\int_t^b\frac{h^{(1)}(\tau)f(\tau)}{(h(\tau)-h(t))^{1-\alpha(t)}}d\tau, \tag{5.20}$$

the left-sided Riemann-Liouville type fractional derivative of variable order with respect to another function is defined as

$$\left({}_{RLqs}^{SR}D_{a+,h}^{\alpha(\cdot)}f\right)(t)=\frac{1}{\Gamma(1-\alpha(t))}\left(\frac{1}{h^{(1)}(t)}\frac{d}{dt}\right)\int_a^t\frac{h^{(1)}(\tau)f(\tau)}{(h(t)-h(\tau))^{\alpha(t)}}d\tau, \tag{5.21}$$

the right-sided Riemann-Liouville type fractional derivative of variable order with respect to another function as

$$\left({}_{RLqs}D_{b-,h}^{\alpha(\cdot)}f\right)(t)=-\frac{1}{\Gamma(1-\alpha(t))}\left(-\frac{1}{h^{(1)}(t)}\frac{d}{dt}\right)\int_t^b\frac{h^{(1)}(\tau)f(\tau)}{(h(\tau)-h(t))^{\alpha(t)}}d\tau, \tag{5.22}$$

the left-sided Riemann-Liouville type fractional derivative of variable order with respect to another function is defined as

$$\left({}_{RLqs}D_{a+,h}^{\alpha(\cdot)}f\right)(t)=\left(\frac{1}{h^{(1)}(t)}\frac{d}{dt}\right)\left(\frac{1}{\Gamma(1-\alpha(t))}\int_a^t\frac{h^{(1)}(\tau)f(\tau)}{(h(t)-h(\tau))^{\alpha(t)}}d\tau\right), \tag{5.23}$$

the right-sided Riemann-Liouville type fractional derivative of variable order with respect to another function as

$$\left({}_{RLqs}D^{\alpha(\cdot)}_{b-,h}f\right)(t) = \left(-\frac{1}{h^{(1)}(t)}\frac{d}{dt}\right)\left(\frac{1}{\Gamma(1-\alpha(t))}\int_t^b \frac{h^{(1)}(\tau)f(\tau)}{(h(\tau)-h(t))^{\alpha(t)}}d\tau\right), \quad (5.24)$$

the left-sided Liouville-Sonine fractional derivative of variable order with respect to another function as

$${}_{LSqs}D^{\alpha(\cdot)}_{a+}f(t) = \frac{1}{\Gamma(1-\alpha(t))}\int_a^t \frac{h^{(1)}(\tau)}{(h(t)-h(\tau))^{\alpha(t)}}\left(\left(\frac{1}{h^{(1)}(\tau)}\frac{d}{d\tau}\right)f_h^{(1)}(\tau)\right)d\tau, \quad (5.25)$$

and the right-sided Liouville-Sonine fractional derivative of variable order with respect to another function as

$${}_{LSqs}D^{\alpha(\cdot)}_{b-,h}f(t) = \frac{1}{\Gamma(1-\alpha(t))}\int_t^b \frac{h^{(1)}(\tau)}{(h(\tau)-h(t))^{\alpha(t)}}\left(\left(-\frac{1}{h^{(1)}(\tau)}\frac{d}{d\tau}\right)f_h^{(1)}(\tau)\right)d\tau. \quad (5.26)$$

More generally, when $\kappa < \alpha(t) < \kappa+1$, the left-sided Riemann-Liouville type fractional derivative of variable order with respect to another function is defined as

$$\left({}_{RLqs}D^{\alpha(\cdot)}_{a+,h}f\right)(t) = \left(\frac{1}{h^{(1)}(t)}\frac{d}{dt}\right)^{\kappa}\left(\frac{1}{\Gamma(\kappa-\alpha(t))}\int_a^t \frac{h^{(1)}(\tau)f(\tau)}{(h(t)-h(\tau))^{\alpha(t)-\kappa+1}}d\tau\right), \quad (5.27)$$

the right-sided Riemann-Liouville type fractional derivative of variable order with respect to another function as

$$\left({}_{RLqs}D^{\alpha(\cdot)}_{b-,h}f\right)(t) = \left(-\frac{1}{h^{(1)}(t)}\frac{d}{dt}\right)^{\kappa}\left(\frac{1}{\Gamma(\kappa-\alpha(t))}\int_t^b \frac{h^{(1)}(\tau)f(\tau)}{(h(\tau)-h(t))^{\alpha(t)-\kappa+1}}d\tau\right), \quad (5.28)$$

the left-sided Liouville-Sonine fractional derivative of variable order with respect to another function as

$${}_{LSqs}D^{\alpha(\cdot)}_{a+,h}f(t) = \frac{1}{\Gamma(\kappa-\alpha(t))}\int_a^t \frac{h^{(1)}(\tau)}{(h(t)-h(\tau))^{\alpha(t)-\kappa+1}}\left(\left(\frac{1}{h^{(1)}(\tau)}\frac{d}{d\tau}\right)^{\kappa}f_h^{(\kappa)}(\tau)\right)d\tau, \quad (5.29)$$

and the right-sided Liouville-Sonine fractional derivative of variable order with respect to another function as

$${}_{LSqs}D^{\alpha(\cdot)}_{b-,h}f(t) = \frac{1}{\Gamma(1-\alpha(t))}\int_t^b \frac{h^{(1)}(\tau)}{(h(\tau)-h(t))^{\alpha(t)}}\left(\left(-\frac{1}{h^{(1)}(\tau)}\frac{d}{d\tau}\right)^{\kappa}f_h^{(\kappa)}(\tau)\right)d\tau. \quad (5.30)$$

5.2 Riesz type fractional derivatives of variable order with respect to another function

In this section, we extend the Riesz type fractional derivatives of variable order (see [313]), proposed by Zayernouri and Karniadakis in 2015, and give the Riesz type fractional derivatives of variable order with respect to another function.

Definition 5.4 *The Riesz type fractional derivative of variable order based on the Riemann-Liouville fractional derivative of variable order with respect to another function is defined as*

$$\begin{aligned}
&\left({}_{ZK}^{Ri}D_{[a,b],h}^{\alpha(\cdot)}f\right)(t)\\
&=-\frac{1}{2\cos(\pi\alpha(t)/2)}\times\left(\left({}_{RLqs}D_{a+,h}^{\alpha(\cdot)}f\right)(t)+\left({}_{RLqs}D_{b-,h}^{\alpha(\cdot)}f\right)(t)\right)\\
&=-\frac{1}{2\cos(\pi\alpha(t)/2)}\times\frac{1}{\Gamma(1-\alpha(t))}\left(\frac{1}{h^{(1)}(t)}\frac{d}{dt}\right)\int_a^t\frac{h^{(1)}(\tau)f(\tau)}{(h(t)-h(\tau))^{\alpha(t)}}d\tau\\
&+\frac{1}{2\cos(\pi\alpha(t)/2)}\times\frac{1}{\Gamma(1-\alpha(t))}\left(\frac{1}{h^{(1)}(t)}\frac{d}{dt}\right)\int_t^b\frac{h^{(1)}(\tau)f(\tau)}{(h(\tau)-h(t))^{\alpha(t)}}d\tau,
\end{aligned}\tag{5.31}$$

where

$$\left({}_{RLqs}D_{a+,h}^{\alpha(\cdot)}f\right)(t)=\frac{1}{\Gamma(1-\alpha(t))}\left(\frac{1}{h^{(1)}(t)}\frac{d}{dt}\right)\int_a^t\frac{h^{(1)}(\tau)f(\tau)}{(h(t)-h(\tau))^{\alpha(t)}}d\tau \tag{5.32}$$

and

$$\left({}_{RLqs}D_{b-,h}^{\alpha(\cdot)}f\right)(t)=\frac{1}{\Gamma(1-\alpha(t))}\left(-\frac{1}{h^{(1)}(t)}\frac{d}{dt}\right)\int_t^b\frac{h^{(1)}(\tau)f(\tau)}{(h(\tau)-h(t))^{\alpha(t)}}d\tau. \tag{5.33}$$

More generally, the Riesz type fractional derivative of variable order with respect to another function based on the Riemann-Liouville fractional derivative of variable order with respect to another function is defined as

$$\begin{aligned}
&\left({}_{GZK}^{Ri}D_{[a,b],h}^{\alpha(\cdot)}f\right)(t)\\
&=-\frac{1}{2\cos(\pi\alpha(t)/2)}\times\left(\left({}_{RLq}D_{a+,h}^{\alpha(\cdot)}f\right)(t)+\left({}_{RLq}D_{b-,h}^{\alpha(\cdot)}f\right)(t)\right)\\
&=-\frac{1}{2\cos(\pi\alpha(t)/2)}\times\left(\frac{1}{h^{(1)}(t)}\frac{d}{dt}\right)\int_a^t\frac{h^{(1)}(\tau)}{\Gamma(1-\alpha(\tau,t))}\frac{f(\tau)}{(h(t)-h(\tau))^{\alpha(\tau,t)}}d\tau\\
&+\frac{1}{2\cos(\pi\alpha(t)/2)}\times\left(\frac{1}{h^{(1)}(t)}\frac{d}{dt}\right)\int_t^b\frac{h^{(1)}(\tau)}{\Gamma(1-\alpha(\tau,t))}\frac{f(\tau)}{(h(\tau)-h(t))^{\alpha(\tau,t)}}d\tau,
\end{aligned}\tag{5.34}$$

where

$$\left({}_{RLq}D^{\alpha(\cdot)}_{a+,h}f\right)(t) = \left(\frac{1}{h^{(1)}(t)}\frac{d}{dt}\right)\int_a^t \frac{h^{(1)}(\tau)}{\Gamma(1-\alpha(\tau,t))}\frac{f(\tau)}{(h(t)-h(\tau))^{\alpha(\tau,t)}}d\tau, \tag{5.35}$$

and

$$\left({}_{RLq}D^{\alpha(\cdot)}_{b-,h}f\right)(t) = \left(-\frac{1}{h^{(1)}(t)}\frac{d}{dt}\right)\int_t^b \frac{h^{(1)}(\tau)}{\Gamma(1-\alpha(\tau,t))}\frac{f(\tau)}{(h(\tau)-h(t))^{\alpha(\tau,t)}}d\tau. \tag{5.36}$$

In particular, when $\alpha(\tau,t) = \alpha(t)$, the Riesz type fractional derivative of variable order with respect to another function based on the Riemann-Liouville type fractional derivative of variable order with respect to another function is defined as

$$\begin{aligned}
&\left({}^{Ri}_{SZK}D^{\alpha(\cdot)}_{[a,b],h}f\right)(t) \\
&= -\frac{1}{2\cos(\pi\alpha(t)/2)} \times \left(\left({}_{RL}D^{\alpha(\cdot)}_{a+,h}f\right)(t) + \left({}_{RL}D^{\alpha(\cdot)}_{b-,h}f\right)(t)\right) \\
&= -\frac{1}{2\cos(\pi\alpha(t)/2)} \times \left(\frac{1}{h^{(1)}(t)}\frac{d}{dt}\right)\int_a^t \frac{h^{(1)}(\tau)}{\Gamma(1-\alpha(\tau))}\frac{f(\tau)}{(h(t)-h(\tau))^{\alpha(\tau)}}d\tau \\
&+\frac{1}{2\cos(\pi\alpha(t)/2)} \times \left(\frac{1}{h^{(1)}(t)}\frac{d}{dt}\right)\int_t^b \frac{h^{(1)}(\tau)}{\Gamma(1-\alpha(\tau))}\frac{f(\tau)}{(h(\tau)-h(t))^{\alpha(\tau)}}d\tau,
\end{aligned} \tag{5.37}$$

where

$$\left({}_{RL}D^{\alpha(\cdot)}_{a+,h}f\right)(t) = \left(\frac{1}{h^{(1)}(t)}\frac{d}{dt}\right)\int_a^t \frac{h^{(1)}(\tau)}{\Gamma(1-\alpha(\tau))}\frac{f(\tau)}{(h(t)-h(\tau))^{\alpha(\tau)}}d\tau \tag{5.38}$$

and

$$\left({}_{RL}D^{\alpha(\cdot)}_{b-,h}f\right)(t) = \left(-\frac{1}{h^{(1)}(t)}\frac{d}{dt}\right)\int_t^b \frac{h^{(1)}(\tau)}{\Gamma(1-\alpha(\tau))}\frac{f(\tau)}{(h(\tau)-h(t))^{\alpha(\tau)}}d\tau. \tag{5.39}$$

Definition 5.5 *The Riesz type fractional derivative of variable order with respect to another function based on the Liouville-Sonine fractional derivative of variable order with respect to another function is defined as*

$$\begin{aligned}
&\left({}^{Ri}_{LSqs}D^{\alpha(\cdot)}_{[a,b],h}f\right)(t) \\
&= -\frac{1}{2\cos(\pi\alpha(t)/2)} \times \left(\left({}_{LSqs}D^{\alpha(\cdot)}_{a+,h}f\right)(t) + \left({}_{LSqs}D^{\alpha(\cdot)}_{b-,h}f\right)(t)\right) \\
&= -\frac{1}{2\cos(\pi\alpha(t)/2)} \times \frac{1}{\Gamma(1-\alpha(t))}\int_a^t \frac{h^{(1)}(\tau)}{(h(t)-h(\tau))^{\alpha(t)}}\left(\frac{1}{h^{(1)}(\tau)}\frac{d}{d\tau}\right)f_h^{(1)}(\tau)\,d\tau \\
&+\frac{1}{2\cos(\pi\alpha(t)/2)} \times \frac{1}{\Gamma(1-\alpha(t))}\int_t^b \frac{h^{(1)}(\tau)}{(h(\tau)-h(t))^{\alpha(t)}}\left(\frac{1}{h^{(1)}(\tau)}\frac{d}{d\tau}\right)f_h^{(1)}(\tau)\,d\tau,
\end{aligned} \tag{5.40}$$

where

$$\left({}_{LSqs}D^{\alpha(\cdot)}_{a+,h}f\right)(t)=\frac{1}{\Gamma\left(1-\alpha\left(t\right)\right)}\int_a^t\frac{h^{(1)}\left(\tau\right)}{\left(h\left(t\right)-h\left(\tau\right)\right)^{\alpha(t)}}\left(\frac{1}{h^{(1)}\left(\tau\right)}\frac{d}{d\tau}\right)f_h^{(1)}\left(\tau\right)d\tau \tag{5.41}$$

and

$$\left({}_{LSqs}D^{\alpha(\cdot)}_{b-}f\right)(t)=\frac{1}{\Gamma\left(1-\alpha\left(t\right)\right)}\int_t^b\frac{h^{(1)}\left(\tau\right)}{\left(h\left(\tau\right)-h\left(t\right)\right)^{\alpha(t)}}\left(-\frac{1}{h^{(1)}\left(\tau\right)}\frac{d}{d\tau}\right)f_h^{(1)}\left(\tau\right)d\tau. \tag{5.42}$$

More generally, the Riesz type fractional derivative of variable order with respect to another function based on the Liouville-Sonine fractional derivative of variable order with respect to another function is defined as

$$\begin{aligned}
&\left({}^{Ri}_{GLSq}D^{\alpha(\cdot)}_{[a,b],h}f\right)(t)\\
&=-\tfrac{1}{2\cos(\pi\alpha(t)/2)}\times\left(\left({}_{LSq}D^{\alpha(\cdot)}_{a+,h}f\right)(t)+\left({}_{LSq}D^{\alpha(\cdot)}_{b-,h}f\right)(t)\right)\\
&=-\tfrac{1}{2\cos(\pi\alpha(t)/2)}\times\int_a^t\tfrac{h^{(1)}(\tau)}{\Gamma(1-\alpha(\tau,t))}\tfrac{1}{(h(t)-h(\tau))^{\alpha(\tau,t)}}\left(\tfrac{1}{h^{(1)}(\tau)}\tfrac{d}{d\tau}\right)f_h^{(1)}\left(\tau\right)d\tau\\
&+\tfrac{1}{2\cos(\pi\alpha(t)/2)}\times\int_t^b\tfrac{h^{(1)}(\tau)}{\Gamma(1-\alpha(\tau,t))}\tfrac{1}{(h(\tau)-h(t))^{\alpha(\tau,t)}}\left(\tfrac{1}{h^{(1)}(\tau)}\tfrac{d}{d\tau}\right)f_h^{(1)}\left(\tau\right)d\tau,
\end{aligned} \tag{5.43}$$

where

$$\left({}_{LSq}D^{\alpha(\cdot)}_{a+,h}f\right)(t)=\int_a^t\frac{1}{\Gamma\left(1-\alpha\left(\tau,t\right)\right)}\frac{h^{(1)}\left(\tau\right)}{\left(h\left(t\right)-h\left(\tau\right)\right)^{\alpha(\tau,t)}}\left(\frac{1}{h^{(1)}\left(\tau\right)}\frac{d}{d\tau}\right)f_h^{(1)}\left(\tau\right)d\tau \tag{5.44}$$

and

$$\left({}_{RLq}D^{\alpha(\cdot)}_{b-,h}f\right)(t)=\int_t^b\frac{1}{\Gamma\left(1-\alpha\left(\tau,t\right)\right)}\frac{h^{(1)}\left(\tau\right)}{\left(h\left(\tau\right)-h\left(t\right)\right)^{\alpha(\tau,t)}}\left(-\frac{1}{h^{(1)}\left(\tau\right)}\frac{d}{d\tau}\right)f_h^{(1)}\left(\tau\right)d\tau. \tag{5.45}$$

In particular, when $\alpha\left(\tau,t\right)=\alpha\left(t\right)$, the Riesz type fractional derivative of variable order with respect to another function based on the Liouville-Sonine fractional derivative of variable order with respect to another function is defined as

$$\begin{aligned}
&\left({}^{Ri}_{GLC}D^{\alpha(\cdot)}_{[a,b],h}f\right)(t)\\
&=-\tfrac{1}{2\cos(\pi\alpha(t)/2)}\times\left(\left({}_{GLC}D^{\alpha(\cdot)}_{a+,h}f\right)(t)+\left({}_{GLC}D^{\alpha(\cdot)}_{b-,h}f\right)(t)\right)\\
&=-\tfrac{1}{2\cos(\pi\alpha(t)/2)}\times\int_a^t\tfrac{1}{\Gamma(1-\alpha(\tau))}\tfrac{h^{(1)}(\tau)}{(h(t)-h(\tau))^{\alpha(\tau)}}\left(\tfrac{1}{h^{(1)}(\tau)}\tfrac{d}{d\tau}\right)f_h^{(1)}\left(\tau\right)d\tau\\
&+\tfrac{1}{2\cos(\pi\alpha(t)/2)}\times\int_t^b\tfrac{1}{\Gamma(1-\alpha(\tau))}\tfrac{h^{(1)}(\tau)}{(h(\tau)-h(t))^{\alpha(\tau)}}\left(\tfrac{1}{h^{(1)}(\tau)}\tfrac{d}{d\tau}\right)f_h^{(1)}\left(\tau\right)d\tau,
\end{aligned} \tag{5.46}$$

where

$$\left({}_{GLC}D_{a+}^{\alpha(\cdot)}f\right)(t)=\int_a^t\frac{1}{\Gamma(1-\alpha(\tau))}\frac{h^{(1)}(\tau)}{(h(t)-h(\tau))^{\alpha(\tau)}}\left(\frac{1}{h^{(1)}(\tau)}\frac{d}{d\tau}\right)f_h^{(1)}(\tau)\,d\tau \tag{5.47}$$

and

$$\left({}_{GLC}D_{b-}^{\alpha(\cdot)}f\right)(t)=\int_t^b\frac{1}{\Gamma(1-\alpha(\tau))}\frac{h^{(1)}(\tau)}{(h(\tau)-h(t))^{\alpha(\tau)}}\left(-\frac{1}{h^{(1)}(\tau)}\frac{d}{d\tau}\right)f_h^{(1)}(\tau)\,d\tau. \tag{5.48}$$

5.3 Hilfer type fractional derivatives of variable order with respect to another function

In this section, we introduce the Hilfer type fractional derivatives of variable order with respect to another function.

Definition 5.6 *Let* $0<\alpha(\tau,t)<1$, $0\le\beta\le1$, $h^{(1)}(t)>0$,$f\in L_1(a,b)$ *and* $-\infty<a<b<\infty$. *The left-sided variable-order Hilfer type fractional derivative with respect to another function is defined as*

$${}_{Sq}^{Hi}D_{a+,h}^{\alpha(\cdot),\beta}f(t)=\left({}_{RLq}^{HL}I_{a+,h}^{\beta(1-\alpha(\cdot))}\left(\frac{1}{h^{(1)}(t)}\frac{d}{dt}\right)\left({}_{RLq}^{HL}I_{a+,h}^{(1-\beta)(1-\alpha(\cdot))}f\right)\right)(t), \tag{5.49}$$

and the right-sided variable-order Hilfer type fractional derivative with respect to another function as

$${}_{Sq}^{Hi}D_{b-,h}^{\alpha(\cdot),\beta}f(t)=\left({}_{RLq}^{HL}I_{b-,h}^{\beta(1-\alpha(\cdot))}\left(\frac{1}{h^{(1)}(t)}\frac{d}{dt}\right)\left({}_{RLq}^{HL}I_{b-,h}^{(1-\beta)(1-\alpha(\cdot))}f\right)\right)(t), \tag{5.50}$$

where

$$\left({}_{RLq}^{HL}I_{a+,h}^{\beta(1-\alpha(\cdot))}f\right)(t)=\int_a^t\frac{h^{(1)}(\tau)}{\Gamma(\beta(1-\alpha(\tau,t)))}\frac{f(\tau)}{(h(t)-h(\tau))^{1-\beta(1-\alpha(\tau,t))}}d\tau, \tag{5.51}$$

$$\left({}_{RLq}^{HL}I_{b-,h}^{\beta(1-\alpha(\cdot))}f\right)(t)=\int_t^b\frac{h^{(1)}(\tau)}{\Gamma(\beta(1-\alpha(\tau,t)))}\frac{f(\tau)}{(h(\tau)-h(t))^{1-\beta(1-\alpha(\tau,t))}}d\tau, \tag{5.52}$$

$$\left({}_{RLq}^{HL}I_{a+,h}^{(1-\beta)(1-\alpha(\cdot))}\right)f(t)=\int_{a}^{t}\frac{h^{(1)}(\tau)}{\Gamma((1-\beta)(1-\alpha(\tau,t)))}\frac{f(\tau)}{(h(t)-h(\tau))^{1-(1-\beta)(1-\alpha(\tau,t))}}d\tau \tag{5.53}$$

and

$$\left({}_{RLq}^{HL}I_{b-,h}^{(1-\beta)(1-\alpha(\cdot))}f\right)(t)=\int_{t}^{b}\frac{h^{(1)}(\tau)}{\Gamma((1-\beta)(1-\alpha(\tau,t)))}\frac{f(\tau)}{(h(\tau)-h(t))^{1-(1-\beta)(1-\alpha(\tau,t))}}d\tau. \tag{5.54}$$

Definition 5.7 *Let $0<\alpha(t)<1$, $0\leq\beta\leq1$, $h^{(1)}(t)>0$, $f\in L_1(a,b)$ and $-\infty<a<b<\infty$. The left-sided variable-order Hilfer type fractional derivative with respect to another function is defined as*

$${}_{S}^{Hi}D_{a+,h}^{\alpha(\cdot),\beta}f(t)=\left({}_{RL}^{HL}I_{a+,h}^{\beta(1-\alpha(\cdot))}\left(\frac{1}{h^{(1)}(t)}\frac{d}{dt}\right)\left({}_{RL}^{HL}I_{a+,h}^{(1-\beta)(1-\alpha(\cdot))}f\right)\right)(t), \tag{5.55}$$

and the right-sided variable-order Hilfer type fractional derivative with respect to another function as

$${}_{S}^{Hi}D_{b-,h}^{\alpha(\cdot),\beta}f(t)=\left({}_{RL}^{HL}I_{b-,h}^{\beta(1-\alpha(\cdot))}\left(-\frac{1}{h^{(1)}(t)}\frac{d}{dt}\right)\left({}_{RL}^{HL}I_{b-,h}^{(1-\beta)(1-\alpha(\cdot))}f\right)\right)(t), \tag{5.56}$$

where

$$\left({}_{RL}^{HL}I_{a+,h}^{\beta(1-\alpha(\cdot))}f\right)(t)=\int_{a}^{t}\frac{h^{(1)}(\tau)}{\Gamma(\beta(1-\alpha(\tau)))}\frac{f(\tau)}{(h(t)-h(\tau))^{1-\beta(1-\alpha(\tau))}}d\tau, \tag{5.57}$$

$$\left({}_{RL}^{HL}I_{b-,h}^{\beta(1-\alpha(\cdot))}f\right)(t)=\int_{t}^{b}\frac{h^{(1)}(\tau)}{\Gamma(\beta(1-\alpha(\tau)))}\frac{f(\tau)}{(h(\tau)-h(t))^{1-\beta(1-\alpha(\tau))}}d\tau, \tag{5.58}$$

$$\left({}_{RL}^{HL}I_{a+,h}^{(1-\beta)(1-\alpha(\cdot))}\right)f(t)=\int_{a}^{t}\frac{h^{(1)}(\tau)}{\Gamma((1-\beta)(1-\alpha(\tau)))}\frac{f(\tau)}{(h(t)-h(\tau))^{1-(1-\beta)(1-\alpha(\tau))}}d\tau \tag{5.59}$$

and

$$\left({}_{RL}^{HL}I_{b-,h}^{(1-\beta)(1-\alpha(\cdot))}f\right)(t)=\int_{t}^{b}\frac{h^{(1)}(\tau)}{\Gamma((1-\beta)(1-\alpha(\tau)))}\frac{f(\tau)}{(h(\tau)-h(t))^{1-(1-\beta)(1-\alpha(\tau))}}d\tau. \tag{5.60}$$

Definition 5.8 *Let $0<\alpha(t)<1$, $0\leq\beta\leq1$, $h^{(1)}(t)>0$, $f\in L_1(a,b)$ and $-\infty<a<b<\infty$. The left-sided variable-order Hilfer type fractional derivative with respect to another function is defined as*

$$ {}_{Sqs}^{Hi}D_{a+,h}^{\alpha(\cdot),\beta} f(t) = \left({}_{RLqs}^{HL}I_{a+,h}^{\beta(1-\alpha(\cdot))} \left(\frac{1}{h^{(1)}(t)} \frac{d}{dt} \right) \left({}_{RLqs}^{HL}I_{a+,h}^{(1-\beta)(1-\alpha(\cdot))} f \right) \right)(t), \tag{5.61} $$

and the right-sided variable-order Hilfer type fractional derivative with respect to another function as

$$ {}_{Sqs}^{Hi}D_{b-,h}^{\alpha(\cdot),\beta} f(t) = \left({}_{RLqs}^{HL}I_{b-,h}^{\beta(1-\alpha(\cdot))} \left(-\frac{1}{h^{(1)}(t)} \frac{d}{dt} \right) \left({}_{RLqs}^{HL}I_{b-,h}^{(1-\beta)(1-\alpha(\cdot))} f \right) \right)(t), \tag{5.62} $$

where

$$ \left({}_{RLqs}^{HL}I_{a+,h}^{\beta(1-\alpha(\cdot))} f \right)(t) = \frac{1}{\Gamma(\beta(1-\alpha(t)))} \int_a^t \frac{h^{(1)}(\tau) f(\tau)}{(h(t)-h(\tau))^{1-\beta(1-\alpha(t))}} d\tau, \tag{5.63} $$

and

$$ \left({}_{RLqs}^{HL}I_{b-,h}^{\beta(1-\alpha(\cdot))} f \right)(t) = \frac{1}{\Gamma(\beta(1-\alpha(t)))} \int_t^b \frac{h^{(1)}(\tau) f(\tau)}{(h(\tau)-h(t))^{1-\beta(1-\alpha(t))}} d\tau, \tag{5.64} $$

$$ \left({}_{RLqs}^{HL}I_{a+,h}^{(1-\beta)(1-\alpha(\cdot))} \right) f(t) = \frac{1}{\Gamma((1-\beta)(1-\alpha(t)))} \int_a^t \frac{h^{(1)}(\tau) f(\tau)}{(h(t)-h(\tau))^{1-(1-\beta)(1-\alpha(t))}} d\tau, \tag{5.65} $$

and

$$ \left({}_{RLqs}^{HL}I_{b-,h}^{(1-\beta)(1-\alpha(\cdot))} f \right)(t) = \frac{1}{\Gamma((1-\beta)(1-\alpha(t)))} \int_t^b \frac{h^{(1)}(\tau) f(\tau)}{(h(\tau)-h(t))^{1-(1-\beta)(1-\alpha(t))}} d\tau. \tag{5.66} $$

5.4 Tempered fractional derivatives of variable order with respect to another function

In this section, we propose the variable-order general fractional derivatives and the variable-order tempered fractional integrals with respect to another function.

Definition 5.9 *Let $\kappa < \alpha(t) < \kappa + 1$, $-\infty < a < b < \infty$ and $\lambda \in \mathbb{R}$. The left-sided variable-order Riemann-Liouville type tempered fractional derivative with respect to another function is defined as*

$$ \left({}_{Cp}^{RL}D_{a+,h}^{\alpha(\cdot),\lambda} f \right)(t) = \left(\frac{1}{h^{(1)}(t)} \frac{d}{dt} \right)^{\kappa} \int_a^t \frac{1}{\Gamma(\kappa-\alpha(\tau))} \frac{e^{-\lambda(h(t)-h(\tau))} h^{(1)}(\tau)}{(h(t)-h(\tau))^{\alpha(\tau)-\kappa+1}} f(\tau)\, d\tau, \tag{5.67} $$

and the right-sided variable-order Riemann-Liouville type tempered fractional derivative with respect to another function as

$$\left({}^{RL}_{Cp}D^{\alpha(\cdot),\lambda}_{b-,h}f\right)(t)=\left(-\frac{1}{h^{(1)}(t)}\frac{d}{dt}\right)^{\kappa}\int_t^b\frac{1}{\Gamma(\kappa-\alpha(\tau))}\frac{h^{(1)}(\tau)e^{-\lambda(h(\tau)-h(t))}}{(h(\tau)-h(t))^{\alpha(\tau)-\kappa+1}}f(\tau)\,d\tau. \tag{5.68}$$

Definition 5.10 *Let $\alpha(t)>0$, $\kappa=[\alpha(t)]+1$, $-\infty<a<b<\infty$and $\lambda\in\mathbb{R}$. The left-sided variable-order Liouville-Sonine-Caputo type tempered fractional derivative with respect to another function is defined as*

$$\left({}^{LSC}_{Cp}D^{\alpha(\cdot),\lambda}_{a+,h}f\right)(t)=\int_a^t\frac{1}{\Gamma(\kappa-\alpha(\tau))}\frac{h^{(1)}(\tau)e^{-\lambda(h(t)-h(\tau))}}{(h(t)-h(\tau))^{\alpha(\tau)-\kappa+1}}\left(\left(\frac{1}{h^{(1)}(\tau)}\frac{d}{d\tau}\right)^{\kappa}f_h^{(\kappa)}(\tau)\right)d\tau, \tag{5.69}$$

and the right-sided variable-order Liouville-Sonine-Caputo type tempered fractional derivative with respect to another function as

$$\left({}^{LSC}_{Cp}D^{\alpha(\cdot),\lambda}_{b-,h}f\right)(t)=\int_t^b\frac{1}{\Gamma(\kappa-\alpha(\tau))}\frac{h^{(1)}(\tau)e^{-\lambda(h(\tau)-h(t))}}{(h(\tau)-h(t))^{\alpha(\tau)-\kappa+1}}\left(\left(-\frac{1}{h^{(1)}(\tau)}\frac{d}{d\tau}\right)^{\kappa}f_h^{(\kappa)}(\tau)\right)d\tau, \tag{5.70}$$

respectively.

Definition 5.11 *Let $\kappa+1>\alpha(t)>\kappa$, $-\infty<a<b<\infty$ and $\lambda\in\mathbb{R}$. The left-sided variable-order tempered fractional integral with respect to another function is defined as*

$$\left({}^{RL}_{Cp}I^{\alpha(\cdot),\lambda}_{a+,h}f\right)(t)=\int_a^t\frac{1}{(h(t)-h(\tau))^{1-\alpha(\tau)}}E^{\alpha(\tau)-\kappa}_{1,\alpha(\tau)}\left(-\lambda\left(h(t)-h(\tau)\right)\right)f(\tau)\,d\tau, \tag{5.71}$$

and the right-sided variable-order tempered fractional integral with respect to another function as

$$\left({}^{RL}_{Cp}I^{\alpha(\cdot),\lambda}_{b-,h}f\right)(t)=\int_t^b\frac{1}{(h(\tau)-h(t))^{1-\alpha(\tau)}}E^{\alpha(\tau)-\kappa}_{1,\alpha(\tau)}\left(-\lambda\left(h(\tau)-h(t)\right)\right)f(\tau)\,d\tau. \tag{5.72}$$

Definition 5.12 *Let $\kappa<\alpha(\tau,t)<\kappa+1$, $-\infty<a<b<\infty$, $h^{(1)}(t)>0$ and $\lambda\in\mathbb{R}$. The left-sided variable-order Riemann-Liouville type tempered fractional derivative with respect to another function is defined as*

$$\left({}^{RL}_{Cpq}D^{\alpha(\cdot),\lambda}_{a+,h}f\right)(t)=\left(\frac{1}{h^{(1)}(t)}\frac{d}{dt}\right)^{\kappa}\int_a^t\frac{1}{\Gamma(\kappa-\alpha(\tau,t))}\frac{e^{-\lambda(h(t)-h(\tau))}h^{(1)}(\tau)}{(h(t)-h(\tau))^{\alpha(\tau,t)-\kappa+1}}f(\tau)\,d\tau, \tag{5.73}$$

and the right-sided variable-order Riemann-Liouville type tempered fractional derivative with respect to another function as

$$\left({}^{RL}_{Cpq}D^{\alpha(\cdot),\lambda}_{b-,h}f\right)(t)=\left(-\frac{1}{h^{(1)}(t)}\frac{d}{dt}\right)^{\kappa}\int_t^b\frac{1}{\Gamma(\kappa-\alpha(\tau,t))}\frac{h^{(1)}(\tau)e^{-\lambda(h(\tau)-h(t))}}{(h(\tau)-h(t))^{\alpha(\tau,t)-\kappa+1}}f(\tau)\,d\tau. \tag{5.74}$$

Definition 5.13 *Let* $\kappa < \alpha(t) < \kappa + 1$, $-\infty < a < b < \infty$, $h^{(1)}(t) > 0$ *and* $\lambda \in \mathbb{R}$. *The left-sided variable-order Riemann-Liouville type tempered fractional derivative with respect to another function is defined as*

$$\left({}_{Cpqs}^{RL}D_{a+,h}^{\alpha(\cdot),\lambda} f\right)(t) = \left(\frac{1}{h^{(1)}(t)}\frac{d}{dt}\right)^{\kappa}\left(\frac{1}{\Gamma(\kappa-\alpha(t))}\int_a^t \frac{e^{-\lambda(h(t)-h(\tau))}h^{(1)}(\tau)}{(h(t)-h(\tau))^{\alpha(t)-\kappa+1}} f(\tau)\,d\tau\right), \tag{5.75}$$

and the right-sided variable-order Riemann-Liouville type tempered fractional derivative with respect to another function as

$$\left({}_{Cpqs}^{RL}D_{b-,h}^{\alpha(\cdot),\lambda} f\right)(t) = \left(-\frac{1}{h^{(1)}(t)}\frac{d}{dt}\right)^{\kappa}\left(\frac{1}{\Gamma(\kappa-\alpha(t))}\int_t^b \frac{h^{(1)}(\tau)e^{-\lambda(h(\tau)-h(t))}}{(h(\tau)-h(t))^{\alpha(t)-\kappa+1}} f(\tau)\,d\tau\right). \tag{5.76}$$

5.5 General fractional derivatives of variable order via Mittag-Leffler function with respect to another function

In this section, we present the general fractional derivatives and general fractional integrals of variable order via Mittag-Leffler function with respect to another function, which was proposed in [324].

Definition 5.14 *Let* $\alpha(t) > 0$, $\kappa = [\alpha(t)] + 1$, $\kappa < \alpha(t) < \kappa + 1$ *and* $\lambda \in \mathbb{R}$. *The left-sided Liouville-Sonine-Caputo type variable-order general fractional derivative via Mittag-Leffler function with respect to another function is defined as*

$${}_{Ml}^{LSC}D_{a+,h}^{\alpha(\cdot),\kappa,\lambda} f(t) = \int_a^t E_{\alpha(\tau)}\left(-\lambda(h(t)-h(\tau))^{\alpha(\tau)}\right)\left(\left(\frac{1}{h^{(1)}(\tau)}\frac{d}{d\tau}\right)^{\kappa} f_h^{(\kappa)}(\tau)\right)d\tau, \tag{5.77}$$

and the right-sided Liouville-Sonine-Caputo type variable-order general fractional derivative via Mittag-Leffler function with respect to another function as

$${}_{Ml}^{LSC}D_{b-,h}^{\alpha(\cdot),\kappa,\lambda} f(t) = \int_t^b E_{\alpha(\tau)}\left(-\lambda(h(\tau)-h(t))^{\alpha(\tau)}\right)\left(\left(-\frac{1}{h^{(1)}(\tau)}\frac{d}{d\tau}\right)^{\kappa} f_h^{(\kappa)}(\tau)\right)d\tau. \tag{5.78}$$

Definition 5.15 *Let* $\alpha(t) > 0$, $\kappa = [\alpha(t)] + 1$, $\kappa < \alpha(t) < \kappa + 1$ *and* $\lambda \in \mathbb{R}$. *The left-sided Riemann-Liouville type variable-order general fractional derivative via Mittag-Leffler function with respect to another function is defined by*

$${}_{Ml}^{RL}D_{a+,h}^{\alpha(\cdot),\kappa,\lambda} f(t) = \left(\frac{1}{h^{(1)}(t)}\frac{d}{dt}\right)^{\kappa}\int_a^t E_{\alpha(\tau)}\left(-\lambda(h(t)-h(\tau))^{\alpha(\tau)}\right) f(\tau)\,d\tau, \tag{5.79}$$

and the right-sided Riemann-Liouville type variable-order general fractional derivative via Mittag-Leffler function with respect to another function by

$$ {}_{Ml}^{RL}D_{b-,h}^{\alpha(\cdot),\kappa,\lambda} f(t) = \left(-\frac{1}{h^{(1)}(t)}\frac{d}{dt}\right)^{\kappa} \int_{t}^{b} E_{\alpha(\tau)}\left(-\lambda\left(h(\tau)-h(t)\right)^{\alpha(\tau)}\right) f(\tau)\, d\tau. \tag{5.80}$$

Definition 5.16 *Let $\alpha(t) > 0$, $1+\kappa > \alpha(t) > \kappa$, $\kappa = [\alpha(t)] + 1$, $-\infty < a < b < \infty$ and $\lambda \in \mathbb{R}$. The left-sided variable-order fractional integral via Mittag-Leffler function with respect to another function is defined as*

$$ {}_{Ml}^{RL}I_{a+,h}^{\alpha(\cdot),\kappa,\lambda} f(t) = \int_{a}^{t} \frac{1}{\left(h(t)-h(\tau)\right)^{\alpha(\tau)+1-\kappa}} E_{\alpha(\tau),\kappa-\alpha(\tau)}^{-1}\left(-\lambda\left(h(t)-h(\tau)\right)^{\alpha(\tau)}\right) f(\tau)\, d\tau, \tag{5.81}$$

and the right-sided variable-order fractional integral via Mittag-Leffler function with respect to another function as

$$ {}_{Ml}^{RL}I_{b-,h}^{\alpha(\cdot),\kappa,\lambda} f(t) = \int_{t}^{b} \frac{1}{\left(h(\tau)-h(t)\right)^{\alpha(\tau)+1-\kappa}} E_{\alpha(\tau),\kappa-\alpha(\tau)}^{-1}\left(-\lambda\left(h(\tau)-h(t)\right)^{\alpha(\tau)}\right) f(\tau)\, d\tau. \tag{5.82}$$

5.6 Applications

In this section, we discuss the mathematical models for the relaxation and the law of deformation based on the variable-order fractional derivatives with respect to another function.

5.6.1 Relaxation models based on variable-order fractional derivatives with respect to another function

Now, the variable-order fractional derivative and variable-order general fractional derivatives with respect to another function are to model the relaxation models in complex phenomenon. To start with the ideas, we present the open problems on the relaxation models based on the variable-order fractional derivatives with respect to another function.

Example 5.1 *The mathematical model of the relaxation involving the Riemann-Liouville type fractional derivative of variable order with respect to another function can be given as*

$$\left({}_{RL}D_{0+,h}^{\alpha(\cdot)} f\right)(t) + \gamma f(t) = 0, \tag{5.83}$$

which becomes

$$\left(\frac{1}{h^{(1)}(t)}\frac{d}{dt}\right)\int_0^t \frac{h^{(1)}(\tau)}{\Gamma(1-\alpha(\tau))}\frac{f(\tau)}{(h(t)-h(\tau))^{\alpha(\tau)}}d\tau+\gamma f(t)=0, \tag{5.84}$$

where γ *is the relaxation constant.*

Example 5.2 *The mathematical model of the relaxation involving the Liouville-Sonine type fractional derivative of variable order with respect to another function can be given as*

$$_{LS}D_{a+,h}^{\alpha(\cdot)}f(t)+\gamma f(t)=0, \tag{5.85}$$

which implies that

$$\int_0^t \frac{1}{\Gamma(1-\alpha(\tau))}\frac{h^{(1)}(\tau)}{(h(t)-h(\tau))^{\alpha(\tau)}}\left(\left(\frac{1}{h^{(1)}(\tau)}\frac{d}{d\tau}\right)f_h^{(1)}(\tau)\right)d\tau+\gamma f(t)=0, \tag{5.86}$$

where γ *is the relaxation constant.*

Example 5.3 *The mathematical model of the relaxation involving the Riemann-Liouville type fractional derivative of variable order with respect to another function can be given as*

$$\left(\frac{1}{h^{(1)}(t)}\frac{d}{dt}\right)\int_0^t \frac{h^{(1)}(\tau)}{\Gamma(1-\alpha(\tau,t))}\frac{f(\tau)}{(h(t)-h(\tau))^{\alpha(\tau,t)}}d\tau+\gamma f(t)=0, \tag{5.87}$$

where γ *is the relaxation constant.*

Example 5.4 *The mathematical model of the relaxation based on the Hilfer type fractional derivative of variable order with respect to another function is written as follows:*

$$\left({}_{RLq}^{HL}I_{0+,h}^{\beta(1-\alpha(\cdot))}\left(\frac{1}{h^{(1)}(t)}\frac{d}{dt}\right)\left({}_{RLq}^{HL}I_{0+,h}^{(1-\beta)(1-\alpha(\cdot))}f\right)\right)(t)+\gamma f(t)=0, \tag{5.88}$$

where γ *is the relaxation constant and* β *is the constant.*

Example 5.5 *The mathematical model of the relaxation based on the Hilfer type fractional derivative of variable order with respect to another function is written as follows:*

$$\left({}_{RLq}^{HL}I_{0+,h}^{\beta(1-\alpha(\cdot))}\left(\frac{1}{h^{(1)}(t)}\frac{d}{dt}\right)\left({}_{RLq}^{HL}I_{0+,h}^{(1-\beta)(1-\alpha(\cdot))}f\right)\right)(t)+\gamma f(t)=0, \tag{5.89}$$

where γ *is the relaxation constant and* β *is the constant.*

Example 5.6 *The mathematical model of the relaxation based on the variable-order Riemann-Liouville type tempered fractional derivative with respect to another function is presented as follows:*

$$\left(\frac{1}{h^{(1)}(t)}\frac{d}{dt}\right)\int\limits_0^t \frac{h^{(1)}(\tau)}{\Gamma(1-\alpha(\tau))}\frac{e^{-\lambda(h(t)-h(\tau))}}{(h(t)-h(\tau))^{\alpha(\tau)}} f(\tau)\,d\tau + \gamma f(t) = 0, \quad (5.90)$$

where γ is the relaxation constant and λ is the constant.

Example 5.7 *The mathematical model of the relaxation based on the variable-order Liouville-Sonine type tempered fractional derivative with respect to another function is represented in the form:*

$$\int\limits_0^t \frac{h^{(1)}(\tau)}{\Gamma(1-\alpha(\tau))}\frac{e^{-\lambda(h(t)-h(\tau))}}{(h(t)-h(\tau))^{\alpha(\tau)}}\left(\left(\frac{1}{h^{(1)}(\tau)}\frac{d}{d\tau}\right) f_h^{(1)}(\tau)\right) d\tau + \gamma f(t) = 0, \quad (5.91)$$

where γ is the relaxation constant and λ is the constant.

Example 5.8 *The mathematical model of the relaxation based on the variable-order Riemann-Liouville type tempered fractional derivative with respect to another function is suggested as follows:*

$$\left(\frac{1}{h^{(1)}(t)}\frac{d}{dt}\right)\int\limits_0^t \frac{h^{(1)}(\tau)}{\Gamma(1-\alpha(\tau,t))}\frac{e^{-\lambda(h(t)-h(\tau))}}{(h(t)-h(\tau))^{\alpha(\tau,t)}} f(\tau)\,d\tau + \gamma f(t) = 0, \quad (5.92)$$

where γ is the relaxation constant and λ is the constant.

Example 5.9 *The mathematical model of the relaxation based on the variable-order Liouville-Sonine type tempered fractional derivative with respect to another function is represented in the form:*

$$\int\limits_0^t \frac{h^{(1)}(\tau)}{\Gamma(1-\alpha(\tau,t))}\frac{e^{-\lambda(h(t)-h(\tau))}}{(h(t)-h(\tau))^{\alpha(\tau,t)}}\left(\left(\frac{1}{h^{(1)}(\tau)}\frac{d}{d\tau}\right) f_h^{(1)}(\tau)\right) d\tau + \gamma f(t) = 0, \quad (5.93)$$

where γ is the relaxation constant and λ is the constant.

Example 5.10 *The mathematical model of the relaxation based on the Liouville-Sonine type variable-order general fractional derivative via the Mittag-Leffler function with respect to another function is given as*

$$\int_0^t E_{\alpha(\tau)}\left(-\lambda\left(h\left(t\right)-h\left(\tau\right)\right)^{\alpha(\tau)}\right)\left(\left(\frac{1}{h^{(1)}\left(\tau\right)}\frac{d}{d\tau}\right)f_h^{(1)}\left(\tau\right)\right)h^{(1)}\left(\tau\right)d\tau+\gamma f\left(t\right)=0, \tag{5.94}$$

where γ is the relaxation constant and λ is the constant.

Example 5.11 *The mathematical model of the relaxation based on the Riemann-Liouville type variable-order general fractional derivative via the Mittag-Leffler function with respect to another function is presented as follows:*

$$\left(\frac{1}{h^{(1)}\left(t\right)}\frac{d}{dt}\right)\int_0^t E_{\alpha(\tau)}\left(-\lambda\left(h\left(t\right)-h\left(\tau\right)\right)^{\alpha(\tau)}\right)h^{(1)}\left(\tau\right)f\left(\tau\right)d\tau+\gamma f\left(t\right)=0, \tag{5.95}$$

where γ is the relaxation constant and λ is the constant.

Example 5.12 *The mathematical model of the relaxation based on the Liouville-Sonine type variable-order general fractional derivative via the Mittag-Leffler function with respect to another function is suggested as follows:*

$$\int_0^t E_{\alpha(t,\tau)}\left(-\lambda\left(h\left(t\right)-h\left(\tau\right)\right)^{\alpha(t,\tau)}\right)\left(\left(\frac{1}{h^{(1)}\left(\tau\right)}\frac{d}{d\tau}\right)f_h^{(1)}\left(\tau\right)\right)h^{(1)}\left(\tau\right)d\tau+\gamma f(t)=0, \tag{5.96}$$

where γ is the relaxation constant and λ is the constant.

Example 5.13 *The mathematical model of the relaxation based on the Riemann-Liouville type variable-order general fractional derivative via the Mittag-Leffler function with respect to another function is reported as follows:*

$$\left(\frac{1}{h^{(1)}\left(t\right)}\frac{d}{dt}\right)\int_0^t E_{\alpha(t,\tau)}\left(-\lambda\left(h\left(t\right)-h\left(\tau\right)\right)^{\alpha(t,\tau)}\right)h^{(1)}\left(\tau\right)f\left(\tau\right)d\tau+\gamma f\left(t\right)=0, \tag{5.97}$$

where γ is the relaxation constant and λ is the constant.

Example 5.14 *The mathematical model of the relaxation based on the Liouville-Sonine type variable-order general fractional derivative via the Mittag-Leffler function with respect to another function is proposed as*

$$\int_0^t E_{\alpha(t)}\left(-\lambda\left(h(t)-h(\tau)\right)^{\alpha(t)}\right)\left(\left(\frac{1}{h^{(1)}(\tau)}\frac{d}{d\tau}\right)f_h^{(1)}(\tau)\right)h^{(1)}(\tau)\,d\tau+\gamma f(t)=0, \tag{5.98}$$

where γ is the relaxation constant and λ is the constant.

Example 5.15 *The mathematical model of the relaxation based on the Riemann-Liouville type variable-order general fractional derivative via the Mittag-Leffler function with respect to another function is written as follows:*

$$\left(\frac{1}{h^{(1)}(t)}\frac{d}{dt}\right)\int_0^t E_{\alpha(t)}\left(-\lambda\left(h(t)-h(\tau)\right)^{\alpha(t)}\right)h^{(1)}(\tau)\,f(\tau)\,d\tau+\gamma f(t)=0, \tag{5.99}$$

where γ is the relaxation constant and λ is the constant.

5.6.2 Rheological models in variable-order fractional derivatives with respect to another function

Here, we address the mathematical models of the rheological materials based on the variable-order fractional derivatives with respect to another function. The open problems for the rheological models in variable-order fractional derivatives with respect to another function are presented as follows.

Example 5.16 *The mathematical model for the law of deformation involving the Riemann-Liouville type fractional derivative of variable order with respect to another function can be given as*

$$\sigma(t)=\xi\left({}_{RL}D_{0+,h}^{\alpha(\cdot)}f\right)(t), \tag{5.100}$$

which becomes

$$\sigma(t)=\xi\left(\frac{1}{h^{(1)}(t)}\frac{d}{dt}\right)\int_0^t \frac{h^{(1)}(\tau)}{\Gamma(1-\alpha(\tau))}\frac{f(\tau)}{\left(h(t)-h(\tau)\right)^{\alpha(\tau)}}d\tau, \tag{5.101}$$

where ξ is the material constant, $\sigma(t)$ is the stress, and $\varepsilon(t)$ is the strain.

Example 5.17 *The mathematical model for the law of deformation involving the Liouville-Sonine type fractional derivative of variable order with respect to another function can be represented as*

$$\sigma(t)=\xi_{LS}D_{a+,h}^{\alpha(\cdot)}f(t), \tag{5.102}$$

which can be written as

$$\sigma(t)=\xi\int_0^t \frac{1}{\Gamma(1-\alpha(\tau))}\frac{h^{(1)}(\tau)}{(h(t)-h(\tau))^{\alpha(\tau)}}\left(\left(\frac{1}{h^{(1)}(\tau)}\frac{d}{d\tau}\right)f_h^{(1)}(\tau)\right)d\tau, \tag{5.103}$$

where ξ is the material constant, $\sigma(t)$ is the stress, and $\varepsilon(t)$ is the strain.

Example 5.18 *The mathematical model for the law of deformation involving the Riemann-Liouville type fractional derivative of variable order with respect to another function can be represented as*

$$\sigma(t)=\xi\left(\frac{1}{h^{(1)}(t)}\frac{d}{dt}\right)\int_0^t \frac{h^{(1)}(\tau)}{\Gamma(1-\alpha(\tau,t))}\frac{f(\tau)}{(h(t)-h(\tau))^{\alpha(\tau,t)}}d\tau, \tag{5.104}$$

where ξ is the material constant, $\sigma(t)$ is the stress, and $\varepsilon(t)$ is the strain.

Example 5.19 *The mathematical model for the law of deformation involving the Liouville-Sonine type fractional derivative of variable order with respect to another function can be given as*

$$\sigma(t)=\xi\int_0^t \frac{1}{\Gamma(1-\alpha(\tau,t))}\frac{h^{(1)}(\tau)}{(h(t)-h(\tau))^{\alpha(\tau,t)}}\left(\left(\frac{1}{h^{(1)}(\tau)}\frac{d}{d\tau}\right)f_h^{(1)}(\tau)\right)d\tau, \tag{5.105}$$

where ξ is the material constant, $\sigma(t)$ is the stress, and $\varepsilon(t)$ is the strain.

Example 5.20 *The mathematical model for the law of deformation based on the Hilfer type fractional derivative of variable order with respect to another function is written as follows*

$$\sigma(t)=\xi\left({}_{RLq}^{HL}I_{0+,h}^{\beta(1-\alpha(\cdot))}\left(\frac{1}{h^{(1)}(t)}\frac{d}{dt}\right)\left({}_{RLq}^{HL}I_{0+,h}^{(1-\beta)(1-\alpha(\cdot))}f\right)\right)(t), \tag{5.106}$$

where β and ξ are the material constants, $\sigma(t)$ is the stress, and $\varepsilon(t)$ is the strain.

Example 5.21 *The mathematical model for the law of deformation based on the Hilfer type fractional derivative of variable order with respect to another function can be expressed as*

$$\sigma(t)=\xi\left({}_{RLq}^{HL}I_{0+,h}^{\beta(1-\alpha(\cdot))}\left(\frac{1}{h^{(1)}(t)}\frac{d}{dt}\right)\left({}_{RLq}^{HL}I_{0+,h}^{(1-\beta)(1-\alpha(\cdot))}f\right)\right)(t), \tag{5.107}$$

where β and ξ are the material constants, $\sigma(t)$ is the stress, and $\varepsilon(t)$ is the strain.

Example 5.22 *The mathematical model for the law of deformation based on the variable-order Riemann-Liouville type tempered fractional derivative with respect to another function is represented in the form*

$$\sigma(t) = \xi\left(\frac{1}{h^{(1)}(t)}\frac{d}{dt}\right)\int_0^t \frac{h^{(1)}(\tau)}{\Gamma(1-\alpha(\tau))}\frac{e^{-\lambda(h(t)-h(\tau))}}{(h(t)-h(\tau))^{\alpha(\tau)}} f(\tau)\,d\tau, \quad (5.108)$$

where λ and ξ are the material constants, $\sigma(t)$ is the stress, and $\varepsilon(t)$ is the strain.

Example 5.23 *The mathematical model for the law of deformation based on the variable-order Liouville-Sonine type tempered fractional derivative with respect to another function is represented as*

$$\sigma(t) = \xi\int_0^t \frac{h^{(1)}(\tau)}{\Gamma(1-\alpha(\tau))}\frac{e^{-\lambda(h(t)-h(\tau))}}{(h(t)-h(\tau))^{\alpha(\tau)}}\left(\left(\frac{1}{h^{(1)}(\tau)}\frac{d}{d\tau}\right) f_h^{(1)}(\tau)\right) d\tau, \quad (5.109)$$

where λ and ξ are the material constants, $\sigma(t)$ is the stress, and $\varepsilon(t)$ is the strain.

Example 5.24 *The mathematical model for the law of deformation based on the variable-order Riemann-Liouville type tempered fractional derivative with respect to another function is represented as*

$$\sigma(t) = \xi\left(\frac{1}{h^{(1)}(t)}\frac{d}{dt}\right)\int_0^t \frac{h^{(1)}(\tau)}{\Gamma(1-\alpha(\tau,t))}\frac{e^{-\lambda(h(t)-h(\tau))}}{(h(t)-h(\tau))^{\alpha(\tau,t)}} f(\tau)\,d\tau, \quad (5.110)$$

where λ and ξ are the material constants, $\sigma(t)$ is the stress, and $\varepsilon(t)$ is the strain.

Example 5.25 *The mathematical model for the law of deformation based on the variable-order Liouville-Sonine type tempered fractional derivative with respect to another function is suggested as*

$$\sigma(t) = \xi\int_0^t \frac{h^{(1)}(\tau)}{\Gamma(1-\alpha(\tau,t))}\frac{e^{-\lambda(h(t)-h(\tau))}}{(h(t)-h(\tau))^{\alpha(\tau,t)}}\left(\left(\frac{1}{h^{(1)}(\tau)}\frac{d}{d\tau}\right) f_h^{(1)}(\tau)\right) d\tau, \quad (5.111)$$

where λ and ξ are the material constants, $\sigma(t)$ is the stress, and $\varepsilon(t)$ is the strain.

Example 5.26 *The mathematical model for the law of deformation based on the Liouville-Sonine type variable-order general fractional derivative via the Mittag-Leffler function with respect to another function is given as*

$$\sigma(t) = \xi \int_0^t E_{\alpha(\tau)} \left(-\lambda \left(h(t) - h(\tau) \right)^{\alpha(\tau)} \right) \left(\left(\frac{1}{h^{(1)}(\tau)} \frac{d}{d\tau} \right) f_h^{(1)}(\tau) \right) h^{(1)}(\tau) \, d\tau, \tag{5.112}$$

where λ *and* ξ *are the material constants,* $\sigma(t)$ *is the stress, and* $\varepsilon(t)$ *is the strain.*

Example 5.27 *The mathematical model for the law of deformation based on the Riemann-Liouville type variable-order general fractional derivative via the Mittag-Leffler function with respect to another function is presented as*

$$\sigma(t) = \xi \left(\frac{1}{h^{(1)}(t)} \frac{d}{dt} \right) \int_0^t E_{\alpha(\tau)} \left(-\lambda \left(h(t) - h(\tau) \right)^{\alpha(\tau)} \right) h^{(1)}(\tau) f(\tau) \, d\tau, \tag{5.113}$$

where λ *and* ξ *are the material constants,* $\sigma(t)$ *is the stress, and* $\varepsilon(t)$ *is the strain.*

Example 5.28 *The mathematical model for the law of deformation based on the Liouville-Sonine type variable-order general fractional derivative via the Mittag-Leffler function with respect to another function is represented in the form*

$$\sigma(t) = \xi \int_0^t E_{\alpha(t,\tau)} \left(-\lambda \left(h(t) - h(\tau) \right)^{\alpha(t,\tau)} \right) \left(\left(\frac{1}{h^{(1)}(\tau)} \frac{d}{d\tau} \right) f_h^{(1)}(\tau) \right) h^{(1)}(\tau) \, d\tau, \tag{5.114}$$

where λ *and* ξ *are the material constants,* $\sigma(t)$ *is the stress, and* $\varepsilon(t)$ *is the strain.*

Example 5.29 *The mathematical model for the law of deformation based on the Riemann-Liouville type variable-order general fractional derivative via the Mittag-Leffler function with respect to another function takes the form*

$$\sigma(t) = \xi \left(\frac{1}{h^{(1)}(t)} \frac{d}{dt} \right) \int_0^t E_{\alpha(t,\tau)} \left(-\lambda \left(h(t) - h(\tau) \right)^{\alpha(t,\tau)} \right) h^{(1)}(\tau) f(\tau) \, d\tau, \tag{5.115}$$

where λ *and* ξ *are the material constants,* $\sigma(t)$ *is the stress, and* $\varepsilon(t)$ *is the strain.*

Example 5.30 *The mathematical model for the law of deformation based on the Liouville-Sonine type variable-order general fractional derivative via the Mittag-Leffler function with respect to another function can be expressed as*

$$\sigma(t) = \xi \int_0^t E_{\alpha(t)}\left(-\lambda\left(h(t) - h(\tau)\right)^{\alpha(t)}\right)\left(\left(\frac{1}{h^{(1)}(\tau)}\frac{d}{d\tau}\right)f_h^{(1)}(\tau)\right)h^{(1)}(\tau)\,d\tau, \tag{5.116}$$

where λ *and* ξ *are the material constants,* $\sigma(t)$ *is the stress, and* $\varepsilon(t)$ *is the strain.*

Example 5.31 *The mathematical model for the law of deformation based on the Riemann-Liouville type variable-order general fractional derivative via the Mittag-Leffler function with respect to another function is expressed as*

$$\sigma(t) = \xi\left(\frac{1}{h^{(1)}(t)}\frac{d}{dt}\right)\int_0^t E_{\alpha(t)}\left(-\lambda\left(h(t) - h(\tau)\right)^{\alpha(t)}\right)h^{(1)}(\tau)\,f(\tau)\,d\tau, \tag{5.117}$$

where λ *and* ξ *are the material constants,* $\sigma(t)$ *is the stress, and* $\varepsilon(t)$ *is the strain.*

Appendix A

Laplace Transforms of the Functions

Laplace gave the integral transform, which is the well-known Laplace transform of the function defined as (see [10, 152, 236])

$$L\left[f(t)\right] := f(s) = \int_0^\infty e^{-st} f(t)\,dt \tag{A.1}$$

and the corresponding inverse Laplace transform is as follows (see [10, 152, 236]):

$$f(t) = L^{-1}\left[f(s)\right] = \frac{1}{2\pi i}\lim_{T\to\infty}\int_{r-iT}^{r+iT} e^{st} f(s)\,ds, \tag{A.2}$$

where $r \in \mathrm{R}$ and $s = r + iT$.

The relation between Laplace and Fourier transforms is given as follows (see [152]):

$$f(s) = f(i\omega), \tag{A.3}$$

where $\omega = is$.

For information on the Laplace transforms of the functions, see [236, 327].

The functions	Laplace transforms
$f(t) + g(t)$	$f(s) + g(s)$
$tf(t)$	$-f^{(1)}(s)$
$t^\kappa f(t)$	$(-1)^\kappa f^{(\kappa)}(s)$
$f^{(1)}(t)$	$sf(s) - f(0)$
$f^{(\kappa)}(t)$	$s^\kappa f(s) - \sum_{j=1}^{\kappa} s^{\kappa-j} f^{(j-1)}(0)$
$\int_0^t f(t)\,dt$	$\frac{f(s)}{s}$
$f(at)\ (\lambda > 0)$	$f(s/\lambda)/\lambda$
$\int_0^t f(\tau)\,g(t-\tau)\,d\tau$	$f(s)\,g(s)$

The functions	Laplace transforms of the functions
The exponential decay $e^{-\lambda t}$	$s^{-1}\left(1+\lambda s^{-1}\right)^{-1}$
The unit step function $\vartheta(\tau)$	s^{-1}
The unit impulse function $\delta(\tau)$	1
$t^{-\nu}/\Gamma(1-\nu)$	s^{ν}
$t^{\nu}/\Gamma(1+\nu)$	$s^{-\nu}$
t^{-1}	s
t	s^{-1}
$t^{1/\kappa}/\Gamma(1+1/\kappa)$	$s^{-(1+1/\kappa)}$
$t^{\nu}e^{-\lambda t}/\Gamma(1+\nu)$	$(s+\lambda)^{-(\nu+1)}$
The sine function $\sin(\lambda t)$	$\frac{\lambda}{s^2+\lambda^2}$
The cosine function $\cos(\lambda t)$	$\frac{s}{s^2+\lambda^2}$
The hyperbolic sine $\sinh(\lambda t)$	$\frac{\lambda}{s^2-\lambda^2}$
The hyperbolic cosine $\cosh(\lambda t)$	$\frac{s}{s^2-\lambda^2}$
$t^{\nu}\sinh(-\lambda t)/\Gamma(1+\nu)$	$\frac{1}{2}\left[(s+\lambda)^{-(\nu+1)}+(s-\lambda)^{-(\nu+1)}\right]$
$t^{\nu}\cosh(-\lambda t)/\Gamma(1+\nu)$	$\frac{1}{2}\left[(s+\lambda)^{-(\nu+1)}-(s-\lambda)^{-(\nu+1)}\right]$
$t^{\nu}\sin(-\lambda t)/\Gamma(1+\nu)$	$\frac{1}{2}\left[(s+i\lambda)^{-(\nu+1)}+(s-i\lambda)^{-(\nu+1)}\right]$
$t^{\nu}\cos(\lambda t)/\Gamma(1+\nu)$	$\frac{1}{2}\left[(s+i\lambda)^{-(\nu+1)}-(s-i\lambda)^{-(\nu+1)}\right]$

The special functions	Laplace transforms
$E_{\nu}(\lambda t^{\nu})$	$s^{-1}\left(1-\lambda s^{-\nu}\right)^{-1}$
$t^{\upsilon-1}E_{\nu,\upsilon}(\lambda t^{\nu})$	$s^{-\upsilon}\left(1-\lambda s^{-\nu}\right)^{-1}$
$t^{\mu+\upsilon-1}E_{\nu,\mu+\upsilon}(\lambda t^{\nu})$	$s^{-(\mu+\upsilon)}\left(1-\lambda s^{-\nu}\right)^{-1}$
$t^{\upsilon-1}E_{1,\upsilon}(\lambda t)$	$s^{-\upsilon}\left(1-\lambda s^{-1}\right)^{-1}$
$t^{\upsilon-1}E^{\varphi}_{\nu,\upsilon}(\lambda t^{\nu})$	$s^{-\upsilon}\left(1-\wp s^{-\nu}\right)^{-\varphi}$
$t^{\mu+\upsilon-1}E^{\varphi}_{\nu,\mu+\upsilon}(\lambda t^{\nu})$	$s^{-\mu+\upsilon}\left(1-\lambda s^{-\nu}\right)^{-\varphi}$
$t^{\upsilon-1}E^{\varphi+\varpi}_{\nu,\upsilon}(\lambda t^{\nu})$	$s^{-\upsilon}\left(1-\lambda s^{-\nu}\right)^{-(\varphi+\varpi)}$
$t^{\mu+\upsilon-1}E^{\varphi+\varpi}_{\nu,\mu+\upsilon}(\lambda t^{\nu})$	$s^{-(\mu+\upsilon)}\left(1-\lambda s^{-\nu}\right)^{-(\varphi+\varpi)}$
$t^{\upsilon-1}E^{\varphi}_{1,\upsilon}(\lambda t)$	$s^{-\upsilon}\left(1-\lambda s^{-1}\right)^{-\varphi}$
$E^{\varphi}_{\nu,1}(\lambda t^{\nu})$	$s^{-1}\left(1-\wp s^{-\nu}\right)^{-\varphi}$
$E^{\varphi}_{1,1}(\lambda t)$	$s^{-1}\left(1-\wp s^{-1}\right)^{-\varphi}$

Appendix B

Fourier Transforms of the Functions

Fourier gave the integral transform, which is the well-known Fourier transform of the function defined as (see [12, 236])

$$F\left[f\left(t\right)\right] = f\left(\omega\right) = \int_{-\infty}^{\infty} e^{-i\omega t} f\left(t\right) dt \tag{B.1}$$

and the corresponding inverse Fourier transform is as follows (see [236]):

$$f\left(t\right) = F^{-1}\left[f\left(\omega\right)\right] = \int_{-\infty}^{\infty} e^{i\omega t} f\left(\omega\right) d\omega. \tag{B.2}$$

The relation between the Fourier and Laplace transforms is given as follows (see [236]):

$$f\left(\omega\right) = f\left(-is\right), \tag{B.3}$$

where $\omega = -is$.

For information on the Fourier transforms of the functions, see [152, 236, 328].

The functions	Fourier transforms
$f(t)+g(t)$	$f(\omega)+g(\omega)$
$tf(t)$	$-if^{(1)}(\omega)$
$t^{\kappa}f(t)$	$(-i)^{\kappa}f^{(\kappa)}(\omega)$
$f^{(1)}(t)$	$i\omega f(\omega)$
$f^{(\kappa)}(t)$	$(i\omega)^{\kappa}f(\omega)$
$f(t-\lambda)$	$e^{-i\lambda\omega}f(s)$
$\int_{-\infty}^{t} f(t)\,dt$	$\frac{f(\omega)}{i\omega}$
$e^{-i\lambda t}f(t)$	$f(\omega-\lambda)$
$f(at)\,(\lambda>0)$	$f(\omega/\lambda)/\lambda$
$\int_{-\infty}^{t} f(\tau)\,g(t-\tau)\,d\tau$	$f(\omega)\,g(\omega)$
The exponential decay $e^{-i\lambda t}$	$2\pi\delta(\omega-\lambda)$
1	$2\pi\delta\tau(\omega)$
The unit impulse function $\delta(\tau)$	1
$t^{-\nu}/\Gamma(1-\nu)$	$(-i\omega)^{\nu}$
$t^{\nu}/\Gamma(1+\nu)$	$(-i\omega)^{-\nu}$
t	$(-i\omega)^{-1}$
$t^{1/\kappa}/\Gamma(1+1/\kappa)$	$(i\omega)^{-(1+1/\kappa)}$
$t^{\nu}e^{-\lambda t}/\Gamma(1+\nu)$	$(i\omega+\lambda)^{-(\nu+1)}$

The special functions	Fourier transforms
$E_{\nu}(\lambda t^{\nu})$	$(i\omega)^{-1}\left(1-\lambda(i\omega)^{-\nu}\right)^{-1}$
$t^{\upsilon-1}E_{\nu,\upsilon}(\lambda t^{\nu})$	$(i\omega)^{-\upsilon}\left(1-\lambda(i\omega)^{-\nu}\right)^{-1}$
$t^{\mu+\upsilon-1}E_{\nu,\mu+\upsilon}(\lambda t^{\nu})$	$(i\omega)^{-(\mu+\upsilon)}\left(1-\lambda(i\omega)^{-\nu}\right)^{-1}$
$t^{\upsilon-1}E_{1,\upsilon}(\lambda t)$	$(i\omega)^{-\upsilon}\left(1-\lambda(i\omega)^{-1}\right)^{-1}$
$t^{\upsilon-1}E^{\varphi}_{\nu,\upsilon}(\lambda t^{\nu})$	$(i\omega)^{-\upsilon}\left(1-\wp(i\omega)^{-\nu}\right)^{-\varphi}$
$t^{\mu+\upsilon-1}E^{\varphi}_{\nu,\mu+\upsilon}(\lambda t^{\nu})$	$(i\omega)^{-\mu+\upsilon}\left(1-\lambda(i\omega)^{-\nu}\right)^{-\varphi}$
$t^{\upsilon-1}E^{\varphi+\varpi}_{\nu,\upsilon}(\lambda t^{\nu})$	$(i\omega)^{-\upsilon}\left(1-\lambda(i\omega)^{-\nu}\right)^{-(\varphi+\varpi)}$
$t^{\mu+\upsilon-1}E^{\varphi+\varpi}_{\nu,\mu+\upsilon}(\lambda t^{\nu})$	$(i\omega)^{-(\mu+\upsilon)}\left(1-\lambda(i\omega)^{-\nu}\right)^{-(\varphi+\varpi)}$
$t^{\upsilon-1}E^{\varphi}_{1,\upsilon}(\lambda t)$	$(i\omega)^{-\upsilon}\left(1-\lambda(i\omega)^{-1}\right)^{-\varphi}$
$E^{\varphi}_{\nu,1}(\lambda t^{\nu})$	$(i\omega)^{-1}\left(1-\wp(i\omega)^{-\nu}\right)^{-\varphi}$
$E^{\varphi}_{1,1}(\lambda t)$	$(i\omega)^{-1}\left(1-\wp(i\omega)^{-1}\right)^{-\varphi}$

Appendix C

Mellin Transforms of the Functions

Mellin gave the integral transform, which is the well-known Mellin transform of the function, defined as (see [236, 255, 329])

$$M\left[f\left(t\right)\right] := f\left(\varpi\right) = \int_{0}^{\infty} t^{\varpi-1} f\left(t\right) dt, \tag{C.1}$$

and the corresponding inverse Mellin transform is as follows (see [255, 236, 329]):

$$f\left(t\right) = M^{-1}\left[f\left(\varpi\right)\right] = \frac{1}{2\pi i} \lim_{T\to\infty} \int_{r-iT}^{r+iT} t^{-\varpi} f\left(\varpi\right) d\varpi, \tag{C.2}$$

where $r \in \mathrm{R}$ and $\varpi = r + iT$.

For information on the Mellin transforms of the functions, see [236, 255].

The functions	Mellin transforms
$f(t)+g(t)$	$f(\varpi)+g(\varpi)$
$tf(t)$	$f(\varpi+1)$
$\frac{f(t)}{t}$	$f(\varpi-1)$
$\int_0^t f(t)\,dt$	$-\varpi^{-1}f(\varpi)$
$\int_t^\infty f(t)\,dt$	$\varpi^{-1}f(\varpi)$
$\int_0^t f(\tau)\,g\left(\frac{t}{\tau}\right)\frac{d\tau}{\tau}$	$f(\varpi)\,g(\varpi)$
$(t/a)^{\nu}\ (t>a)$	$(\nu+\varpi)^{-1}a^{\varpi}$
$t^{\nu}\ (t>a)$	$-(\nu+\varpi)^{-1}a^{\nu+\varpi}$
$-1\ (t>a)$	$\varpi^{-1}a^{\varpi}$
$e^{-\lambda t}$	$\Gamma(\varpi)\,\lambda^{-\varpi}$
$E_{\nu}(-\lambda t)$	$\frac{\Gamma(\varpi)\Gamma(1-\varpi)}{\Gamma(1-\nu\varpi)}\lambda^{-\varpi}$
$E_{\nu,\upsilon}(-\lambda t)$	$\frac{\Gamma(\varpi)\Gamma(1-\varpi)}{\Gamma(\upsilon-\nu\varpi)}\lambda^{-\varpi}$
$E_{\nu,\mu+\upsilon}(-\lambda t)$	$\frac{\Gamma(\varpi)\Gamma(1-\varpi)}{\Gamma(\mu+\upsilon-\nu\varpi)}\lambda^{-\varpi}$
$E_{1,\upsilon}(-\lambda t)$	$\frac{\Gamma(\varpi)\Gamma(1-\varpi)}{\Gamma(\upsilon-\varpi)}\lambda^{-\varpi}$
$E^{\varphi}_{\nu,\upsilon}(-\lambda t)$	$\frac{\Gamma(\varpi)\Gamma(\varphi-\varpi)}{\Gamma(\varphi)\Gamma(\upsilon-\nu\varpi)}\lambda^{-\varpi}$
$E^{\varphi}_{\nu,\mu+\upsilon}(-\lambda t)$	$\frac{\Gamma(\varpi)\Gamma(\varphi-\varpi)}{\Gamma(\varphi)\Gamma(\mu+\upsilon-\nu\varpi)}\lambda^{-\varpi}$
$E^{\varphi+\varpi}_{\nu,\upsilon}(-\lambda t)$	$\frac{\Gamma(\varpi)\Gamma(\varphi+\varpi-\varpi)}{\Gamma(\varphi+\varpi)\Gamma(\upsilon-\nu\varpi)}\lambda^{-\varpi}$
$E^{\varphi+\varpi}_{\nu,\mu+\upsilon}(-\lambda t)$	$\frac{\Gamma(\varpi)\Gamma(\varphi+\varpi-\varpi)}{\Gamma(\varphi)\Gamma(\mu+\upsilon-\nu\varpi)}\lambda^{-\varpi}$
$E^{\varphi}_{1,\upsilon}(-\lambda t)$	$\frac{\Gamma(\varpi)\Gamma(\varphi-\varpi)}{\Gamma(\varphi)\Gamma(\upsilon-\varpi)}\lambda^{-\varpi}$
$E^{\varphi}_{\nu,1}(-\lambda t)$	$\frac{\Gamma(\varpi)\Gamma(\varphi-\varpi)}{\Gamma(\varphi)\Gamma(1-\nu\varpi)}\lambda^{-\varpi}$
$E^{\varphi}_{1,1}(-\lambda t)$	$\frac{\Gamma(\varpi)\Gamma(\varphi-\varpi)}{\Gamma(\varphi)\Gamma(1-\varpi)}\lambda^{-\varpi}$

Appendix D

The Special Functions and Their Expansions

D.1 The one-parametric special functions via Mittag-Leffler functions

We now present the definitions and properties of the subcosine, subcosine, hyperbolic subcosine and hyperbolic subcosine functions via Mittag-Leffler functions and their Laplace transforms.

Let $\kappa \in \mathbb{N}$, $i = \sqrt{-1}$, $\lambda \in \mathbb{C}$, $\alpha \in \mathbb{C}$, $Re\,(\lambda) \geq 0$ and $Re\,(\alpha) \geq 0$. The Mittag-Leffler function with one-parameter constant λ is defined as [152]

$$E_\alpha\left(\lambda t^\alpha\right) = \sum_{\kappa=0}^{\infty} \frac{\lambda^\kappa t^{\kappa\alpha}}{\Gamma\left(\kappa\alpha + 1\right)}, \tag{D.1}$$

such that

$$\Im\left(\alpha\right) = \sum_{\kappa=0}^{\infty} \frac{1}{\Gamma\left(\kappa\alpha + 1\right)}, \tag{D.2}$$

which, for $\alpha = 1$, leads to

$$\sum_{\kappa=0}^{\infty} \frac{1}{\Gamma\left(\kappa + 1\right)} = e = \Im\left(1\right). \tag{D.3}$$

The subsine function, denoted by $LSin_\alpha\left(t^\alpha\right)$, is defined as [330, 331]

$$LSin_\alpha\left(t^\alpha\right) = \sum_{\kappa=0}^{\infty} \frac{\left(-1\right)^\kappa t^{(2\kappa+1)\alpha}}{\Gamma\left(\left(2\kappa + 1\right)\alpha + 1\right)}, \tag{D.4}$$

which implies that

$$LSin_\alpha\left(\lambda t^\alpha\right) = \sum_{\kappa=0}^{\infty} \frac{\left(-1\right)^\kappa \lambda^{2\kappa+1} t^{(2\kappa+1)\alpha}}{\Gamma\left(\left(2\kappa + 1\right)\alpha + 1\right)}, \tag{D.5}$$

the subcosine function, denoted by $LCos_\alpha\left(t^\alpha\right)$, as [330, 331]

$$LCos_\alpha\left(t^\alpha\right) = \sum_{\kappa=0}^{\infty} \frac{\left(-1\right)^\kappa t^{2\kappa\alpha}}{\Gamma\left(2\kappa\alpha + 1\right)}, \tag{D.6}$$

which implies that

$$LCos_{\alpha}\left(\lambda t^{\alpha}\right)=\sum_{\kappa=0}^{\infty}\frac{(-1)^{\kappa}\lambda^{2\kappa}t^{2\kappa\alpha}}{\Gamma\left(2\kappa\alpha+1\right)}, \tag{D.7}$$

the hyperbolic subsine function, denoted by $LSinh_{\alpha}\left(t^{\alpha}\right)$, as

$$LSinh_{\alpha}\left(t^{\alpha}\right)=\sum_{\kappa=0}^{\infty}\frac{t^{(2\kappa+1)\alpha}}{\Gamma\left(\left(2\kappa+1\right)\alpha+1\right)}, \tag{D.8}$$

which implies that

$$LSinh_{\alpha}\left(\lambda t^{\alpha}\right)=\sum_{\kappa=0}^{\infty}\frac{\lambda^{2\kappa+1}t^{(2\kappa+1)\alpha}}{\Gamma\left(\left(2\kappa+1\right)\alpha+1\right)}, \tag{D.9}$$

and the hyperbolic subcosine function, denoted by $LCosh_{\alpha}\left(t^{\alpha}\right)$, as

$$LCosh_{\alpha}\left(t^{\alpha}\right)=\sum_{\kappa=0}^{\infty}\frac{t^{2\kappa\alpha}}{\Gamma\left(2\kappa\alpha+1\right)}, \tag{D.10}$$

which implies that

$$LCosh_{\alpha}\left(\lambda t^{\alpha}\right)=\sum_{\kappa=0}^{\infty}\frac{\lambda^{2\kappa}t^{2\kappa\alpha}}{\Gamma\left(2\kappa\alpha+1\right)}. \tag{D.11}$$

In fact, the Mittag-Leffler function with one-parameter constant can be written as [330, 331]

$$E_{\alpha}\left(\lambda t^{\alpha}\right)=LCosh_{\alpha}\left(\lambda t^{\alpha}\right)+LSinh_{\alpha}\left(\lambda t^{\alpha}\right), \tag{D.12}$$

which leads to [330, 331]

$$E_{\alpha}\left(i\lambda t^{\alpha}\right)=LCos_{\alpha}\left(\lambda t^{\alpha}\right)+iLSin_{\alpha}\left(\lambda t^{\alpha}\right), \tag{D.13}$$

and

$$E_{\alpha}\left((i\lambda t)^{\alpha}\right)=\sum_{\kappa=0}^{\infty}\frac{(-1)^{\kappa\alpha}(\lambda t)^{2\kappa\alpha}}{\Gamma(2\kappa\alpha+1)}+i^{\alpha}\sum_{\kappa=0}^{\infty}\frac{(-1)^{\kappa\alpha}(\lambda t)^{(2\kappa+1)\alpha}}{\Gamma((2\kappa+1)\alpha+1)}, \tag{D.14}$$

which from Eq. (D.1) leads to

$$E_{\alpha}\left((i\lambda t)^{\alpha}\right)=RLCos_{\alpha}\left(\lambda t^{\alpha}\right)+i^{\alpha}RLSin_{\alpha}\left(\lambda t^{\alpha}\right), \tag{D.15}$$

where

$$RLSin_{\alpha}\left((\lambda t)^{\alpha}\right):=\sum_{\kappa=0}^{\infty}\frac{(-1)^{\kappa\alpha}\left(\lambda t\right)^{(2\kappa+1)\alpha}}{\Gamma\left(\left(2\kappa+1\right)\alpha+1\right)} \tag{D.16}$$

and

$$RLCos_{\alpha}\left((\lambda t)^{\alpha}\right) := \sum_{\kappa=0}^{\infty} \frac{(-1)^{\kappa\alpha}(\lambda t)^{2\kappa\alpha}}{\Gamma(2\kappa\alpha+1)}. \tag{D.17}$$

With the use of the formulation

$$L\left[\frac{t^{\kappa\alpha}}{\Gamma(\kappa\alpha+1)}\right] = \frac{1}{s^{\kappa\alpha}}, \tag{D.18}$$

we have [332]

$$L\left\{E_{\alpha}\left(-\lambda t^{\alpha}\right)\right\} = s^{-1}\left(1+\lambda s^{-\alpha}\right)^{-1} \quad \left(\left|\lambda s^{-\alpha}\right|<1\right) \tag{D.19}$$

such that

$$L\left\{LSin_{\alpha}\left(\lambda t^{\alpha}\right)\right\} = \lambda s^{-(1+\alpha)}\left(1+\lambda^{2}s^{-2\alpha}\right)^{-1} \quad \left(\left|\lambda s^{-\alpha}\right|<1\right), \tag{D.20}$$

$$L\left\{LCos_{\alpha}\left(\lambda t^{\alpha}\right)\right\} = s^{-1}\left(1+\lambda^{2}s^{-2\alpha}\right)^{-1} \quad \left(\left|\lambda s^{-\alpha}\right|<1\right), \tag{D.21}$$

$$L\left\{LSinh_{\alpha}\left(\lambda t^{\alpha}\right)\right\} = \lambda s^{-(1+\alpha)}\left(1-\lambda^{2}s^{-2\alpha}\right)^{-1} \quad \left(\left|\lambda s^{-\alpha}\right|<1\right), \tag{D.22}$$

and

$$L\left\{LCosh_{\alpha}\left(\lambda t^{\alpha}\right)\right\} = s^{-1}\left(1-\lambda^{2}s^{-2\alpha}\right)^{-1} \quad \left(\left|\lambda s^{-\alpha}\right|<1\right). \tag{D.23}$$

Therefore, there are

$$LSin_{\alpha}\left(\lambda t^{\alpha}\right) = \frac{E_{\alpha}\left(i\lambda t^{\alpha}\right)-E_{\alpha}\left(-i\lambda t^{\alpha}\right)}{2i}, \tag{D.24}$$

$$LCos_{\alpha}\left(\lambda t^{\alpha}\right) = \frac{E_{\alpha}\left(i\lambda t^{\alpha}\right)+E_{\alpha}\left(-i\lambda t^{\alpha}\right)}{2}, \tag{D.25}$$

$$LSinh_{\alpha}\left(\lambda t^{\alpha}\right) = \frac{E_{\alpha}\left(\lambda t^{\alpha}\right)-E_{\alpha}\left(-\lambda t^{\alpha}\right)}{2} \tag{D.26}$$

and

$$LCosh_{\alpha}\left(\lambda t^{\alpha}\right) = \frac{E_{\alpha}\left(\lambda t^{\alpha}\right)+E_{\alpha}\left(-\lambda t^{\alpha}\right)}{2}. \tag{D.27}$$

There is [330, 331]

$$E_{\alpha}\left(it^{\alpha}\right) = LCos_{\alpha}\left(t^{\alpha}\right)+iLSin_{\alpha}\left(t^{\alpha}\right). \tag{D.28}$$

In this case, there are [333]

$$LSin_{\alpha}\left(t^{\alpha}\right) = \frac{E_{\alpha}\left(it^{\alpha}\right)-E_{\alpha}\left(-it^{\alpha}\right)}{2i}, \tag{D.29}$$

$$LCos_{\alpha}\left(t^{\alpha}\right) = \frac{E_{\alpha}\left(it^{\alpha}\right)+E_{\alpha}\left(-it^{\alpha}\right)}{2}, \tag{D.30}$$

$$LSinh_\alpha\left(t^\alpha\right) = \frac{E_\alpha\left(t^\alpha\right) - E_\alpha\left(-t^\alpha\right)}{2} \tag{D.31}$$

and

$$LCosh_\alpha\left(t^\alpha\right) = \frac{E_\alpha\left(t^\alpha\right) + E_\alpha\left(-t^\alpha\right)}{2}. \tag{D.32}$$

Thus, we have

$$E_\alpha\left(-it^\alpha\right) = LCos_\alpha\left(t^\alpha\right) - iLSin_\alpha\left(t^\alpha\right) \tag{D.33}$$

and

$$E_\alpha\left(-i\lambda t^\alpha\right) = LCos_\alpha\left(\lambda t^\alpha\right) - iLSin_\alpha\left(\lambda t^\alpha\right) \tag{D.34}$$

such that

$$E_\alpha\left(-it^\alpha\right) = Re\left(E_\alpha\left(-it^\alpha\right)\right) + iIm\left(E_\alpha\left(-it^\alpha\right)\right), \tag{D.35}$$

where

$$Re\left(E_\alpha\left(-it^\alpha\right)\right) = Cos_\alpha\left(-t^\alpha\right) \tag{D.36}$$

and

$$Im\left(E_\alpha\left(-it^\alpha\right)\right) = LSin_\alpha\left(-t^\alpha\right). \tag{D.37}$$

In a similar way, we have

$$E_\alpha\left(it^\alpha\right) = LCos_\alpha\left(t^\alpha\right) + iLSin_\alpha\left(t^\alpha\right) \tag{D.38}$$

and

$$E_\alpha\left(i\lambda t^\alpha\right) = LCos_\alpha\left(\lambda t^\alpha\right) + iLSin_\alpha\left(\lambda t^\alpha\right) \tag{D.39}$$

such that

$$E_\alpha\left(it^\alpha\right) = Re\left(E_\alpha\left(it^\alpha\right)\right) + iIm\left(E_\alpha\left(it^\alpha\right)\right), \tag{D.40}$$

where

$$Re\left(E_\alpha\left(it^\alpha\right)\right) = Cos_\alpha\left(t^\alpha\right) \tag{D.41}$$

and

$$Im\left(E_\alpha\left(it^\alpha\right)\right) = LSin_\alpha\left(t^\alpha\right). \tag{D.42}$$

In this case, we find that

$$LCos_\alpha\left(-\lambda t^\alpha\right) = LCos_\alpha\left(\lambda t^\alpha\right) \tag{D.43}$$

and

$$LSin_\alpha\left(-\lambda t^\alpha\right) = -LSin_\alpha\left(\lambda t^\alpha\right). \tag{D.44}$$

We obtain the following results:

$$\vartheta_1\left(\alpha\right) := \sum_{\kappa=0}^{\infty} \frac{1}{\Gamma\left(2\kappa\alpha + 1\right)}, \tag{D.45}$$

$$\vartheta_2\left(\alpha\right) := \sum_{\kappa=0}^{\infty} \frac{t^{(2\kappa+1)\alpha}}{\Gamma\left(\left(2\kappa + 1\right)\alpha + 1\right)}, \tag{D.46}$$

$$\vartheta_3(\alpha) := \sum_{\kappa=0}^{\infty} \frac{(-1)^{\kappa}}{\Gamma(2\kappa\alpha + 1)} \tag{D.47}$$

and

$$\vartheta_4(\alpha) := \sum_{\kappa=0}^{\infty} \frac{(-1)^{\kappa}}{\Gamma((2\kappa + 1)\alpha + 1)}. \tag{D.48}$$

D.2 The one-parametric special functions via Plotnikov functions

The Plotnikov subsine function, suggested by Plotnikov in 1979, is defined as [334]

$$PSin_{\alpha}(t^{\alpha}) = \sum_{\kappa=0}^{\infty} \frac{(-1)^{\kappa} t^{(2-\alpha)\kappa+1}}{\Gamma((2-\alpha)\kappa + 2)}, \tag{D.49}$$

which implies that

$$PSin_{\alpha}(\lambda t^{\alpha}) = \sum_{\kappa=0}^{\infty} \frac{(-1)^{\kappa} \lambda^{2\kappa+1} t^{(2-\alpha)\kappa+1}}{\Gamma((2-\alpha)\kappa + 2)}, \tag{D.50}$$

Plotnikov subcosine function, denoted by $PCos_{\alpha}(t^{\alpha})$, as [334]

$$PCos_{\alpha}(t^{\alpha}) = \sum_{\kappa=0}^{\infty} \frac{(-1)^{\kappa} t^{(2-\alpha)\kappa}}{\Gamma((2-\alpha)\kappa + 1)}, \tag{D.51}$$

which implies that

$$PCos_{\alpha}(\lambda t^{\alpha}) = \sum_{\kappa=0}^{\infty} \frac{(-1)^{\kappa} \lambda^{2\kappa} t^{(2-\alpha)\kappa}}{\Gamma((2-\alpha)\kappa + 1)}, \tag{D.52}$$

the hyperbolic subsine function, denoted by $PSinh_{\alpha}(t^{\alpha})$, as

$$PSinh_{\alpha}(t^{\alpha}) = \sum_{\kappa=0}^{\infty} \frac{t^{(2-\alpha)\kappa+1}}{\Gamma((2-\alpha)\kappa + 2)}, \tag{D.53}$$

which implies that

$$PSinh_{\alpha}(\lambda t^{\alpha}) = \sum_{\kappa=0}^{\infty} \frac{\lambda^{2\kappa+1} t^{(2-\alpha)\kappa+1}}{\Gamma((2-\alpha)\kappa + 2)}, \tag{D.54}$$

the hyperbolic subcosine function, denoted by $PCosh_{\alpha}(t^{\alpha})$, as

$$PCosh_{\alpha}(t^{\alpha}) = \sum_{\kappa=0}^{\infty} \frac{t^{(2-\alpha)\kappa}}{\Gamma((2-\alpha)\kappa + 1)}, \tag{D.55}$$

which implies that

$$PCosh_{\alpha}\left(\lambda t^{\alpha}\right)=\sum_{\kappa=0}^{\infty}\frac{\lambda^{2\kappa}t^{(2-\alpha)\kappa}}{\Gamma\left((2-\alpha)\kappa+1\right)}. \tag{D.56}$$

Finding the Laplace transforms of the above formulations, we have

$$L\left[PSin_{\alpha}\left(\lambda t^{\alpha}\right)\right]=\lambda s^{-2}\left(1+\lambda^{2}s^{2-\alpha}\right)^{-1}\quad\left(\left|\lambda^{2}s^{2-\alpha}\right|<1\right), \tag{D.57}$$

$$L\left[PCos_{\alpha}\left(\lambda t^{\alpha}\right)\right]=s^{-1}\left(1+\lambda^{2}s^{2-\alpha}\right)^{-1}\quad\left(\left|\lambda^{2}s^{2-\alpha}\right|<1\right), \tag{D.58}$$

$$L\left[PSinh_{\alpha}\left(\lambda t^{\alpha}\right)\right]=\lambda s^{-2}\left(1-\lambda^{2}s^{2-\alpha}\right)^{-1}\quad\left(\left|\lambda^{2}s^{2-\alpha}\right|<1\right) \tag{D.59}$$

and

$$L\left[PCosh_{\alpha}\left(\lambda t^{\alpha}\right)\right]=s^{-1}\left(1-\lambda^{2}s^{2-\alpha}\right)^{-1}\quad\left(\left|\lambda^{2}s^{2-\alpha}\right|<1\right). \tag{D.60}$$

D.3 The one-parametric special functions via Miller-Ross functions

The Miller-Ross function with one-parameter constant λ can be written as follows [1]:

$$\mathrm{M}_{\alpha}\left(\lambda t^{\alpha}\right)=\sum_{\kappa=0}^{\infty}\frac{\lambda^{\kappa}t^{\kappa+\alpha}}{\Gamma(\kappa+1+\alpha)}. \tag{D.61}$$

The subsine function of Miller-Ross type, denoted by $MRSin_{\alpha}\left(t^{\alpha}\right)$, is defined as [1]

$$MRSin_{\alpha}\left(t^{\alpha}\right)=\sum_{\kappa=0}^{\infty}\frac{(-1)^{\kappa}t^{2\kappa+1+\alpha}}{\Gamma(2\kappa+2+\alpha)}, \tag{D.62}$$

which implies that

$$MRSin_{\alpha}\left(\lambda t^{\alpha}\right)=\sum_{\kappa=0}^{\infty}\frac{(-1)^{\kappa}\lambda^{2\kappa+1}t^{2\kappa+1+\alpha}}{\Gamma\left(2\kappa+2+\alpha\right)}, \tag{D.63}$$

and the subcosine function of Miller-Ross type, denoted by $MRCos_{\alpha}\left(t^{\alpha}\right)$, as

$$MRCos_{\alpha}\left(t^{\alpha}\right)=\sum_{\kappa=0}^{\infty}\frac{(-1)^{\kappa}t^{2\kappa+\alpha}}{\Gamma(2\kappa+1+\alpha)}, \tag{D.64}$$

which implies that

$$MRCos_{\alpha}\left(\lambda t^{\alpha}\right)=\sum_{\kappa=0}^{\infty}\frac{(-1)^{\kappa}\lambda^{2\kappa}t^{2\kappa+\alpha}}{\Gamma\left(2\kappa+1+\alpha\right)}, \tag{D.65}$$

the hyperbolic subsine function of Miller-Ross type, denoted by $MRSin_{\alpha}(t^{\alpha})$, as [1]

$$MRSinh_{\alpha}(t^{\alpha}) = \sum_{\kappa=0}^{\infty} \frac{t^{2\kappa+1+\alpha}}{\Gamma(2\kappa+2+\alpha)}, \tag{D.66}$$

which implies that

$$MRSinh_{\alpha}(\lambda t^{\alpha}) = \sum_{\kappa=0}^{\infty} \frac{\lambda^{2\kappa+1} t^{2\kappa+1+\alpha}}{\Gamma(2\kappa+2+\alpha)}, \tag{D.67}$$

and the hyperbolic subcosine function of Miller-Ross type, denoted by $MRCosh_{\alpha}(t^{\alpha})$, as [1]

$$MRCosh_{\alpha}(t^{\alpha}) = \sum_{\kappa=0}^{\infty} \frac{t^{2\kappa+\alpha}}{\Gamma(2\kappa+1+\alpha)}, \tag{D.68}$$

which implies that

$$MRCosh_{\alpha}(\lambda t^{\alpha}) = \sum_{\kappa=0}^{\infty} \frac{\lambda^{2\kappa} t^{2\kappa+\alpha}}{\Gamma(2\kappa+1+\alpha)}. \tag{D.69}$$

In this case, we have

$$\mathrm{M}_{\alpha}(i\lambda t^{\alpha}) = MRCos_{\alpha}(\lambda t^{\alpha}) + iMRSin_{\alpha}(\lambda t^{\alpha}) \tag{D.70}$$

and

$$\mathrm{M}_{\alpha}(\lambda t^{\alpha}) = MRCos_{\alpha}(\lambda t^{\alpha}) + MRSin_{\alpha}(\lambda t^{\alpha}) \tag{D.71}$$

such that

$$MRSin_{\alpha}(\lambda t^{\alpha}) = \frac{\mathrm{M}_{\alpha}(i\lambda t^{\alpha}) - \mathrm{M}_{\alpha}(-i\lambda t^{\alpha})}{2i}, \tag{D.72}$$

$$MRCos_{\alpha}(\lambda t^{\alpha}) = \frac{\mathrm{M}_{\alpha}(i\lambda t^{\alpha}) + \mathrm{M}_{\alpha}(-i\lambda t^{\alpha})}{2}, \tag{D.73}$$

$$MRSinh_{\alpha}(\lambda t^{\alpha}) = \frac{\mathrm{M}_{\alpha}(\lambda t^{\alpha}) - \mathrm{M}_{\alpha}(-\lambda t^{\alpha})}{2} \tag{D.74}$$

and

$$MRCosh_{\alpha}(\lambda t^{\alpha}) = \frac{\mathrm{M}_{\alpha}(\lambda t^{\alpha}) + \mathrm{M}_{\alpha}(-\lambda t^{\alpha})}{2}. \tag{D.75}$$

With the use of the formulation

$$L\left[\frac{t^{\kappa\alpha}}{\Gamma(\kappa\alpha+1)}\right] = \frac{1}{s^{\kappa\alpha}}, \tag{D.76}$$

we have (see [1])

$$L\{\mathrm{M}_{\alpha}(\lambda t^{\alpha})\} = s^{-(\alpha+1)}\left(1-\lambda s^{-1}\right)^{-1}\left(\left|\lambda s^{-1}\right|<1\right) \tag{D.77}$$

such that (see [1])

$$L\left\{MRSin_{\alpha}\left(t^{\alpha}\right)\right\} = \lambda s^{-(2+\alpha)}\left(1+\lambda^{2}s^{-2}\right)^{-1}\left(\left|\lambda s^{-1}\right|<1\right), \tag{D.78}$$

$$L\left\{MRSinh_{\alpha}\left(t^{\alpha}\right)\right\} = \lambda s^{-(2+\alpha)}\left(1-\lambda^{2}s^{-2}\right)^{-1}\left(\left|\lambda s^{-1}\right|<1\right), \tag{D.79}$$

$$L\left\{MRCos_{\alpha}\left(\lambda t^{\alpha}\right)\right\} = s^{-1}\left(1+\lambda^{2}s^{-2}\right)^{-1}\left(\left|\lambda s^{-1}\right|<1\right) \tag{D.80}$$

and

$$L\left\{MRCosh_{\alpha}\left(\lambda t^{\alpha}\right)\right\} = s^{-1}\left(1-\lambda^{2}s^{-2}\right)^{-1}\left(\left|\lambda s^{-1}\right|<1\right). \tag{D.81}$$

In this case, we have

$$\mu_0(\alpha) = \sum_{\kappa=0}^{\infty}\frac{1}{\Gamma(\kappa+1+\alpha)}, \tag{D.82}$$

$$\mu_1(\alpha) = \sum_{\kappa=0}^{\infty}\frac{(-1)^{\kappa}}{\Gamma(2\kappa+2+\alpha)}, \tag{D.83}$$

$$\mu_2(\alpha) = \sum_{\kappa=0}^{\infty}\frac{(-1)^{\kappa}}{\Gamma(2\kappa+1+\alpha)}, \tag{D.84}$$

$$\mu_3(\alpha) = \sum_{\kappa=0}^{\infty}\frac{t^{2\kappa+1+\alpha}}{\Gamma(2\kappa+2+\alpha)}, \tag{D.85}$$

and

$$\mu_4(\alpha) = \sum_{\kappa=0}^{\infty}\frac{1}{\Gamma(2\kappa+1+\alpha)}. \tag{D.86}$$

D.4 The one-parametric special functions via Rabotnov functions

The Rabotnov function with one-parameter constant λ can be written as follows [217]:

$$\Re_{\alpha}\left((\iota t)^{\alpha}\right) = \sum_{\kappa=0}^{\infty}\frac{(\iota t)^{(\kappa+1)(\alpha+1)-1}}{\Gamma((\kappa+1)(\alpha+1))}, \tag{D.87}$$

which can be written as

$$\Re_{\alpha}\left(\lambda t^{\alpha}\right) = \sum_{\kappa=0}^{\infty}\frac{\lambda^{\kappa}t^{(\kappa+1)(\alpha+1)-1}}{\Gamma((\kappa+1)(\alpha+1))}, \tag{D.88}$$

where $\lambda = \iota^{\alpha}$.

The corresponding subsine function, denoted by $\Re Sin_{\alpha}(t^{\alpha})$, is defined as

$$\Re Sin_{\alpha}(t^{\alpha}) = \sum_{\kappa=0}^{\infty} \frac{(-1)^{\kappa} t^{2(\kappa+1)(\alpha+1)-1}}{\Gamma(2(\kappa+1)(\alpha+1))}, \tag{D.89}$$

which implies that

$$\Re Sin_{\alpha}(\lambda t^{\alpha}) = \sum_{\kappa=0}^{\infty} \frac{(-1)^{\kappa} \lambda^{2\kappa+1} t^{2(\kappa+1)(\alpha+1)-1}}{\Gamma(2(\kappa+1)(\alpha+1))}, \tag{D.90}$$

the subcosine function, denoted by $\Re Cos_{\alpha}(t^{\alpha})$, as

$$\Re Cos_{\alpha}(t^{\alpha}) = \sum_{\kappa=0}^{\infty} \frac{(-1)^{\kappa} t^{(2\kappa+1)(\alpha+1)-1}}{\Gamma((2\kappa+1)(\alpha+1))}, \tag{D.91}$$

which implies that

$$\Re Cos_{\alpha}(\lambda t^{\alpha}) = \sum_{\kappa=0}^{\infty} \frac{(-1)^{\kappa} \lambda^{2\kappa} t^{(2\kappa+1)(\alpha+1)-1}}{\Gamma((2\kappa+1)(\alpha+1))}. \tag{D.92}$$

the hyperbolic subsine function, denoted by $\Re Sinh_{\alpha}(t^{\alpha})$, is defined as

$$\Re Sinh_{\alpha}(t^{\alpha}) = \sum_{\kappa=0}^{\infty} \frac{t^{2(\kappa+1)(\alpha+1)-1}}{\Gamma(2(\kappa+1)(\alpha+1))}, \tag{D.93}$$

which implies that

$$\Re Sinh_{\alpha}(\lambda t^{\alpha}) = \sum_{\kappa=0}^{\infty} \frac{\lambda^{2\kappa+1} t^{2(\kappa+1)(\alpha+1)-1}}{\Gamma(2(\kappa+1)(\alpha+1))}, \tag{D.94}$$

and the hyperbolic subcosine function, denoted by $\Re Cosh_{\alpha}(t^{\alpha})$, as

$$\Re Cosh_{\alpha}(t^{\alpha}) = \sum_{\kappa=0}^{\infty} \frac{t^{(2\kappa+1)(\alpha+1)-1}}{\Gamma((2\kappa+1)(\alpha+1))}, \tag{D.95}$$

which implies that

$$\Re Cosh_{\alpha}(\lambda t^{\alpha}) = \sum_{\kappa=0}^{\infty} \frac{\lambda^{2\kappa} t^{(2\kappa+1)(\alpha+1)-1}}{\Gamma((2\kappa+1)(\alpha+1))}. \tag{D.96}$$

With the use of the formulation

$$L\left[\frac{t^{\kappa\alpha}}{\Gamma(\kappa\alpha+1)}\right] = \frac{1}{s^{\kappa\alpha}}, \tag{D.97}$$

we have (see [237])

$$L\{\Re_{\alpha}(\lambda t^{\alpha})\} = s^{-(\alpha 1)}\left(1 - \lambda s^{-(\alpha+1)}\right)^{-1} \left(\left|\lambda s^{-(\alpha+1)}\right| < 1\right) \tag{D.98}$$

such that

$$L\left\{\Re Sin_{\alpha}\left(\lambda t^{\alpha}\right)\right\} = s^{-(\alpha 1)}\left(1-\lambda s^{-(\alpha+1)}\right)^{-1}\left(\left|\lambda^{2} s^{-2(\alpha+1)}\right|<1\right), \tag{D.99}$$

$$L\left\{\Re Sinh_{\alpha}\left(\lambda t^{\alpha}\right)\right\} = \lambda s^{-2(1+\alpha)}\left(1+\lambda^{2} s^{-2(\alpha+1)}\right)^{-1}\left(\left|\lambda^{2} s^{-2(\alpha+1)}\right|<1\right), \tag{D.100}$$

$$L\left\{\Re Cos_{\alpha}\left(\lambda t^{\alpha}\right)\right\} = \lambda s^{-2(1+\alpha)}\left(1-\lambda^{2} s^{-2(\alpha+1)}\right)^{-1}\left(\left|\lambda^{2} s^{-2(\alpha+1)}\right|<1\right) \tag{D.101}$$

and

$$L\left\{\Re Cosh_{\alpha}\left(\lambda t^{\alpha}\right)\right\} = s^{-(1+\alpha)}\left(1+\lambda^{2} s^{-2(\alpha+1)}\right)^{-1}\left(\left|\lambda^{2} s^{-2(\alpha+1)}\right|<1\right). \tag{D.102}$$

Thus, we have

$$\Re_{\alpha}\left(\lambda t^{\alpha}\right) = \Re Cos_{\alpha}\left(\lambda t^{\alpha}\right) + \Re Sin_{\alpha}\left(\lambda t^{\alpha}\right) \tag{D.103}$$

and

$$\Re_{\alpha}\left(i\lambda t^{\alpha}\right) = \Re Cos_{\alpha}\left(\lambda t^{\alpha}\right) + i\Re Sin_{\alpha}\left(\lambda t^{\alpha}\right) \tag{D.104}$$

such that

$$\Re Sin_{\alpha}\left(t^{\alpha}\right) = \frac{\Re_{\alpha}\left(it^{\alpha}\right) - \Re_{\alpha}\left(-it^{\alpha}\right)}{2i}, \tag{D.105}$$

$$\Re Cos_{\alpha}\left(t^{\alpha}\right) = \frac{\Re_{\alpha}\left(it^{\alpha}\right) + \Re_{\alpha}\left(-it^{\alpha}\right)}{2}, \tag{D.106}$$

$$\Re Sinh_{\alpha}\left(t^{\alpha}\right) = \frac{\Re_{\alpha}\left(t^{\alpha}\right) - \Re_{\alpha}\left(-t^{\alpha}\right)}{2}, \tag{D.107}$$

$$\Re Cosh_{\alpha}\left(t^{\alpha}\right) = \frac{\Re_{\alpha}\left(t^{\alpha}\right) + \Re_{\alpha}\left(-t^{\alpha}\right)}{2}, \tag{D.108}$$

$$\Re Sin_{\alpha}\left(\lambda t^{\alpha}\right) = \frac{\Re_{\alpha}\left(i\lambda t^{\alpha}\right) - \Re_{\alpha}\left(-i\lambda t^{\alpha}\right)}{2i}, \tag{D.109}$$

$$\Re Cos_{\alpha}\left(\lambda t^{\alpha}\right) = \frac{\Re_{\alpha}\left(i\lambda t^{\alpha}\right) + \Re_{\alpha}\left(-i\lambda t^{\alpha}\right)}{2}, \tag{D.110}$$

$$\Re Sinh_{\alpha}\left(\lambda t^{\alpha}\right) = \frac{\Re_{\alpha}\left(\lambda t^{\alpha}\right) - \Re_{\alpha}\left(-\lambda t^{\alpha}\right)}{2} \tag{D.111}$$

and

$$\Re Cosh_{\alpha}\left(\lambda t^{\alpha}\right) = \frac{\Re_{\alpha}\left(\lambda t^{\alpha}\right) + \Re_{\alpha}\left(-\lambda t^{\alpha}\right)}{2}. \tag{D.112}$$

D.5 The one-parametric special functions via one-parametric Lorenzo-Hartley functions

The Lorenzo-Hartley function with the constant λ is defined as [335, 336]

$$F_\alpha\left(\lambda t^\alpha\right)=\sum_{\kappa=0}^{\infty}\frac{\lambda^\kappa t^{(\kappa+1)\alpha-1}}{\Gamma\left((\kappa+1)\alpha\right)}. \tag{D.113}$$

The corresponding subsine function, denoted by $FSin_\alpha\left(t^\alpha\right)$, is defined as

$$FSin_\alpha\left(t^\alpha\right)=\sum_{\kappa=0}^{\infty}\frac{(-1)^\kappa t^{2(\kappa+1)\alpha-1}}{\Gamma\left(2(\kappa+1)\alpha\right)}, \tag{D.114}$$

which implies that

$$FSin_\alpha\left(\lambda t^\alpha\right)=\sum_{\kappa=0}^{\infty}\frac{(-1)^\kappa \lambda^{2\kappa+1}t^{2(\kappa+1)\alpha-1}}{\Gamma\left(2(\kappa+1)\alpha\right)}, \tag{D.115}$$

the subcosine function, denoted by $FCos_\alpha\left(t^\alpha\right)$, as

$$FCos_\alpha\left(t^\alpha\right)=\sum_{\kappa=0}^{\infty}\frac{(-1)^\kappa t^{(2\kappa+1)\alpha-1}}{\Gamma\left((2\kappa+1)\alpha\right)}, \tag{D.116}$$

which implies that

$$FCos_\alpha\left(\lambda t^\alpha\right)=\sum_{\kappa=0}^{\infty}\frac{(-1)^\kappa \lambda^{2\kappa}t^{(2\kappa+1)\alpha-1}}{\Gamma\left((2\kappa+1)\alpha\right)}. \tag{D.117}$$

the hyperbolic subsine function, denoted by $FSinh_\alpha\left(t^\alpha\right)$, is defined as

$$FSinh_\alpha\left(t^\alpha\right)=\sum_{\kappa=0}^{\infty}\frac{t^{2(\kappa+1)\alpha-1}}{\Gamma\left(2(\kappa+1)\alpha\right)}, \tag{D.118}$$

which implies that

$$FSinh_\alpha\left(\lambda t^\alpha\right)=\sum_{\kappa=0}^{\infty}\frac{\lambda^\kappa t^{2(\kappa+1)\alpha-1}}{\Gamma\left(2(\kappa+1)\alpha\right)}, \tag{D.119}$$

and the hyperbolic subcosine function, denoted by $FCosh_\alpha\left(t^\alpha\right)$, as

$$FCosh_\alpha\left(t^\alpha\right)=\sum_{\kappa=0}^{\infty}\frac{\lambda^\kappa t^{(2\kappa+1)\alpha-1}}{\Gamma\left((2\kappa+1)\alpha\right)}, \tag{D.120}$$

which implies that

$$FCosh_{\alpha}(\lambda t^{\alpha}) = \sum_{\kappa=0}^{\infty} \frac{\lambda^{\kappa} t^{(2\kappa+1)\alpha-1}}{\Gamma((2\kappa+1)\alpha)}. \tag{D.121}$$

With the use of the formulation

$$L\left[\frac{t^{\kappa\alpha}}{\Gamma(\kappa\alpha+1)}\right] = \frac{1}{s^{\kappa\alpha}}, \tag{D.122}$$

we have

$$L\{F_{\alpha}(\lambda t^{\alpha})\}(|\lambda s^{-\alpha}| < 1) \tag{D.123}$$

such that

$$L\{FSin_{\alpha}(\lambda t^{\alpha})\}\left(\left|\lambda^{2} s^{-2\alpha}\right| < 1\right), \tag{D.124}$$

$$L\{FSinh_{\alpha}(\lambda t^{\alpha})\}\left(\left|\lambda^{2} s^{-2\alpha}\right| < 1\right), \tag{D.125}$$

$$L\{FCos_{\alpha}(\lambda t^{\alpha})\}\left(\left|\lambda^{2} s^{-2\alpha}\right| < 1\right) \tag{D.126}$$

and

$$L\{FCosh_{\alpha}(\lambda t^{\alpha})\}\left(\left|\lambda^{2} s^{-2\alpha}\right| < 1\right). \tag{D.127}$$

Thus, we have

$$F_{\alpha}(\lambda t^{\alpha}) = FCosh_{\alpha}(\lambda t^{\alpha}) + FSinh_{\alpha}(\lambda t^{\alpha}) \tag{D.128}$$

and

$$F_{\alpha}(i\lambda t) = FCos_{\alpha}(\lambda t^{\alpha}) + iFSin_{\alpha}(\lambda t^{\alpha}) \tag{D.129}$$

such that

$$FSin_{\alpha}(t^{\alpha}) = \frac{F_{\alpha}(it^{\alpha}) - F_{\alpha}(-it^{\alpha})}{2i}, \tag{D.130}$$

$$FCos_{\alpha}(t^{\alpha}) = \frac{F_{\alpha}(it^{\alpha}) + F_{\alpha}(-it^{\alpha})}{2}, \tag{D.131}$$

$$FSinh_{\alpha}(t^{\alpha}) = \frac{F_{\alpha}(t^{\alpha}) - F_{\alpha}(-t^{\alpha})}{2}, \tag{D.132}$$

$$FCosh_{\alpha}(t^{\alpha}) = \frac{F_{\alpha}(t^{\alpha}) + F_{\alpha}(-t^{\alpha})}{2}, \tag{D.133}$$

$$FSin_{\alpha}(\lambda t^{\alpha}) = \frac{F_{\alpha}(i\lambda t^{\alpha}) - F_{\alpha}(-i\lambda t^{\alpha})}{2i}, \tag{D.134}$$

$$FCos_{\alpha}(\lambda t^{\alpha}) = \frac{F_{\alpha}(i\lambda t^{\alpha}) + F_{\alpha}(-i\lambda t^{\alpha})}{2}, \tag{D.135}$$

$$FSinh_{\alpha}(\lambda t^{\alpha}) = \frac{F_{\alpha}(\lambda t^{\alpha}) - F_{\alpha}(-\lambda t^{\alpha})}{2} \tag{D.136}$$

and

$$FCosh_{\alpha}(\lambda t^{\alpha}) = \frac{F_{\alpha}(\lambda t^{\alpha}) + F_{\alpha}(-\lambda t^{\alpha})}{2}. \tag{D.137}$$

D.6 The three-parametric special functions via Prabhakar functions

The generalized Prabhakar function with one-parameter constant λ can be written as follows [133, 236]:

$$t^{\upsilon-1}E^{\varphi}_{\alpha,\upsilon}\left(\lambda t^{\alpha}\right)=\sum_{\kappa=0}^{\infty}\begin{pmatrix}-\varphi\\ \kappa\end{pmatrix}\frac{(-1)^{\kappa}\lambda^{\kappa}t^{\kappa\alpha+\upsilon-1}}{\Gamma(\kappa\alpha+\upsilon)},\tag{D.138}$$

where

$$\frac{(\varphi)_{\kappa}}{\Gamma\left(\kappa+1\right)}=\begin{pmatrix}-\varphi\\ \kappa\end{pmatrix}(-1)^{\kappa},\tag{D.139}$$

and the Prabhakar function is given as (see [133])

$$E^{\varphi}_{\alpha,\upsilon}\left(t^{\alpha}\right)=\sum_{\kappa=0}^{\infty}\frac{(\varphi)_{\kappa}}{\Gamma(\kappa+1)}\frac{t^{\kappa\alpha}}{\Gamma(\kappa\alpha+\upsilon)}.\tag{D.140}$$

The supersine function, denoted by $Supersin^{\varphi}_{\alpha,\upsilon}\left(t^{\alpha}\right)$, is defined as

$$Supersin^{\varphi}_{\alpha,\upsilon}\left(t^{\alpha}\right)=\sum_{\kappa=0}^{\infty}\frac{\Gamma(\varphi+2\kappa+1)}{\Gamma(2\kappa+2)\Gamma(\varphi)}\frac{(-1)^{\kappa}t^{(2\kappa+1)\alpha+\upsilon-1}}{\Gamma((2\kappa+1)\alpha+\upsilon)}\tag{D.141}$$

which implies that

$$Supersin^{\varphi}_{\alpha,\upsilon}\left(\lambda t^{\alpha}\right)=\sum_{\kappa=0}^{\infty}\frac{\Gamma(\varphi+2\kappa+1)}{\Gamma(2\kappa+2)\Gamma(\varphi)}\frac{(-1)^{\kappa}\lambda^{2\kappa+1}t^{(2\kappa+1)\alpha+\upsilon-1}}{\Gamma((2\kappa+1)\alpha+\upsilon)},\tag{D.142}$$

the supercosine function, denoted by $Supercos^{\varphi}_{\alpha,\upsilon}\left(t^{\alpha}\right)$, as

$$Supercos^{\varphi}_{\alpha,\upsilon}\left(t^{\alpha}\right)=\sum_{\kappa=0}^{\infty}\frac{\Gamma(\varphi+2\kappa)}{\Gamma(2\kappa+1)\Gamma(\varphi)}\frac{(-1)^{\kappa}t^{2\kappa\alpha+\upsilon-1}}{\Gamma(2\kappa\alpha+\upsilon)},\tag{D.143}$$

which yields that

$$Supercos^{\varphi}_{\alpha,\upsilon}\left(\lambda t^{\alpha}\right)=\sum_{\kappa=0}^{\infty}\frac{\Gamma(\varphi+2\kappa)}{\Gamma(2\kappa+1)\Gamma(\varphi)}\frac{(-1)^{\kappa}\lambda^{2\kappa}t^{2\kappa\alpha+\upsilon-1}}{\Gamma(2\kappa\alpha+\upsilon)}.\tag{D.144}$$

The hyperbolic supersine function, denoted by $Supersinh^{\varphi}_{\alpha,\upsilon}\left(t^{\alpha}\right)$, as

$$Supersinh^{\varphi}_{\alpha,\upsilon}\left(t^{\alpha}\right)=\sum_{\kappa=0}^{\infty}\frac{\Gamma(\varphi+2\kappa+1)}{\Gamma(2\kappa+2)\Gamma(\varphi)}\frac{t^{(2\kappa+1)\alpha+\upsilon-1}}{\Gamma((2\kappa+1)\alpha+\upsilon)},\tag{D.145}$$

which implies that

$$Supersinh^{\varphi}_{\alpha,\upsilon}\left(\lambda t^{\alpha}\right)=\sum_{\kappa=0}^{\infty}\frac{\Gamma(\varphi+2\kappa+1)}{\Gamma(2\kappa+2)\Gamma(\varphi)}\frac{\lambda^{2\kappa+1}t^{(2\kappa+1)\alpha+\upsilon-1}}{\Gamma((2\kappa+1)\alpha+\upsilon)},\tag{D.146}$$

and the hyperbolic supercosine function, denoted by $Supercosh_{\alpha,\upsilon}^{\varphi}(t^{\alpha})$, as

$$Supercosh_{\alpha,\upsilon}^{\varphi}(t^{\alpha}) = \sum_{\kappa=0}^{\infty} \frac{\Gamma(\varphi+2\kappa)}{\Gamma(2\kappa+1)\Gamma(\varphi)} \frac{t^{2\kappa\alpha+\upsilon-1}}{\Gamma(2\kappa\alpha+\upsilon)}, \tag{D.147}$$

which leads to

$$Supercosh_{\alpha,\upsilon}^{\varphi}(\lambda t^{\alpha}) = \sum_{\kappa=0}^{\infty} \frac{\Gamma(\varphi+2\kappa)}{\Gamma(2\kappa+1)\Gamma(\varphi)} \frac{\lambda^{2\kappa} t^{2\kappa\alpha+\upsilon-1}}{\Gamma(2\kappa\alpha+\upsilon)}. \tag{D.148}$$

Thus, we have

$$t^{\upsilon-1}E_{\alpha,\upsilon}^{\varphi}(\lambda t^{\alpha}) = Supercosh_{\alpha,\upsilon}^{\varphi}(\lambda t^{\alpha}) + Supersinh_{\alpha,\upsilon}^{\varphi}(\lambda t^{\alpha}) \tag{D.149}$$

and

$$t^{\upsilon-1}E_{\alpha,\upsilon}^{\varphi}(i\lambda t^{\alpha}) = Supercos_{\alpha,\upsilon}^{\varphi}(\lambda t^{\alpha}) + iSupersin_{\alpha,\upsilon}^{\varphi}(\lambda t^{\alpha}) \tag{D.150}$$

such that

$$Supersin_{\alpha,\upsilon}^{\varphi}(t^{\alpha}) = \frac{t^{\upsilon-1}E_{\alpha,\upsilon}^{\varphi}(it^{\alpha}) - t^{\upsilon-1}E_{\alpha,\upsilon}^{\varphi}(-it^{\alpha})}{2i}, \tag{D.151}$$

$$Supercos_{\alpha,\upsilon}^{\varphi}(t^{\alpha}) = \frac{t^{\upsilon-1}E_{\alpha,\upsilon}^{\varphi}(-i\lambda t^{\alpha}) + t^{\upsilon-1}E_{\alpha,\upsilon}^{\varphi}(-i\lambda t^{\alpha})}{2}, \tag{D.152}$$

$$Supersinh_{\alpha,\upsilon}^{\varphi}(t^{\alpha}) = \frac{t^{\upsilon-1}E_{\alpha,\upsilon}^{\varphi}(t^{\alpha}) - t^{\upsilon-1}E_{\alpha,\upsilon}^{\varphi}(-t^{\alpha})}{2}, \tag{D.153}$$

$$Supercosh_{\alpha,\upsilon}^{\varphi}(t^{\alpha}) = \frac{t^{\upsilon-1}E_{\alpha,\upsilon}^{\varphi}(t^{\alpha}) + t^{\upsilon-1}E_{\alpha,\upsilon}^{\varphi}(-t^{\alpha})}{2}, \tag{D.154}$$

$$Supersin_{\alpha,\upsilon}^{\varphi}(\lambda t^{\alpha}) = \frac{t^{\upsilon-1}E_{\alpha,\upsilon}^{\varphi}(i\lambda t^{\alpha}) - t^{\upsilon-1}E_{\alpha,\upsilon}^{\varphi}(-i\lambda t^{\alpha})}{2i}, \tag{D.155}$$

$$Supercos_{\alpha,\upsilon}^{\varphi}(\lambda t^{\alpha}) = \frac{t^{\upsilon-1}E_{\alpha,\upsilon}^{\varphi}(i\lambda t^{\alpha}) + t^{\upsilon-1}E_{\alpha,\upsilon}^{\varphi}(-i\lambda t^{\alpha})}{2}, \tag{D.156}$$

$$Supersinh_{\alpha,\upsilon}^{\varphi}(\lambda t^{\alpha}) = \frac{t^{\upsilon-1}E_{\alpha,\upsilon}^{\varphi}(\lambda t^{\alpha}) - t^{\upsilon-1}E_{\alpha,\upsilon}^{\varphi}(-\lambda t^{\alpha})}{2} \tag{D.157}$$

and

$$Supercosh_{\alpha,\upsilon}^{\varphi}(\lambda t^{\alpha}) = \frac{t^{\upsilon-1}E_{\alpha,\upsilon}^{\varphi}(\lambda t^{\alpha}) + t^{\upsilon-1}E_{\alpha,\upsilon}^{\varphi}(-\lambda t^{\alpha})}{2}. \tag{D.158}$$

Making use of

$$L\left[\frac{t^{\kappa\alpha}}{\Gamma(\kappa\alpha+1)}\right] = \frac{1}{s^{\kappa\alpha}}, \tag{D.159}$$

we give

$$L\left\{t^{\upsilon-1}E_{\alpha,\upsilon}^{\varphi}(\lambda t^{\alpha})\right\} = s^{-\upsilon}\left(1-\lambda s^{-\alpha}\right)^{-\varphi}, \tag{D.160}$$

$$L\left\{Supersin_{\alpha,\upsilon}^{\varphi}\left(\lambda t^{\alpha}\right)\right\} = \frac{is^{-\upsilon}}{2}\frac{\left(1-i\lambda s^{-\alpha}\right)^{\varphi}-\left(1+i\lambda s^{-\alpha}\right)^{\varphi}}{\left(1+\lambda^{2}s^{-2\alpha}\right)^{\varphi}}, \tag{D.161}$$

$$L\left\{Supersinh_{\alpha,\upsilon}^{\varphi}\left(\lambda t^{\alpha}\right)\right\} = \frac{s^{-\upsilon}}{2}\frac{\left(1-\lambda s^{-\alpha}\right)^{\varphi}-\left(1+\lambda s^{-\alpha}\right)^{\varphi}}{\left(1-\lambda^{2}s^{-2\alpha}\right)^{\varphi}}, \tag{D.162}$$

$$L\left\{Supercos_{\alpha,\upsilon}^{\varphi}\left(\lambda t^{\alpha}\right)\right\} = \frac{s^{-\upsilon}}{2}\frac{\left(1+i\lambda s^{-\alpha}\right)^{\varphi}+\left(1-i\lambda s^{-\alpha}\right)^{\varphi}}{\left(1+\lambda^{2}s^{-2\alpha}\right)^{\varphi}}, \tag{D.163}$$

and

$$L\left\{Supercosh_{\alpha,\upsilon}^{\varphi}\left(\lambda t^{\alpha}\right)\right\} = \frac{s^{-\upsilon}}{2}\frac{\left(1+\lambda s^{-\alpha}\right)^{\varphi}+\left(1-\lambda s^{-\alpha}\right)^{\varphi}}{\left(1-\lambda^{2}s^{-2\alpha}\right)^{\varphi}}, \tag{D.164}$$

where

$$\frac{1}{2}\left[(1+\lambda t)^{-\varphi}+(\lambda t-1)^{-\varphi}\right]=\sum_{\kappa=0}^{\infty}\binom{-\varphi}{2\kappa}(\lambda t)^{-\varphi-2\kappa}\left(|\lambda t|<1\right), \tag{D.165}$$

$$\frac{1}{2}\left[(1+\lambda t)^{-\varphi}+(1-\lambda t)^{-\varphi}\right]=\sum_{\kappa=0}^{\infty}\binom{-\varphi}{2\kappa}(\lambda t)^{2\kappa}\left(|\lambda t|<1\right), \tag{D.166}$$

$$\frac{1}{2}\left[(1+\lambda t)^{-\varphi}-(\lambda t-1)^{-\varphi}\right]=\sum_{\kappa=0}^{\infty}\binom{-\varphi}{2\kappa+1}(\lambda t)^{-\varphi-2\kappa-1}\left(|\lambda t|<1\right), \tag{D.167}$$

$$\frac{1}{2}\left[(1+\lambda t)^{-\varphi}-(1-\lambda t)^{-\varphi}\right]=\sum_{\kappa=0}^{\infty}\binom{-\varphi}{2\kappa+1}(\lambda t)^{2\kappa+1}\left(|\lambda t|<1\right), \tag{D.168}$$

$$(1+\lambda t)^{-\varphi}=\sum_{\kappa=0}^{\infty}\binom{-\varphi}{\kappa}(\lambda t)^{-\varphi-\kappa}, \tag{D.169}$$

$$(\lambda t-1)^{-\varphi}=\sum_{\kappa=0}^{\infty}\binom{-\varphi}{\kappa}(\lambda t)^{\kappa}(-1)^{-\varphi-\kappa}\left(|\lambda t|<1\right) \tag{D.170}$$

and

$$(1-\lambda t)^{-\varphi}=\sum_{\kappa=0}^{\infty}\binom{-\varphi}{\kappa}(-\lambda t)^{-\varphi-\kappa}\left(|\lambda t|<1\right), \tag{D.171}$$

which are reduced by (see [337])

$$\frac{1}{2}\left[(1+t)^{-\varphi}+(t-1)^{-\varphi}\right]=\sum_{\kappa=0}^{\infty}\binom{-\varphi}{2\kappa}t^{-\varphi-2\kappa}\quad\left(|t|<1\right), \tag{D.172}$$

$$\frac{1}{2}\left[(1+t)^{-\varphi}+(1-t)^{-\varphi}\right]=\sum_{\kappa=0}^{\infty}\binom{-\varphi}{2\kappa}t^{2\kappa}\quad\left(|t|<1\right), \tag{D.173}$$

$$\frac{1}{2}\left[(1+t)^{-\varphi}-(t-1)^{-\varphi}\right]=\sum_{\kappa=0}^{\infty}\binom{-\varphi}{2\kappa+1}t^{-\varphi-2\kappa-1}\ (|t|<1) \tag{D.174}$$

and

$$(1+t)^{-\varphi}=\sum_{\kappa=0}^{\infty}\binom{-\varphi}{\kappa}t^{\kappa}\ (|t|<1)\,. \tag{D.175}$$

Let us consider the first case as follows.

The generalized Prabhakar function with one-parameter constant λ is

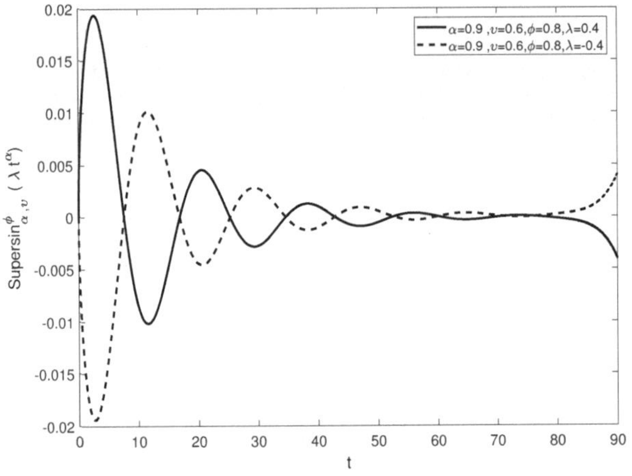

FIGURE D.1: The plot of the subsine function $Supersin^{\varphi}_{\alpha,\upsilon}\left(\lambda t^{\alpha}\right)$.

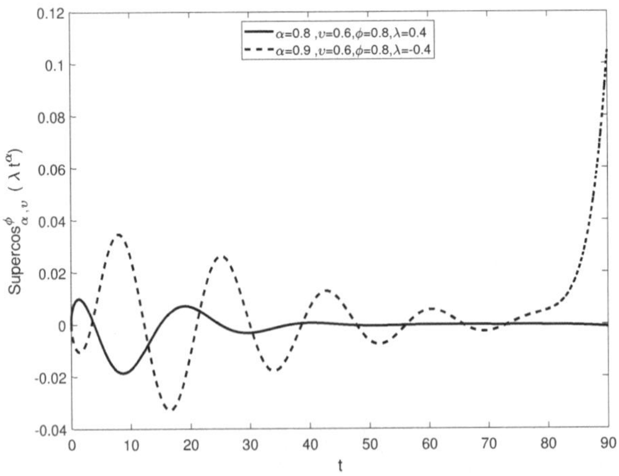

FIGURE D.2: The plot of the subcosine function $Supercos^{\varphi}_{\alpha,\upsilon}\left(\lambda t^{\alpha}\right)$.

given as [236]

$$E_{\alpha,1}^{\varphi}\left(\lambda t^{\alpha}\right)=\sum_{\kappa=0}^{\infty}\left(\begin{array}{c}-\varphi\\ \kappa\end{array}\right)\frac{(-1)^{\kappa}\lambda^{\kappa}t^{\kappa\alpha}}{\Gamma(\kappa\alpha+1)}. \tag{D.176}$$

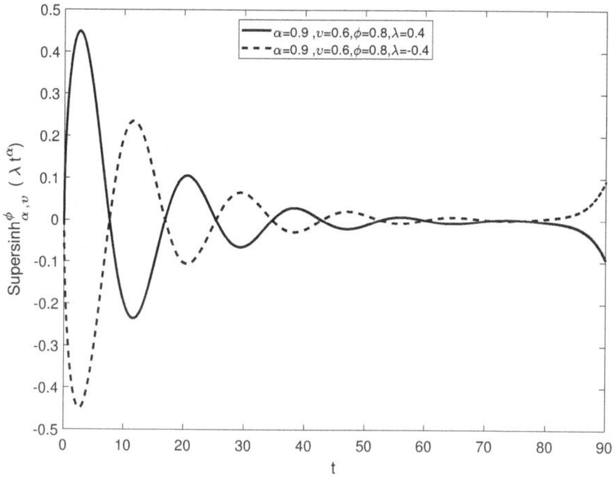

FIGURE D.3: The plot of the hyperbolic subsine function $Supersinh_{\alpha,\upsilon}^{\varphi}\left(\lambda t^{\alpha}\right)$.

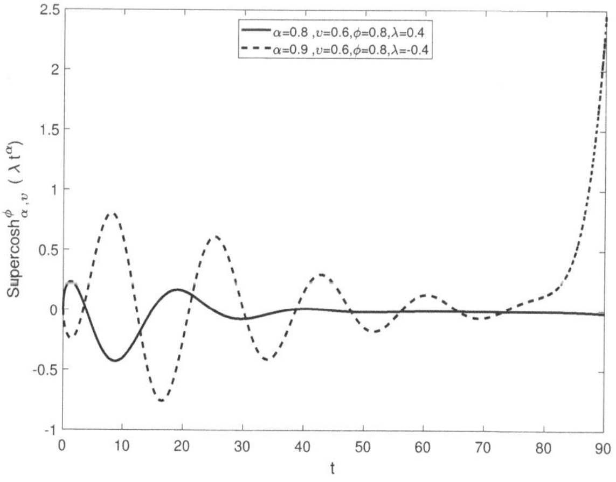

FIGURE D.4: The plot of the hyperbolic subcosine function $Supercosh_{\alpha,\upsilon}^{\varphi}\left(\lambda t^{\alpha}\right)$.

The supersine function, denoted by $Supersin_{\alpha,1}^{\varphi}(t^{\alpha})$, is defined as

$$Supersin_{\alpha,1}^{\varphi}(t^{\alpha}) = \sum_{\kappa=0}^{\infty} \frac{\Gamma(\varphi+2\kappa+1)}{\Gamma(2\kappa+2)\Gamma(\varphi)} \frac{(-1)^{\kappa} t^{(2\kappa+1)\alpha}}{\Gamma((2\kappa+1)\alpha+1)}, \tag{D.177}$$

which implies that

$$Supersin_{\alpha,1}^{\varphi}(\lambda t^{\alpha}) = \sum_{\kappa=0}^{\infty} \frac{\Gamma(\varphi+2\kappa+1)}{\Gamma(2\kappa+2)\Gamma(\varphi)} \frac{(-1)^{\kappa} \lambda^{2\kappa+1} t^{(2\kappa+1)\alpha}}{\Gamma((2\kappa+1)\alpha+1)}, \tag{D.178}$$

the supercosine function, denoted by $Supercos_{\alpha,1}^{\varphi}(t^{\alpha})$, as

$$Supercos_{\alpha,1}^{\varphi}(t^{\alpha}) = \sum_{\kappa=0}^{\infty} \frac{\Gamma(\varphi+2\kappa)}{\Gamma(2\kappa+1)\Gamma(\varphi)} \frac{(-1)^{\kappa} t^{2\kappa\alpha}}{\Gamma(2\kappa\alpha+1)}, \tag{D.179}$$

which yields that

$$Supercos_{\alpha,1}^{\varphi}(\lambda t^{\alpha}) = \sum_{\kappa=0}^{\infty} \frac{\Gamma(\varphi+2\kappa)}{\Gamma(2\kappa+1)\Gamma(\varphi)} \frac{(-1)^{\kappa} \lambda^{2\kappa} t^{2\kappa\alpha}}{\Gamma(2\kappa\alpha+1)}, \tag{D.180}$$

the hyperbolic supersine function, denoted by $Supersinh_{\alpha,1}^{\varphi}(t^{\alpha})$, as

$$Supersinh_{\alpha,1}^{\varphi}(t^{\alpha}) = \sum_{\kappa=0}^{\infty} \frac{\Gamma(\varphi+2\kappa+1)}{\Gamma(2\kappa+2)\Gamma(\varphi)} \frac{t^{(2\kappa+1)\alpha}}{\Gamma((2\kappa+1)\alpha+1)}, \tag{D.181}$$

which leads to

$$Supersinh_{\alpha,1}^{\varphi}(\lambda t^{\alpha}) = \sum_{\kappa=0}^{\infty} \frac{\Gamma(\varphi+2\kappa+1)}{\Gamma(2\kappa+2)\Gamma(\varphi)} \frac{\lambda^{2\kappa+1} t^{(2\kappa+1)\alpha}}{\Gamma((2\kappa+1)\alpha+1)}, \tag{D.182}$$

and the hyperbolic supercosine function, denoted by $Supercosh_{\alpha,1}^{\varphi}(t^{\alpha})$, as

$$Supercosh_{\alpha,1}^{\varphi}(t^{\alpha}) = \sum_{\kappa=0}^{\infty} \frac{\Gamma(\varphi+2\kappa)}{\Gamma(2\kappa+1)\Gamma(\varphi)} \frac{t^{2\kappa\alpha}}{\Gamma(2\kappa\alpha+1)}, \tag{D.183}$$

which implies that

$$Supercosh_{\alpha,1}^{\varphi}(\lambda t^{\alpha}) = \sum_{\kappa=0}^{\infty} \frac{\Gamma(\varphi+2\kappa)}{\Gamma(2\kappa+1)\Gamma(\varphi)} \frac{\lambda^{2\kappa} t^{2\kappa\alpha}}{\Gamma(2\kappa\alpha+1)}. \tag{D.184}$$

Thus, we have

$$E_{\alpha,1}^{\varphi}(\lambda t^{\alpha}) = Supercosh_{\alpha,1}^{\varphi}(\lambda t^{\alpha}) + Supersinh_{\alpha,1}^{\varphi}(\lambda t^{\alpha}) \tag{D.185}$$

and

$$E_{\alpha,1}^{\varphi}(i\lambda t^{\alpha}) = Supercos_{\alpha,1}^{\varphi}(\lambda t^{\alpha}) + iSupersin_{\alpha,1}^{\varphi}(\lambda t^{\alpha}) \tag{D.186}$$

such that

$$Supersin_{\alpha,1}^{\varphi}(t^{\alpha}) = \frac{E_{\alpha,1}^{\varphi}(it^{\alpha}) - E_{\alpha,1}^{\varphi}(-it^{\alpha})}{2i}, \tag{D.187}$$

$$Supercos_{\alpha,1}^{\varphi}\left(t^{\alpha}\right)=\frac{E_{\alpha,1}^{\varphi}\left(-i\lambda t^{\alpha}\right)+E_{\alpha,1}^{\varphi}\left(-i\lambda t^{\alpha}\right)}{2}, \tag{D.188}$$

$$Supersinh_{\alpha,1}^{\varphi}\left(t^{\alpha}\right)=\frac{E_{\alpha,1}^{\varphi}\left(t^{\alpha}\right)-E_{\alpha,1}^{\varphi}\left(-t^{\alpha}\right)}{2}, \tag{D.189}$$

$$Supercosh_{\alpha,1}^{\varphi}\left(t^{\alpha}\right)=\frac{E_{\alpha,\upsilon}^{\varphi}\left(t^{\alpha}\right)+E_{\alpha,1}^{\varphi}\left(-t^{\alpha}\right)}{2}, \tag{D.190}$$

$$Supersin_{\alpha,1}^{\varphi}\left(\lambda t^{\alpha}\right)=\frac{E_{\alpha,1}^{\varphi}\left(i\lambda t^{\alpha}\right)-E_{\alpha,1}^{\varphi}\left(-i\lambda t^{\alpha}\right)}{2i}, \tag{D.191}$$

$$Supercos_{\alpha,1}^{\varphi}\left(\lambda t^{\alpha}\right)=\frac{E_{\alpha,1}^{\varphi}\left(i\lambda t^{\alpha}\right)+E_{\alpha,1}^{\varphi}\left(-i\lambda t^{\alpha}\right)}{2}, \tag{D.192}$$

$$Supersinh_{\alpha,1}^{\varphi}\left(\lambda t^{\alpha}\right)=\frac{E_{\alpha,1}^{\varphi}\left(\lambda t^{\alpha}\right)-E_{\alpha,1}^{\varphi}\left(-\lambda t^{\alpha}\right)}{2} \tag{D.193}$$

and

$$Supercosh_{\alpha,1}^{\varphi}\left(\lambda t^{\alpha}\right)=\frac{E_{\alpha,1}^{\varphi}\left(\lambda t^{\alpha}\right)+E_{\alpha,1}^{\varphi}\left(-\lambda t^{\alpha}\right)}{2}. \tag{D.194}$$

Therefore, we obtain

$$L\left\{E_{\alpha,1}^{\varphi}\left(\lambda t^{\alpha}\right)\right\}=s^{-1}\left(1-\lambda s^{-\alpha}\right)^{-\varphi}\left(\left|\lambda s^{-\alpha}\right|<1; Re\left(\varphi\right)>0\right), \tag{D.195}$$

$$L\left\{Supersin_{\alpha,1}^{\varphi}\left(\lambda t^{\alpha}\right)\right\}=\frac{is^{-1}}{2}\frac{\left(1-i\lambda s^{-\alpha}\right)^{\varphi}-\left(1+i\lambda s^{-\alpha}\right)^{\varphi}}{\left(1+\lambda^{2}s^{-2\alpha}\right)^{\varphi}}\left(\left|\lambda^{2}s^{-2\alpha}\right|<1\right), \tag{D.196}$$

$$L\left\{Super\sinh_{\alpha,1}^{\varphi}\left(\lambda t^{\alpha}\right)\right\}=\frac{s^{-1}}{2}\frac{\left(1-\lambda s^{-\alpha}\right)^{\varphi}-\left(1+\lambda s^{-\alpha}\right)^{\varphi}}{\left(1-\lambda^{2}s^{-2\alpha}\right)^{\varphi}}\left(\left|\lambda^{2}s^{-2\alpha}\right|<1\right), \tag{D.197}$$

$$L\left\{Supercos_{\alpha,1}^{\varphi}\left(\lambda t^{\alpha}\right)\right\}=\frac{s^{-1}}{2}\frac{\left(1+i\lambda s^{-\alpha}\right)^{\varphi}+\left(1-i\lambda s^{-\alpha}\right)^{\varphi}}{\left(1+\lambda^{2}s^{-2\alpha}\right)^{\varphi}}\left(\left|\lambda^{2}s^{-2\alpha}\right|<1\right), \tag{D.198}$$

and

$$L\left\{Super\cosh_{\alpha,1}^{\varphi}\left(\lambda t^{\alpha}\right)\right\}=\frac{s^{-1}}{2}\frac{\left(1+\lambda s^{-\alpha}\right)^{\varphi}+\left(1-\lambda s^{-\alpha}\right)^{\varphi}}{\left(1-\lambda^{2}s^{-2\alpha}\right)^{\varphi}}\left(\left|\lambda^{2}s^{-2\alpha}\right|<1\right). \tag{D.199}$$

Let us consider the second case as follows:

The generalized Kilbas-Saigo-Saxena type function with one-parameter constant λ can be written as follows:

$$t^{\upsilon-1}E_{1,\upsilon}^{\varphi}\left(\lambda t\right)=\sum_{\kappa=0}^{\infty}\begin{pmatrix}-\varphi\\ \kappa\end{pmatrix}\frac{(-1)^{\kappa}\lambda^{\kappa}t^{\kappa+\upsilon-1}}{\Gamma(\kappa+\upsilon)}, \tag{D.200}$$

where the Kilbas-Saigo-Saxena function is defined as (see [97])

$$t^{\upsilon-1}E_{1,\upsilon}^{\varphi}\left(t\right)=\sum_{\kappa=0}^{\infty}\frac{(\varphi)_{\kappa}}{\Gamma(\kappa+1)}\frac{t^{\kappa}}{\Gamma(\kappa+\upsilon)}. \tag{D.201}$$

The supersine function, denoted by $Supersin_{1,\upsilon}^{\varphi}(t)$, is defined as

$$Supersin_{1,\upsilon}^{\varphi}(t) = \sum_{\kappa=0}^{\infty} \frac{\Gamma(\varphi+2\kappa+1)}{\Gamma(2\kappa+2)\Gamma(\varphi)} \frac{(-1)^{\kappa} t^{2\kappa+\upsilon}}{\Gamma(2\kappa+1+\upsilon)}, \tag{D.202}$$

which yields that

$$Supersin_{1,\upsilon}^{\varphi}(\lambda t) = \sum_{\kappa=0}^{\infty} \frac{\Gamma(\varphi+2\kappa+1)}{\Gamma(2\kappa+2)\Gamma(\varphi)} \frac{(-1)^{\kappa} \lambda^{2\kappa+1} t^{2\kappa+\upsilon}}{\Gamma(2\kappa+1+\upsilon)}, \tag{D.203}$$

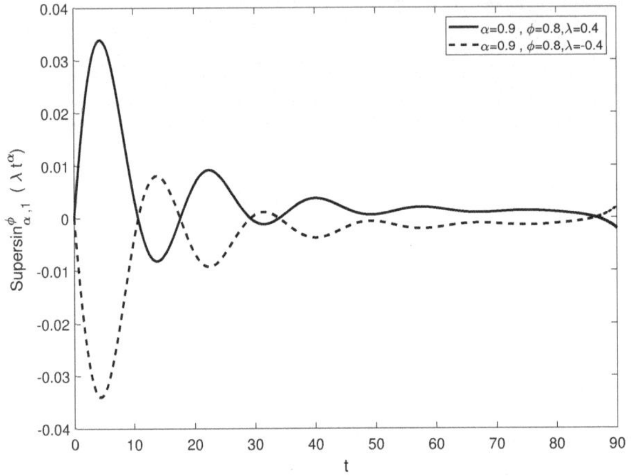

FIGURE D.5: The plot of the subsine function $Supersin_{\alpha,1}^{\varphi}(\lambda t^{\alpha})$.

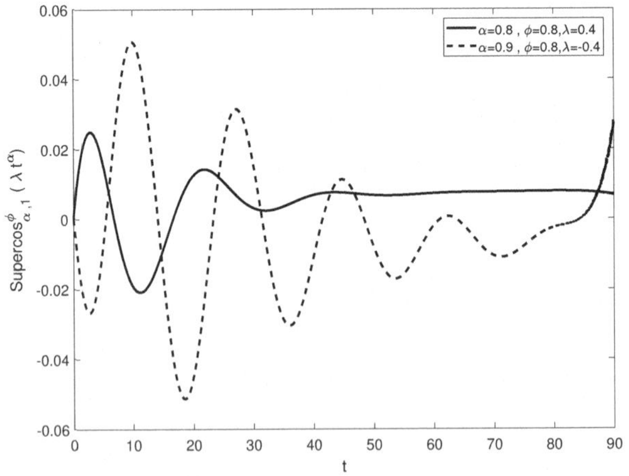

FIGURE D.6: The plot of the subcosine function $Supercos_{\alpha,1}^{\varphi}(\lambda t^{\alpha})$.

the supercosine function, denoted by $Supercos_{1,\upsilon}^{\varphi}(t)$, as

$$Supercos_{1,\upsilon}^{\varphi}(t) = \sum_{\kappa=0}^{\infty} \frac{\Gamma(\varphi+2\kappa)}{\Gamma(2\kappa+1)\Gamma(\varphi)} \frac{(-1)^{\kappa} t^{2\kappa+\upsilon-1}}{\Gamma(2\kappa+\upsilon)}, \tag{D.204}$$

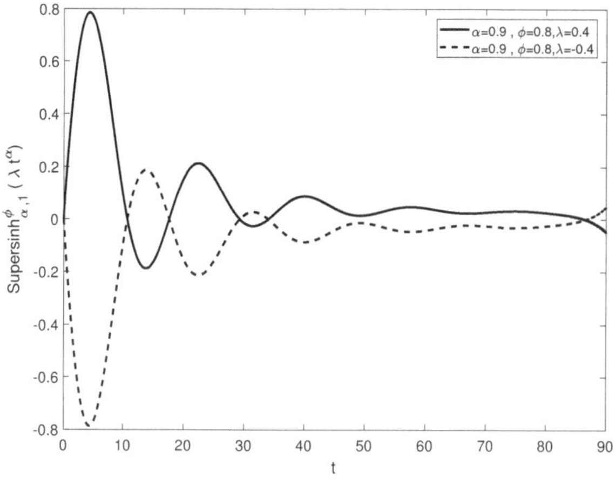

FIGURE D.7: The plot of the hyperbolic subsine function $Supersinh_{\alpha,1}^{\varphi}(\lambda t^{\alpha})$.

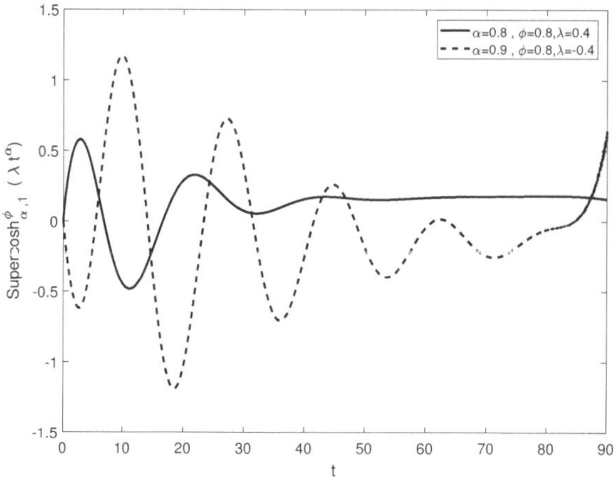

FIGURE D.8: The plot of the hyperbolic subcosine function $Supercosh_{\alpha,1}^{\varphi}(\lambda t^{\alpha})$.

which yields that

$$Supercos_{1,\upsilon}^{\varphi}(\lambda t)=\sum_{\kappa=0}^{\infty}\frac{\Gamma(\varphi+2\kappa)}{\Gamma(2\kappa+1)\Gamma(\varphi)}\frac{(-1)^{\kappa}\lambda^{2\kappa}t^{2\kappa+\upsilon-1}}{\Gamma(2\kappa+\upsilon)}, \tag{D.205}$$

the hyperbolic supersine function, denoted by $Supersinh_{1,\upsilon}^{\varphi}(t)$, as

$$Supersinh_{1,\upsilon}^{\varphi}(t)=\sum_{\kappa=0}^{\infty}\frac{\Gamma(\varphi+2\kappa+1)}{\Gamma(2\kappa+2)\Gamma(\varphi)}\frac{t^{2\kappa+\upsilon}}{\Gamma(2\kappa+1+\upsilon)}, \tag{D.206}$$

which implies that

$$Supersinh_{1,\upsilon}^{\varphi}(\lambda t)=\sum_{\kappa=0}^{\infty}\frac{\Gamma(\varphi+2\kappa+1)}{\Gamma(2\kappa+2)\Gamma(\varphi)}\frac{\lambda^{2\kappa+1}t^{2\kappa+\upsilon}}{\Gamma(2\kappa+1+\upsilon)}, \tag{D.207}$$

and the hyperbolic supercosine function, denoted by $Supercosh_{1,\upsilon}^{\varphi}(t)$, as

$$Supercosh_{1,\upsilon}^{\varphi}(t)=\sum_{\kappa=0}^{\infty}\frac{\Gamma(\varphi+2\kappa)}{\Gamma(2\kappa+1)\Gamma(\varphi)}\frac{t^{2\kappa+\upsilon-1}}{\Gamma(2\kappa+\upsilon)}, \tag{D.208}$$

which leads to

$$Supercosh_{1,\upsilon}^{\varphi}(\lambda t)=\sum_{\kappa=0}^{\infty}\frac{\Gamma(\varphi+2\kappa)}{\Gamma(2\kappa+1)\Gamma(\varphi)}\frac{\lambda^{2\kappa}t^{2\kappa+\upsilon-1}}{\Gamma(2\kappa+\upsilon)}. \tag{D.209}$$

Thus, we have

$$t^{\upsilon-1}E_{1,\upsilon}^{\varphi}(\lambda t)=Supercosh_{1,\upsilon}^{\varphi}(\lambda t)+Supersinh_{1,\upsilon}^{\varphi}(\lambda t) \tag{D.210}$$

and

$$t^{\upsilon-1}E_{1,\upsilon}^{\varphi}(i\lambda t)=Supercos_{1,\upsilon}^{\varphi}(\lambda t^{\alpha})+iSupersin_{1,\upsilon}^{\varphi}(\lambda t) \tag{D.211}$$

such that

$$Supersin_{1,\upsilon}^{\varphi}(t)=\frac{t^{\upsilon-1}E_{1,\upsilon}^{\varphi}(it)-t^{\upsilon-1}E_{1,\upsilon}^{\varphi}(-it)}{2i}, \tag{D.212}$$

$$Supercos_{1,\upsilon}^{\varphi}(t)=\frac{t^{\upsilon-1}E_{1,\upsilon}^{\varphi}(-i\lambda t)+t^{\upsilon-1}E_{1,\upsilon}^{\varphi}(-i\lambda t)}{2}, \tag{D.213}$$

$$Supersinh_{1,\upsilon}^{\varphi}(t)=\frac{t^{\upsilon-1}E_{1,\upsilon}^{\varphi}(t)-t^{\upsilon-1}E_{1,\upsilon}^{\varphi}(-t)}{2}, \tag{D.214}$$

$$Supercosh_{1,\upsilon}^{\varphi}(t)=\frac{t^{\upsilon-1}E_{1,\upsilon}^{\varphi}(t)+t^{\upsilon-1}E_{1,\upsilon}^{\varphi}(-t)}{2}, \tag{D.215}$$

$$Supersin_{1,\upsilon}^{\varphi}(\lambda t)=\frac{t^{\upsilon-1}E_{1,\upsilon}^{\varphi}(i\lambda t)-t^{\upsilon-1}E_{1,\upsilon}^{\varphi}(-i\lambda t)}{2i}, \tag{D.216}$$

$$Supercos_{1,\upsilon}^{\varphi}(\lambda t)=\frac{t^{\upsilon-1}E_{1,\upsilon}^{\varphi}(i\lambda t)+t^{\upsilon-1}E_{1,\upsilon}^{\varphi}(-i\lambda t)}{2}, \tag{D.217}$$

$$Supersinh_{1,\upsilon}^{\varphi}(\lambda t)=\frac{t^{\upsilon-1}E_{1,\upsilon}^{\varphi}(\lambda t)-t^{\upsilon-1}E_{1,\upsilon}^{\varphi}(-\lambda t)}{2} \quad (D.218)$$

and

$$Supercosh_{1,\upsilon}^{\varphi}(\lambda t)=\frac{t^{\upsilon-1}E_{1,\upsilon}^{\varphi}(\lambda t)+t^{\upsilon-1}E_{1,\upsilon}^{\varphi}(-\lambda t)}{2}. \quad (D.219)$$

Thus, we have

$$\begin{array}{l} L\left\{t^{\upsilon-1}E_{1,\upsilon}^{\varphi}(\lambda t)\right\}=s^{-\upsilon}\left(1-\lambda s^{-1}\right)^{-\varphi} \\ \left(\left|\lambda s^{-1}\right|<1; Re(\varphi)>0; Re(\upsilon)>0\right) \end{array} \quad (D.220)$$

such that

$$\begin{array}{l} L\left\{Supersin_{1,\upsilon}^{\varphi}(\lambda t)\right\}=\frac{is^{-\upsilon}}{2}\frac{\left(1-i\lambda s^{-1}\right)^{\varphi}-\left(1+i\lambda s^{-1}\right)^{\varphi}}{\left(1+\lambda^{2}s^{-2}\right)^{\varphi}} \\ \left(\left|\lambda^{2}s^{-2}\right|<1; Re(\varphi)>0; Re(\upsilon)>0\right), \end{array} \quad (D.221)$$

$$\begin{array}{l} L\left\{Supersinh_{1,\upsilon}^{\varphi}(\lambda t)\right\}=\frac{s^{-\upsilon}}{2}\frac{\left(1-\lambda s^{-1}\right)^{\varphi}-\left(1+\lambda s^{-1}\right)^{\varphi}}{\left(1-\lambda^{2}s^{-2}\right)^{\varphi}} \\ \left(\left|\lambda^{2}s^{-2}\right|<1; Re(\varphi)>0; Re(\upsilon)>0\right), \end{array} \quad (D.222)$$

$$\begin{array}{l} L\left\{Supercos_{1,\upsilon}^{\varphi}(\lambda t)\right\}=\frac{s^{-\upsilon}}{2}\frac{\left(1+i\lambda s^{-1}\right)^{\varphi}+\left(1-i\lambda s^{-1}\right)^{\varphi}}{\left(1+\lambda^{2}s^{-2}\right)^{\varphi}} \\ \left(\left|\lambda^{2}s^{-2}\right|<1; Re(\varphi)>0; Re(\upsilon)>0\right) \end{array} \quad (D.223)$$

and

$$\begin{array}{l} L\left\{Supercosh_{1,\upsilon}^{\varphi}(\lambda t)\right\}=\frac{s^{-\upsilon}}{2}\frac{\left(1+\lambda s^{-1}\right)^{\varphi}+\left(1-\lambda s^{-1}\right)^{\varphi}}{\left(1-\lambda^{2}s^{-2}\right)^{\varphi}} \\ \left(\left|\lambda^{2}s^{-2}\right|<1; Re(\varphi)>0; Re(\upsilon)>0\right). \end{array} \quad (D.224)$$

Let us consider the third case as follows.

The generalized Prabhakar function with one-parameter constant λ is given as [236]

$$E_{1,1}^{\varphi}(\lambda t)=\sum_{\kappa=0}^{\infty}\binom{-\varphi}{\kappa}\frac{(-1)^{\kappa}\lambda^{\kappa}t^{\kappa}}{\Gamma(\kappa+1)}. \quad (D.225)$$

The supersine function, denoted by $Supersin_{1,1}^{\varphi}(t)$, is defined as

$$Supersin_{1,1}^{\varphi}(t)=\sum_{\kappa=0}^{\infty}\frac{\Gamma(\varphi+2\kappa+1)}{\Gamma(2\kappa+2)\Gamma(\varphi)}\frac{(-1)^{\kappa}t^{2\kappa+1}}{\Gamma(2\kappa+2)}, \quad (D.226)$$

which implies that

$$Supersin_{1,1}^{\varphi}(\lambda t^{\alpha})=\sum_{\kappa=0}^{\infty}\frac{\Gamma(\varphi+2\kappa+1)}{\Gamma(2\kappa+2)\Gamma(\varphi)}\frac{(-1)^{\kappa}\lambda^{2\kappa+1}t^{2\kappa+1}}{\Gamma(2\kappa+2)}, \quad (D.227)$$

the supercosine function, denoted by $Supercos_{1,1}^{\varphi}(t)$, as

$$Supercos_{1,1}^{\varphi}(t)=\sum_{\kappa=0}^{\infty}\frac{\Gamma(\varphi+2\kappa)}{\Gamma(2\kappa+1)\Gamma(\varphi)}\frac{(-1)^{\kappa}t^{2\kappa}}{\Gamma(2\kappa+1)}, \quad (D.228)$$

which yields that

$$Supercos_{1,1}^{\varphi}(\lambda t) = \sum_{\kappa=0}^{\infty} \frac{\Gamma(\varphi+2\kappa)}{\Gamma(2\kappa+1)\Gamma(\varphi)} \frac{(-1)^{\kappa}\lambda^{2\kappa}t^{2\kappa}}{\Gamma(2\kappa+1)}, \tag{D.229}$$

the hyperbolic supersine function, denoted by $Supersinh_{1,1}^{\varphi}(t)$, as

$$Supersinh_{1,1}^{\varphi}(t) = \sum_{\kappa=0}^{\infty} \frac{\Gamma(\varphi+2\kappa+1)}{\Gamma(2\kappa+2)\Gamma(\varphi)} \frac{t^{2\kappa+1}}{\Gamma(2\kappa+2)}, \tag{D.230}$$

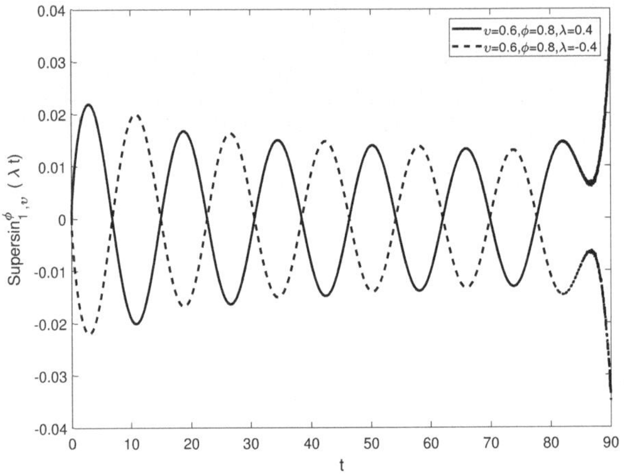

FIGURE D.9: The plot of the subsine function $Supersinh_{1,\upsilon}^{\varphi}(\lambda t)$.

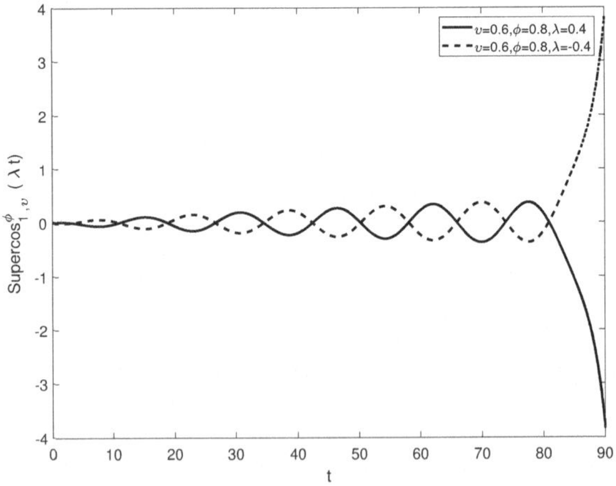

FIGURE D.10: The plot of the subcosine function $Supercosh_{1,\upsilon}^{\varphi}(\lambda t)$.

which leads to

$$Supersinh_{1,1}^{\varphi}(\lambda t) = \sum_{\kappa=0}^{\infty} \frac{\Gamma(\varphi+2\kappa+1)}{\Gamma(2\kappa+2)\Gamma(\varphi)} \frac{\lambda^{2\kappa+1}t^{2\kappa+1}}{\Gamma(2\kappa+2)}, \tag{D.231}$$

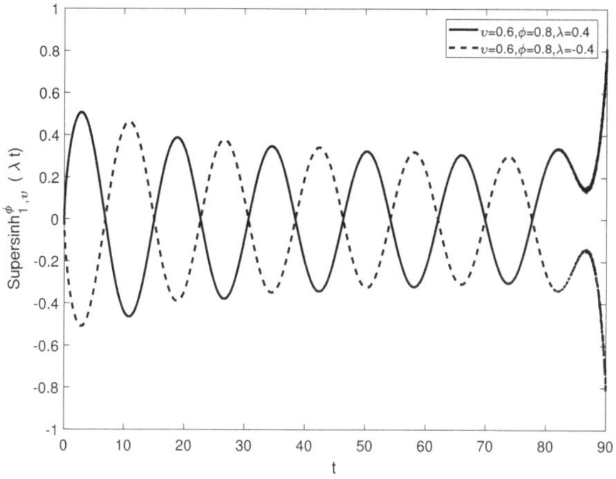

FIGURE D.11: The plot of the hyperbolic subsine function $Supersinh_{1,\upsilon}^{\varphi}(\lambda t)$.

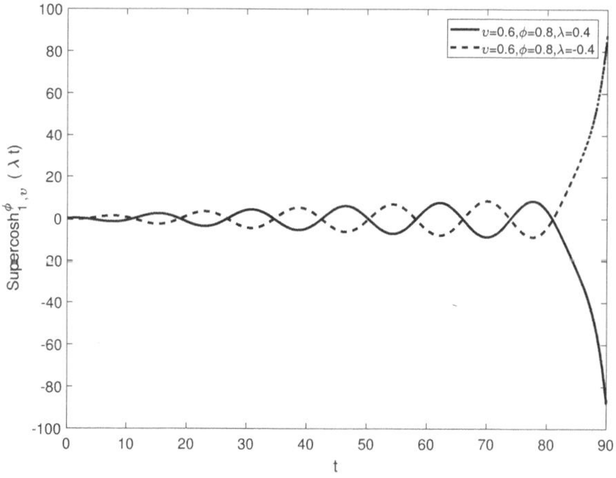

FIGURE D.12: The plot of the hyperbolic subcosine function $Supercosh_{1,\upsilon}^{\varphi}(\lambda t)$.

and the hyperbolic supercosine function, denoted by $Supercosh_{1,1}^{\varphi}(t)$, as

$$Supercosh_{1,1}^{\varphi}(t) = \sum_{\kappa=0}^{\infty} \frac{\Gamma(\varphi+2\kappa)}{\Gamma(2\kappa+1)\Gamma(\varphi)} \frac{t^{2\kappa}}{\Gamma(2\kappa+1)}, \tag{D.232}$$

which implies that

$$Supercosh_{1,1}^{\varphi}(\lambda t) = \sum_{\kappa=0}^{\infty} \frac{\Gamma(\varphi+2\kappa)}{\Gamma(2\kappa+1)\Gamma(\varphi)} \frac{\lambda^{2\kappa} t^{2\kappa}}{\Gamma(2\kappa+1)}. \tag{D.233}$$

According to the above results, we have

$$E_{1,1}^{\varphi}(\lambda t) = Supercosh_{1,1}^{\varphi}(\lambda t) + Supersinh_{1,1}^{\varphi}(\lambda t) \tag{D.234}$$

and

$$E_{1,1}^{\varphi}(i\lambda t) = Supercos_{1,1}^{\varphi}(\lambda t) + iSupersin_{1,1}^{\varphi}(\lambda t) \tag{D.235}$$

such that

$$Supersin_{1,1}^{\varphi}(t) = \frac{E_{1,1}^{\varphi}(it) - E_{1,1}^{\varphi}(-it)}{2i}, \tag{D.236}$$

$$Supercos_{1,1}^{\varphi}(t) = \frac{E_{1,1}^{\varphi}(-i\lambda t) + E_{1,1}^{\varphi}(-i\lambda t)}{2}, \tag{D.237}$$

$$Supersinh_{1,1}^{\varphi}(t) = \frac{E_{1,1}^{\varphi}(t) - E_{1,1}^{\varphi}(-t)}{2}, \tag{D.238}$$

$$Supercosh_{1,1}^{\varphi}(t) = \frac{E_{1,1}^{\varphi}(t) + E_{1,1}^{\varphi}(-t)}{2}, \tag{D.239}$$

$$Supersin_{1,1}^{\varphi}(\lambda t) = \frac{E_{1,1}^{\varphi}(i\lambda t) - E_{1,1}^{\varphi}(-i\lambda t)}{2i}, \tag{D.240}$$

$$Supercos_{1,1}^{\varphi}(\lambda t) = \frac{E_{1,1}^{\varphi}(i\lambda t) + E_{1,1}^{\varphi}(-i\lambda t)}{2}, \tag{D.241}$$

$$Supersinh_{1,1}^{\varphi}(\lambda t) = \frac{E_{1,1}^{\varphi}(\lambda t) - E_{1,1}^{\varphi}(-\lambda t)}{2} \tag{D.242}$$

and

$$Supercosh_{1,1}^{\varphi}(\lambda t) = \frac{E_{1,1}^{\varphi}(\lambda t) + E_{1,1}^{\varphi}(-\lambda t)}{2}. \tag{D.243}$$

Making use of

$$L\left[\frac{t^{\kappa\alpha}}{\Gamma(\kappa\alpha+1)}\right] = \frac{1}{s^{\kappa\alpha}}, \tag{D.244}$$

we present

$$L\left\{E_{1,1}^{\varphi}(\lambda t)\right\} = s^{-1}\left(1-\lambda s^{-1}\right)^{-\varphi} \quad \left(\left|\lambda s^{-1}\right| < 1; Re(\varphi) > 0\right), \tag{D.245}$$

$$L\left\{Supersin_{1,1}^{\varphi}(\lambda t)\right\} = \frac{is^{-1}}{2} \frac{\left(1-i\lambda s^{-1}\right)^{\varphi} - \left(1+i\lambda s^{-1}\right)^{\varphi}}{\left(1+\lambda^2 s^{-2}\right)^{\varphi}} \left(\left|\lambda^2 s^{-2}\right| < 1\right), \tag{D.246}$$

$$L\left\{Super\sinh_{1,1}^{\varphi}(\lambda t)\right\} = \frac{s^{-1}}{2}\frac{\left(1-\lambda s^{-1}\right)^{\varphi}-\left(1+\lambda s^{-1}\right)^{\varphi}}{(1-\lambda^2 s^{-2})^{\varphi}}\left(\left|\lambda^2 s^{-2}\right|<1\right), \quad \text{(D.247)}$$

$$L\left\{Supercos_{1,1}^{\varphi}(\lambda t)\right\} = \frac{s^{-1}}{2}\frac{\left(1+i\lambda s^{-1}\right)^{\varphi}+\left(1-i\lambda s^{-1}\right)^{\varphi}}{(1+\lambda^2 s^{-2})^{\varphi}}\left(\left|\lambda^2 s^{-2}\right|<1\right) \quad \text{(D.248)}$$

and

$$L\left\{Super\cosh_{1,1}^{\varphi}(\lambda t)\right\} = \frac{s^{-1}}{2}\frac{\left(1+\lambda s^{-1}\right)^{\varphi}+\left(1-\lambda s^{-1}\right)^{\varphi}}{(1-\lambda^2 s^{-2})^{\varphi}}\left(\left|\lambda^2 s^{-2}\right|<1\right). \quad \text{(D.249)}$$

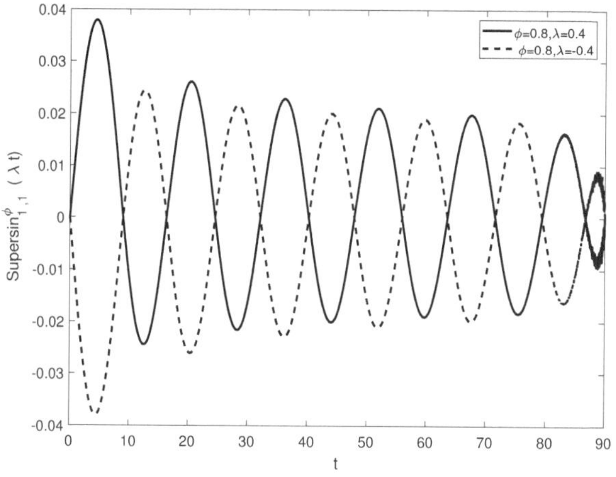

FIGURE D.13: The plot of the subsine function $Supersin_{1,1}^{\varphi}(\lambda t)$.

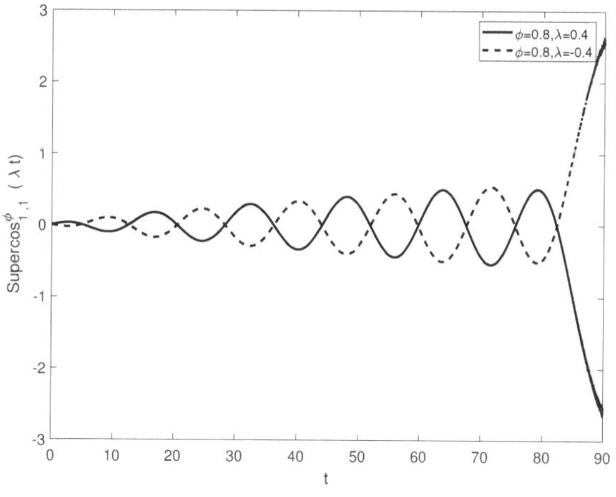

FIGURE D.14: The plot of the subcosine function $Supercos_{1,1}^{\varphi}(\lambda t)$.

D.7 The two-parametric special functions via Wiman functions

The generalized Wiman function with one-parameter constant λ is defined as [236]

$$t^{\upsilon-1}E_{\alpha,\upsilon}\left(\lambda t^{\alpha}\right)=\sum_{\kappa=0}^{\infty}\frac{\lambda^{\kappa}t^{\kappa\alpha+\upsilon-1}}{\Gamma(\kappa\alpha+\upsilon)},\tag{D.250}$$

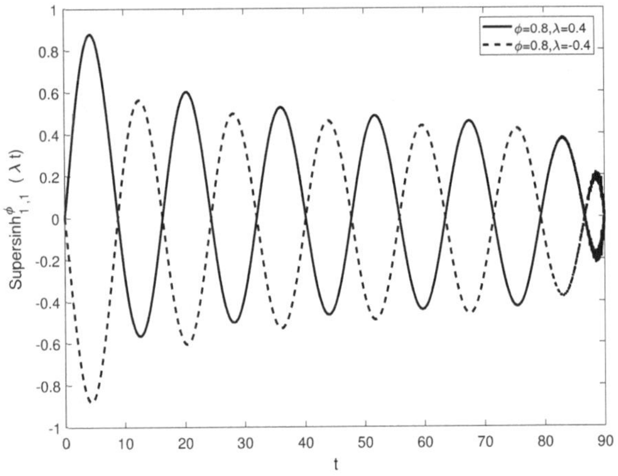

FIGURE D.15: The plot of the hyperbolic subsine function $Supersinh^{\varphi}_{1,1}\left(\lambda t\right)$.

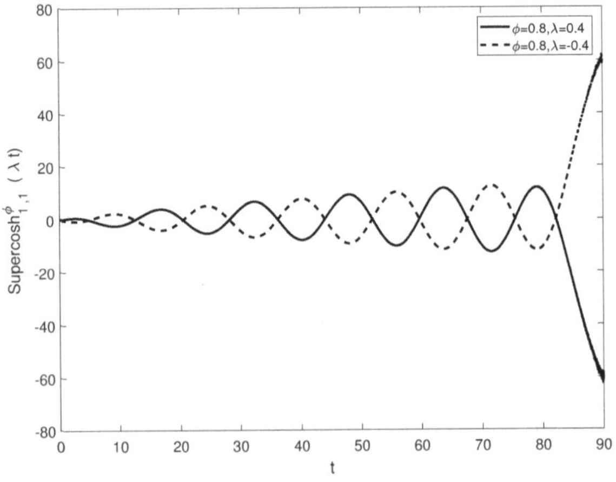

FIGURE D.16: The plot of the hyperbolic subcosine function $Supercosh^{\varphi}_{1,1}\left(\lambda t\right)$.

and the Wiman function as (see [208])

$$E_{\alpha,\upsilon}\left(t^{\alpha}\right)=\sum_{\kappa=0}^{\infty}\frac{t^{\kappa\alpha}}{\Gamma\left(\kappa\alpha+\upsilon\right)}. \tag{D.251}$$

The supersine function, denoted by $Supersin_{\alpha,\upsilon}\left(t^{\alpha}\right)$, is defined as

$$Supersin_{\alpha,\upsilon}\left(t^{\alpha}\right)=\sum_{\kappa=0}^{\infty}\frac{\left(-1\right)^{\kappa}t^{(2\kappa+1)\alpha+\upsilon-1}}{\Gamma\left(\left(2\kappa+1\right)\alpha+\upsilon\right)} \tag{D.252}$$

which implies that

$$Supersin_{\alpha,\upsilon}\left(\lambda t^{\alpha}\right)=\sum_{\kappa=0}^{\infty}\frac{\left(-1\right)^{\kappa}\lambda^{2\kappa+1}t^{(2\kappa+1)\alpha+\upsilon-1}}{\Gamma\left(\left(2\kappa+1\right)\alpha+\upsilon\right)}, \tag{D.253}$$

the supercosine function, denoted by $Supercos_{\alpha,\upsilon}\left(t^{\alpha}\right)$, as

$$Supercos_{\alpha,\upsilon}\left(t^{\alpha}\right)=\sum_{\kappa=0}^{\infty}\frac{\left(-1\right)^{\kappa}t^{2\kappa\alpha+\upsilon-1}}{\Gamma\left(2\kappa\alpha+\upsilon\right)}, \tag{D.254}$$

which implies that

$$Supercos_{\alpha,\upsilon}\left(\lambda t^{\alpha}\right)=\sum_{\kappa=0}^{\infty}\frac{\left(-1\right)^{\kappa}\lambda^{2\kappa}t^{2\kappa\alpha+\upsilon-1}}{\Gamma\left(2\kappa\alpha+\upsilon\right)}, \tag{D.255}$$

the hyperbolic supersine function, denoted by $Supersinh_{\alpha,\upsilon}\left(t^{\alpha}\right)$, as

$$Supersinh_{\alpha,\upsilon}\left(t^{\alpha}\right)=\sum_{\kappa=0}^{\infty}\frac{t^{(2\kappa+1)\alpha+\upsilon-1}}{\Gamma\left(\left(2\kappa+1\right)\alpha+\upsilon\right)} \tag{D.256}$$

which implies that

$$Supersinh_{\alpha,\upsilon}\left(\lambda t^{\alpha}\right)=\sum_{\kappa=0}^{\infty}\frac{\lambda^{2\kappa+1}t^{(2\kappa+1)\alpha+\upsilon-1}}{\Gamma\left(\left(2\kappa+1\right)\alpha+\upsilon\right)}, \tag{D.257}$$

and the hyperbolic supercosine function, denoted by $Supercosh_{\alpha,\upsilon}\left(t^{\alpha}\right)$, as

$$Supercosh_{\alpha,\upsilon}\left(t^{\alpha}\right)=\sum_{\kappa=0}^{\infty}\frac{t^{2\kappa\alpha+\upsilon-1}}{\Gamma\left(2\kappa\alpha+\upsilon\right)}, \tag{D.258}$$

which implies that

$$Supercosh_{\alpha,\upsilon}\left(\lambda t^{\alpha}\right)=\sum_{\kappa=0}^{\infty}\frac{\lambda^{2\kappa}t^{2\kappa\alpha+\upsilon-1}}{\Gamma\left(2\kappa\alpha+\upsilon\right)}. \tag{D.259}$$

Thus, we have

$$t^{\upsilon-1}E_{\alpha,\upsilon}(\lambda t^{\alpha}) = Supercosh_{\alpha,\upsilon}(\lambda t^{\alpha}) + Supersinh_{\alpha,\upsilon}(\lambda t^{\alpha}) \tag{D.260}$$

and

$$t^{\upsilon-1}E_{\alpha,\upsilon}(i\lambda t^{\alpha}) = Supercos_{\alpha,\upsilon}(\lambda t^{\alpha}) + iSupersin_{\alpha,\upsilon}(\lambda t^{\alpha}) \tag{D.261}$$

such that

$$Supersin_{\alpha,\upsilon}(t^{\alpha}) = \frac{t^{\upsilon-1}E_{\alpha,\upsilon}(it^{\alpha}) - t^{\upsilon-1}E_{\alpha,\upsilon}(-it^{\alpha})}{2i}, \tag{D.262}$$

$$Supercos_{\alpha,\upsilon}(t^{\alpha}) = \frac{t^{\upsilon-1}E_{\alpha,\upsilon}(-i\lambda t^{\alpha}) + t^{\upsilon-1}E_{\alpha,\upsilon}(-i\lambda t^{\alpha})}{2}, \tag{D.263}$$

$$Supersinh_{\alpha,\upsilon}(t^{\alpha}) = \frac{t^{\upsilon-1}E_{\alpha,\upsilon}(t^{\alpha}) - t^{\upsilon-1}E_{\alpha,\upsilon}(-t^{\alpha})}{2}, \tag{D.264}$$

$$Supercosh_{\alpha,\upsilon}(t^{\alpha}) = \frac{t^{\upsilon-1}E_{\alpha,\upsilon}(t^{\alpha}) + t^{\upsilon-1}E_{\alpha,\upsilon}(-t^{\alpha})}{2}, \tag{D.265}$$

$$Supersin_{\alpha,\upsilon}(\lambda t^{\alpha}) = \frac{t^{\upsilon-1}E_{\alpha,\upsilon}(i\lambda t^{\alpha}) - t^{\upsilon-1}E_{\alpha,\upsilon}(-i\lambda t^{\alpha})}{2i}, \tag{D.266}$$

$$Supercos_{\alpha,\upsilon}(\lambda t^{\alpha}) = \frac{t^{\upsilon-1}E_{\alpha,\upsilon}(i\lambda t^{\alpha}) + t^{\upsilon-1}E_{\alpha,\upsilon}(-i\lambda t^{\alpha})}{2}, \tag{D.267}$$

$$Supersinh_{\alpha,\upsilon}(\lambda t^{\alpha}) = \frac{t^{\upsilon-1}E_{\alpha,\upsilon}(\lambda t^{\alpha}) - t^{\upsilon-1}E_{\alpha,\upsilon}(-\lambda t^{\alpha})}{2} \tag{D.268}$$

and

$$Supercosh_{\alpha,\upsilon}(\lambda t^{\alpha}) = \frac{t^{\upsilon-1}E_{\alpha,\upsilon}(\lambda t^{\alpha}) + t^{\upsilon-1}E_{\alpha,\upsilon}(-\lambda t^{\alpha})}{2}. \tag{D.269}$$

With the use of

$$L\left[\frac{t^{\kappa\alpha}}{\Gamma(\kappa\alpha+1)}\right] = \frac{1}{s^{\kappa\alpha}}, \tag{D.270}$$

we have

$$L\left\{t^{\upsilon-1}E_{\alpha,\upsilon}(\lambda t^{\alpha})\right\} = s^{-\upsilon}(1-\lambda s^{-\alpha}) \quad (|\lambda s^{-\alpha}| < 1; Re(\upsilon) > 0) \tag{D.271}$$

such that

$$L\{Supersin_{\alpha,\upsilon}(\lambda t^{\alpha})\} = \lambda s^{-(\upsilon+\alpha)}\left(1+\lambda^2 s^{-2\alpha}\right)^{-1} \quad \left(\left|\lambda^2 s^{-2\alpha}\right| < 1\right), \tag{D.272}$$

$$L\{Supersinh_{\alpha,\upsilon}(\lambda t^{\alpha})\} = \lambda s^{-(\upsilon+\alpha)}\left(1-\lambda^2 s^{-2\alpha}\right)^{-1} \left(\left|\lambda^2 s^{-2\alpha}\right| < 1\right), \tag{D.273}$$

$$L\{Supercos_{\alpha,\upsilon}(\lambda t^{\alpha})\} = s^{-\upsilon}\left(1+\lambda^2 s^{-2\alpha}\right)^{-1} \left(\left|\lambda^2 s^{-2\alpha}\right| < 1\right), \tag{D.274}$$

and

$$L\{Supercosh_{\alpha,\upsilon}(\lambda t^{\alpha})\} = s^{-\upsilon}\left(1-\lambda^2 s^{-2\alpha}\right)^{-1} \left(\left|\lambda^2 s^{-2\alpha}\right| < 1\right). \tag{D.275}$$

D.8 The two-parametric special functions via generalized Wiman functions

The generalized Wiman function with one-parameter constant λ is defined as (see [236])

$$t^{\upsilon-1}E_{1,\upsilon}(\lambda t) = \sum_{\kappa=0}^{\infty} \frac{\lambda^{\kappa} t^{\kappa+\upsilon-1}}{\Gamma(\kappa+\upsilon)}. \tag{D.276}$$

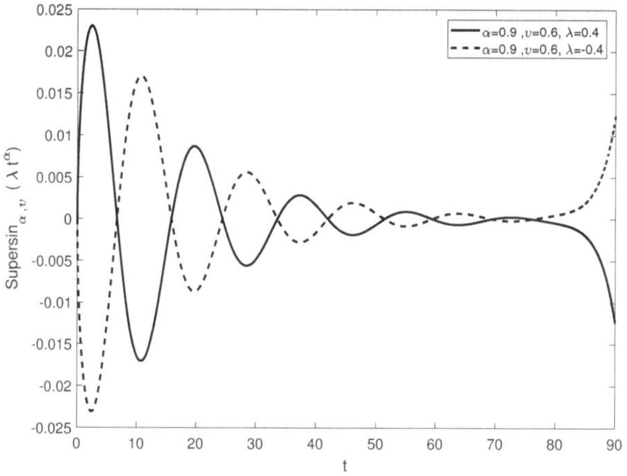

FIGURE D.17: The plot of the subsine function $Supersin_{\alpha,\upsilon}(\lambda t^{\alpha})$.

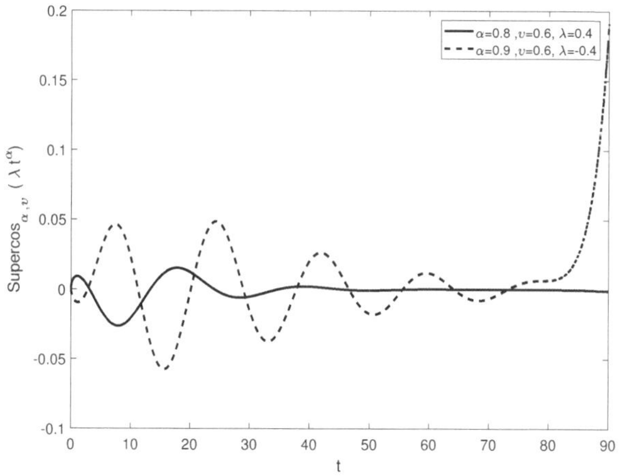

FIGURE D.18: The plot of the subcosine function $Supercos_{\alpha,\upsilon}(\lambda t^{\alpha})$.

The supersine function, denoted by $Supersin_{1,\upsilon}(t)$, is defined as

$$Supersin_{1,\upsilon}(t) = \sum_{\kappa=0}^{\infty} \frac{(-1)^{\kappa} t^{2\kappa+\upsilon}}{\Gamma(2\kappa+1+\upsilon)}, \tag{D.277}$$

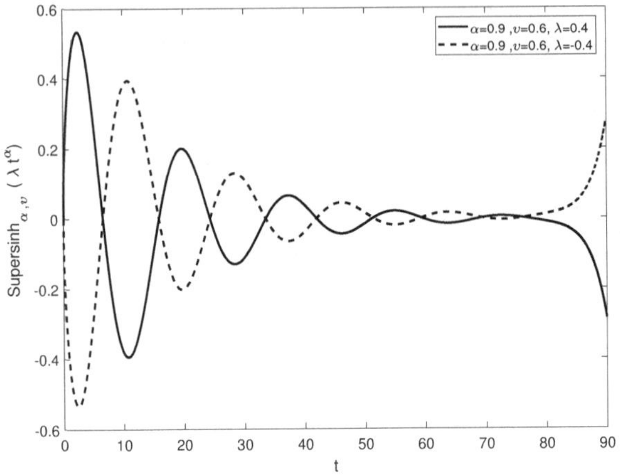

FIGURE D.19: The plot of the hyperbolic subsine function $Supersinh_{\alpha,\upsilon}(\lambda t^{\alpha})$.

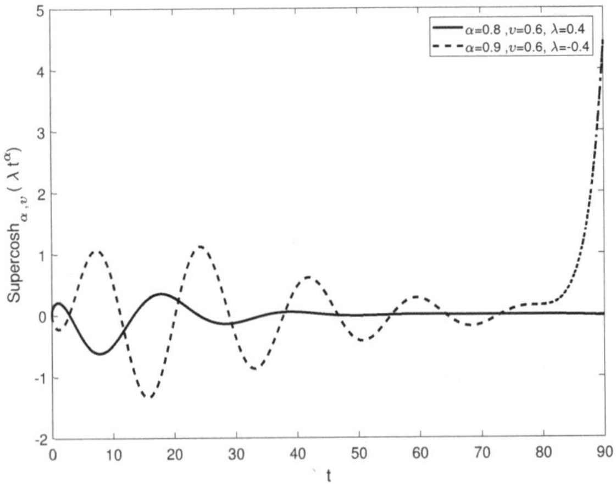

FIGURE D.20: The plot of the hyperbolic subcosine function $Supercosh_{\alpha,\upsilon}(\lambda t^{\alpha})$.

which implies that

$$Supersin_{1,\upsilon}(\lambda t) = \sum_{\kappa=0}^{\infty} \frac{(-1)^{\kappa} \lambda^{2\kappa+1} t^{2\kappa+\upsilon}}{\Gamma(2\kappa+1+\upsilon)}, \tag{D.278}$$

the supercosine function, denoted by $Supercos_{1,\upsilon}(t)$, as

$$Supercos_{1,\upsilon}(t) = \sum_{\kappa=0}^{\infty} \frac{(-1)^{\kappa} t^{2\kappa+\upsilon-1}}{\Gamma(2\kappa+\upsilon)}, \tag{D.279}$$

which implies that

$$Supercos_{1,\upsilon}(\lambda t) = \sum_{\kappa=0}^{\infty} \frac{(-1)^{\kappa} \lambda^{2\kappa} t^{2\alpha+\upsilon-1}}{\Gamma(2\kappa+\upsilon)}, \tag{D.280}$$

the hyperbolic supersine function, denoted by $Supersinh_{1,\upsilon}(t)$, as

$$Supersinh_{1,\upsilon}(t) = \sum_{\kappa=0}^{\infty} \frac{t^{2\kappa+\upsilon}}{\Gamma(2\kappa+1+\upsilon)}, \tag{D.281}$$

which implies that

$$Supersinh_{1,\upsilon}(\lambda t) = \sum_{\kappa=0}^{\infty} \frac{\lambda^{2\kappa+1} t^{2\kappa+\upsilon}}{\Gamma(2\kappa+1+\upsilon)}, \tag{D.282}$$

and the hyperbolic supercosine function, denoted by $Supercosh_{1,\upsilon}(t)$, as

$$Supercosh_{1,\upsilon}(t) = \sum_{\kappa=0}^{\infty} \frac{t^{2\kappa+\upsilon-1}}{\Gamma(2\kappa+\upsilon)}, \tag{D.283}$$

which implies that

$$Supercosh_{1,\upsilon}(\lambda t) = \sum_{\kappa=0}^{\infty} \frac{\lambda^{2\kappa} t^{2\kappa+\upsilon-1}}{\Gamma(2\kappa+\upsilon)}. \tag{D.284}$$

Thus, we have

$$t^{\upsilon-1} E_{1,\upsilon}(\lambda t) = Supercosh_{1,\upsilon}(\lambda t) + Supersinh_{1,\upsilon}(\lambda t) \tag{D.285}$$

and

$$t^{\upsilon-1} E_{1,\upsilon}(i\lambda t) = Supercos_{1,\upsilon}(\lambda t) + iSupersin_{1,\upsilon}(\lambda t) \tag{D.286}$$

such that

$$Supersin_{1,\upsilon}(t) = \frac{t^{\upsilon-1} E_{1,\upsilon}(it) - t^{\upsilon-1} E_{1,\upsilon}(-it)}{2i}, \tag{D.287}$$

$$Supercos_{1,\upsilon}(t) = \frac{t^{\upsilon-1}E_{1,\upsilon}(-i\lambda t) + t^{\upsilon-1}E_{1,\upsilon}(-i\lambda t)}{2}, \tag{D.288}$$

$$Supersinh_{1,\upsilon}(t) = \frac{t^{\upsilon-1}E_{1,\upsilon}(t) - t^{\upsilon-1}E_{1,\upsilon}(-t)}{2}, \tag{D.289}$$

$$Supercosh_{1,\upsilon}(t) = \frac{t^{\upsilon-1}E_{1,\upsilon}(t) + t^{\upsilon-1}E_{1,\upsilon}(-t)}{2}, \tag{D.290}$$

$$Supersin_{1,\upsilon}(\lambda t) = \frac{t^{\upsilon-1}E_{1,\upsilon}(i\lambda t) - t^{\upsilon-1}E_{1,\upsilon}(-i\lambda t)}{2i}, \tag{D.291}$$

$$Supercos_{1,\upsilon}(\lambda t) = \frac{t^{\upsilon-1}E_{1,\upsilon}(i\lambda t) + t^{\upsilon-1}E_{1,\upsilon}(-i\lambda t)}{2}, \tag{D.292}$$

$$Supersinh_{1,\upsilon}(\lambda t) = \frac{t^{\upsilon-1}E_{1,\upsilon}(\lambda t) - t^{\upsilon-1}E_{1,\upsilon}(-\lambda t)}{2} \tag{D.293}$$

and

$$Supercosh_{1,\upsilon}(\lambda t) = \frac{t^{\upsilon-1}E_{1,\upsilon}(\lambda t) + t^{\upsilon-1}E_{1,\upsilon}(-\lambda t)}{2}. \tag{D.294}$$

In view of

$$L\left[\frac{t^{\kappa\alpha}}{\Gamma(\kappa\alpha+1)}\right] = \frac{1}{s^{\kappa\alpha}}, \tag{D.295}$$

we have

$$L\left\{t^{\upsilon-1}E_{1,\upsilon}(\lambda t)\right\} = s^{-\upsilon}\left(1-\lambda s^{-1}\right) \quad \left(\left|\lambda s^{-1}\right|<1; Re(\upsilon)>0\right) \tag{D.296}$$

such that

$$L\left\{Supersin_{1,\upsilon}(\lambda t)\right\} = \lambda s^{-(\upsilon+1)}\left(1+\lambda^2 s^{-2}\right)^{-1} \quad \left(\left|\lambda^2 s^{-2}\right|<1\right), \tag{D.297}$$

$$L\left\{Supersinh_{1,\upsilon}(\lambda t)\right\} = \lambda s^{-(\upsilon+1)}\left(1-\lambda^2 s^{-2}\right)^{-1}\left(\left|\lambda^2 s^{-2}\right|<1\right), \tag{D.298}$$

$$L\left\{Supercos_{1,\upsilon}(\lambda t)\right\} = s^{-\upsilon}\left(1+\lambda^2 s^{-2}\right)^{-1}\left(\left|\lambda^2 s^{-2}\right|<1\right), \tag{D.299}$$

and

$$L\left\{Supercosh_{1,\upsilon}(\lambda t)\right\} = s^{-\upsilon}\left(1-\lambda^2 s^{-2}\right)^{-1}\left(\left|\lambda^2 s^{-2}\right|<1\right). \tag{D.300}$$

D.9 The two-parametric special functions via two-parametric Lorenzo-Hartley functions

The two-parametric Lorenzo-Hartley function is defined as (see [338])

$$R_{\alpha,\upsilon}(\lambda t^{\alpha}) = \sum_{\kappa=0}^{\infty} \frac{\lambda^{\kappa} t^{(\kappa+1)\alpha-1-\upsilon}}{\Gamma((\kappa+1)\alpha-\upsilon)}. \tag{D.301}$$

The corresponding subsine function, denoted by $RSin_\alpha(t^\alpha)$, is defined as

$$RSin_{\alpha,v}(t^\alpha) = \sum_{\kappa=0}^{\infty} \frac{(-1)^\kappa t^{2(\kappa+1)\alpha-1-v}}{\Gamma(2(\kappa+1)\alpha - v)}, \tag{D.302}$$

which implies that

$$RSin_{\alpha,v}(\lambda t^\alpha) = \sum_{\kappa=0}^{\infty} \frac{(-1)^\kappa \lambda^{2\kappa+1} t^{2(\kappa+1)\alpha-1-v}}{\Gamma(2(\kappa+1)\alpha - v)}, \tag{D.303}$$

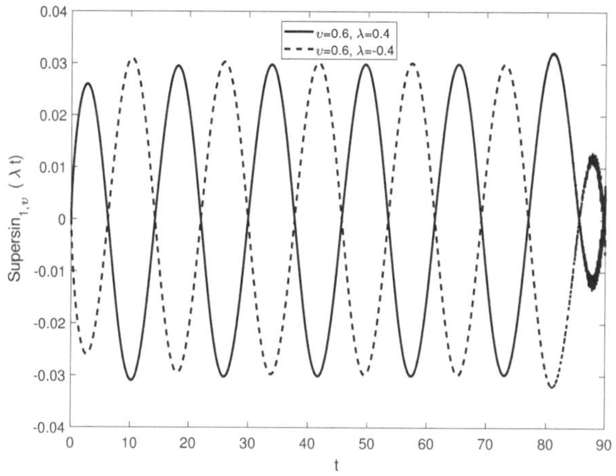

FIGURE D.21: The plot of the subsine function $Supersin_{1,v}(\lambda t)$.

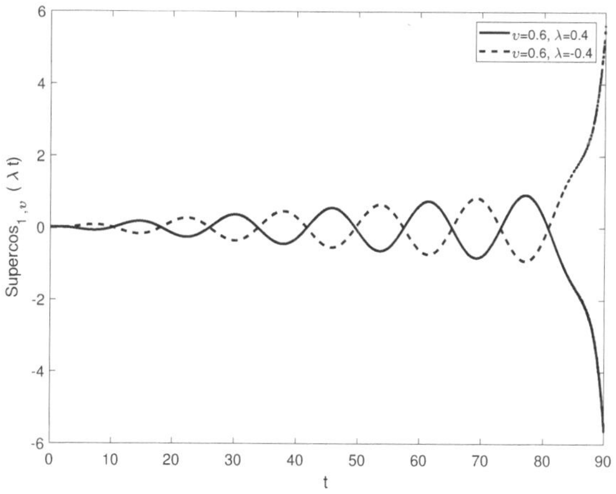

FIGURE D.22: The plot of the subcosine function $Supercos_{1,v}(\lambda t)$.

the subcosine function, denoted by $RCos_{\alpha}(t^{\alpha})$, as

$$RCos_{\alpha,\upsilon}(t^{\alpha}) = \sum_{\kappa=0}^{\infty} \frac{(-1)^{\kappa} t^{(2\kappa+1)\alpha-1-\upsilon}}{\Gamma((2\kappa+1)\alpha-\upsilon)}, \tag{D.304}$$

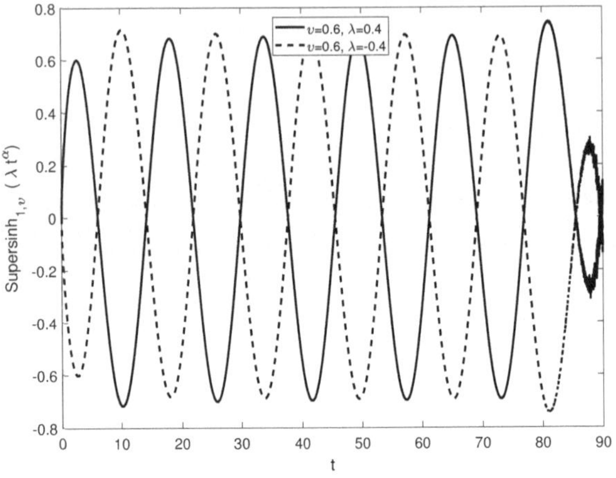

FIGURE D.23: The plot of the hyperbolic subsine function $Supersinh_{1,\upsilon}(\lambda t)$.

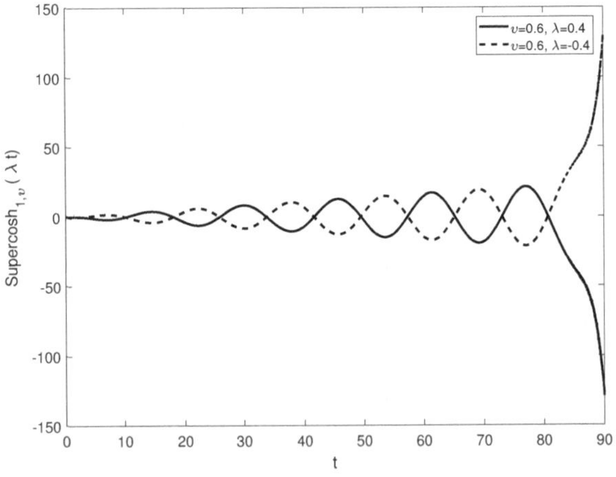

FIGURE D.24: The plot of the hyperbolic subcosine function $Supercosh_{1,\upsilon}(\lambda t)$.

which implies that

$$RCos_{\alpha,\upsilon}\left(\lambda t^{\alpha}\right)=\sum_{\kappa=0}^{\infty}\frac{(-1)^{\kappa}\,\lambda^{2\kappa}t^{(2\kappa+1)\alpha-1-\upsilon}}{\Gamma\left((2\kappa+1)\,\alpha-\upsilon\right)}. \tag{D.305}$$

the hyperbolic subsine function, denoted by $RSinh_{\alpha}\left(t^{\alpha}\right)$, is defined as

$$RSinh_{\alpha,\upsilon}\left(t^{\alpha}\right)=\sum_{\kappa=0}^{\infty}\frac{t^{2(\kappa+1)\alpha-1-\upsilon}}{\Gamma\left(2\left(\kappa+1\right)\alpha-\upsilon\right)}, \tag{D.306}$$

which implies that

$$RSinh_{\alpha,\upsilon}\left(\lambda t^{\alpha}\right)=\sum_{\kappa=0}^{\infty}\frac{\lambda^{\kappa}t^{2(\kappa+1)\alpha-1-\upsilon}}{\Gamma\left(2\left(\kappa+1\right)\alpha-\upsilon\right)}, \tag{D.307}$$

and the hyperbolic subcosine function, denoted by $RCosh_{\alpha}\left(t^{\alpha}\right)$, as

$$RCosh_{\alpha,\upsilon}\left(t^{\alpha}\right)=\sum_{\kappa=0}^{\infty}\frac{\lambda^{\kappa}t^{(2\kappa+1)\alpha-1-\upsilon}}{\Gamma\left((2\kappa+1)\,\alpha-\upsilon\right)}, \tag{D.308}$$

which implies that

$$RCosh_{\alpha,\upsilon}\left(\lambda t^{\alpha}\right)=\sum_{\kappa=0}^{\infty}\frac{\lambda^{\kappa}t^{(2\kappa+1)\alpha-1-\upsilon}}{\Gamma\left((2\kappa+1)\,\alpha-\upsilon\right)}. \tag{D.309}$$

Making use of

$$L\left[\frac{t^{\kappa\alpha}}{\Gamma\left(\kappa\alpha+1\right)}\right]=\frac{1}{s^{\kappa\alpha}}, \tag{D.310}$$

we have

$$L\left\{R_{\alpha,\upsilon}\left(\lambda t^{\alpha}\right)\right\}=s^{-(\alpha-\upsilon)}\left(1-\lambda s^{-\alpha}\right)^{-1}\quad\left(\left|\lambda s^{-\alpha}\right|<1\right) \tag{D.311}$$

such that

$$L\left\{RSin_{\alpha,\upsilon}\left(\lambda t^{\alpha}\right)\right\}=\lambda s^{-(2\alpha-\upsilon)}\left(1+\lambda^{2}s^{-2\alpha}\right)^{-1}\quad\left(\left|\lambda^{2}s^{-2\alpha}\right|<1\right), \tag{D.312}$$

$$L\left\{RSinh_{\alpha,\upsilon}\left(\lambda t^{\alpha}\right)\right\}=\lambda s^{-(2\alpha-\upsilon)}\left(1-\lambda^{2}s^{-2\alpha}\right)^{-1}\quad\left(\left|\lambda^{2}s^{-2\alpha}\right|<1\right), \tag{D.313}$$

$$L\left\{RCos_{\alpha,\upsilon}\left(\lambda t^{\alpha}\right)\right\}=s^{-(\alpha-\upsilon)}\left(1+\lambda^{2}s^{-2\alpha}\right)^{-1}\quad\left(\left|\lambda^{2}s^{-2\alpha}\right|<1\right), \tag{D.314}$$

and

$$L\left\{RCosh_{\alpha,\upsilon}\left(\lambda t^{\alpha}\right)\right\}=s^{-\alpha+\upsilon}\left(1-\lambda^{2}s^{-2\alpha}\right)^{-1}\quad\left(\left|\lambda^{2}s^{-2\alpha}\right|<1\right). \tag{D.315}$$

Therefore, we have

$$R_{\alpha,\upsilon}\left(\lambda t^{\alpha}\right)=RCosh_{\alpha,\upsilon}\left(\lambda t^{\alpha}\right)+RSinh_{\alpha,\upsilon}\left(\lambda t^{\alpha}\right) \tag{D.316}$$

and

$$R_{\alpha,\upsilon}\left(i\lambda t^{\alpha}\right) = RCos_{\alpha,\upsilon}\left(\lambda t^{\alpha}\right) + iRSin_{\alpha,\upsilon}\left(\lambda t^{\alpha}\right) \tag{D.317}$$

such that

$$RSin_{\alpha,\upsilon}\left(t^{\alpha}\right) = \frac{R_{\alpha,\upsilon}\left(it^{\alpha}\right) - R_{\alpha,\upsilon}\left(-it^{\alpha}\right)}{2i}, \tag{D.318}$$

$$RCos_{\alpha,\upsilon}\left(t^{\alpha}\right) = \frac{R_{\alpha,\upsilon}\left(it^{\alpha}\right) + R_{\alpha,\upsilon}\left(-it^{\alpha}\right)}{2}, \tag{D.319}$$

$$RSinh_{\alpha,\upsilon}\left(t^{\alpha}\right) = \frac{R_{\alpha,\upsilon}\left(t^{\alpha}\right) - R_{\alpha,\upsilon}\left(-t^{\alpha}\right)}{2}, \tag{D.320}$$

$$RCosh_{\alpha,\upsilon}\left(t^{\alpha}\right) = \frac{R_{\alpha,\upsilon}\left(t^{\alpha}\right) + R_{\alpha,\upsilon}\left(-t^{\alpha}\right)}{2}, \tag{D.321}$$

$$RSin_{\alpha,\upsilon}\left(\lambda t^{\alpha}\right) = \frac{R_{\alpha,\upsilon}\left(i\lambda t^{\alpha}\right) - R_{\alpha,\upsilon}\left(-i\lambda t^{\alpha}\right)}{2i}, \tag{D.322}$$

$$RCos_{\alpha,\upsilon}\left(\lambda t^{\alpha}\right) = \frac{R_{\alpha,\upsilon}\left(i\lambda t^{\alpha}\right) + R_{\alpha,\upsilon}\left(-i\lambda t^{\alpha}\right)}{2}, \tag{D.323}$$

$$RSinh_{\alpha,\upsilon}\left(\lambda t^{\alpha}\right) = \frac{R_{\alpha,\upsilon}\left(\lambda t^{\alpha}\right) - R_{\alpha,\upsilon}\left(-\lambda t^{\alpha}\right)}{2} \tag{D.324}$$

and

$$RCosh_{\alpha,\upsilon}\left(\lambda t^{\alpha}\right) = \frac{R_{\alpha,\upsilon}\left(\lambda t^{\alpha}\right) + R_{\alpha,\upsilon}\left(-\lambda t^{\alpha}\right)}{2}. \tag{D.325}$$

D.10 The two-parametric special functions via two-parametric Gorenflo-Mainardi functions

The two-parametric Gorenflo-Mainardi function is defined as (see [214]; also see [225])

$$G_{\alpha,\upsilon}\left(\lambda t^{\alpha}\right) = \sum_{\kappa=0}^{\infty} \frac{\lambda^{\kappa} t^{\kappa\alpha-1+\upsilon}}{\Gamma\left(\kappa\alpha+\upsilon\right)}. \tag{D.326}$$

The corresponding subsine function, denoted by $GSin_{\alpha}\left(t^{\alpha}\right)$, is defined as

$$GSin_{\alpha,\upsilon}\left(t^{\alpha}\right) = \sum_{\kappa=0}^{\infty} \frac{(-1)^{\kappa} t^{(2\kappa+1)\alpha-1+\upsilon}}{\Gamma\left((2\kappa+1)\alpha+\upsilon\right)}, \tag{D.327}$$

which implies that

$$GSin_{\alpha,\upsilon}\left(\lambda t^{\alpha}\right) = \sum_{\kappa=0}^{\infty} \frac{(-1)^{\kappa} \lambda^{2\kappa+1} t^{(2\kappa+1)\alpha-1+\upsilon}}{\Gamma\left((2\kappa+1)\alpha+\upsilon\right)}, \tag{D.328}$$

the subcosine function, denoted by $GCos_{\alpha}(t^{\alpha})$, as

$$GCos_{\alpha,v}(t^{\alpha}) = \sum_{\kappa=0}^{\infty} \frac{(-1)^{\kappa} t^{2\kappa\alpha-1+v}}{\Gamma(2\kappa\alpha+v)}, \tag{D.329}$$

which implies that

$$GCos_{\alpha,v}(\lambda t^{\alpha}) = \sum_{\kappa=0}^{\infty} \frac{(-1)^{\kappa} \lambda^{2\kappa} t^{2\kappa\alpha-1+v}}{\Gamma(2\kappa\alpha+v)}. \tag{D.330}$$

the hyperbolic subsine function, denoted by $GSinh_{\alpha}(t^{\alpha})$, is defined as

$$GSinh_{\alpha,v}(t^{\alpha}) = \sum_{\kappa=0}^{\infty} \frac{t^{(2\kappa+1)\alpha-1+v}}{\Gamma((2\kappa+1)\alpha+v)}, \tag{D.331}$$

which implies that

$$GSinh_{\alpha,v}(\lambda t^{\alpha}) = \sum_{\kappa=0}^{\infty} \frac{\lambda^{\kappa} t^{(2\kappa+1)\alpha-1+v}}{\Gamma((2\kappa+1)\alpha+v)}, \tag{D.332}$$

and the hyperbolic subcosine function, denoted by $GCosh_{\alpha}(t^{\alpha})$, as

$$RCosh_{\alpha,v}(t^{\alpha}) = \sum_{\kappa=0}^{\infty} \frac{\lambda^{\kappa} t^{2\kappa\alpha-1+v}}{\Gamma(2\kappa\alpha+v)}, \tag{D.333}$$

which implies that

$$GCosh_{\alpha,v}(\lambda t^{\alpha}) = \sum_{\kappa=0}^{\infty} \frac{\lambda^{\kappa} t^{2\kappa\alpha-1+v}}{\Gamma(2\kappa\alpha+v)}. \tag{D.334}$$

In view of

$$L\left[\frac{t^{\kappa\alpha}}{\Gamma(\kappa\alpha+1)}\right] = \frac{1}{s^{\kappa\alpha}}, \tag{D.335}$$

we have

$$L\{G_{\alpha,v}(\lambda t^{\alpha})\} = s^{-v}(1-\lambda s^{-\alpha})^{-1} \quad (|\lambda s^{-\alpha}|<1) \tag{D.336}$$

such that

$$L\{GSin_{\alpha,v}(\lambda t^{\alpha})\} = \lambda s^{-(v+\alpha)}\left(1+\lambda^2 s^{-2\alpha}\right)^{-1} \quad \left(\left|\lambda^2 s^{-2\alpha}\right|<1\right), \tag{D.337}$$

$$L\{GSinh_{\alpha,v}(\lambda t^{\alpha})\} = \lambda s^{-(v+\alpha)}\left(1-\lambda^2 s^{-2\alpha}\right)^{-1} \quad \left(\left|\lambda^2 s^{-2\alpha}\right|<1\right), \tag{D.338}$$

$$L\{GCos_{\alpha,v}(\lambda t^{\alpha})\} = s^{-v}\left(1+\lambda^2 s^{-2\alpha}\right)^{-1} \quad \left(\left|\lambda^2 s^{-2\alpha}\right|<1\right), \tag{D.339}$$

and

$$L\{GCosh_{\alpha,v}(\lambda t^{\alpha})\} = s^{-v}\left(1-\lambda^2 s^{-2\alpha}\right)^{-1} \quad \left(\left|\lambda^2 s^{-2\alpha}\right|<1\right). \tag{D.340}$$

In this case, we have

$$G_{\alpha,\upsilon}(\lambda t^{\alpha}) = GCosh_{\alpha,\upsilon}(\lambda t^{\alpha}) + GSinh_{\alpha,\upsilon}(\lambda t^{\alpha}) \tag{D.341}$$

and

$$G_{\alpha,\upsilon}(i\lambda t^{\alpha}) = GCos_{\alpha,\upsilon}(\lambda t^{\alpha}) + iGSin_{\alpha,\upsilon}(\lambda t^{\alpha}) \tag{D.342}$$

such that

$$GSin_{\alpha,\upsilon}(t^{\alpha}) = \frac{G_{\alpha,\upsilon}(it^{\alpha}) - G_{\alpha,\upsilon}(-it^{\alpha})}{2i}, \tag{D.343}$$

$$GCos_{\alpha,\upsilon}(t^{\alpha}) = \frac{G_{\alpha,\upsilon}(it^{\alpha}) + G_{\alpha,\upsilon}(-it^{\alpha})}{2}, \tag{D.344}$$

$$GSinh_{\alpha,\upsilon}(t^{\alpha}) = \frac{G_{\alpha,\upsilon}(t^{\alpha}) - G_{\alpha,\upsilon}(-t^{\alpha})}{2}, \tag{D.345}$$

$$GCosh_{\alpha,\upsilon}(t^{\alpha}) = \frac{G_{\alpha,\upsilon}(t^{\alpha}) + G_{\alpha,\upsilon}(-t^{\alpha})}{2}, \tag{D.346}$$

$$GSin_{\alpha,\upsilon}(\lambda t^{\alpha}) = \frac{G_{\alpha,\upsilon}(i\lambda t^{\alpha}) - G_{\alpha,\upsilon}(-i\lambda t^{\alpha})}{2i}, \tag{D.347}$$

$$GCos_{\alpha,\upsilon}(\lambda t^{\alpha}) = \frac{G_{\alpha,\upsilon}(i\lambda t^{\alpha}) + G_{\alpha,\upsilon}(-i\lambda t^{\alpha})}{2}, \tag{D.348}$$

$$GSinh_{\alpha,\upsilon}(\lambda t^{\alpha}) = \frac{G_{\alpha,\upsilon}(\lambda t^{\alpha}) - G_{\alpha,\upsilon}(-\lambda t^{\alpha})}{2}. \tag{D.349}$$

and

$$GCosh_{\alpha,\upsilon}(\lambda t^{\alpha}) = \frac{G_{\alpha,\upsilon}(\lambda t^{\alpha}) + G_{\alpha,\upsilon}(-\lambda t^{\alpha})}{2}. \tag{D.350}$$

Bibliography

[1] K. S. Miller and B. Ross. *An Introduction to the Fractional Calculus and Fractional Differential Equations.* John Wiley & Sons. New York Press, 1993.

[2] K. Oldham and J. Spanier. *The Fractional Calculus Theory and Applications of Differentiation and Integration to Arbitrary Order (Vol. 111).* Elsevier, 1974.

[3] S. G. Samko, A. A. Kilbas, and O. I. Marichev. *Fractional Integrals and Derivatives and Some of Their Applications.* Nauka i Tekhnika, Minsk, 1987 (in Russian).

[4] G. W. Leibniz. *Letter from Hanover, Germany, September 30, 1695 to G. A. L'Hospital. Leibnizen Mathematische Schrifien,* volume 2, pages 301–302, Olms Verlag, Hildesheim, Germany, 1962, (First published in 1849).

[5] G. W. Leibniz. *Letter from Hanover, Germany, to Johann Bernoulli, December 28, 1695, in Mathematische Schriften, reprinted 1962.* Olms Verlag, Hildesheim, Germany, 3, 226.

[6] G. W. Leibniz. *Letter from Hanover, Germany, May 28, 1697 to J. Wallis, Leibnizen Mathematische Schrifien,* volume 4, page 25. Olms Verlag, Hildesheim, Germany, 1962, (First published in 1859).

[7] L. Euler. De Progressionibus Transcentibus, sev Quarum Termini Algebraice Dari Nequeunt. *Comment. Acad. Sci. Imperialis Petropolitanae,* Volume 5, pages 38–57, 1738.

[8] J. L. Lagrange. Sur une nouvelle espèce de calcul rélatif à la différentiation & à l'intégration des quantités variables. *Oeuvres de Lagrange,* Volume 3, pages 441–476. Gauthier-Villars, Paris, 1849. 1738, (First appeared in Nouv. Mem. Acad. Roy. Sci. Belles Lett. Berlin 3, 185–206. 1772).

[9] A. V. Letnikov. Theory of Differentiation of Fractional Order. *Mat. Sb,* 3, pages 1–68, 1868.

[10] P. S. Laplace. *Theorie Analytique des Probability.* Courcier, Paris, 1820, (First appeared in 1812).

[11] S. F. Lacroix. *Traité du Calcul Differentiel et du Calcul Intégral*, Volume 3, pages 409–410, Courtier, Paris, 1819.

[12] J. B. J. Fourier. *Theorie Analytique de la Chaleur*. Oeuvres de Fourier, Volume 1, page 508. Didot, Paris, 1822.

[13] N. H. Abel. Solution de quelques problèmes à l'aide d'intégrales définites. *Oeuvres Completes*, Volume 1, pages 16–18. Grondahl, Christiania, Norway, 1881. (This paper first appeared in Mag. Naturvidenkaberne, 1823).

[14] N. H. Abel. Auflösung einer mechanischen Aufgabe. *J. fiir reine undangew. Math*, 1, 153–157, 1826.

[15] J. Liouville. Memoir sur quelques Questions de Gomtrie et de Mecanique, et sur un nouveau genre de Calcul pour resoudre ces Questions, *Journal de L'École Polytechnique*, tome XIII, XXIe cahier, pages 1–69, 1832.

[16] J. Liouville. Memoire sur le Calcul des different idles a indices quelconques. *Journal de L'École Polytechnique*, 13(21), 71–162, 1832.

[17] J. Liouville. Memoire sur le changement de la variable dans le calcul des differentielles a indices quelconques. *Journal de L'École Polytechnique*, 15(24), 17–54, 1835.

[18] B. Riemann. Versuch einer allgemeinen Auffassung der Integration und Differentiation, 14 Janvier 1847. *Bernhard Riemanns Gesammelte Mathematische Werke*, pages 353–362, 1892.

[19] N. Y. Sonin. On differentiation with arbitrary index. *Moscow Matem, Sbornik*, 6(1), 1–38, 1869.

[20] A. K. Grünwald. Ueber begrenzte Derivationen und deren Anwendung. *Zeitschrift fur Mathematik und Physik*, 12, 441–480, 1867.

[21] P. L. Butzer and U. Westphal. An access to fractional differentiation via fractional difference quotients. *Fractional Calculus and Its Applications*. Springer, Berlin, Heidelberg, pages 116–145, 1975.

[22] E. L. Post. Generalized differentiation. *Transactions of the American Mathematical Society*, 32(4), 723–781, 1930.

[23] A. V. Letnikov. Theory of differentiation of fractional order. *Matematicheskii Sbornik*, 3, 1–68, 1868.

[24] N. Sonine. Sur la differentiation a indice quelconque. *Matematicheskii Sbornik*, 6(1), 1–38, 1872.

[25] N. Sonine. Recherches sur les fonctions cylindriques et le developpement des fonctions continues en series. *Mathematische Annalen*, 16(1), 1–80, 1880.

[26] N. Sonine. Sur la generalisation d'une formule d'Abel. *Acta Mathematica*, 4(1), 171–176, 1884.

[27] S. G. Samko and R. P. Cardoso. Integral equations of the first kind of Sonine type. *International Journal of Mathematics and Mathematical Sciences*, 2003(57), 3609–3632, 2003.

[28] B. Choudhary. An extension of Abel's integral equation. *Journal of Mathematical Analysis and Applications*, 44(1), 113–130, 1973.

[29] N. Sonine. Sur la reduction d'une integrale multiple. *Matematicheskii Sbornik*, 14(4), 527–536, 1890.

[30] J. Hadamard. Essai sur l'étude des fonctions données par leur développement de Taylor. *J. math. Pures et Appl*, 4, 101–186, 1892.

[31] V. Volterra. Sulla inversione degli integrali definiti, I. Atti Rule Accad. *delle Sci. Torifl*, 31, 311–323, 1896.

[32] M. Boocher. *An Introduction to the Study of Integral Equations.* Cambridge University Press, 1909.

[33] H. Weyl. Bemerkungen zum Begriff des Differential quotienten gebrochener Ordnung. *Vierteljal & rechrift tier Ntdrforchentlen Geellchaft in Zirich*, 62(1-2), 296–302, 1917.

[34] A. Marchaud. Sur les denvees et sur les differences des fonctions de variables reeiles. *Journal de Mathematiques Pures et Appliques*, 6(4), 337–425, 1927.

[35] L. Tonelli. Su un problema di Abel. *Mathematische Annalen*, 99(1), 183–199, 1928.

[36] J. D. Tamarkin. On integrable solutions of Abel's integral equation. *Annals of Mathematics*, 2(31), 219–229, 1930.

[37] E. Hille and J. D. Tamarkin. On the theory of linear integral equations. *Annals of Mathematics*, 3(1), 479–528, 1930.

[38] Y. Watanabe. Notes on the generalized derivative of Riemann-Liouville and its application to Leibnitz's formula, I and IT. *Tohoku Mathematical Journal*, 34, 8–27 and 28–41, 1931.

[39] T. Sato. On Abel's integral equation. *Mem. Coli. Sci. Kroto*, 18, 63–78, 1935.

[40] E. R. Love. Fractional integration and almost periodic functions. *Proceedings of the London Mathematical Society*, Ser. 2, 44, No. 5, 363–397, 1938.

[41] E. R. Love and L. C. Young. In fractional integration by parts. *Proceedings of the London Mathematical Society*, Ser. 2, 44, 1–35, 1938.

[42] E. Hille. Notes on linear transformations. II. Analyticity of semi-groups. *Annals of Mathematics*, 40(1), 1–47, 1939.

[43] A. Erdèlyi and H. Kober. Some remarks on Hankel transforms. *The Quarterly Journal of Mathematics*, 11(1), 212–221, 1940.

[44] J. Cossar. A theorem on Cesaro summability. *Journal of the London Mathematical Society*, volume 16, 56–68, 1941.

[45] M. Riesz. L'integrale de Riemann-Liouville et le probleme de Cauchy. *Acta Mathematica*, 81(1), 1–222, 1949.

[46] A. Erdèlyi. On some functional transformations. *Univ. e Politec. Torino, Rend. Sem. Mat*, 10, 217–234, 1951.

[47] W. Feller. On a generalization of Marcel Riesz' potentials and the semi-groups generated by them. *Comm. Sem. Mathem. Universite de Lund*, 72–81, 1952.

[48] E. Hille and R. S. Phillips. Functional analysis and semi-groups. *American Mathematical Soc.*, volume 31, 1996, (First published in 1957).

[49] Y. W. Chen. Entire solutions of a class of differential equations of mixed type. *Communications on Pure and Applied Mathematics*, 14(3), 229–255, 1961.

[50] A. S. Peters. Certain dual integral equations and Sonine's integrals. *New York University, Institute of Mathematical Sciences*, Report, No.285, 1961.

[51] J. Burlak. A pair of dual integral equations occurring in diffraction theory. *Proceedings of the Edinburgh Mathematical Society*, 13(2), 179–187, 1962.

[52] I. N. Sneddon. Dual integral equations with trigonometrical kernels. *Glasgow Mathematical Journal*, 5(3), 147–152, 1962.

[53] K. N. Srivastava. Dual integral equations with Bessel function and trigonometrical kernels. *Proceedings of the Edinburgh Mathematical Society*, 13(4), 351–354, 1963.

[54] K. N. Srivastava. Integral equations involving a confluent hypergeometric function as kernel. *Journal d'Analyse Mathematique*, 13(1), 391–397, 1964.

[55] J. C. Cooke. The solution of triple integral equations in operational form. *The Quarterly Journal of Mechanics and Applied Mathematics*, 18(1), 57–72, 1965.

[56] K. D. Sakalyuk. The generalized Abel integral equation with inner coefficients (Russian). *Kisinev. Gos. Ucen. Zap.*, 82, 60–68, 1965.

[57] J. S. Lowndes. Dual Integral Equations with Trigonometric Kernels. *Proceedings of the Edinburgh Mathematical Society*, 15(1), 73–74, 1966.

[58] R. K. Saxena. On fractional integration operators. *Mathematische Zeitschrift*, 96(4), 288–291, 1967.

[59] G. K. Kalisch. On fractional integrals of pure imaginary order in L_p. *Proceedings of the American Mathematical Society*, 136–139, 1967.

[60] B. P. Parashar. Domain and range of fractional integration operators. *Math. Japon*, 12(2), 141–145, 1967.

[61] M. Caputo. Linear models of dissipation whose Q is almost frequency independent II. *Geophysical Journal International*, 13(5), 529–539, 1967.

[62] L. S. Bosanquet. Some extensions of M. Riesz's mean value theorem. *Indian J. Math*, 9, 65–90, 1967.

[63] M. M. Dzherbashyan. A generalized Riemann-Liouville operator and some applications of it (Russian). *Dokl. Akad. Nav.k SSSR*, 177(4), 767–770, 1967.

[64] H. M. Srivastava. Fractional integration and inversion formulae associated with the generalized Whittaker transform. *Pacific Journal of Mathematics*, 26(2), 375–377, 1968.

[65] T. Flett. Mean values of power series. *Pacific Journal of Mathematics*, 25(3), 463–494, 1968.

[66] M. Dzhrbashyan and A. Nersesyan. Drobnye proizvodnye i zadacha cauchy dlya differencialnykh uravneniy drobnogo poryadka. *Izvestiya Akademii Nauk Armyanskoj SSR, ser. Matematika*, 3(1), 3–29, 1968.

[67] S. L. Kalla. Integral operators involving Fox's H-function I. *Acta Mexicana Cienc, Tecn*, 3, 117–122, 1969.

[68] W. Smit and H. De Vries. Rheological models containing fractional derivatives. *In Rheologica Acta*, 9(4), 525–534, 1970.

[69] J. S. Lowndes. A generalisation of the Erdlyi-Kober operators. *Proceedings of the Edinburgh Mathematical Society*, 17(2), 139–148, 1970.

[70] T. J. Osler. The fractional derivative of a composite function. *SIAM Journal on Mathematical Analysis*, 1(2), 288–293, 1970.

[71] L. M. B. C. Campos. On generalizations of the series of Taylor, Lagrange, Laurent and Teixeira. *International Journal of Mathematics and Mathematical Sciences*, 13(4), 687–708, 1990.

[72] T. J. Osler. Leibniz rule for fractional derivatives generalized and an application to infinite series. *SIAM Journal on Applied Mathematics*, 18(3), 658–674, 1970.

[73] T. J. Osler. Taylor's series generalized for fractional derivatives and applications. *SIAM Journal on Mathematical Analysis*, 2(1), 37–48, 1971.

[74] M. Mikols. On the recent trends in the development, theory and applications of fractional calculus. *Fractional Calculus and Its Applications*, pages 357–375, Springer, Berlin, Heidelberg, 1975.

[75] S. G. Samko, A. A. Kilbas, and O. I. Marichev. *Fractional Integrals and Derivatives. Theory and Applications.* Gordon and Breach, Yverdon, 1993.

[76] E. R. Love. Fractional derivatives of imaginary order. *Journal of the London Mathematical Society*, 2(2), 241–259, 1971.

[77] B. Ross. A brief history and exposition of the fundamental theory of fractional calculus. *Fractional Calculus and Its Applications.* Pages 1–36, Springer, Berlin, Heidelberg, 1975.

[78] S. Z. Rafal'son. Fourier-Laguerre coefficients. *Matematika, Izvestiya Vysshikh Uchebnykh Zavedenii*, 11, 93–98, 1971.

[79] T. R. Prabhakar. Hypergeometric integral equations of a general kind and fractional integration. *SIAM Journal on Mathematical Analysis*, 3(3), 422–425, 1972.

[80] B. Choudhary. An extension of Abel's integral equation. *Journal of Mathematical Analysis and Applications*, 44(1), 113–130, 1973.

[81] I. N. Sneddon. The use in mathematical physics of Erdlyi-Kober operators and of some of their generalizations. *Fractional Calculus and Its Applications*, pages 37–79, Springer, Berlin, Heidelberg, 1975.

[82] V. Malovichko. On a generalized hypergeometric function and some integral operators that contain it. *Mat. fiz, Kiev: Naukova Dumlra, Vyp*, 19, 99–103, 1976

[83] V. A. Krasnov. The fractional derivative of a function with respect to another function. *Taskent. Gos. Univ. Naucn. Trudy No. 548 Voprosy Mat*, 142, 5660, 1977. (Russian)

[84] M. Saigo. A remark on integral operators involving the Gauss hypergeometric functions. *Mathematical reports of College of General Education, Kyushu University*, 11(2), 135–143, 1978.

[85] R. P. Srivastav. On dual integral equations with trigonometric kernels. *Quarterly of Applied Mathematics*, 35(4), 524–526, 1978.

[86] L. Gearhart. The Weyl semigroup and left translation invariant subspaces. *Journal of Mathematical Analysis and Applications*, 67(1), 75–91, 1979.

[87] A. C. McBride. *Fractional Calculus and Integral Transforms of Generalized Functions*. Pitman. London, 1979.

[88] K. Skornik. On fractional integrals and derivatives of one class of generalized functions (Russian). *Dokl. Akad. Nad SSSR*, 254, No. 5, 1085–1087, 1980.

[89] K. Skornik. On tempered integrals and derivatives of non-negative orders. *Annales Polonici Mathematici*, 1(40), pages 47–57, 1981.

[90] R. Estrada and R. P. Kanwal. Distributional solutions of singular integral equations. *The Journal of Integral Equations*, 8(1), 41–85, 1985.

[91] V. K. Tuan. On the factorization of integral transformations of convolution type in the space L_2. *Dokl. Akad. Nauk Armenyan SSR*, 83, 7–10, 1986.

[92] A. I. Peschanskii. Description of a space of fractional integrals of curvilinear convolution type. *Izvestiya Vysshikh Uchebnykh Zavedenii. Matematika*, 7, 29–39, 1989.

[93] S. G. Samko and B. Ross. Integration and differentiation to a variable fractional order. *Integral Transforms and Special Functions*, 1(4), 277–300, 1993.

[94] R. Hilfer. *Applications Of Fractional Calculus In Physics*, 30(1): 122–173, 2005. World Scientific, River Edge, New Jersey, 2000.

[95] C. F. Coimbra. Mechanics with variable-order differential operators. *Annalen der Physik*, 12(11-12), 692–703, 2003.

[96] S. G. Samko and R. P. Cardoso. Integral equations of the first kind of Sonine type. *International Journal of Mathematics and Mathematical Sciences*, 2003(57), 3609–3632, 2003.

[97] A. A. Kilbas, M. Saigo, and R. K. Saxena. Generalized Mittag-Leffler function and generalized fractional calculus operators. *Integral Transforms and Special Functions*, 15(1), 31–49, 2004.

[98] O. P. Agrawal. Fractional variational calculus in terms of Riesz fractional derivatives. *Journal of Physics A: Mathematical and Theoretical*, 40(24), 6287, 2007.

[99] J. Gajda and M. Magdziarz. Fractional Fokker-Planck equation with tempered stable waiting times: Langevin picture and computer simulation. *Physical Review E*, 82(1), 011117, 2010.

[100] M. M. Meerschaert, F. Sabzikar, M. S. Phanikumar, and A. Zeleke. Tempered fractional time series model for turbulence in geophysical flows. *Journal of Statistical Mechanics: Theory and Experiment*, 2014(9), 09023, 2014.

[101] R. Garra et al. Hilfer-Prabhakar derivatives and some applications. *Applied Mathematics and Computation*, 242, 576–589, 2014.

[102] M. Caputo and M. Fabrizio. A new definition of fractional derivative without singular kernel. *Progress in Fractional Differentiation and Applications*, 1(2), 1–13, 2015.

[103] M. Zayernouri, M. Ainsworth, and G. E. Karniadakis. Tempered Fractional Sturm-Liouville EigenProblems. *SIAM Journal on Scientific Computing*, 37(4), 1777–1800, 2015.

[104] X. J. Yang, H. M. Srivastava, and T. J. Machado. A new fractional derivative without singular kernel: Application to the modelling of the steady heat flow. *Thermal Science*, 20(2), 753–756, 2016.

[105] A. Atangana and D. Baleanu. New fractional derivatives with nonlocal and non-singular kernel: Theory and application to heat transfer model. *Thermal Science*, 20(2), 763–769, 2016.

[106] F. Sabzikar, M. M. Meerschaert, and J. Chen. Tempered fractional calculus. *Journal of Computational Physics*, 293, 14–28, 2015.

[107] R. G. Buschman. Decomposition of an integral operator by use of Mikusiski calculus. *SIAM Journal on Mathematical Analysis*, 3(1), 83–85, 1972.

[108] C. Li and W. Deng. High order schemes for the tempered fractional diffusion equations. *Advances in Computational Mathematics*, 42(3), 543–572, 2016.

[109] A. Cartea and D. del-Castillo-Negrete. Fluid limit of the continuous-time random walk with general Levy jump distribution functions. *Physical Review E*, 76(4), 041105, 2007.

[110] X. J. Yang. Fractional derivatives of constant and variable orders applied to anomalous relaxation models in heat-transfer problems. *Thermal Science*, 21(3), 1161–1171, 2016.

[111] C. Li and W. Deng. High order schemes for the tempered fractional diffusion equations. *Advances in Computational Mathematics*, 42(3), 543–572, 2016.

[112] C. Torres. Tempered fractional differential equation: variational approach. *Mathematical Methods in the Applied Sciences*, 40(13), 4962–4973, 2017.

[113] H. Sun, X. Hao, Y. Zhang, and D. Baleanu. Relaxation and diffusion models with non-singular kernels. *Physica A: Statistical Mechanics and Its Applications*, 468, 590–596, 2017.

[114] X. J. Yang and J. T. Machado. A new fractional operator of variable order: application in the description of anomalous diffusion. *Physica A: Statistical Mechanics and Its Applications*, 481, 276–283, 2017.

[115] M. Dehghan, M. Abbaszadeh, and W. Deng. Fourth-order numerical method for the space-time tempered fractional diffusion-wave equation. *Applied Mathematics Letters*, 73, 120–127, 2017.

[116] X. J. Yang, J. T. Machado, and D. Baleanu. Anomalous diffusion models with general fractional derivatives within the kernels of the extended Mittag-Leffler type functions. *Romanian Reports in Physics*, 69(4), 115, 2017.

[117] X. J. Yang, F. Gao, J. A. T. Machado, and D. Baleanu. A new fractional derivative involving the normalized sinc function without singular kernel. *The European Physical Journal Special Topics*, 226(16-18), 3567-3575.

[118] J. Sousa and E. C. de Oliveira. A new fractional derivative of variable order with non-singular kernel and fractional differential equations. *Computational and Applied Mathematics*, DOI: 10.1007/s40314-018-0639-x, 2018.

[119] H. Holmgren. On differential kalkylen: Med indices af hvad natur somhelst. *Kongl. Stenh Vetendap Ahd. Handl. Stockholm*, 5, 11, 1–83, 1865.

[120] H. T. Davis. The application of fractional operators to functional equations. *American Journal of Mathematics*, 49(1), 123–142, 1927.

[121] G. H. Hardy and J. E. Littlewood. Some properties of fractional integrals. I. *Mathematische Zeitschrift*, 27(1), 565–606, 1928.

[122] J. C. Burkill. Fractional orders of integrability. *Journal of the London Mathematical Society*, 1(3), 220–226, 1936.

[123] E. T. Copson. XXII.–Some Applications of Marcel Riesz's Integrals of Fractional Order. *Proceedings of the Royal Society of Edinburgh Section A: Mathematics*, 61(3), 260–272, 1943.

[124] A. Zygmund. A theorem on fractional derivatives. *Duke Mathematical Journal*, 12(3), 455–464, 1945.

[125] M. A. Bassam. Some properties of Holmgren-Riesz transform. *Annali della Scuola Normale Superiore di Pisa-Classe di Scienze*, 15(1-2), 1–24, 1961.

[126] B. Kuttner. On differences of fractional order. *Proceedings of the London Mathematical Society*, 3(1), 453–466, 1957.

[127] R. N. Kesarwani. Fractional integration and certain dual integral equations. *Mathematische Zeitschrift*, 98(1), 83–88, 1967.

[128] G. K. Kalisch. On fractional integrals of pure imaginary order in L_p. *Proceedings of the American Mathematical Society*, 136–139, 1967.

[129] R. O'Neil. Fractional integration in Orlicz spaces.I. *Transactions of the American Mathematical Society*, 115, 300–328, 1965.

[130] H. Kober. New properties of the Weyl extended integral. *Proceedings of the London Mathematical Society*, 3(3), 557–575, 1970.

[131] M. C. Gaer and L. A. Rubel. The fractional derivative via entire functions. *Journal of Mathematical Analysis and Applications*, 34(2), 289–301, 1971.

[132] E. R. Love. Fractional derivatives of imaginary order. *Journal of the London Mathematical Society*, 2(2), 241–259, 1971.

[133] T. R. Prabhakar. Hypergeometric Integral Equations of a general kind and fractional integration. *SIAM Journal on Mathematical Analysis*, 3(3), 422–425, 1972.

[134] B. S. Rubin. Fractional integrals in Holder spaces with weight, and operators of potential type. *Izv. Akad. Nauk Armjan. SSR Ser. Mat*, 9(4), 308324, 1974.

[135] D. R. Adams. A note on Riesz potentials. *Duke Mathematical Journal*, 42(4), 765–778, 1975.

[136] R. C. Koeller. Applications of fractional calculus to the theory of viscoelasticity. *Journal of Applied Mechanics*, 51(2), 299–307, 1984.

[137] P. Ahern and M. Jevtic. Mean modulus and the fractional derivative of an inner function. *Complex Variables and Elliptic Equations*, 3(4), 431–445, 1984.

[138] W. G. Glöckle and T. F. Nonnenmacher. A fractional calculus approach to self-similar protein dynamics. *Biophysical Journal*, 68(1), 46–53, 1995.

[139] R. L. Bagley and P. J. Torvik. On the fractional calculus model of viscoelastic behavior. *Journal of Rheology*, 30(1), 133–155, 1986.

[140] R. C. Koeller. Polynomial operators, Stieltjes convolution, and fractional calculus in hereditary mechanics. *Acta Mechanica*, 58(3-4), 251–264, 1986.

[141] H. M. Srivastava, M. Saigo, and S. Owa. A class of distortion theorems involving certain operators of fractional calculus. *Journal of Mathematical Analysis and Applications*, 131(2), 412–420, 1988.

[142] L. Debnath. Recent applications of fractional calculus to science and engineering. *International Journal of Mathematics and Mathematical Sciences*, 2003(54), 3413–3442, 2003.

[143] B. Ross. The development of fractional calculus 16951900. *Historia Mathematica*, 4(1), 75–89, 1977.

[144] R. P. Agarwal, V. Lakshmikantham, and J. J. Nieto. On the concept of solution for fractional differential equations with uncertainty. *Nonlinear Analysis: Theory, Methods & Applications*, 72(6), 2859–2862, 2010.

[145] R. L. Magin. Fractional calculus models of complex dynamics in biological tissues. *Computers & Mathematics with Applications*, 59(5), 1586–1593, 2010.

[146] N. Engheia. On the role of fractional calculus in electromagnetic theory. *IEEE Antennas and Propagation Magazine*, 39(4), 35–46, 1997.

[147] A. Kilicman and Z. A. A. Al Zhour. Kronecker operational matrices for fractional calculus and some applications. *Applied Mathematics and Computation*, 187(1), 250–265, 2007.

[148] J. T. Machado, V. Kiryakova, and F. Mainardi. Recent history of fractional calculus. *Communications in Nonlinear Science and Numerical Simulation*, 16(3), 1140–1153, 2011.

[149] K. Oldham and J. Spanier. *The Fractional Calculus Theory and Applications of Differentiation and Integration to Arbitrary Order (Vol. 111)*. Elsevier, 1974.

[150] I. Podlubny. *Fractional differential equations: an introduction to fractional derivatives, fractional differential equations, to methods of their solution and some of their applications (Vol. 198)*. Elsevier, 1998.

[151] R. L. Magin. Fractional calculus in bioengineering. *Redding: Begell House*, pages 269–355, 2006.

[152] A. A. Kilbas, H. M. Srivastava, and J. J. Trujillo. *Theory And Applications of Fractional Differential Equations (Vol. 204)*. Elsevier Science Limited, 2006.

[153] J. Sabatier, O. P. Agrawal, and J. T. Machado. *Advances in Fractional Calculus (Vol. 4, No. 9)*. Dordrecht: Springer, 2007.

[154] K. Diethelm. *The Analysis of Fractional Differential Equations: An Application-Oriented Exposition Using Differential Operators of Caputo Type*. Springer, 2010.

[155] C. A. Monje et al. *Fractional-Order Systems and Controls: Fundamentals and Applications*. Springer Science & Business Media, 2010.

[156] V. E. Tarasov. *Fractional Dynamics: Applications of Fractional Calculus to Dynamics of Particles, Fields and Media*. Springer Science & Business Media, 2011.

[157] M. D. Ortigueira. *Fractional Calculus for Scientists and Engineers (Vol. 84)*. Springer Science & Business Media, 2011.

[158] D. Baleanu, J. A. T. Machado, and A. C. Luo (Eds.). *Fractional Dynamics and Control*. Springer Science & Business Media, 2011.

[159] G. A. Anastassiou. *Advances on Fractional Inequalities*. Springer Science & Business Media, 2011.

[160] J. Klafter, S. C. Lim, and R. Metzler. *Fractional Dynamics: Recent Advances*. F. World Scientific, 2012.

[161] M. M. Meerschaert and A. Sikorskii. *Stochastic Models for Fractional Calculus (Vol. 43)*. Walter de Gruyter, 2012.

[162] D. Baleanu, K. Diethelm, E. Scalas, and J. J. Trujillo. *Fractional Calculus: Models and Numerical Methods (Vol. 3)*. World Scientific, 2012.

[163] A. B. Malinowska and D. F. Torres. *Introduction to the Fractional Calculus of Variations*. World Scientific Publishing Company, 2012.

[164] S. Abbas, M. Benchohra, and G. M. N'Guèrèkata. *Topics in Fractional Differential Equations (Vol. 27)*. Springer Science & Business Media, 2012.

[165] X. J. Yang. *Advanced Local Fractional Calculus and Its Applications*. World Science Publisher, New York, 2012.

[166] J. D. Tamarkin, On integrable solutions of Abel's integral equation. *Annals of Mathematics*, 31, 219-229, 1930.

[167] V. V. Uchaikin. *Fractional Derivatives for Physicists and Engineers (Vol. 2)*. Berlin: Springer, 2013.

[168] H. Richard. *Fractional Calculus: An Introduction for Physicists.* World Scientific, 2014.

[169] A. Carpinteri and F. Mainardi (Eds.). *Fractals and Fractional Calculus in Continuum Mechanics (Vol. 378).* Springer, 2014.

[170] T. M. Atanackovic, S. Pilipovic, B. Stankovic, and D. Zorica. *Fractional Calculus with Applications in Mechanics: Wave Propagation, Impact and Variational Principles.* John Wiley & Sons, 2014.

[171] C. Li and F. Zeng. *Numerical Methods for Fractional Calculus (Vol. 24).* CRC Press, 2015.

[172] Z. Yong, W. Jinrong, and Z. Lu. *Basic Theory of Fractional Differential Equations.* World Scientific, 2016.

[173] B. L. Guo, P. Xueke, and H. Fenghui. *Fractional Partial Differential Equations and Their Numerical Solutions.* World Scientific, 2015.

[174] C. F. Lorenzo and T. T. Hartley. *The Fractional Trigonometry: With Applications to Fractional Differential Equations and Science.* John Wiley & Sons, Springer, 2016.

[175] B. J. West. *Fractional Calculus View of Complexity: Tomorrow's Science.* CRC Press, 2016.

[176] B. Ahmad et al. *Hadamard-Type Fractional Differential Equations, Inclusions and Inequalities.* Springer International Publishing, 2017.

[177] J. Tenreiro Machado, V. Kiryakova, and F. Mainardi. A poster about the old history of fractional calculus. *Fractional Calculus and Applied Analysis*, 13(4), 447–454, 2010.

[178] T. Machado, V. Kiryakova, and F. Mainardi. A poster about the recent history of fractional calculus. *Fractional Calculus and Applied Analysis*, 13(3), 329–334, 2010.

[179] J. T. Machado, V. Kiryakova, and F. Mainardi. Recent history of fractional calculus. *Communications in Nonlinear Science and Numerical Simulation*, 16(3), 1140–1153, 2011.

[180] J. T. Machado, A. M. Galhano, and J. J. Trujillo. On development of fractional calculus during the last fifty years. *Scientometrics*, 98(1), 577–582, 2014.

[181] V. S. Kiryakova. *Generalized Fractional Calculus and Applications.* CRC Press, 1993.

[182] A. C. McBride. *Fractional Calculus and Integral Transforms of Generalized Functions.* Pitman, San Francisco, CA, 1979.

[183] H. M. Srivastava and S. Owa. *Univalent Functions, Fractional Calculus, and Their Applications.* Ellis Horwood, New York, Toronto: Halsted Press, 1989.

[184] A. H. Zemanian. *Generalized Integral Transformations.* Interscience, New York, 1968.

[185] A. I. Zayed. *Handbook of Function and Generalized Function Transformations.* CRC Press, 1996.

[186] P. Antosik, J. Mikusiski, and R. Sikorski. *Theory of Distributions: The Sequential Approach.* American Elsevier Pub. Co, New York, 1973.

[187] V. Kiryakova. A brief story about the operators of the generalized fractional calculus. *Fractional Calculus and Applied Analysis*, 11(2), 203–220, 2008.

[188] J. D. Tamarkin, On integrable solutions of Abel's integral equation. *Annals of Mathematics*, 31, 219–229, 1930.

[189] J. T. Machado, F. Mainardi, and V. Kiryakova. Fractional calculus: Quo vadimus? (Where are we going?). *Fractional Calculus and Applied Analysis*, 18(2), 495–526, 2015.

[190] L. Euler. De progressionibus transcendentibus sen quaroum termini generales algebrare dari nequeunt, (1738). *Commentarii academiae scientiarum Petropolitanae*, 5, 36–57, presented to the St. Petersburg Academy on November 28, 1729.

[191] A. M. Legendre. Recherches sur diverses sortes d'integrales definies. *Mem. de l'Institut*, 10, 416-509, 1809.

[192] L. Euler. Evolutio formula integralis, etc. *Commentarii Academiae Scientiarum Petropolitanae*, 16, 91-139, 1771, (First published in 1772).

[193] P. J. Davis. Leonhard Eulers integral: A historical profile of the Gamma function: In memoriam: Milton Abramowitz. *The American Mathematical Monthly*, 66(10), 849–869, 1959.

[194] G. E. Andrews, R. Askey, and R. Roy. *Special functions, Volume 71 of Encyclopedia of Mathematics and Its Applications.* Cambridge University Press, Cambridge, UK, 1999.

[195] L. J. Slater. *Generalized Hypergeometric Functions.* Cambridge University Press, London, UK, 1966.

[196] C. F. Gauss. Disquisitiones generales circa seriem infinitam. *Comm. Soc. Reg. Gott. Gesammelte Werke*, 3, 123–162, 1812.

[197] L. Pochhammer. Ueber Hypergeometrischen Funktionen Hoheren Ordungen. *Journal fr die Reine und Angewandte Mathematik*, 71, 216, 1870.

[198] F. W. Bessel. *Untersuchung des Theils der planetarischen Storungen aus der Bewegung der Sonne entsteht.* Berliner Abh. 1824, 1-52.

[199] G. N. Watson. *A Treatise on the Theory of Bessel Functions.* Cambridge University Press, 1922.

[200] Y. L. Luke. *Integrals of Bessel Functions.* McGraw-Hill, New York, 1962.

[201] F. Bowman. *Introduction to Bessel Functions.* Dover, New York, 1958.

[202] E. E. Kummer. Uber die hypergeometrische Reihe $F(a; b; x)$. *Journal fr die Reine und Angewandte Mathematik*, 15, 39–83, 127–172, 1836.

[203] R. Kohlrausch. Theorie des elektrischen Rckstandes in der Leidener Flasche. *Annalen der Physik*, 167(2), 179–214, 1854.

[204] G. Williams and D. C. Watts. Non-symmetrical dielectric relaxation behaviour arising from a simple empirical decay function. *Transactions of the Faraday society*, 66, 80–85, 1970.

[205] W. K. Clifford. *On Bessels Functions, Mathematical Papers.* Macmillan and Company, London: 346–349, 1882.

[206] H. M. Macdonald. Zeroes of the Bessel functions. *Proceedings of the London Mathematical Society*, 1(1), 165-179, 1898.

[207] G. M. Mittag-Leffler. Sur La Nouvelle Fonction $E(x)$. *Omptes Rendus de l'Academie des Sciences*, 137, 554–558, 1903.

[208] A. Wiman. Uber Den Fundamental Satz in Der Theorie Der Funktionen $E(x)$. *Acta Mathematica*, 29, 217–234, 1905.

[209] E. W. Barnes. The asymptotic expansion of integral functions defined by generalised hypergeometric series. *Proceedings of the London Mathematical Society*, 2(1), 59–116, 1907.

[210] E. M. Wright. On the coefficients of power series having exponential singularities. *Journal of the London Mathematical Society*, 1(1), 71–79, 1933.

[211] W. N. Bailey. *Generalised Hypergeometric Series.* Cambridge University Press, England: Cambridge, 1935.

[212] C. S. Meijer. Uber Whittakersche bzw. Besselsche Funktionen und deren Produkte. *Nieuw Archief voor Wiskunde (2) (German)*, 18(4), 10–39, 1936.

[213] T. M. MacRobert. I. Induction Proofs of the Relations between certain Asymptotic Expansions and Corresponding Generalised Hypergeometric Series. *Proceedings of the Royal Society of Edinburgh*, 58, 1–13, 1939.

[214] L. J. Slater. *Confluent Hypergeometric Functions*. Cambridge, England: Cambridge University Press, 1960.

[215] L. J. Slater. *Generalized Hypergeometric Functions*. Cambridge, England: Cambridge University Press, 1966.

[216] A. M. Mathai and R. K. Saxena. *Generalized Hypergeometr Functions with Appellations in Statists and Physical Sciences*. Lecture Notes in Mathematics, Springer, New York, 1973.

[217] Y. N. Rabotnov. Equilibrium of an elastic medium with after effect. *Prikl. Matem. i Mekh*, 12(1), 81–91, 1948. (Russian)

[218] R. P. Agarwal. Some Inversion Formulae for the Generalised Hankel Transform. *Bulletin of Calcutta Mathematical Society*, volume 45, 69–73, 1953.

[219] A. Rdelyi, W. Magnus, F. Oberhettinger, and F. G. Tricomi. *Tables of Integral Transforms*. The Bateman Project, McGraw-Hill Book Co., LCC, volume 1, Number 54–6214, 1954.

[220] C. Fox. The G and H functions as symmetrical Fourier kernels. *Transactions of the American Mathematical Society*, 98(3), 395–429, 1961.

[221] T. R. Prabhakar. A Singular Integral Equation with a Generalized Mittag Leffler Function in the Kernel. *Yokohama Mathematical Journal*, 19(1), 7–15, 1971.

[222] R. Gorenflo, A. A. Kilbas, F. Mainardi, S. V. Rogosin. *Mittag-Leffler Functions, Related Topics and Applications*. Berlin: Springer, 2014.

[223] K. S. Miller and B. Ross. *An Introduction to the Fractional Calculus and Fractional Differential Equations*. Wiley, New York, 1993.

[224] R. Gorenflo and F. Mainardi. Fractional oscillations and Mittag-Leffler functions. *International Workshop on the Recent Advances in Applied Mathematics (RAAM96), State of Kuwait, Proceedings, Kuwait University*, 193–208, 1996.

[225] T. T. Hartley and C. F. Lorenzo. The vector linear fractional initialization problem. *NASA*, TP-1999-208919, 1999.

[226] L. C. Andrews. *Special Functions of Mathematics for Engineers*. New York: McGraw-Hill, 1992.

[227] W. Magnus and F. Oberhettinger. *Formulas and Theorems for the Special Functions of Mathematical Physics*. Springer, New York, 1969.

[228] G. N. Watson. *A Treatise on the Theory of Bessel Functions.* Cambridge University Press, London, 1952.

[229] Y. L. Luke. *The Special Functions and Their Approximations.* Academic Press, New York, 1969.

[230] E. D. Rainville. *Special Functions.* Chelsea, New York, 1960.

[231] F. W. J. Olver. *Asymptotics and Special Functions.* Academic, New York, 1974.

[232] W. W. Bell. *Special Functions for Scientists and Engineers.* Van Nostrand, London, 1968.

[233] B. C. Carlson. *Special Functions of Applied Mathematics.* Academic, New York, 1977.

[234] I. S. Gradshteyn and I. M. Ryzhik. *Table of Integrals, Series, and Products.* Academic Press, New York, 1980.

[235] Atangana, A. Derivative with two fractional orders: A new avenue of investigation toward revolution in fractional calculus. *The European Physical Journal Plus*, 131(10): 373, 2016.

[236] R. Gorenflo, A. A. Kilbas, F. Mainardi, and S. V. Rogosin. *Mittag-Leffler Functions, Related Topics and Applications.* Berlin: Springer, volume 2, 2014.

[237] C. F. Lorenzo and T. T. Hartley. *The Fractional Trigonometry: With Applications to Fractional Differential Equations and Science.* John Wiley & Sons, 2016.

[238] N. N. Lebedev. *Special Functions and Their Applications. Revised English Edition.* Translated and edited by Richard A. Silverman. Prentice-Hall, London, UK, 1965.

[239] A. M. Legendre. *Essai sur la Theorie des Nombres.* Paris: Courcier, 1808.

[240] F. E. Prym. Zur Theorie der Gammafunktion. *J. Reine Angew. Math*, 82: 165–172, 1877.

[241] T. J. I. A. Bromwich. *An Introduction to the Theory of Infinite Series.* Macmillan, New York, 1908.

[242] F. G. Tricomi. Asymptotische eigenschaften der unvollständigen gammafunktion. *Mathematische Zeitschrift*, 53(1), 136–148, 1950.

[243] N. M. Temme. Uniform asymptotic expansions of the incomplete gamma functions and the incomplete beta function. *Mathematics of Computation*, 29(132), 1109–1114, 1975.

[244] G. J. O. Jameson. The incomplete gamma functions. *Mathematical Gazette*, 100(548), 298–306, 2016.

[245] K. Weierstraβ. Uber die Theorie der analytischen Facultaten, Schlomilch. O., Analytische Studien, Engelmann, Leipzig 1848. *Journal fur die reine und angewandte Mathematik*, 51, 1–60, 1856.

[246] T. H. Gronwall. An elementary exposition of the theory of the gamma function. *Annals of Mathematics*, 17(3), 124–166, 1916.

[247] L. Euler. Evolutio formulae integralis $\int x^{f-1}(\log x)^{m/n} dx$ integratione a valore $x = 0$ ad $x = 1$ extensa, Nov. *Comm. Acad. Sci. Petr*, 16: 91–139, 1772 (Opera Omnia, Series I, vol. 17, pp. 316–357, 1771).

[248] O. Schlomilch. Analytische Studien *Engelmann, Leipzig*, 1848.

[249] F. W. Newman. On $\Gamma(a)$, especially when a is negative. *The Cambridge and Dublin Mathematical Journal, 3: 57–60*, 1848.

[250] W. Freeden and M. Gutting. *Special Functions of Mathematical (geo-) Physics.* Springer Science & Business Media, 2013.

[251] F. Mainardi. *Fractional Calculus and Waves in Linear Viscoelasticity: An Introduction to Mathematical Models.* World Scientific, 2010.

[252] H. Hankel. Die Euler'schen integrale bei unbeschrankter variabilitat der arguments. *Leipzig*, 1863.

[253] N. M. Temme. *Special Functions: An Introduction to the Classical Functions of Mathematical Physics.* John Wiley & Sons, 2011.

[254] I. N. Sneddon. *Special Functions of Mathematical Physics and Chemistry.* Oliver and Boyd, Edinburgh, 1961.

[255] F. Oberhettinger. *Tables of Mellin Transforms.* Springer Science & Business Media, 2012.

[256] T. Bayes. An assay foward solving a problem in the doctrine of chances. *Philosophical Transactions of the Royal Society of London Series I*, 53, 370-418, 1763.

[257] H. Weiler. The use of incomplete beta functions for prior distributions in binomial sampling. *Technometrics*, 7(3), 335-347, 1965.

[258] S. M. Nikol'skii. *Course of Mathematical Analysis (Russian).* vol. 1–2, Nauka, Moscow, 1983.

[259] A. N. Kolmogorov and S. V. Fomin. *Fundamentals of the Theory of Functions and Functional Analysis.* Nauka, Moscow, 1968.

[260] J. Lützen. *Joseph Liouville 1809–1882: Master of pure and applied mathematics (Vol.15).* Springer Science & Business Media, 2012.

[261] B. Ross. The development of fractional calculus 1695–1900. *Historia Mathematica*, 4(1), 75–89, 1977.

[262] A. M. Mathai, R. K. Saxena, and H. J. Haubold. *The H-Function: theory and applications.* Springer Science & Business Media, 2009.

[263] A. Krug. Theorie der Derivationen Akad. Wiss. Wien Denkschriften Math. *Naturwiss*, 57, 151–228, 1890.

[264] Y. A. Rossikhin and M. V. Shitikova. Application of fractional calculus for dynamic problems of solid mechanics: novel trends and recent results. *Applied Mechanics Reviews*, 63(1), 010801, 2010.

[265] Y. A. Rossikhin. Reflections on two parallel ways in the progress of fractional calculus in mechanics of solids. *Applied Mechanics Reviews*, 63(1), 010701, 2010.

[266] I. M. Sokolov, J. Klafter, and A. Blumen. Fractional kinetics. *Physics Today*, 55(11), 48–54, 2002.

[267] D. Baleanu, J. A. T. Machado, and A. C. Luo (Eds.). *Fractional dynamics and control.* Springer Science & Business Media, 2011.

[268] D. Baleanu, Z. B. Güveç, and J. T. Machado (Eds.). *New trends in nanotechnology and fractional calculus applications (p.C397).* New York: Springer, 2010.

[269] A. N. Gerasimov. A generalization of linear laws of deformation and its application to problems of internal friction. *Akad. Nauk SSSR. Prikl. Mat. Meh*, 12, 251–260, 1948.

[270] M. Caputo and F. Mainardi. Linear models of dissipation in anelastic solids. *La Rivista del Nuovo Cimento (1971–1977)*, 1(2), 161–198, 1971.

[271] M. Caputo and F. Mainardi. A new dissipation model based on memory mechanism. *Pure and Applied Geophysics*, 91(1), 134–147, 1971.

[272] R. Hilfer, Y. Luchko, and Z. Tomovski. Operational method for the solution of fractional differential equations with generalized Riemann-Liouville fractional derivatives. *Fractional Calculus and Applied Analysis*, 12(3), 299–318, 2009.

[273] X. J. Yang, J. T. Machado, and D. Baleanu. Anomalous diffusion models with general fractional derivatives within the kernels of the extended Mittag-Leffler type functions. *Romanian Reports in Physics*, 69(4), 115, 2017.

[274] V. E. Tarasov. Fractional vector calculus and fractional Maxwell's equations. *Annals of Physics*, 323(11), 2756–2778, 2008.

[275] X. J. Yang, D. Baleanu, and H. M. Srivastava. *Local Fractional Integral Transforms and Their Applications.* Academic Press, 2015.

[276] P. L. Butzer and U. Westphal. An access to fractional differentiation via fractional difference quotients. In *Fractional Calculus and Its Applications* (pp. 116–145). Springer, Berlin, Heidelberg, 1975.

[277] U. Westphal. An approach to fractional powers of operators via fractional differences. *Proceedings of the London Mathematical Society*, 3(3), 557–576, 1974.

[278] N. Stuloff. Die Differentiation beliebiger reeller Ordnung. *Mathematische Annalen*, 122(5), 400–410, 1950.

[279] V. E. Tarasov. Lattice model of fractional gradient and integral elasticity: long-range interaction of Grünwald-Letnikov-Riesz type. *Mechanics of Materials*, 70, 106–114, 2014.

[280] R. Herrmann. *Fractional Calculus: An Introduction for Physicists.* World Scientific, 2011.

[281] Katugampola and N. Udita. New approach to a generalized fractional integral. *Applied Mathematics and Computation*, 218(3), 860–865, 2011.

[282] U. N. Katugampola. A new approach to generalized fractional derivatives. *Bulletin of Mathematical Analysis and Applications*, 6(4): 1–15, 2014.

[283] J. Bernoulli. Principia calculi exponentialium seu percurrentium. *Acta Eruditorium, March*, 125–133, 1697.

[284] L. Euler. *Introductio in Analysin Infinitorum.* St. Petersburg, 1748.

[285] L. Euler. De la controverse entre Mrs. Leibniz et Bernoulli sur les logarithmes des nombres négatifs et imaginaires. *Mémoires de l'Académie des Sciences de Berlin*, 5: 139–179, 1749.

[286] L. Euler. *Introduction to Analysis of the Infinite (Introductio in Analysin Infinitorum).* Springer-Verlag, New York, (Translator: D. Blanton), 1988.

[287] M. Chen and W. Deng. A second-order accurate numerical method for the space-time tempered fractional diffusion-wave equation. *Applied Mathematics Letters*, 68, 87–93, 2017.

[288] H. Pollard and D. V. Widder. Inversion of a convolution transform related to heat conduction. *SIAM Journal on Mathematical Analysis*, 1(4), 527–532, 1970.

[289] W. Deng and Z. Zhang. Variational formulation and efficient implementation for solving the tempered fractional problems. *Numerical Methods for Partial Differential Equations*, 34(4), 1224–1257, 2018.

[290] F. Sabzikar, M. M. Meerschaert, and J. Chen. Tempered fractional calculus. *Journal of Computational Physics*, 293, 14–28, 2015.

[291] C. Torres. Tempered fractional differential equation: variational approach. *Mathematical Methods in the Applied Sciences*, 40(13), 4962–4973, 2017.

[292] R. Hilfer. *Three fold introduction to fractional derivatives, Anomalous Transport: Foundations and Applications*. pp. 17–73, 2008.

[293] Z. Tomovski, R. Hilfer, and H. M. Srivastava. Fractional and operational calculus with generalized fractional derivative operators and Mittag-Leffler type functions. *Integral Transforms Spec. Funct*, 21(11), 797–814, 2010.

[294] H. M. Srivastava and Z. Tomovski. Fractional calculus with an integral operator containing a generalized Mittag-Leffler function in the kernel. *Applied Mathematics and Computation*, 211(1), 198–210, 2009.

[295] J. V. D. C. Sousa and E. C. de Oliveira. On the ψ-Hilfer fractional derivative. *Communications in Nonlinear Science and Numerical Simulation*, 60, 72–91, 2018.

[296] X. J. Yang, H. M. Srivastava and T. J. Machado. A new fractional derivative without singular kernel: Application to the modelling of the steady heat flow. *Thermal Science*, 20(2), 753–756, 2016.

[297] X. J. Yang, H. M. Srivastava, D. F. Torres, and A. Debbouche. General fractional-order anomalous diffusion with non-singular power-law kernel. *Thermal Science*, 21(1), S1–S9, 2017.

[298] X. J. Yang. General fractional calculus operators containing the generalized Mittag-Leffler functions applied to anomalous relaxation. *Thermal Science*, 21(S1), 317–326, 2017.

[299] X. J. Yang. Fractional derivatives of constant and variable orders applied to anomalous relaxation models in heat transfer problems. *Thermal Science*, 21(3), 1161–1171, 2017.

[300] X. J. Yang. New general fractional-order rheological models with kernels of Mittag-Leffler functions. *Romanian Reports in Physics*, 69(4), 118, 2017.

[301] X. J. Yang. *General fractional calculus: a new prospective on viscoelasticity*. International workshop on theory and applications of fractional partial differential equations, June 30, 2018, Qingdao, China.

[302] X. J. Yang. *Theoretical studies on general fractional-order viscoelasticity, Ph.D. Thesis.* China University of Mining and Technology, Xuzhou, China, December, 2017.

[303] R. Gorenflo and F. Mainardi. *Fractional oscillations and Mittag-Leffler functions.* Preprint No A-96-14, Fachbereich Mathematik und Informatik, Freie Univ. Berlin, Serie Mathematik, pp. 22, 1996.

[304] X. J. Yang. New rheological problems involving general fractional derivatives with nonsingular power-law kernels. *Proceedings of the Romanian Academy Series A-Mathematics Physics Technical Sciences Information Science*, 19(1), 45–52, 2017.

[305] M. Caputo and M. Fabrizio. Applications of new time and spatial fractional derivatives with exponential kernels. *Progress in Fractional Differentiation and Applications*, 2(2), 1–11, 2016.

[306] D. W. Davidson and R. H. Cole. Dielectric relaxation in glycerol, propylene glycol, and n-propanol. *The Journal of Chemical Physics*, 19(12), 1484–1490, 1951.

[307] S. Havriliak and S. Negami. A complex plane representation of dielectric and mechanical relaxation processes in some polymers. *Polymer*, 8, 161–210, 1967.

[308] A. Jurlewicz, J. Trzmiel and K. Weron. Two-power-law relaxation processes in complex materials. *Acta Physica Polonica B*, 41(5), 1001, 2010.

[309] E. C. de Oliveira, F. Mainardi and J. Vaz. Fractional models of anomalous relaxation based on the Kilbas and Saigo function. *Meccanica*, 49(9), 2049–2060, 2014.

[310] A. V. Chechkin, R. Gorenflo and I. M. Sokolov. Fractional diffusion in inhomogeneous media. *Journal of Physics A: Mathematical and General*, 38(42), 679, 2005.

[311] C. F. Lorenzo and T. T. Hartley. Variable order and distributed order fractional operators. *Nonlinear Dynamics*, 29(1-4), 57–98, 2002.

[312] A. B. Malinowska, T. Odzijewicz and D. F. Torres. *Advanced Methods in the Fractional Calculus of Variations.* Cham: Springer, 2015.

[313] M. Zayernouri and G. E. Karniadakis. Fractional spectral collocation methods for linear and nonlinear variable order FPDEs. *Journal of Computational Physics*, 293, 312–338, 2015.

[314] H. Sun, Y. Zhang, W. Chen, D. M. Reeves. Use of a variable-index fractional-derivative model to capture transient dispersion in heterogeneous media. *Journal of Contaminant Hydrology*, 157, 47–58, 2014.

[315] H. Sun, W. Chen and Y. Chen. Variable-order fractional differential operators in anomalous diffusion modeling. *Physica A: Statistical Mechanics and Its Applications*, 388(21), 4586–4592, 2009.

[316] S. Shen, F. Liu, J. Chen, I. Turner and V. Anh. Numerical techniques for the variable order time fractional diffusion equation. *Applied Mathematics and Computation*, 218(22), 10861–10870, 2012.

[317] Z. Li, H. Wang, R. Xiao and S. Yang. A variable-order fractional differential equation model of shape memory polymers. *Chaos, Solitons & Fractals*, 102, 473–485, 2017.

[318] A. H. Bhrawy and M. A. Zaky. An improved collocation method for multi-dimensional space-time variable-order fractional Schrdinger equations. *Applied Numerical Mathematics*, 111, 197–218, 2017.

[319] A. H. Bhrawy and M. A. Zaky. Numerical simulation for two-dimensional variable-order fractional nonlinear cable equation. *Nonlinear Dynamics*, 80(1-2), 101–116, 2015.

[320] B. P. Moghaddam and J. A. T. Machado. A stable three-level explicit spline finite difference scheme for a class of nonlinear time variable order fractional partial differential equations. *Computers & Mathematics with Applications*, 73(6), 1262–1269, 2017.

[321] B. P. Moghaddam and J. A. T. Machado. Extended algorithms for approximating variable order fractional derivatives with applications. *Journal of Scientific Computing*, 71(3), 1351–1374, 2017.

[322] L. E. Ramirez and C. F. Coimbra. On the variable order dynamics of the nonlinear wake caused by a sedimenting particle. *Physica D: Nonlinear Phenomena*, 240(13), 1111–1118, 2011.

[323] R. Almeida, D. Tavares and D. F. Torres. The Variable-Order Fractional Calculus of Variations. arXiv preprint arXiv:1805.00720, 2018.

[324] J. V. D. C. Sousa and E. C. de Oliveira. Two new fractional derivatives of variable order with non-singular kernel and fractional differential equation. *Computational and Applied Mathematics*, 2018, 37(4), 5375–5394, 2018.

[325] L. E. Ramirez and C. F. Coimbra. A variable order constitutive relation for viscoelasticity. *Annalen der Physik*, 16(7-8), 543–552, 2007.

[326] F. Wu, J. F. Liu and J. Wang. An improved Maxwell creep model for rock based on variable-order fractional derivatives. *Environmental Earth Sciences*, 73(11), 6965–6971, 2015.

[327] F. Oberhettinger and L. Badii. *Tables of Laplace Transforms*. Springer-Verlag, 1973.

[328] L. Debnath and D. Bhatta. *Integral Transforms and Their Applications.* CRC Press, New York, 2014.

[329] H. Mellin. Über die fundamentale Wichtigkeit des Satzes von Cauchy für die Theorien der Gamma-und der hypergeometrischen Functionen. *Officina Typographica Societatis Litterari Fennic*, 21(1), 1–115, 1896.

[330] R. L. Bagley. The Initial Value Problem for Fractional Order Differential Equations with Constant Coefficients. *Air Force Institute of Technology Report*, AFIT-TR-EN-88-1, 1988.

[331] G. Jumarie. Fourier's transform of fractional order via Mittag-Leffler function and modified Riemann-Liouville derivative. *Journal of Applied Mathematics & Informatics*, 26(5-6), 1101–1121, 2008.

[332] I. Podlubny. The Laplace Transform Method for Linear Differential Equations of the Fractional Order. Preprint, arXiv:funct-an/9710005v11997, 1997.

[333] S. Zhang and H. Q. Zhang. Fractional sub-equation method and its applications to nonlinear fractional PDEs. *Physics Letters A*, 375(7), 1069–1073, 2011.

[334] Y. I. Plotnikov. *Steady-state vibrations of plane and axesymmetric stamps on a viscoelastic foundation, Doctoral dissertation, Ph.D. thesis.* Moscow, 1979. (Russian)

[335] T. T. Hartley and C. F. Lorenzo. A solution to the fundamental linear fractional order differential equation. *NASA/TP-1998-208693*, December, 1998.

[336] C. F. Lorenzo and T. T. Hartley. Generalized functions for the fractional calculus. *Critical Reviews in Biomedical Engineering*, 36(1), 39–55, 2008.

[337] J. Riordan. *Combinatorial Identities (Vol. 6).* New York: Wiley, 1968.

[338] Blair, G. S. The role of psychophysics in rheology. *Journal of Colloid Science*, 2(1), 21–32, 1947.

[339] Nutting, P. G. A new general law of deformation. *Journal of the Franklin Institute*, 191(5), 679–685, 1921.

[340] Gemant A. A method of analyzing experimental results obtained from elasto-viscous Bodies. *Physics*, 7(8), 311–317, 1936.

[341] Mainardi, F. and Spada, G. Creep, relaxation and viscosity properties for basic fractional models in rheology. *The European Physical Journal-Special Topics*, 193(1), 133–160, 2011.

[342] Stiassnie, M. On the application of fractional calculus for the formulation of viscoelastic models. *Applied Mathematical Modelling*, 3(4), 300–302, 1979.

[343] Colombaro, I., Garra, R., Giusti, A. and Mainardi, F. Scott-Blair models with time-varying viscosity. *Applied Mathematics Letters*, 86, 57–63, 2018.

[344] Garra, R., & Mainardi, F. On Scott-Blair model with time-varying viscosity in linear viscoelasticity. arXiv preprint arXiv:1803.03408, 2018.

Index